Lecture Notes in Computer Science 9309

Commenced Publication in 1973
Founding and Former Series Editors:
Gerhard Goos, Juris Hartmanis, and Jan van Leeuwen

More information about this series at http://www.springer.com/series/7407

Costas S. Iliopoulos · Simon J. Puglisi
Emine Yilmaz (Eds.)

String Processing and Information Retrieval

22nd International Symposium, SPIRE 2015
London, UK, September 1–4, 2015
Proceedings

Springer

Editors

Costas S. Iliopoulos
King's College London
London
UK

Emine Yilmaz
University College London
London
UK

Simon J. Puglisi
University of Helsinki
Helsinki
Finland

ISSN 0302-9743   ISSN 1611-3349 (electronic)
Lecture Notes in Computer Science
ISBN 978-3-319-23825-8   ISBN 978-3-319-23826-5 (eBook)
DOI 10.1007/978-3-319-23826-5

Library of Congress Control Number: 2015947399

LNCS Sublibrary: SL1 – Theoretical Computer Science and General Issues

Springer Cham Heidelberg New York Dordrecht London

Printed on acid-free paper

Springer International Publishing AG Switzerland is part of Springer Science+Business Media
(www.springer.com)

Preface

From humble beginnings as a regional meeting focused on string algorithms (under the auspices of WSP: South American Workshop on String Processing), the International Symposium on String Processing and Information Retrieval (SPIRE) has, in the last two decades, developed into a vibrant conference at the broad nexus of algorithms and data structures for sequences and graphs, data compression, databases, data mining, and information retrieval.

This volume contains the papers presented at SPIRE 2015, the 22nd International Symposium on String Processing and Information Retrieval, held from August 31 to September 2, 2015 in London, UK, in the Great Hall of King's College London's Strand Campus. There were 90 submissions. Each submission was reviewed by at least 3, and on average 3.1, Program Committee members. The committee decided to accept 34 papers. The program also included 3 invited talks.

The main conference, which spanned the three days during 1–3 September, featured keynote talks by Aristides Gionis (Aalto Univeristy, Finland), Mounia Lalmas (Yahoo! Labs London, UK), and Rajeev Raman (University of Leicester, UK), together with presentations by authors of the 33 accepted papers. The 10th Workshop on Compression, Text, and Algorithms (WCTA 2015) was then held on September 4, the day immediately after the main conference, as has become a recent tradition. WCTA was coordinated this year by Travis Gagie and Tatiana Starikovskaya, and featured a keynote talk from Richard Durbin of the Wellcome Trust Sanger Institute, UK.

We take this opportunity to thank King's College London for its generous sponsorship of SPIRE this year. Our deep thanks also go to all the members of this year's Program Committee and additional reviewers, for the prompt, thorough reviewing and vibrant discussion that made our job as chairs easy. We thank the SPIRE Steering Committee, for giving us the opportunity to host this wonderful community of researchers in London, and finally, the Local Organizing Committee (led by Solon Pissis), for their efforts to ensure that the whole week ran smoothly, and that a relaxed and inspiring time was had by all.

July 2015

Costas S. Iliopoulos
Simon J. Puglisi
Emine Yilmaz

Organization

Program Committee

Sengor Altingovde	
Amihood Amir	Bar-Ilan University, Israel
Leif Azzopardi	University of Glasgow, UK
Golnaz Badkobeh	University of Sheffield, UK
Hideo Bannai	Kyushu University, Japan
Philip Bille	Technical University of Denmark, Denmark
Christina Boucher	Colorado State University, CO, USA
Ben Carterette	University of Delaware, DE, USA
Charles Clarke	University of Waterloo, Canada
Gianluca Demartini	University of Sheffield, UK
Johannes Fischer	TU Dortmund, Germany
Travis Gagie	University of Helsinki, Finland
Paweł Gawrychowski	University of Warsaw, Poland
Simon Gog	Karlsruhe Institute of Technology, Germany
Danny Hermelin	Ben-Gurion University, Israel
Djoerd Hiemstra	University of Twente, The Netherlands
Katja Hofmann	Microsoft Research Cambridge, UK
Costas S. Iliopoulos	King's College London, UK
Jaap Kamps	University of Amsterdam, The Netherlands
Evangelos Kanoulas	University of Amsterdam, The Netherlands
Gabriella Kazai	Lumi, Semion Ltd, UK
Juha Kärkkäinen	University of Helsinki, Finland
Susana Ladra	University of A Coruña, Spain
Gad Landau	University of Haifa, Israel
Zsuzsanna Lipták	University of Verona, Italy
Gonzalo Navarro	University of Chile, Chile
Kunsoo Park	Seoul National University, South Korea
Nadia Pisanti	University of Pisa, Italy
Solon Pissis	King's College London, UK
Simon J. Puglisi	University of Helsinki, Finland
Jakub Radoszewski	University of Warsaw, Poland
Falk Scholer	RMIT, Australia
Marinella Sciortino	University of Palermo, Italy
Jouni Sirén	Wellcome Trust Sanger Institute, UK
Tatiana Starikovskaya	University of Bristol, UK
Torsten Suel	NYU Poly, NY, USA
Yasuo Tabei	Japan Science and Technology Agency, Japan

Rossano Venturini   University of Pisa, Italy
Grace Yang   Georgetown University, DC, USA
Emine Yilmaz   University College London, UK

Additional Reviewers

Abouelhoda, Mohamed   Inenaga, Shunsuke   Rahman, M. Sohel
Amit, Mika   Karimi, Sarvnaz   Raymond, Rob
Belazzougui, Djamal   Kayaaslan, Enver   Reynier, Pierre-Alain
Bilò, Davide   Kempa, Dominik   Rodriguez, Juan
Bingmann, Timo   Keogh, Eamonn   Rosone, Giovanna
Brown, C. Titus   Korkin, Dmitry   Rozenberg, Liat
Christiansen, Anders Roy Kurpicz, Florian   Sacomoto, Gustavo
Cunial, Fabio   Köppl, Dominik   Salmela, Leena
Della Vedova, Gianluca   Larsson, N. Jesper   Seco, Diego
Doerr, Benjamin   Levy, Avivit   Shangsong, Liang
Eisenberg, Estrella   Lewenstein, Moshe   Song, Xuemeng
Epifanio, Chiara   Mantaci, Sabrina   Straszak, Damian
Farruggia, Andrea   Marino, Andrea   Sugimoto, Shiho
Fici, Gabriele   Markov, Ilya   Tischler, German
Ganguly, Debasis   Metke, Alejandro   Tomescu, Alexandru I.
Gasieniec, Leszek   Micale, Giovanni   Turpin, Andrew
Giaquinta, Emanuele   Na, Joong Chae   Valenzuela, Daniel
Grossi, Roberto   Nadalin, Francesca   Välimäki, Niko
Harrison, Thomas   Ozcan, Rifat   Walen, Tomasz
I, Tomohiro   Piatkowski, Marcin   Weimann, Oren
Ilie, Lucian   Pulvirenti, Alfredo

Invited Talks

Untitled Tales

Computational Problems in Mining Urban Data

Aristides Gionis

Department of Computer Science, Aalto University

With the fast growth of smart devices and sensor networks, large amounts of data are collected recording location, activity, and mobility of people living in urban environments. Additionally, data generated on location-aware social media provide rich information about places where people spend their time (shopping malls, café s, parks, etc). The availability of this type of data provides novel opportunities for developing methods for extracting interesting patterns, detecting trends, modelling people's behaviour, and eventually building intelligent systems that improve the interaction of citizens with their cities and help them to utilize better the available resources. In this talk we will review recent work in the area of mining urban data. We formulate and discuss computational problems motivated by applications in detecting events, mining trajectories, finding similar neighbourhoods, and recommending locations.

A Journey into Evaluation: From Retrieval Effectiveness to User Engagement

Mounia Lalmas

Yahoo! Labs London

Building retrieval systems that return results to users that satisfy their information need is one thing; Information Retrieval has a long history in evaluating how effective retrieval systems are. Building a retrieval system that not only returns good results to users, but does so in a way that users will want to use that system again is something more challenging; a positive search experience has been shown to lead to users engaging long-term with the retrieval system. In this talk, I will review state-of-the-art evaluation approaches for search, with respect to retrieval effectiveness but also user satisfaction. I will then focus on those approaches aiming at evaluating user engagement, and describe current works in this area within and outside the search realm. The talk will end with the proposal of a framework incorporating effectiveness evaluation into user engagement in search. An important component of this framework is to consider both within- and across-search session measurement.

Encodings = (Data Structures) - (Data)

Rajeev Raman

Department of Computer Science, University of Leicester

Driven by the increasing need to analyze and search for complex patterns in very large data sets, the area of compressed and succinct data structures has grown rapidly in the last 10-15 years. Such data structures have very low memory requirements, allowing them to fit into the main memory of a computer, which in turn avoids expensive computation on hard disks.

This talk will focus on a sub-topic that has become popular recently: encoding "the data structure" itself. Some data structuring problems involve supporting queries on data, but the queries that need to be supported do not allow the original data to be deduced from the queries. This presents opportunities to obtain space savings even when the data is incompressible, by pre-processing the data, extracting only the information needed to answer the queries, and then deleting the data. The minimum information needed to answer the queries is called the effective entropy of the problem: precisely determining the effective entropy can involve interesting combinatorics.

Contents

Faster Exact Search Using Document Clustering

Jonathan Dimond and Peter Sanders[✉]

Karlsruhe Institute of Technology, Karlsruhe, Germany
sanders@kit.edu

Abstract. We show how full-text search based on inverted indices can be accelerated by clustering the documents without losing results (SeCluD – **Se**arch with **Clu**stered **D**ocuments). We develop a fast multilevel clustering algorithm that uses query cost of conjunctive queries as an objective function. Depending on the inputs we get up to four times faster than non-clustered search. The resulting clusters are also useful for data compression and for distributing the work over many machines.

1 Introduction

Full-text search is one of the enabling techniques of the information society since it is needed for all kinds of search engines. The approach most used in practice uses *inverted indices*: Consider a set D of n *documents*. Each document contains a set of *terms* from a dictionary T. The index stores for each term $t \in T$ a *posting list* of document IDs where it occurs. A typical query asks for documents containing a set of two or more terms. This query can then be answered by intersecting the corresponding posting lists. Unfortunately, for large inputs, the lists become huge incurring substantial energy consumption for full-text search. For example, for web search, thousands of machines can be involved in answering a single query. What helps is that two sorted sequences of document IDs can be intersected in time linear in the size of the *smaller* list [12].

The starting point for this paper was the observation that we can boost the impact of such advanced set intersection algorithms by distributing the documents to clusters where terms are distributed as nonuniformly as possible. Lets consider a simple example to illustrate this effect. Suppose we want to know all documents that contain both term a and term b. Assume we have four clusters with the following number of occurrences of a and b.

Counting $\min(x, y)$ steps of the algorithm from [12] for intersecting lists of size x and y respectively, we get a total of 5 000 steps for performing the intersection in all clusters whereas we get 37 000 steps for intersecting the posting lists in an unclustered scenario – more than a factor five difference. More generally, for conjunctive

Cluster	#a	#b	min
1	2 000	10 000	2 000
2	10 000	1 000	1 000
3	40 000	1 000	1 000
4	1 000	25 000	1 000
Σ	53 000	37 000	5 000

queries where the number of term occurrences in the clusters is not too highly correlated, we expect improved performance from clustering. The main purpose

© Springer International Publishing Switzerland 2015
C. Iliopoulos et al. (Eds.): SPIRE 2015, LNCS 9309, pp. 1–12, 2015.
DOI: 10.1007/978-3-319-23826-5_1

of this paper is to get a first idea how much this approach could help in practice. After introducing basic concepts in Section 2, we formalize this idea in Section 3 and develop an efficient clustering algorithm for the resulting objective function. This algorithm is based on the well known K-Means principle but uses a fast multilevel scheme for initialization that may be of independent interest. In Section 5 we evaluate our approach using large real world instances. Section 6 discusses the results and possible future work.

More Related Work

This paper is based on the diploma thesis of Jonathan Dimond [4], see also the TR [5]. Clustering has been previously proposed for accelerating full-text search, e.g., [8, Section 7.1.6] following the *clustering hypothesis* [14] – documents that are relevant to a query tend to be more similar to each other than documents that are not relevant. Documents similar to each other are clustered together. Queries are then executed via *collection selection*. Document clusters considered irrelevant are disregarded and only clusters relevant to the query are used for searching. Although this yields good scores in standard information retrieval benchmarks, such an approach may lead to unexpected results in practice since unsupervised learning techniques such as document clustering are notoriously unreliable. In many commercial applications one even wants – at least as a first step – a complete and well defined set of results. For example, in SAP HANA [6,13] the default mode of full-text search to is to find *all* documents matching a query. One reason is that the full text query is often only one of several filtering criteria in a complex SQL query. Further speedup techniques not based on document clustering have also been considered including geographical tiering [2], static index pruning [9] and dynamic index pruning [11]. However, all these techniques improve efficiency by disregarding parts of the index.

2   Preliminaries

Let $D = \{d_1, \ldots, d_n\}$ denote the set of documents. Furthermore, let $T = \{t_1, \ldots, t_m\}$ denote the set of terms occurring in D. Given a desired number of clusters k, we want to partition the documents into a set of cluster $C = \{c_1, \ldots, c_k\}$ such that the expected query time is small.

3   Clustering Based Search

3.1   Developing an Objective Function

Of course, query costs depend on the query algorithm and the distribution of queries. In order to come to an easy to handle objective function, we make some assumptions and simplifications here that arguably lead to little loss in precision.

First of all, we focus on exact conjunctive queries involving exactly two terms. Exact conjunctive queries with more than two terms are covered insofar as it is usually a good idea to first intersect the lists for the two most rare terms and often this takes most of the query time. We ignore ranking in this paper for simplicity and because some applications require computing the full set of answers, for example in a relational database context where further filtering may throw away most of the results later.

In our implementation, we assume a two-level query algorithm that first inspects an inverted index where each cluster is viewed as one document and then inspects each cluster containing any document with the search terms. See [4] for details. To simplify the exposition, in this section, we assume that the query cost is the sum of the per cluster query costs. Let $\Phi(x, y)$ denote the cost for intersecting two lists of length x and y respectively. For the Lookup algorithm from [12] we can approximate $\Phi(x, y) = \min(x, y)$. For a particular query (t, u) we then get cost

$$\Psi_C(t, u) := \sum_{i=1}^{k} \Phi(n_i(t), n_i(u))$$

where $n_i(t)$ is the number of documents in cluster i of clustering C containing term t. When the clustering C is clear from the context, we omit the subscript 'C'. If we know the distribution of queries, we can now compute the expected cost of a query:

$$\psi := \mathbf{E}[\Psi] := \sum_{\{t,u\} \in \binom{T}{2}} \mathbf{P}[(t, u)] \Psi(t, u) \tag{1}$$

where $\mathbf{P}[(t, u)]$ denotes the probability of observing query (t, u). Unfortunately, it is unrealistic to assume that we know the probability of all queries so that we need an approximation. We thus only assume that we know the probability of each query term. These probabilities can be estimated from a query log or (less accurately) using statistics on the frequency of the terms in the document collection itself. If we now also assume that terms in a query are chosen independently, we get

$$\psi = \sum_{\{t,u\} \in \binom{T}{2}} \mathbf{P}[t] \mathbf{P}[u] \Psi(t, u) . \tag{2}$$

3.2   The Clustering Algorithm

Our starting point is an adaptation of the well known K-means algorithm to our problem: In each iteration of the algorithm, each document d is added to the cluster where it "fits" best. In order to decide what the best fit is for objective function ψ from Equation (2) we only have to decide how ψ changes when d is added. Hence, let $\delta_j^+(d)$ denote the change in ψ when document d is added to cluster j. Similarly, let $\delta_j^+(t)$ denote the change in ψ when a document is added to cluster j that contains only the single term t. We have $\delta_j^+(d) = \sum_{t \in d} \delta_j^+(t)$. Adding a single term t to cluster j only affects a summand $\mathbf{P}[t] \mathbf{P}[u] \min(n_j(t), n_j(u))$ of ϕ if $n_j(t) < n_j(u)$. In that case, $\min(n_j(t), n_j(u))$ increases by one. Hence, $\delta_j^+(t) = \mathbf{P}[t] \sum_{u \neq t, n_j(t) < n_j(u)} \mathbf{P}[u]$. $\delta_j^+(t)$ can be approximated in constant time using a lookup table that only has to be recomputed after each iteration. Hence, $\delta_j^+(d) = \sum_{t \in d} \delta_j^+(t)$ can be computed in time linear in the document size. Computing this lookup table itself can be done by sorting the terms occurring in cluster j by $n_j(t)$ and then computing a prefix sum over that array. Assuming that the term frequencies are polynomial in the number of terms occurring in a cluster, sorting can be done in linear time. Overall, one iteration of the K-means algorithm then takes time (and space) $\mathcal{O}(kN)$ where $N = \sum_{d \in D} |d|$ the total size of the corpus.

Refinements

The basic clustering algorithm defined above is already quite fast. However a number of additional improvements are critical to scale to really large inputs.

Multilevel Initialization. K-Means algorithms converge much faster if the initial solution is already of high quality. We use a multilevel initialization that may be of independent interest also in other applications. For a scaling factor $\epsilon < 1$, we take a sample of size $\max(k, \epsilon|D|)$ of the documents, cluster it recursively into k clusters (which is trivial for the base case $|D| = k$) and then run the K-means algorithm. The K-means algorithm can lead to oscillations in the clustering that are particularly pronounced when the clusters are small. When D becomes small, we therefore switch to an algorithm that updates the objective function after every assignment of a document.

TopDown Clustering. Since the running time of the K-Means algorithm grows at least linearly with the number of clusters k, we use a hierarchical clustering algorithms that recursively splits the documents: Subproblems with $s > |D|/k$ documents are split into $\min(\chi, sk/|D|)$ pieces where the *splitting factor* χ is a tuning parameter. This way we obtain between k and $2k$ clusters. An important side effect is that this approach balances cluster sizes.

Ignoring Infrequent Terms. Most search time is spent on queries involving long posting lists, i.e., frequent terms. Hence, the rare terms hardly contribute to the overall cost of queries. Therefore, we can ignore rare terms while evaluating the objective function without significantly affecting overall performance. This greatly accelerate the clustering algorithm and simplifies its parallelization.

Parallelization. We use shared memory parallelization. A massively parallel distributed memory implementation is also easy as long as we can afford to store word frequency statistics for each cluster and each frequent term on each node of the system – simply assign a subset of the documents to each node. In connection with the above optimizations on TopDown clustering and ignoring infrequent terms, this seems quite realistic even for huge document collections.

3.3   Query Algorithms

As already explained, we focus on conjunctive queries involving two terms. The most direct way to use the clustering is to run the query on each cluster. When the number of clusters k is large, a possible improvement is to build a *cluster index* listing for each term which clusters contain documents with this term. A query (t, u) will then first intersect the list corresponding to t and u in the cluster index. The query is then only forwarded to the clusters containing both t and u. Note that a cluster index can be viewed as an inverted index for a corpus with k documents where each document is the concatenation of all the documents in a cluster.

We can also use the clustering to reorder the documents: The j-th document in cluster i gets document id $j + \sum_{\ell < i} |c_\ell|$. Beyond this reordering, the clustering is ignored – we use the single-cluster Lookup algorithm. The motivation for this is the observation in [12] that a nonuniform distribution of document IDs in the posting lists accelerates the set intersection. Renumbering makes the distribution less uniform and thus may accelerate the search.

4   Implementation Details

We have made a prototypical implementation using about 4 000 lines of Haskell and 2 000 lines of C where all the time critical parts – clustering and query – are implemented in C. We use OpenMP for parallelization[1]. We switch to document grained updates of the objective function once $|D| < 100k$. The default shrink factor for the multilevel initialization is $\epsilon = 0.1$. The K-means algorithm repeats as long as the objective function value improves by at least 1 %. The splitting factor for hierarchical clustering is set to $\chi = 8$. Only the TC $= 100\,000$ most frequent terms are used for clustering. The Lookup algorithm [12] uses bucket size 16 (8 for the cluster index). This seems a good tradeoff between space and speed requirements.

5   Experiments

All experiments were done on a machine with two octa-core Intel Sandy Bridge Xeon E5-2670 processors with 2.6GHz and 64 GB RAM (i.e., 16 cores and 32 hardware threads). The operating system was SuSE Linux Enterprise Server 11 (kernel version 3.0.42). The compilers used were GHC 7.6.2 and GCC 4.7.2 with optimization level -O3. Clustering is run in parallel. Queries are run sequentially. For the source code refer to https://github.com/jdimond/diplomarbeit.

Table 1 gives the text corpora used for our benchmarks. GOV2 [3] is one of the standard benchmarks used in the literature. GOV2s is the same corpus but each *sentence* is used as one document. The sentences were extracted using the Stanford NLP library [7]. This emulates a corpus with many very small

---

[1] Source are here: https://www.github.com/jdimond/diplomarbeit

Table 1. Dataset Statistics

	GOV2	GOV2s	Wikipedia	pagenstecher.de
Documents	25 205 179	631 975 969	6 096 279	786 474
Terms	38 562 580	25 221 691	12 295 297	573 725
Terms / Document	652.22	18.19	230.54	35.70
Input size (raw)	396.74 GB	396.74 GB	12.37 GB	175.14 MB
Inverted Index size	16.25 GB	32.83 GB	3.09 GB	89.8 MB

Table 2. Query Log Statistics

	AOL	Wikipedia	pagenstecher.de
Queries	29 077 553	11 000 000	13 230
Distinct Terms	1 501 946	1 067 091	981

documents as you may find it in corporate databases and a situation where you are looking for terms occurring together in the same sentence. Wikipedia is the plain text contained in the articles in the English Wikipedia in May 3, 2013. Pagenstecher.de contains user posts from a German online community for car tuning. This corpus was selected because we have an authentic query log for it.

Further query logs used are shown in Table 2. The AOL log contains the two term queries from [10]. We use this log as semi-realistic input for our GOV2 and GOV2s test collections. For the Wikipedia corpus, we generate a synthetic log: for each article reference, we add all pairs of terms in the title of the referenced article to the log. The query logs are split randomly into a set used for clustering and one used for evaluation. Figure 1 shows the distribution of term frequencies in these logs. We can see that all of them, including the synthetic Wikipedia log, show a Zipf-like distribution of term frequencies.

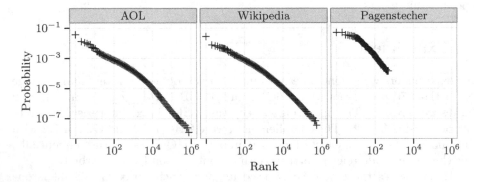

Fig. 1. Probability of a term appearing in a query as a function of its rank on a log-log scale. A sample of 100 terms with exponentially growing ranks is plotted.

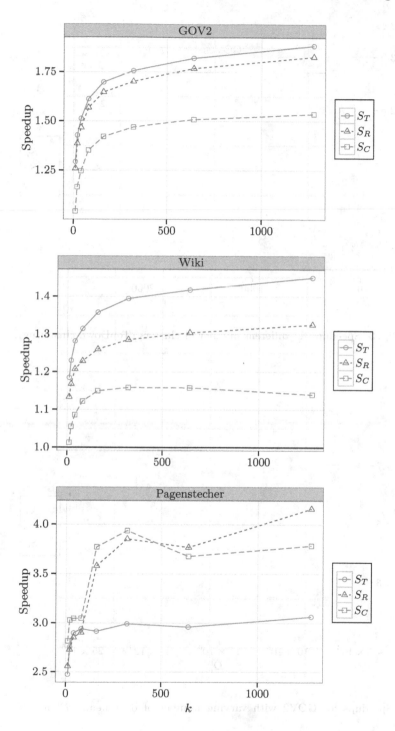

Fig. 2. Speedups for different number of clusters (flat clustering).

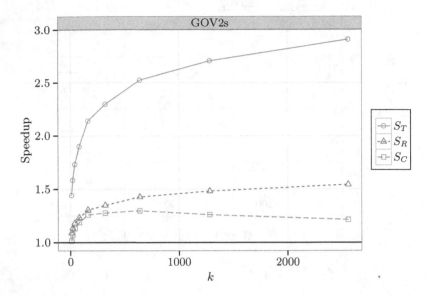

Fig. 3. Speedups for different number of clusters (TopDown clustering).

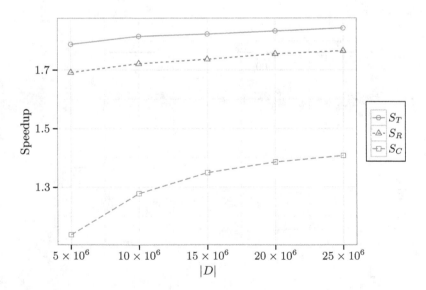

Fig. 4. Speedups for GOV2 with varying number of documents $|D|$ for $k = 2500$ clusters.

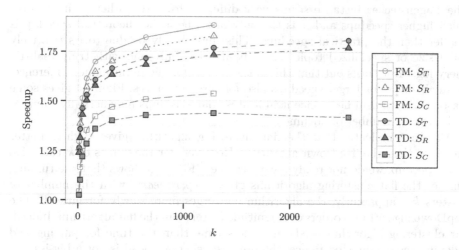

Fig. 5. Speedups comparison for flat (FM) and TopDown (TD) clustering (GOV2).

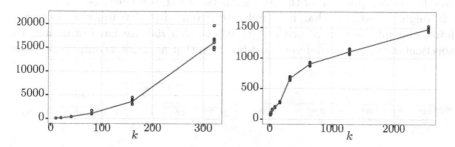

Fig. 6. Clustering times [s] for flat (left) and TopDown (right) clustering for 10 independent trials.

Our main concern are speedups over the single cluster case. We distinguish the "theoretical" speedup S_T predicted by the function ψ from Equation (2), the speedup S_C obtained using the cluster index from Section 3.3, and the speedup S_R using the reordering algorithm from Section 3.3. Figures 2 and 3 compare these values for varying number k of clusters. Most of the time, S_T overestimates the speedup but it seems strongly correlated with the actual behavior – which is all we need for making it a useful objective function for the clustering algorithm. Generally, increasing the number of clusters helps to increase speedup. However, for the Wikipedia instance using the cluster index, the speedup decreases for $k > 500$ – it is clear that eventually, overheads for the additional indirection start to show. A surprise is that the reordering algorithm achieves much better performance than the cluster index. Overall, we achieve speedups between 1.3 and 4 which is not overwhelming but certainly significant and possibly useful.

The Pagenstecher instance seems very different from the others – it achieves much higher speedups and it is the only one where the theoretical speedup is smaller than the practical speedups. This may simply be due to its relatively small size or specialized topic but it is also the only one with truly realistic query logs. If it turns out that this is the reason for the performance difference, we might hope for larger speedups also for large instances. Figure 4 gives some reason for optimism here since it indicates that speedups may actually *increase* with growing number of documents.

Figure 5 indicates that the flat clustering algorithm gives slightly better speedups than the TopDown algorithm. However, for the values of k that give good speedup, we do not really have a choice – Figure 6 shows that the running time of the flat clustering algorithms grows *superlinearly* with the number of clusters k – apparently, the algorithm converges more slowly for large k. The TopDown algorithm is orders of magnitude faster than the flat algorithm. Indeed, our clustering algorithm needs much less time than the time for parsing and indexing the documents. Hence, the preprocessing overhead is not a big issue.

Usually, preprocessing techniques also come with a penalty for storing the preprocessed information. However, in our case the contrary is true – clustering allows better compression of the posting lists, see [4,5] for details.

It might be argued that it is risky that our objective function is tied to a particular intersection algorithm. To assess this risk we have evaluated the theoretical speedup for a different cost function that assumes a comparison based

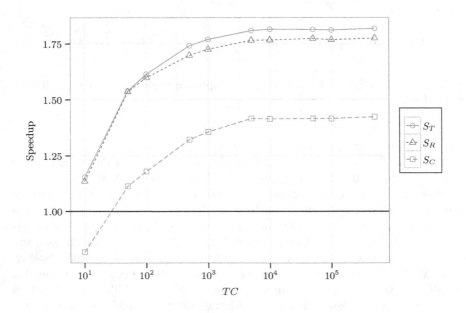

Fig. 7. Speedups for GOV2 with varying TC for $k = 2\,500$.

intersection algorithm with running time $\Phi(x, y) = x \log \frac{y}{x}$ for $x > y$ (e.g., [1]). Experiments in [4,5] indicate that this gives very similar results – indeed, the theoretical speedups are even higher than for the lookup algorithm even though $\min(x, y)$ was used as the objective function for clustering.

In Figure 7 we investigate how the number of terms (TC) we use for clustering influences speedup. All three speedup measure show that even the 10 000 most frequent terms would be enough for the GOV2 input. This is good news since this allows faster clustering algorithm. In particular, a massively parallel clustering algorithm can probably afford to replicate all lookup tables over all processors.

6   Conclusions

We have demonstrated that document clustering can significantly accelerate conjunctive queries while still giving exact results. Using a multilevel hierarchical clustering algorithm we were able to do high quality clustering even faster than the time needed for parsing and indexing the documents. Our approach of tailoring the objective function for clustering to the actual performance of the query algorithm might also be useful in other situations like clustering for inexact search or in order to compress data.

Several ideas suggest themselves for further improving the results. Since the approach is most useful for big inputs, scaling to even bigger corpora is interesting. This seems feasible since a massively parallel implementation is relatively easy. We have not done so yet since it naturally requires a lot of resources.

Equally interesting are efforts to increase the quality of the clustering. Currently, we only look at the impact of adding documents to a cluster. However, for small clusters (e.g., during initialization) it also matters how removing a document from a cluster affects query performance. We believe that our lookup-table approach can take this into account without big performance penalties. Compromises between the pure round based K-means algorithm and the variant with document-wise updates seem possible which might improve convergence speed and overall quality. Further tuning of the TopDown algorithm seems promising. Probably one wants to use larger splitting factors at least at the top of the recursion tree in order to improve quality.

Since using the clustering for reordering posting lists was very successful, this should also be considered more closely. In particular, we would like to have a more symmetric version of the lookup algorithm for list intersection. Currently, the algorithm traverses the shorter list while making lookups in the longer list. Actually, we would like to have a more adaptive algorithm that scans the list which is more dense at the current position. For example, when a lookup finds an empty bucket, we might switch to the other list. It is quite clear that clustering for reordering should (recursively) go all the way down to clusters consisting only of a few documents. To make this efficient, we need to dynamize our cutoffs – we only want to consider terms frequent in the remaining documents.

Last but not least, we have to investigate what happens for other types of queries. Perhaps most interesting are top-K queries where we are only interested in the most relevant results using some scoring function. Once more, we want to exploit the information inherent in the clustering to gain performance but the clustering must not influence the result. The hope might be that if a cluster c contains only few documents with a query term t then often c will contain even less *relevant* documents with term t so that only a fraction of t's posting list actually needs to be considered.

References

1. Baeza-Yates, R.: A fast set intersection algorithm for sorted sequences. In: Sahinalp, S.C., Muthukrishnan, S.M., Dogrusoz, U. (eds.) CPM 2004. LNCS, vol. 3109, pp. 400–408. Springer, Heidelberg (2004)
2. Cambazoglu, B.B., Plachouras, V., Baeza-Yates, R.: Quantifying performance and quality gains in distributed web search engines. In: 32nd ACM SIGIR Conference on Research and Development in Information Retrieval, pp. 411–418. ACM (2009)
3. Clarke, C., Soboroff, I., Craswell, N.: Gov2 test collection (2004)
4. Dimond, J.: Faster full text search through document clustering. Diploma thesis, Kalsruhe Institute of Technology (2013)
5. Dimond, J., Sanders, P.: Faster exact search using document clustering (2014). CoRR, abs/1411.1220
6. Färber, F., et al.: SAP HANA Database: Data management for modern business applications. SIGMOD Rec. **40**(4), 45–51 (2012)
7. Manning, C.D., et al.: The Stanford CoreNLP natural language processing toolkit. In: 52nd Meeting of the Association for Computational Linguistics: System Demonstrations, pp. 55–60 (2014)
8. Manning, C.D., Raghavan, P., Schütze, H.: Introduction to Information Retrieval. Cambridge University Press (2008)
9. Moura, D., et al.: Improving web search efficiency via a locality based static pruning method. In: 14th World Wide Web Conference, pp. 235–244. ACM (2005)
10. Pass, G., Chowdhury, A., Torgeson, C.: A picture of search. In: 1st International Conference on Scalable Information Systems, p. 1. Citeseer (2006)
11. Persin, M.: Document filtering for fast ranking. In 17th ACM SIGIR Conference on Research and Development in Information Retrieval, pp. 339–348 (1994)
12. Sanders, P., Transier, F.: Intersection in integer inverted indices. In: Workshop on Algorithm Engineering and Experiments (ALENEX) (2007)
13. Transier, F., Sanders, P.: Engineering basic algorithms of an in-memory text search engine. ACM Trans. Inf. Syst. **29**(1) (2010)
14. Van Rijsbergen, C.: Information retrieval. Butterworths (1979)

Fast Online Lempel-Ziv Factorization in Compressed Space

Alberto Policriti[1,2] and Nicola Prezza[1(✉)]

[1] Department of Mathematics and Computer Science,
University of Udine, Udine, Italy
prezza.nicola@spes.uniud.it
[2] Institute of Applied Genomics, Udine, Italy

Abstract. Let T be a text of length n on an alphabet Σ of size σ, and let H_0 be the zero-order empirical entropy of T. We show that the LZ77 factorization of T can be computed in $nH_0 + o(n \log \sigma) + \mathcal{O}(\sigma \log n)$ bits of working space with an online algorithm running in $\mathcal{O}(n \log n)$ time. Previous space-efficient online solutions either work in compact space and $\mathcal{O}(n \log n)$ time, or in succinct space and $\mathcal{O}(n \log^3 n)$ time.

Keywords: Lempel-Ziv · Compression · BWT

1 Introduction and Related Work

Let $T = a_1 a_2 ... a_{n-1}\$$ be a length-n text on an alphabet Σ of size σ, with $\$ \in \Sigma$ being a symbol appearing only at the end of T (in this work we will implicitly assume that the input text ends with $\$$). The Lempel-Ziv factorization—LZ77 for brevity—of T [18] is a sequence

$$\mathcal{Z} = \langle pos_1, len_1, c_1 \rangle ... \langle pos_i, len_i, c_i \rangle ... \langle pos_z, len_z, c_z \rangle$$

where $0 \leq pos_i, len_i < n$, $c_i \in \Sigma$ for $i = 1, ..., z$, and:

1. $T = \omega_1 c_1 ... \omega_z c_z$, with $\omega_i = \epsilon$ if $len_i = 0$ and $\omega_i = T[pos_i, ..., pos_i + len_i - 1]$ otherwise.
2. For any $i = 1, ..., z$ with $len_i > 0$, it follows that $pos_i < \sum_{j=1}^{i-1}(len_j + 1)$.
3. For any $i = 1, ..., z$, ω_i must be the *longest* prefix of $\omega_i c_i ... \omega_z c_z$ that occurs in a previous position of T.

The Lempel-Ziv factorization is an important tool in text compression, being its size z closely related with the number of repetitions in the processed string. Moreover, by augmenting it with additional (light) structures, one can obtain fast and high-order compressed full-text indexes [9,12]. Structures based on LZ77 have been shown to be competitive in terms of space on repetitive text collections with respect to BWT-based self indexes [9], and a careful combination of the two techniques stands at the basis of some of the most time-and-space efficient repetition-aware indexes [1].

C. Iliopoulos et al. (Eds.): SPIRE 2015, LNCS 9309, pp. 13–20, 2015.
DOI: 10.1007/978-3-319-23826-5_2

The Lempel-Ziv factorization can be computed in linear time and $\mathcal{O}(n \log n)$ bits of working space by using suffix trees or suffix arrays [2,3,7]. Recent results—building up on the FM index [4] data structure—reduced space to *compact* ($\mathcal{O}(n \log \sigma)$ bits), while retaining linear running time [13]. The best space bound to date is achieved by the algorithm discussed in [8], which builds the LZ77 factorization of the text in $\mathcal{O}(n \log^{1+\epsilon} n)$ time ($\epsilon > 0$) and $n(H_k+2)+o(n \log \sigma)$ bits of space (although the $\mathcal{O}(n)$ term prevents space from being *fully* compressed).

A line on this research is focused on the *online* computation of the LZ factorization. Okanohara et al. [14] showed that this task can be carried out in $\mathcal{O}(n \log^3 n)$ time using only $(1 + o(1))n \log \sigma + \mathcal{O}(n)$ bits of working space. Starikovskaya in [16] reduced the running time to $\mathcal{O}(n \log^2 n)$, while slightly increasing the working space to $\mathcal{O}(n \log \sigma)$ bits. Finally, Yamamoto et al. in [17] obtained $\mathcal{O}(n \log n)$ running time within $\mathcal{O}(n \log \sigma)$ bits of working space by using Directed Acyclic Word Graphs (DAWGs).

In this paper, we improve upon the space of all the above discussed solutions by describing an *online* algorithm that computes the LZ77 factorization of a length-n string in $\mathcal{O}(n \log n)$ time using only $nH_0 + o(n \log \sigma) + \mathcal{O}(\sigma \log n)$ bits of working space, H_0 being the empirical zero-order entropy of the input text. If one is interested in computing only the phrase boundaries, then running time can be improved to $\mathcal{O}(n \log n/ \log \log n)$. Our basic structure is a dynamic FM index over the reversed text, updated by inserting T-characters from the first to the last.

2   Notation

With T-, L- and F-*positions* we will denote positions on the text T and on the L (last) and F (first) column of the BWT matrix, respectively. Indices start from 0, and we will assume that the text length n and the alphabet $\Sigma = \{0, ..., \sigma - 1\}$ are known beforehand. The only restriction we pose on the alphabet size is $\sigma \leq n$ (which is always true after a re-mapping of the symbols). $BWT(T)$ will denote the Burrows-Wheeler transform of string T, and, when clear from the context, we will refer to it simply as BWT. With $\langle l, r \rangle$ we will denote the right-open BWT interval $[l, r)$. $BWT.F(c)$, $c \in \Sigma$, will denote the starting F-position of the block corresponding to character c in the BWT matrix. Letting $W \in \Sigma^*$, the *interval of* W will be the interval $[l, r)$ of rows prefixed by W in the BWT matrix ($r = l$ if W does not occur in T). Letting S be a dynamic string representation on the alphabet Σ, $S[i]$ will indicate the i-th character of S, $S.rank(c, i)$, $c \in \Sigma$, $0 \leq i \leq |S|$ the number of characters equal to c in S before position i *excluded*, and $S.insert(c, i)$, $c \in \Sigma$, $0 \leq i \leq |S|$ the insertion of a character c in S at position i.

3   Fast Online LZ-Factorization in Compressed Space

Our result builds upon a recent insight by Navarro and Nekrich on the optimal representation of dynamic strings [11]: there exists a data structure that permits to represent a sequence $S[0, n-1]$ over an alphabet $\Sigma = \{0, ..., \sigma - 1\}$ in $nH_0 + o(n \log \sigma) + \mathcal{O}(\sigma \log n)$ bits of space and that supports queries (access, rank, select) and updates (insertions and deletions) in $\mathcal{O}(\log n / \log \log n)$ time. The bound is worst-case for the queries and amortized for the updates.

We use the optimal sequence representation of Navarro and Nekrich to build a dynamic FM index taking $nH_0 + o(n \log \sigma) + \mathcal{O}(\sigma \log n)$ bits of space that supports (amortized) $\mathcal{O}(\log n / \log \log n)$-time left-extension of the text with an arbitrary character, $\mathcal{O}(\log n / \log \log n)$-time LF function computation, and $\mathcal{O}(\log^2 n / \log \sigma)$-time locate. Our algorithm scans the text from its first to last character, building the dynamic FM index of the reversed text. At each step (i.e. text character), we (1) update the BWT interval of the current LZ phrase and (2) insert a new text character in the index. Each time the BWT interval becomes empty, we have reached the end of the current LZ phrase and we use a locate query to compute the LZ-factor.

3.1   Dynamic FM Index

The principal component of our dynamic FM index is a dynamic BWT. There is a simple and well-known algorithm that permits to update the Burrows-Wheeler transform $BWT(S)$ of a sequence $S = s_1 s_2 ... s_{u-1}\#$, $\# \notin \Sigma$ being a character [1] lexicographically smaller than all $s \in \Sigma$, by left-extending S with a character $c \in \Sigma$ (see, for example, section 10.3 of [11]). Letting j be such that $BWT(S)[j] = \#$ and $r = BWT(S).rank(c, j)$, we update $BWT(S)$ by:

1) $BWT(S)[j] \leftarrow c$ and
2) $BWT(S).insert(\#, BWT(S).F(c) + r)$.

Let T^R denote the reversed text. In our algorithm, we index the sequence $S = T^R\#$. By using the dynamic sequence representation of [11], we can build $BWT(T^R\#)$ online in overall $\mathcal{O}(n \log n / \log \log n)$ time and $nH_0 + o(n \log \sigma) + \mathcal{O}(\sigma \log n)$ bits of space by inserting characters in the order $\#, T[0], ..., T[n-1]$ with the above procedure. In the following paragraphs, we will denote with BWT the Burrow-Wheeler transform of the current suffix of $S = T^R\#$.

The second ingredient we need in order to compute the LZ77 factorization of T is a dynamic suffix array sampling to support fast locate. The main challenge is to add such functionality without asymptotically increasing space usage. Let $\gamma > 0$ be the sample rate, and $m = \lceil n/\gamma \rceil$ be the number of stored suffix array pointers. To this end, we employ two structures:

1. A compressed dynamic bitvector B to mark with a "1" sampled F-positions.
2. A dynamic sequence representation $SA[0, m-1]$ over the alphabet $[0, n-1]$ taking compact space ($\mathcal{O}(m \log n)$ bits) and supporting $\mathcal{O}(\log n)$-time access and insert operations.

---

[1] Note that we use two *different* terminator symbols—$\$ \in \Sigma$ and $\# \notin \Sigma$—to mark the end of the forward (LZ77 algorithm) and reverse (BWT algorithm) text, respectively. Our algorithm will therefore work on texts of the form $\#W\$$, $W \in \Sigma^*$.

We use a sample rate of $\gamma = \log_\sigma n \log \log n$. For component (1), we use again the dynamic sequence representation of Navarro and Nekrich. We remind the reader that the size of a zero-order compressed bitvector B' with b bits set is $nH_0(B') \leq b \log(n/b) + b \log e$. Since B has $m = \lceil n/\gamma \rceil = \lceil \frac{n}{\log_\sigma n \log \log n} \rceil$ bits set, it follows easily that B takes overall $nH_0(B) + o(n) + \mathcal{O}(\log n) = o(n)$ bits of space.

For component (2), we use a simple balanced tree (e.g. a red-black tree or a B-tree with constant fanout) where we store suffix array samples in the leafs and we augment each internal node with the size of the corresponding subtree. Access and insert in position i are then implemented by descending the tree according to the subtree-size counters, accessing/inserting the suffix array pointer in the leafs, and (in the case of insert) updating $\mathcal{O}(\log m)$ subtree-size counters. The tree takes overall $\mathcal{O}(m \log n) = o(n \log \sigma)$ bits of space, and access/insert operations take $\mathcal{O}(\log m) = \mathcal{O}(\log n)$ time. Structures B and SA take overall $o(n \log \sigma)$ bits of space.

Implementing *Extend*. With $BWT.extend(c) \in \{0, ..., |BWT|\}$, $c \in \Sigma \cup \{\#\}$, we will denote the function that:

1. updates the BWT of the current S suffix by left-extending it with a new character c
2. updates the suffix array samples, and
3. returns the L-position of character $\#$ *after* the left-extension has taken place.

To avoid updating the already inserted suffix array pointers at each text extension, in structure SA we enumerate S-positions starting from the last. In this sense, $S[n] = \#$ corresponds to SA-position 0, and $S[0]$ corresponds to SA-position n (remember that $|S| = |T^R\#| = n + 1$). Suppose we have built the structures for the length-$(i - 1)$ suffix of S and that we want to left-extend it with the new character $S[n - i + 1]$. Let j be such that $BWT[j] = \#$, $r = BWT.rank(S[n - i + 1], j)$, and $k = BWT.F(S[n - i + 1]) + r$. Operation $BWT.extend(S[n - i + 1])$ is implemented as follows:

1. We update BWT with the new text character $S[n - i + 1]$ as described at the beginning of this section.
2. If $i \mod \gamma = 0$, then we insert a new suffix array pointer in SA and mark with a "1" the corresponding F-position in B: $SA.insert(i-1, B.rank(1, k))$ and $B.insert(1, k)$.
3. Otherwise ($i \mod \gamma \neq 0$), we mark with a "0" the new suffix F-position in B: $B.insert(0, k)$.

Step (1) takes $\mathcal{O}(\log n / \log \log n)$ amortized time. The insertion of a bit in B takes $\mathcal{O}(\log n / \log \log n)$ time, and the insertion of a suffix array pointer in SA takes $\mathcal{O}(\log n)$ time. Since we update SA every $\log_\sigma n \log \log n$ left-extensions, *extend* takes overall $\mathcal{O}(\log n / \log \log n)$ amortized time.

Implementing *Locate*. Let BWT be the Burrows-Wheeler transform of the current S suffix. Operation $BWT.locate(i)$ returns the S-position (enumerated from right to left) corresponding to the F-position i. We implement this operation as usual, i.e. by backward-navigating the current S suffix until a sampled F-position or the first suffix position is found:

1. If i is such that $BWT[i] =' \#'$, then we return $|BWT| - 1$.
2. Otherwise:
   (a) If $B[i] = 1$, then we return $SA[B.rank(1, i)]$.
   (b) If $B[i] = 0$, then we return $BWT.locate(i') - 1$, where $i' = BWT.F(c) + BWT.rank(c, i)$ and $c = BWT[i]$.

Since we use a sample rate of $\log_\sigma n \log\log n$ and access and rank operations on BWT take $\mathcal{O}(\log n / \log\log n)$ time, after $\mathcal{O}(\log^2 n / \log \sigma)$ time we find a marked F-position. Then, extracting the suffix array pointer from structure SA takes $\mathcal{O}(\log n)$ time. Since we assume $\sigma \leq n$, *locate* takes overall $\mathcal{O}(\log^2 n / \log \sigma)$ time.

Implementing LF Function. With $BWT.LF(\langle l, r\rangle, c), 0 \leq l < |BWT|, 0 \leq r \leq |BWT|, c \in \Sigma \cup \{\#\}$, we will denote function LF applied to BWT intervals: if $\langle l, r\rangle$ is the interval of a string $W \in \Sigma^*$ in BWT, then $BWT.LF(\langle l, r\rangle, c)$ returns the interval $\langle l', r'\rangle$ of cW in BWT. LF requires a constant number of rank and access operations on BWT, so it takes overall $\mathcal{O}(\log n / \log\log n)$ time.

3.2 Main Algorithm

The extension step of our algorithm is described in Algorithm 1. The algorithm takes as input one T character c, and outputs either the LZ factor ended by c or nothing if c does not end a factor. In Algorithm 1, variables BWT (the dynamic BWT described in section 3.1), $\langle l, r\rangle$ (right-open BWT interval of the current phrase), *len* (length of the current phrase), and i (L-position of character $\#$) are global, and are initialized at the beginning as $BWT \leftarrow' \#', \langle l, r\rangle \leftarrow \langle 0, 1\rangle$, $len \leftarrow 0$, and $i \leftarrow 0$.

First of all, in line 1 we perform one backward-search step using function LF. The new BWT interval $\langle l', r'\rangle$ is nonempty if and only if the current phrase $Wc, W \in \Sigma^*$, does appear previously in the text. If this is the case (lines 16-19), then we increment the current phrase length (line 17), left-extend the current S suffix (line 18), and update the BWT interval of cW^R (line 19) by incrementing its right bound r'. This step is always needed since in line 18 the new S suffix (prefixed by cW^R) falls inside the closed interval $[l', r']$.

Otherwise, if Wc does not occur previously and $len = |W| > 0$ (lines 2-8), then Wc is a new LZ factor and interval $\langle l, r\rangle$ holds all occurrences of W^R seen until now in the *reversed* text. Notice, however, that $\langle l, r\rangle$ holds also the *current* occurrence of W^R (i.e. $i \in [l, r)$) in addition to at least one previous occurrence (i.e. $r - l \geq 2$). We must therefore be careful to output a *previous* occurrence of W^R: in lines 4-8 we locate either l or $r - 1$, depending on which one is different than i. Moreover, we must subtract *len* from the located text position since *locate* returns an occurrence of W^R in the *reversed* text, and position 0 is reserved for the terminator character #. After locating the occurrence, we can extend the BWT with character c (line 12), reset the BWT interval to the full range $\langle 0, |BWT| \rangle$ (line 13), reset phrase length to zero (line 14), and return the factor.

The last case to consider is when Wc does not occur previously and $len = |W| = 0$ (lines 9 and 10). Then, this is the first occurrence of c in the text and we simply output a factor $\langle null, 0, c \rangle$ after extending the BWT with character c and resetting the global variables as described above (lines 13-14).

Algorithm 1. add_character(c)

---

    input : Character $c \in \Sigma$ (right-extending current T prefix)
    output: A factor $\langle pos, len, c \rangle$ if c ends a factor. Nothing otherwise.

1    $\langle l', r' \rangle \leftarrow BWT.LF(\langle l, r \rangle, c)$;   /* backward search step */

2 **if** $l' \geq r'$ **then**

3   **if** $len > 0$ **then**

4   **if** $i = l$ **then**

5                $occ \leftarrow r - 1$;

6   **else**

7                $occ \leftarrow l$;

8            $P \leftarrow BWT.locate(occ) - len$;   /* locate a previous occurrence */

9   **else**

10           $P \leftarrow null$;   /* first occurrence of c */

11      $L \leftarrow len$;   /* length of current phrase (c excluded) */

12      $BWT.extend(c)$;   /* insert character c in the BWT */

13      $\langle l, r \rangle \leftarrow \langle 0, |BWT| \rangle$;   /* reset interval */

14      $len \leftarrow 0$;   /* reset phrase length */

15   **return** $\langle P, L, c \rangle$;   /* return LZ factor */

16 **else**

17      $len \leftarrow len + 1$;   /* increase current phrase length */

18      $i \leftarrow BWT.extend(c)$;   /* insert character c in the BWT */

19      $\langle l, r \rangle \leftarrow \langle l', r' + 1 \rangle$;   /* new suffix falls inside $[l', r')$ */

From the analysis carried out in section 3.1 it is clear that, excluding *locate*, all steps in Algorithm 1 take (amortized) $\mathcal{O}(\log n / \log \log n)$ time. Notice that we call *locate* once per phrase. It is known that the number z of LZ77 phrases satisfies $z \in \mathcal{O}(n / \log_\sigma n)$ [10]. Since the cost of a single *locate* query is $\mathcal{O}(\log^2 n / \log \sigma)$, in Algorithm 1 *locate* takes $\mathcal{O}(\log n)$ amortized time. We can state our final result:

Theorem 1. *Let $T \in \Sigma^n$. By calling Algorithm 1 on $T[0], ..., T[n-1]$, we build the LZ77 factorization of T online in $nH_0 + o(n \log \sigma) + \mathcal{O}(\sigma \log n)$ bits of working space and $\mathcal{O}(n \log n)$ time.*

Notice that, if we wish to compute only the LZ phrase boundaries, then we do not need *locate*, and the LZ factorization can be built using a simplified version of Algorithm 1 in $\mathcal{O}(n \log n / \log \log n)$ time.

4 Conclusions

In this paper, we presented an online algorithm for computing the LZ77 factorization of a text in $nH_0 + o(n \log \sigma) + \mathcal{O}(\sigma \log n)$ bits of working space and $\mathcal{O}(n \log n)$ time. To our knowledge, ours is the first solution of this problem reaching *fully compressed* working space. Moreover, we obtain this result while being as fast as the fastest online LZ77-construction algorithms described in literature.

Solving this task in small space is of great importance in areas such as LZ-based self-indexing, where computing the LZ77 parse of the text is a spatial bottleneck during index construction. Ideally, it would be desirable being able to solve the problem in $\mathcal{O}(z)$ words of working space (result easily reachable with LZ78), considering that for repetitive text collections z can be *exponentially* smaller than n. One first improvement over our approach could be to obtain high-order compressed space, e.g. by using techniques similar to those employed in [6,11,15]. However, this strategy would still not perform well over highly repetitive text collections—being H_k not sensitive to long repetitions—and being entropy-based techniques usually affected by an $o(n)$ spatial term that could be exponentially larger than z. Alternatively, one could consider using a run-length compressed BWT. Yet, this approach would also require a more sparse SA sampling, which in the most efficient implementations [1,5] is based on the LZ parse itself.

References

1. Belazzougui, D., Cunial, F., Gagie, T., Prezza, N., Raffinot, M.: Composite repetition-aware data structures. In: Cicalese, F., Porat, E., Vaccaro, U. (eds.) CPM 2015. LNCS, vol. 9133, pp. 26–39. Springer, Heidelberg (2015)
2. Crochemore, M., Ilie, L.: Computing longest previous factor in linear time and applications. Information Processing Letters **106**(2), 75–80 (2008)
3. Crochemore, M., Ilie, L., Smyth, W.F.: A simple algorithm for computing the Lempel-Ziv factorization. In: 18th Data Compression Conference (DCC 2008), pp. 482–488. IEEE Computer Society Press, Los Alamitos (2008)
4. Ferragina, P., Manzini, G.: Opportunistic data structures with applications. In: Proceedings of the 41st Annual Symposium on Foundations of Computer Science, 2000, pp. 390–398. IEEE (2000)
5. Ferragina, P., Manzini, G.: Indexing compressed text. Journal of the ACM (JACM) **52**(4), 552–581 (2005)

6. Ferragina, P., Manzini, G., Mäkinen, V., Navarro, G.: An alphabet-friendly FM-index. In: Apostolico, A., Melucci, M. (eds.) SPIRE 2004. LNCS, vol. 3246, pp. 150–160. Springer, Heidelberg (2004)
7. Kärkkäinen, J., Kempa, D., Puglisi, S.J.: Linear time Lempel-Ziv factorization: simple, fast, small. In: Fischer, J., Sanders, P. (eds.) CPM 2013. LNCS, vol. 7922, pp. 189–200. Springer, Heidelberg (2013)
8. Kreft, S., Navarro, G.: Self-index based on LZ77 (Ph.D. thesis) (2011). arXiv preprint arXiv:1112.4578
9. Kreft, S., Navarro, G.: Self-indexing based on LZ77. In: Giancarlo, R., Manzini, G. (eds.) CPM 2011. LNCS, vol. 6661, pp. 41–54. Springer, Heidelberg (2011)
10. Lempel, A., Ziv, J.: On the complexity of finite sequences. IEEE Transactions on Information Theory 22(1), 75–81 (1976)
11. Navarro, G., Nekrich, Y.: Optimal dynamic sequence representations. SIAM Journal on Computing 43(5), 1781–1806 (2014)
12. Navarro, G., Raffinot, M.: Practical and flexible pattern matching over Ziv-Lempel compressed text. Journal of Discrete Algorithms 2(3), 347–371 (2004)
13. Ohlebusch, E., Gog, S.: Lempel-Ziv factorization revisited. In: Giancarlo, R., Manzini, G. (eds.) CPM 2011. LNCS, vol. 6661, pp. 15–26. Springer, Heidelberg (2011)
14. Okanohara, D., Sadakane, K.: An online algorithm for finding the longest previous factors. In: Halperin, D., Mehlhorn, K. (eds.) ESA 2008. LNCS, vol. 5193, pp. 696–707. Springer, Heidelberg (2008)
15. Policriti, A., Gigante, N., Prezza, N.: Average linear time and compressed space construction of the Burrows-Wheeler transform. In: Dediu, A.-H., Formenti, E., Martín-Vide, C., Truthe, B. (eds.) LATA 2015. LNCS, vol. 8977, pp. 587–598. Springer, Heidelberg (2015)
16. Starikovskaya, T.: Computing Lempel-Ziv factorization online. In: Rovan, B., Sassone, V., Widmayer, P. (eds.) MFCS 2012. LNCS, vol. 7464, pp. 789–799. Springer, Heidelberg (2012)
17. Yamamoto, J., I, T., Bannai, H., Inenaga, S., Takeda, M.: Faster compact online Lempel-Ziv factorization. In: 31st International Symposium on Theoretical Aspects of Computer Science (STACS 2014). Leibniz International Proceedings in Informatics (LIPIcs), vol. 25, pp. 675–686. Schloss Dagstuhl-Leibniz-Zentrum fuer Informatik, Dagstuhl (2014)
18. Ziv, J., Lempel, A.: A universal algorithm for sequential data compression. IEEE Transactions on information theory 23(3), 337–343 (1977)

Adaptive Computation of the Swap-Insert Correction Distance

Jérémy Barbay[1](✉) and Pablo Pérez-Lantero[2](✉)

[1] Departamento de Ciencias de la Computación, Universidad de Chile,
Santiago, Chile
jeremy@barbay.cl
[2] Escuela de Ingeniería Civil en Informática, Universidad de Valparaíso,
Valparaíso, Chile
pplantero@yahoo.com

Abstract. The Swap-Insert Correction distance from a string S of length n to another string L of length $m \geq n$ on the alphabet $[1..d]$ is the minimum number of insertions, and swaps of pairs of adjacent symbols, converting S into L. Contrarily to other correction distances, computing it is NP-Hard in the size d of the alphabet. We describe an algorithm computing this distance in time within $O(d^2 nmg^{d-1})$, where there are n_α occurrences of α in S, m_α occurrences of α in L, and where $g = \max_{\alpha \in [1..d]} \min\{n_\alpha, m_\alpha - n_\alpha\}$ measures the difficulty of the instance. The difficulty g is bounded by above by various terms, such as the length of the shortest string S, and by the maximum number of occurrences of a single character in S. The latter bound yields a running time within $O(d(n+m) + (d/(d-1)^{d-2}) \cdot n^d(m-n))$ in the worst case over instances of fixed lengths n and m for S and L, which further simplifies to within $O(n^d(m-n)+m)$ when d is fixed, the state of the art for this problem. This illustrates how, in many cases, the correction distance between two strings can be easier to compute than in the worst case scenario.

Keywords: Adaptive · Dynamic programming · Edit distance · Insert · Swap

1 Introduction

Given two strings S and L on the alphabet $\Sigma = [1..d]$ and a list of correction operations on strings, the STRING-TO-STRING CORRECTION distance is the minimum number of operations required to transform the string S into the string L. Introduced in 1974 by Wagner and Fischer [7], this concept has many applications, from suggesting corrections for typing mistakes, to decomposing the changes between two consecutive versions into a minimum number of correction steps, for example within a control version system.

J. Barbay and P. Pérez-Lantero—Partially supported by Millennium Nucleus Information and Coordination in Networks ICM/FIC RC130003.

Each distinct set of correction operators yields a distinct correction distance on strings. For instance, Wagner and Fischer [7] showed that for the three following operations, the `insertion` of a symbol at some arbitrary position, the `deletion` of a symbol at some arbitrary position, and the `substitution` of a symbol at some arbitrary position, there is a dynamic program solving this problem in time within $O(nm)$ when S is of length n and L of length m. Similar complexity results, all polynomial, hold for many other different subsets of the natural correction operators, with one striking exception: Wagner [6] proved the NP-hardness of the SWAP-INSERT CORRECTION distance, denoted $\delta(S, L)$ through this paper, i.e. the correction distance when restricted to the operators `insertion` and `swap` (or, by symmetry, to the operators `deletion` and `swap`).

The SWAP-INSERT CORRECTION distance's difficulty attracted special interest, with two results of importance: Abu-Khzam et al. [1] described an algorithm computing $\delta(S, L)$ in time within $O(1.6181^{\delta(S,L)}m)$, and Meister [4] described an algorithm computing $\delta(S, L)$ in time polynomial in the input size when S and L are strings on a finite alphabet.

The complexity of Meister's result [4], polynomial in m of degree $2d + 1$, is a very pessimistic approximation of the computational complexity of the distance. At one extreme, the SWAP-INSERT CORRECTION distance between two strings which are very similar (e.g. only a finite number of symbols need to be swapped or inserted) can be computed in time linear in n and d. At the other extreme, the SWAP-INSERT CORRECTION distance of strings which are completely different (e.g. their effective alphabets are disjoint) can also be computed in linear time (it is then close to $n + m$). Even when S and L are quite different, $\delta(S, L)$ can be "easy" to compute: when mostly swaps are involved to transform S into L (i.e. S and L are almost of the same length), and when mostly insertions are involved to transform S into L (i.e. many symbols present in L are absent from S).

Hypothesis: We consider whether the SWAP-INSERT CORRECTION distance $\delta(S, L)$ can be computed in time polynomial in the length of the input strings for a constant alphabet size, while still taking advantage of cases such as those described above, where the distance $\delta(S, L)$ can be computed much faster.

Our Results: After a short review of previous results and techniques in Section 2, we present such an algorithm in Section 3, in four steps: the intuition behind the algorithm in Section 3.1, the formal description of the dynamic program in Section 3.2, and the formal analysis of its complexity in Section 3.3. In the latter, we define the local imbalance $g_\alpha = \min\{n_\alpha, m_\alpha - n_\alpha\}$ for each symbol $\alpha \in \Sigma$, summarized by the global imbalance measure $g = \max_{\alpha \in \Sigma} g_\alpha$, and prove that our algorithm runs in time within $O(d^2 g^{d-1} nm)$ in the worst case over instances where d, n, m and g are fixed.

We discuss in Section 4 some implied results, and some questions left open. Additional details are deferred to the full version [2].

2   Background

In 1974, motivated by the problem of correcting typing and transmission errors, Wagner and Fischer [7] introduced the STRING-TO-STRING CORRECTION problem, which is to compute the minimum number of corrections required to change the source string S into the target string L. They considered the following operators: the insertion of a symbol at some arbitrary position, the deletion of a symbol at some arbitrary position, and the substitution of a symbol at some arbitrary position. They described a dynamic program solving this problem in time within $O(nm)$ when S is of length n and L of length m. The worst case among instances of fixed input size $n + m$ is when $n = m/2$, which yields a complexity within $O(n^2)$.

In 1975, Lowrance and Wagner [8] extended the STRING-TO-STRING COR-RECTION distance to the cases where one considers not only the insertion, deletion, and substitution operators, but also the swap operator, which exchanges the positions of two contiguous symbols. Not counting the identity, fifteen different variants arise when considering any given subset of those four correction operators. Thirteen of those variants can be computed in polynomial time [6–8]. The two remaining distances, the computation of the SWAP-INSERT CORRECTION distance and its symmetric the SWAP-DELETE CORRECTION distance, are equivalent by symmetry, and are NP-hard to compute [6], hence our interest. All our results on the computation of the SWAP-INSERT CORRECTION distance from S to L directly imply the same results on the computation of the SWAP-DELETE CORRECTION distance from L to S.

In 2013, Spreen [5] observed that Wagner's NP-hardness proof [6] was based on unbounded alphabet sizes (i.e. the SWAP-INSERT CORRECTION problem is NP-hard when the size d of the alphabet is part of the input), and suggested that this problem might be tractable for fixed alphabet sizes. He described some polynomial-time algorithms for various special cases when the alphabet is binary, and some more general properties.

In 2014, Meister [4] extended Spreen's work [5] to an algorithm computing the SWAP-INSERT CORRECTION distance from a string S of length n to another string L of length m on any fixed alphabet size $d \geq 2$, in time polynomial in n and m. The algorithm is explicitly based on finding an injective function $\varphi : [1..n] \to [1..m]$ such that $\varphi(i) = j$ if and only if $S[i] = L[j]$, and the total number of crossings is minimized. Two positions $i < i'$ of S define a crossing if and only if $\varphi(i) > \varphi(i')$. Such a number of crossings equals the number of swaps, and the number of insertions is always equal to $m - n$. Meister proved that the time complexity of this algorithm is equal to $(m + 1)^{2d+1} \cdot (n + 1)^2$ times some function polynomial in n and m.

3   Algorithm

We describe the intuition behind our algorithm in Section 3.1, the high level description of the dynamic program in Section 3.2, and the formal analysis of its complexity in Section 3.3.

3.1   High Level Description

The algorithm runs through S and L from left to right, building a mapping from the characters of S to a subset of the characters of L, using the fact that, for each distinct character, the mapping function on positions is monotone. The Dynamic Programming matrix has size $n_1 \times \cdots \times n_d < n^d$.

For every string $X \in \{S, L\}$, let $X[i]$ denote the i-th symbol of X from left to right ($i \in [1..|X|]$), and $X[i..j]$ denote the substring of X from the i-th symbol to the j-th symbol ($1 \leq i \leq j \leq |X|$). For every $1 \leq j < i \leq n$, let $X[i..j]$ denote the empty string. Given any symbol $\alpha \in \Sigma$, let $rank(X, i, \alpha)$ denote the number of occurrences of the symbol α in the substring $X[1..i]$, and $select(X, k, \alpha)$ denote the value $j \in [1..|X|]$ such that the k-th occurrence of α in X is precisely at position j, if j exists. If j does not exist, then $select(X, k, \alpha)$ is *null*.

The algorithm runs through S and L simultaneously from left to right, skipping positions where the current symbol of S equals the current symbol of L, and otherwise branching out between two options to correct the current symbol of S: inserting a symbol equal to the current symbol of L in the current position of S, or moving (by applying many swaps) the first symbol of the part not scanned of S equal to the current symbol of L, to the current position in S.

More formally, the computation of $\delta(S, L)$ can be reduced to the application of four rules:

- **if S is empty**: We just return the length $|L|$ of L, since insertions are the only possible operations to perform in S.
- **if some $\alpha \in \Sigma$ appears more times in S than in L**: We return $+\infty$, since **delete** operations are not allowed to make S and L match.
- **if S and L are not empty, $S[1] = L[1]$**: We return $\delta(S[2..|S|], L[2..|L|])$.
- **if S and L are not empty, $S[1] \neq L[1]$**: We compute two distances: the distance $d_{ins} = 1 + \delta(S, L[2..|L|])$ corresponding to an **insertion** of the symbol $L[1]$ at the first position of S, and the distance $d_{swaps} = (r - 1) + \delta(S', L[2..|L|])$ corresponding to perform $r - 1$ **swaps** to bring to the first position of S the first symbol of S equal to $L[1]$. In this case, r denotes the position of such a symbol, and S' the string resulting from S by removing that symbol. We then return $\min\{d_{ins}, d_{swaps}\}$.

There can be several overlapping subproblems in the recursive definition of $\delta(S, L)$ described above, which calls for *dynamic programming* [3] and *memoization*. In any call $\delta(S', L')$ in the recursive computation of $\delta(S, L)$, the string L' is always a substring $L[j..|J|]$ for some $j \in [1..|J|]$, and can thus be replaced by such an index j, but this is not always the case for the string S'. Observe that S' is a substring $S[i..|S|]$ for some $i \in [1..|S|]$ with (eventually) some symbols removed. Furthermore, if for some symbol $\alpha \in \Sigma$ precisely c_α symbols α of $S[i..|S|]$ have been removed, then those symbols are precisely the first c_α symbols α from left to right. We can then represent S' by the index i and a counter c_α for each symbol $\alpha \in \Sigma$ of how many symbols α of $S[i..|S|]$ are removed (i.e. ignored). In the above fourth rule, the position r is equivalent to the position of the $(c_{L[1]} + 1)$-th occurrence of the symbol $L[1]$ in $S[i..|S|]$. To quickly compute r, the functions $rank$ and $select$ will be used.

Let $\mathbb{W} = \prod_{\alpha=1}^{d}[0..n_\alpha]$ denote the domain of such vectors of counters, where for any $\bar{c} = (c_1, c_2, \ldots, c_d) \in \mathbb{W}$, c_α denotes the counter for $\alpha \in \Sigma$. Using the ideas described above, the algorithm recursively computes the extension $DIST(i, j, \bar{c})$ of $\delta(S, L)$, defined for each $i \in [1..n + 1]$, $j \in [1..m + 1]$, and $\bar{c} = (c_1, c_2, \ldots, c_d) \in \mathbb{W}$, as the value of $\delta(S[i..n]_{\bar{c}}, L[j..m])$, where $S[i..n]_{\bar{c}}$ is the string obtained from $S[i..n]$ by removing (i.e. ignoring) for each $\alpha \in \Sigma$ the first c_α occurrences of α from left to right.

Given this definition, $\delta(S, L) = DIST(1, 1, \bar{0})$, where $\bar{0}$ denotes the vector $(0, \ldots, 0) \in \mathbb{W}$. Given i, j, and $\bar{c}$, $DIST(i, j, \bar{c}) < +\infty$ if and only if for each symbol $\alpha \in \Sigma$ the number of considered (i.e. not removed or ignored) α symbols in $S[i..n]$ is at most the number of α symbols in $L[j..m]$. That is, $count(S, i, \alpha) - c_\alpha \leq count(L, j, \alpha)$ for all $\alpha \in \Sigma$, where $count(X, i, \alpha) = rank(X, |X|, \alpha) - rank(|X|, i - 1, \alpha)$ is the number of symbols α in the string $X[i..|X|]$. In the following, we show how to compute $DIST(i, j, \bar{c})$ recursively for every i, j, and $\bar{c}$. For a given $\alpha \in \Sigma$, let $\overline{w}_\alpha \in \mathbb{W}$ be the vector whose components are all equal to zero except the α-th component that is equal to 1.

3.2   Recursive Computation of $DIST(i, j, \bar{c})$

We will use the following observation which considers the swap operations performed in the optimal transformation from a short string S of length n to a larger string L of length m.

Observation 1 ([1,5]). *The swap operations used in any optimal solution satisfy the following properties: two equal symbols cannot be swapped; each symbol is always swapped in the same direction in the string; and if some symbol is moved from some position to another by performing swaps operations, then no symbol equal to it can be inserted afterwards between these two positions.*

The following lemma deals with the basic case where $S[i..n]$ and $L[j..m]$ start with the same symbol, i.e. $S[i] = L[j]$. When the beginnings of both strings are the same, matching those two symbols seems like an obvious choice in order to minimize the distance, but one must be careful to check first if the first symbol from $S[i..n]$ has not been scheduled to be "swapped" to an earlier position, in which case it must be ignored and skipped:

Lemma 1. *Given two strings S and L over the alphabet Σ, for any positions $i \in [1..n]$ in S and $j \in [1..m]$ in L, for any vector of counters $\bar{c} = (c_1, \ldots, c_d) \in \mathbb{W}$ and for any symbol $\alpha \in \Sigma$,*

$$\left.\begin{array}{r} S[i] = L[j] = \alpha \\ c_\alpha = 0 \end{array}\right\} \implies DIST(i, j, \bar{c}) = DIST(i + 1, j + 1, \bar{c}).$$

Proof. Given strings X, Y in the alphabet Σ, and an integer k, Abu-Khzam et al. [1, Corollary1] proved that if $X[1] = Y[1]$, then:

$$\delta(X, Y) \leq k \text{ if and only if } \delta(X[2..|X|], Y[2..|Y|]) \leq k.$$

Given that one option to transform X into Y with the minimum number of operations is to transform $X[2..|X|]$ into $Y[2..|Y|]$ with the minimum number of operations (matching $X[1]$ with $Y[1]$), we have:

$$\delta(X,Y) \leq \delta(X[2..|X|], Y[2..|Y|]).$$

By selecting $k = \delta(X,Y)$, we obtain the equality

$$\delta(X,Y) = \delta(X[2..|X|], Y[2..|Y|]).$$

Then, since the symbol $\alpha = S[i]$ must be considered (because $c_\alpha = 0$), and $S[i] = L[j]$, we can apply the above statement for $X = S[i..n]_{\bar{c}}$ and $Y = L[j..m]$ to obtain the next equalities:

$$DIST(i,j,\bar{c}) = \delta(X,Y) = \delta(X[2..|X|], Y[2..|Y|]) = DIST(i+1, j+1, \bar{c}).$$

The result thus follows.   □

The second simplest case is when the first available symbol of $S[i..n]$ is already matched (through swaps) to a symbol from $L[1..j-1]$. The following lemma shows how to simply skip such a symbol:

Lemma 2. *Given S and L over the alphabet Σ, for any positions $i \in [1..n]$ in S and $j \in [1..m]$ in L, and for any vector of counters $\bar{c} = (c_1, \ldots, c_d) \in \mathbb{W}$ and for any symbol $\alpha \in \Sigma$,*

$$\left. \begin{array}{c} S[i] = \alpha \\ c_\alpha > 0 \end{array} \right\} \implies DIST(i,j,\bar{c}) = DIST(i+1, j, \bar{c} - \overline{w}_\alpha).$$

Proof. Since $c_\alpha > 0$, the first c_α symbols α of $S[i..n]$ have been ignored, thus $S[i]$ is ignored. Then, $DIST(i,j,\bar{c})$ must be equal to $DIST(i+1, j, \bar{c} - \overline{w}_\alpha)$, case in which $c_\alpha - 1$ symbols α of $S[i+1..n]$ are ignored.   □

The most important case is when the first symbols of $S[i..n]$ and $L[j..m]$ do not match: the minimum "path" from S to L can then start either by an **insertion** or a **swap** operation.

Lemma 3. *Given S and L over the alphabet Σ, for any positions $i \in [1..n]$ in S and $j \in [1..m]$ in L, and for any vector of counters $\bar{c} = (c_1, \ldots, c_d) \in \mathbb{W}$, note $\alpha, \beta \in \Sigma$ the symbols $\alpha = S[i]$ and $\beta = L[j]$, r the position $r = select(S, rank(S, i, \beta) + c_\beta + 1, \beta)$ in S of the $(c_\beta + 1)$-th symbol β of $S[i..n]$, and Δ the number $\sum_{\theta=1}^{d} \min\{c_\theta, rank(S, r, \theta) - rank(S, i-1, \theta)\}$ of symbols ignored in $S[i..r]$.*
If $\alpha \neq \beta$ and $c_\alpha = 0$, then $DIST(i,j,\bar{c}) = \min\{d_{ins}, d_{swaps}\}$, where

$$d_{ins} = \begin{cases} DIST(i, j+1, \bar{c}) + 1 & \text{if } c_\beta = 0 \\ +\infty & \text{if } c_\beta > 0 \end{cases}$$

and

$$d_{swaps} = \begin{cases} (r-i) - \Delta + DIST(i, j+1, \bar{c} + \overline{w}_\beta) & \text{if } r \neq 0 \\ +\infty & \text{if } r = 0. \end{cases}$$

Proof. Let $S'[1..n'] = S[i..n]_{\overline{c}}$. Given that $\alpha \neq \beta$ and $c_\alpha = 0$, there are two possibilities for $DIST(i, j, \overline{c})$: (1) transform $S'[1..n']$ into $L[j+1..m]$ with the minimum number of operations, and after that insert a symbol β at the first position of the resulting $S'[1..n']$; or (2) swap the first symbol β in $S'[2..n']$ from left to right from its current position r' to the position 1 performing $r'-1$ **swaps**, and then transform the resulting $S'[2..n']$ into $L[j+1..m]$ with the minimum number of operations. Observe that option (1) can be performed if and only if there is no symbol β ignored in $S[i..n]$ (see Observation 1). If this is the case, then $DIST(i, j, \overline{c}) = DIST(i, j+1, \overline{c}) + 1$. Option (2) can be used if and only if there is a non-ignored symbol β in $S[i..n]$, where the first one from left to right is precisely at position $r = select(S, rank(S, i, \beta) + c_\beta + 1, \beta)$. In such a case $r' = (r - i + 1) - \Delta$, where $\Delta = \sum_{\theta=1}^{d} \min\{c_\theta, rank(S, r, \theta) - rank(S, i-1, \theta)\}$ is the total number of ignored symbols in the string $S[i..r]$. Hence, the number of swaps counts to $r' - 1 = (r - i) - \Delta$. Then, the correctness of d_{ins}, d_{swaps}, and the result follow. $\qquad\square$

The next two lemmas deal with the cases where one string is completely processed. When L has been completely processed, either the remaining symbols in S have all previously been matched via **swaps** and the distance equals zero, or there is no sequence of operations correcting S into L:

Lemma 4. *Given S and L over the alphabet Σ, for any positions $i \in [1..n+1]$ in S and $j \in [1..m]$ in L, for any vector of counters $\overline{c} = (c_1, \ldots, c_d) \in \mathbb{W}$,*

$$DIST(i, m+1, \overline{c}) = \begin{cases} 0 & \text{if } c_1 + \ldots + c_d = n - i + 1 \text{ and} \\ +\infty & \text{otherwise.} \end{cases}$$

Proof. Note that $DIST(i, m+1, \overline{c})$ is the minimum number of operations to transform the string $S[i..n]$ into the empty string $L[m+1..m]$. This number is null if and only if all the $n - i + 1$ symbols of $S[i..n]$ have been ignored, that is, $c_1 + \ldots + c_d = n - i + 1$. If not all the symbols have been ignored, then such a transformation does not exist and $DIST(i, m+1, \overline{c}) = +\infty$. $\qquad\square$

When S has been completely processed, there are only insertions left to perform: the distance can be computed in constant time, and the list of corrections in linear time.

Lemma 5. *Given S and L over the alphabet Σ, for any position $j \in [1..m+1]$ in L, and for any vector of counters $\overline{c} = (c_1, \ldots, c_d) \in \mathbb{W}$,*

$$DIST(n+1, j, \overline{c}) = \begin{cases} m - j + 1 & \text{if } \overline{c} = \overline{0} \text{ and} \\ +\infty & \text{otherwise.} \end{cases}$$

Proof. Note that $DIST(i, m+1, \overline{c})$ is the minimum number of operations to transform the empty string $S[n+1..n]$ into the string $L[j..m]$. If $\overline{c} = \overline{0}$, then $DIST(n+1, j, \overline{c}) < +\infty$ and the transformation consists of only insertions which are $m - j + 1$. If $\overline{c} \neq \overline{0}$, then $DIST(n+1, j, \overline{c}) = +\infty$. $\qquad\square$

Algorithm $DIST(i, j, \bar{c} = (c_1, \ldots, c_d))$
1.   **if** $DIST(i, j, \bar{c}) = +\infty$ **then**
2.   **return** $+\infty$
3.   **else if** $i = n + 1$ **then**
4.   (* insertions *)
5.   **return** $m - j + 1$
6.   **else if** $j = m + 1$ **then**
7.   (* skip all symbols since they were ignored *)
8.   **return** 0
9.   **else**
10. $\alpha \leftarrow S[i], \beta \leftarrow L[j]$
11.   **if** $c_\alpha > 0$ **then**
12.   (* skip $S[i]$, it was ignored *)
13.   **return** $DIST(i + 1, j, \bar{c} - \overline{w}_\alpha)$
14.   **else if** $\alpha = \beta$ **then**
15.   (* $S[i]$ and $L[j]$ match *)
16.   **return** $DIST(i + 1, j + 1, \bar{c})$
17.   **else**
18. $d_{ins} \leftarrow +\infty, d_{swaps} \leftarrow +\infty$
19.   **if** $c_\beta = 0$ **then**
20.   (* insert a β at index i *)
21. $d_{ins} \leftarrow 1 + DIST(i, j + 1, \bar{c})$
22. $r \leftarrow select(S, rank(S, i, \beta) + c_\beta + 1, \beta)$
23.   **if** $r \neq null$ **then**
24. $\Delta \leftarrow \sum_{\theta=1}^{d} \min\{c_\theta, rank(S, r, \theta) - rank(S, i - 1, \theta)\}$
25.   (* swaps *)
26. $d_{swaps} \leftarrow (r - i) - \Delta + DIST(i, j + 1, \bar{c} + \overline{w}_\beta)$
27.   **return** $\min\{d_{ins}, d_{swaps}\}$

Fig. 1. Informal algorithm to compute $DIST(i, j, \bar{c})$: Lemma 4 and Lemma 5 guarantee the correctness of lines 1 to 8; Lemma 2 guarantees the correctness of lines 11 to 13; Lemma 1 guarantees the correctness of lines 14 to 16; and Lemma 3 guarantees the correctness of lines 18 to 27.

3.3   Complexity Analysis

Combining Lemmas 1 to 5, the value of $DIST(1, 1, \bar{0})$ can be computed recursively, as shown in the algorithm of Figure 1. We analyze the formal complexity of this algorithm in Theorem 1, in the finest model that we can define, taking into account the relation for each symbol $\alpha \in \Sigma$ between the number n_α of occurrences of α in S and the number m_α of occurrences of α in L.

Theorem 1. *Given two strings S and L over the alphabet Σ, for each symbol $\alpha \in \Sigma$, note n_α the number of occurrences of α in S and m_α the number of occurrences of m in L, their sums $n = n_1 + \cdots + n_d$ and $m = m_1 + \cdots + m_d$, and $g_\alpha = \min\{n_\alpha, m_\alpha - n_\alpha\}$ a measure of how far n_α is from $m_\alpha/2$. There is an algorithm computing the* SWAP-INSERT CORRECTION *distance $\delta(S, L)$ in time* *within $O(d + m)$ if S and L have no symbol in common, and otherwise in time within*

$$O\left(d(n + m) + d^2 n \cdot \sum_{\alpha=1}^{d} (m_\alpha - g_\alpha) \cdot \prod_{\alpha \in \Sigma_+} (g_\alpha + 1)\right),$$

where $\Sigma_+ = \{\alpha \in \Sigma : g_\alpha > 0\}$ if $g_\alpha = 0$ for any $\alpha \in \Sigma$, and $\Sigma_+ = \Sigma \setminus \{\arg\min_{\alpha \in \Sigma} g_\alpha\}$ otherwise.

Proof. Observe first that there is a reordering of $\Sigma = [1..d]$ such that $0 < g_1 \leq g_2 \leq \cdots \leq g_s$ and $g_{s+1} = g_{s+2} = \cdots = g_d$ for some index $s \in [0..d]$, and we assume such an ordering from now on. Note also that given any string $X \in \{S, L\}$, a simple 2-dimensional array using space within $O(d \cdot |X|)$ can be computed in time within $O(d \cdot |X|)$, to support the queries $rank(X, i, \alpha)$ and $select(X, k, \alpha)$ in constant time for all values of $i \in [1..n]$, $k \in [1..|X|]$, and $\alpha \in \Sigma$.

The case where the two strings S and L have no symbol in common is easy: the distance is then $+\infty$. The algorithm detects this case by testing if $g_\alpha = 0$ for all $\alpha \in \Sigma$, in time within $O(d + m)$.

Consider the algorithm of Figure 1, and let $i \in [1..n]$, $j \in [1..m]$, and $\bar{c} = (c_1, \ldots, c_d)$ be parameters such that $DIST(i, j, \bar{c}) < +\infty$.

At least one of the $c_1, \ldots, c_d$ is equal to zero: in the first entry $DIST(1, 1, \bar{0})$ all the counters $c_1, c_2, \ldots, c_d$ are equal to zero, and any counter is incremented only at line 26, in which another counter must be equal to zero because of the lines 11 and 14.

The number of **insertions** counted in line 21, in previous calls to the function $DIST$ in the recursion path from $DIST(1, 1, \bar{0})$ to $DIST(i, j, \bar{c})$, is equal to $j - i - (c_1 + \cdots + c_d)$. Let t_α denote the number of such insertions for the symbol $\alpha \in \Sigma$. Then, we have

$$j = i + (c_1 + \cdots + c_d) + (t_1 + \cdots + t_d),$$

and for all $\alpha \in \Sigma$, $c_\alpha \leq n_\alpha$, $t_\alpha \leq m_\alpha - n_\alpha$, and

$$c_\alpha + t_\alpha = rank(L, j - 1, \alpha) - rank(S, i - 1, \alpha).$$

Using the above observations, we encode all entries $DIST(i, j, \bar{c})$, for i, j and $\bar{c}$ such that $DIST(i, j, \bar{c}) < +\infty$, into the following table T of $s + 2 \leq d + 2$ dimensions. If we have $s = d$, then

$$T[p, i, k, r_1, \ldots, r_{d-1}] = DIST(i, j, \bar{c} = (c_1, \ldots, c_d)),$$

where

$$c_p = 0,$$
$$(r_1, \ldots, r_{d-1}) = (x_1, \ldots, x_{p-1}, x_{p+1}, \ldots, x_d)$$
$$x_\alpha = \begin{cases} c_\alpha & \text{if } n_\alpha \leq m_\alpha - n_\alpha \\ t_\alpha & \text{if } m_\alpha - n_\alpha < n_\alpha \end{cases} \text{ for every } \alpha \in \Sigma, \text{ and}$$
$$k = (c_1 + \cdots + c_d) + (t_1 + \cdots + t_d) - (r_1 + \cdots + r_{d-1}).$$

Furthermore, given any combination of values $i, j, c_1, \ldots, c_d$ we can switch to the values $p, i, k, r_1, \ldots, r_{d-1}$, and vice versa, in time within $O(d)$. Otherwise, if $s < d$, then

$$T[i, k, r_1, \ldots, r_s] \;=\; DIST(i, j, \bar{c} = (c_1, \ldots, c_d)),$$

where $(r_1, \ldots, r_s) = (x_1, \ldots, x_s)$. Again, given the values $i, j, c_1, \ldots, c_d$ we can switch to the values $i, k, r_1, \ldots, r_s$, and vice versa, in $O(d)$ time.

Since $p \in [1..d]$, $i \in [1..n+1]$, $k \in [0.. \sum_{\alpha=1}^{d}(m_\alpha - g_\alpha)]$, and $r_\alpha \in [0..g_\alpha]$ for every α, the table T can be as large as $d \times (n+1) \times (1 + \sum_{\alpha=1}^{d}(m_\alpha - g_\alpha)) \times (g_2 + 1) \times \cdots \times (g_d + 1)$ if $s = d$, and as large as $(n+1) \times (1 + \sum_{\alpha=1}^{d}(m_\alpha - g_\alpha)) \times (g_1 + 1) \times \cdots \times (g_s + 1)$ if $0 < s < d$. For $s = 0$, no table is needed. The running time of this new algorithm includes the $O(d(n+m)) = O(dm)$ time for processing each of S and L for $rank$ and $select$, and the time to compute $DIST(1, 1, \bar{0})$ which is within $O(d)$ times $n + m$ plus the number of cells of the table T. If $s = d$, the time to compute $DIST(1, 1, \bar{0})$ is within

$$O\left(d(n+m) + d^2 n \cdot \sum_{\alpha=1}^{d}(m_\alpha - g_\alpha) \cdot (g_2 + 1) \cdots (g_d + 1) \right).$$

Otherwise, if $0 \le s < d$, the time to compute $DIST(1, 1, \bar{0})$ is within

$$O\left(d(n+m) + dn \cdot \sum_{\alpha=1}^{d}(m_\alpha - g_\alpha) \cdot (g_1 + 1) \cdots (g_s + 1) \right).$$

The result follows by noting that: if $s = d$, then $\Sigma_+ = \{2, \ldots, d\}$. Otherwise, if $s < d$, then $\Sigma_+ = \{1, \ldots, s\}$. □

The result above, about the complexity in the worst case over instances with $d, n_1, \ldots, n_d, m_1, \ldots, m_d$ fixed, implies results in less precise models, such as in the worst case over instances for d, n, m fixed:

Corollary 1. *Given two strings S and L over the alphabet Σ, of respective sizes n and m, the algorithm analyzed in Theorem 1 computes the* SWAP-INSERT CORRECTION *distance $\delta(S, L)$ in time within*

$$O\left(d(n+m) + d^2 n(m-n) \left(\frac{n}{d-1} + 1 \right)^{d-1} \right),$$

which is within $O\left(n + m + n^d(m-n) \right)$ for alphabets of fixed size d; and within

$$O\left(d(n+m) + d^2 n^2 \left(\frac{m-n}{d-1} + 1 \right)^{d-1} \right),$$

which is within $O\left(n + m + n^2(m-n)^{d-1} \right)$ for alphabets of fixed size d.

4   Discussion

The exact running time of our algorithm is within

$$
O\left(d(n + m) + d^2 n \cdot \sum_{\alpha=1}^{d}(m_\alpha - g_\alpha) \cdot \prod_{\alpha \in \Sigma_+} (g_\alpha + 1) \right),
$$

where n_α and m_α are the respective number of occurrences of symbol $\alpha \in [1..d]$ in S and L respectively; where the vector formed by the values $g_\alpha = \min\{n_\alpha, m_\alpha - n_\alpha\}$ measures the distance between $(n_1, \ldots, n_\sigma)$ and $(m_1, \ldots, m_\sigma)$; and where $\Sigma_+ = \{\alpha \in \Sigma : g_\alpha > 0\}$ if $g_\alpha = 0$ for any $\alpha \in \Sigma$, and $\Sigma_+ = \Sigma \setminus \{\arg\min_{\alpha \in \Sigma} g_\alpha\}$ otherwise.

Summarizing the disequilibrium between the frequency distributions of the symbols in the two strings via the measure $g = \max_{\alpha \in \Sigma} g_\alpha \leq n$, this yields a complexity within $O(d^2 nmg^{d-1})$, which is polynomial in n and m, and exponential only in d of base g. Since this disequilibrium g is smaller than the length n of the smallest string S, this implies a worst case complexity within $O(d^2 mn^d)$ over instances formed by strings of lengths n and m over an alphabet of size d, a result matching the state of the art [4] for this problem.

4.1   Implicit Results

The result from Theorem 1 implies the following additional results:

Weighted Operators: Wagner and Fisher [7] considered variants where the cost c_{ins} of an insertion and the cost c_{swap} of an swap are distinct. In the SWAP-INSERT CORRECTION problem, there are always $n - m$ insertions, and always $\delta(S, L) - n + m$ swaps, which implies the optimality of the algorithm we described in such variants.

Implied Improvements When Only Swaps Are Needed: Abu-Khzam et al. [1] mention an algorithm computing the SWAP STRING-TO-STRING CORRECTION distance (i.e. only swaps are allowed) in time within $O(n^2)$. This is a particular case of the SWAP-INSERT CORRECTION distance, which happens exactly when the two strings are of the same size $n = m$ (and no insertion is neither required nor allowed). In this particular case, our algorithm yields a solution running in time within $O(dm)$, hence improving on Abu-Khzam et al.'s solution [1].

Effective Alphabet: Let d' be the effective alphabet of the instance, i.e. the number of symbols α of $\Sigma = [1..d]$ such that the number of occurrences of α in S is a constant fraction of the number of occurrences of α in L (i.e. $n_\alpha \in \Theta(m_\alpha)$). Our result implies that the real difficulty is d' rather than d, i.e. that even for a large alphabet size d the distance can still be computed in reasonable time if d' is finite.

4.2   Perspectives

Those results suggest various directions for future research:

Further Improvements of the Algorithm: our algorithm can be improved
further using a lazy evaluation of the min operator on line 27, so that
the computation in the second branch of the execution stops any time the
computed distance becomes larger than the distance computed in the first
branch. This would save time in practice, but it would not improve the worst-
case complexity in our analysis, in which both branches are fully explored:
one would require a finer measure of difficulty to express how such a modification could improve the complexity of the algorithm

Further Improvements of the Analysis: The complexity of Abu-Khzam et
al.'s algorithm [1], sensitive to the distance from S to L, is an orthogonal
result to ours. An algorithm simulating both their algorithm and ours in
parallel yields a solution adaptive to both measures, but an algorithm using
both techniques in synergy would outperform both on some instances, while
never performing worse on other instances.

Adaptivity for other Existing Distances: Can other STRING-TO-STRING
CORRECTION distances be computed faster when the number of occurrences
of symbols in both strings are similar for most symbols? Edit distances such
as when only insertions or only deletions are allowed are linear anyway, but
more complex combinations require further studies.

Acknowledgement. The authors would like to thank the anonymous referees of
SPIRE 2015 for insightful comments.

References

1. Abu-Khzam, F.N., Fernau, H., Langston, M.A., Lee-Cultura, S., Stege, U.: Charge
   and reduce: A fixed-parameter algorithm for String-to-String Correction. Discrete
   Optimization (DO) **8**(1), 41–49 (2011)
2. Barbay, J., Pérez-Lantero, P.: Adaptive computation of the Swap-Insert Edition
   Distance. arXiv preprint arXiv:1504.07298 (2015)
3. Cormen, T.H., Leiserson, C.E., Rivest, R. L., Stein, C.: Introduction to Algorithms,
   3rd edn. The MIT Press (2009)
4. Meister, Daniel: Using swaps and deletes to make strings match. Theoretical Computer Science (TCS) **562**, 606–620 (2015)
5. Spreen, T.D.: The Binary String-to-String Correction Problem. Master's thesis, University of Victoria, Canada (2013)
6. Wagner, R.A.: On the complexity of the extended String-to-String Correction Problem. In: Proceedings of the Seventh Annual ACM Symposium on Theory Of Computing (STOC), pp. 218–223. ACM (1975)
7. Wagner, R.A., Fischer, M.J.: The String-to-String Correction Problem. Journal of
   the ACM (JACM) **21**(1), 168–173 (1974)
8. Wagner, R.A., Lowrance, R.: An extension of the String-to-String Correction Problem. Journal of the ACM (JACM) **22**(2), 177–183 (1975)

Transforming XML Streams with References

Sebastian Maneth[1]([✉]), Alberto Ordóñez[2], and Helmut Seidl[3]

[1] University of Edinburgh, Edinburgh, UK
smaneth@inf.ed.ac.uk
[2] Universidade da Coruña, Coruña, Spain
alberto.ordonez@udc.es
[3] TU München, München, Germany
seidl@in.tum.de

Abstract. Many useful XML transformations can be formulated through deterministic top-down tree transducers. If transducers process parts of the input repeatedly or in non-document order, then they cannot be realized over the XML stream with constant or even depth-bounded memory. We show that by enriching streams by *forward references* both in the input and in the output, every such transformation can be compiled into a stream processor with a space consumption depending only on the transducer and the depth of the XML document. References allow to produce DAG-compressed output that is guaranteed to be linear in the size of the input (up to the space required for labels). Our model is designed so that without decompression, the output may again serve as the input of a subsequent transducer.

1 Introduction

In many scenarios data arrives in a stream, e.g., sensor readings, news feeds, or large data that cannot fit in memory. If the streamed data is tree structured, such as XML, further challenges arise because documents have nesting-depth as well as width. One basic question is, which tree transformations can be computed with *constant memory*, i.e., memory only depending on the transformation but not on the size of the input stream. It turns out that even very simple transformations cannot be realized with constant memory, e.g., the transformation that removes all subtrees with a certain root label from a document adhering to a non-recursive DTD (see [11]). Therefore we consider a milder restriction: a tree transformation is *left-depth-bounded memory* (LDBM), if it can be computed with memory only depending on the transformation and on the *left-depth* of the input tree. Similar to the ordinary depth, the left-depth of a ranked tree is defined as the maximal length of a path from the root to a leaf, with the difference that edges from nodes to their *right-most* (last) child are not counted. Thus, monadic trees have left-depth 0, ordinary lists have left-depth 1, and the left-depth of the binary tree representation of an XML document corresponds to the nesting-depth of the document. This is practically relevant since most XML documents are of small nesting-depth.

© Springer International Publishing Switzerland 2015
C. Iliopoulos et al. (Eds.): SPIRE 2015, LNCS 9309, pp. 33–45, 2015.
DOI: 10.1007/978-3-319-23826-5_4

There are two fundamental limitations of LDBM tree translations:

- subtrees must be transformed in the order they arrive;
- subtrees may not be transformed multiple times ("copied").

To see this, consider first the transformation that flips the order of two lists: on input $\mathsf{root}(l_1, l_2)$ it outputs $\mathsf{root}(l_2, l_1)$, where l_1, l_2 are arbitrary lists. The translation is not LDBM because l_1 must be stored in memory. For the same reason, the translation from $\mathsf{root}(l_1)$ to $\mathsf{root}(l_1, l_1)$ is not LDBM. Translations that flip or copy, frequently appear in practice. In this paper we enrich streams by *forward references*. A forward reference is a pointer to a later position in the stream. It is given by a *label* which is defined later in the stream. Labels are considered as abstract data objects (of size "1") that can only be created and compared for equality; we represent them by natural numbers. A valid output for the flip of $\mathsf{root}(l_1, l_2)$ is this output stream with references:

$$\mathsf{root}(\mathsf{ref}\ 1, \mathsf{ref}\ 2)\ 2{:}l_1\ 1{:}l_2.$$

The translation generating this output can be performed in LDBM. Similarly, the copy of the list l_1 can be produced in LDBM by outputting $\mathsf{root}(\mathsf{ref}\ 1, \mathsf{ref}\ 1)\ 1 :$ l_1, corresponding to a DAG-compressed representation of the corresponding tree. A stream processor takes as input an enriched stream (i.e., a serialization of a DAG) and produces as output an enriched stream. Thus, the initial input may be any DAG representing the tree. Note that the minimal DAG of typical XML documents has only 10% of the edges of the original tree [1]. We show: **(1)** Any *deterministic top-down tree transducer* (*dtop*) can be transformed into an LDBM stream processor. The memory and the number of references in the output is bounded by the depth of the input, the transducer, and the number of references in the input. Thus, it is independent of the length m of the input stream. The size of the output is in $\mathcal{O}(m)$, even if the dtop produces exponentially many copies. **(2)** For the same model, the generation of reference chains can be avoided, at the expense of introducing *sets* of labels. The cardinality of those sets is bounded by the maximal sharing in the input and the maximal number of visits of input nodes by the transducer. Our experimental results are summarized at the end of the paper.

Related Work. We are not aware of works that consider tree streams with references. Filiot, Gauwin, Reynier, and Servais [2] show that two large subclasses of visibly pushdown transducers (VPTs) can be streamed with memory only depending on the height of the unranked input tree and on the transducer. For one class, the memory depends exponentially on the height of the input and for the other class it depends quadratically on the input height. The expressive power of VPTs is incomparable to that of dtops, but includes linear size increase dtop translations. They also show that it is decidable for a given VPT, whether or not it can be streamed with height bounded memory.

There are several best-effort implementations of tree transformations that use "as little memory as possible", but do not give guarantees. Most notably, Michael Kay's SAXON system streams XSLT transformations in a best-effort approach. Michael Kay is also the editor of the W3C working draft on XSLT 3.0; the primary purpose of that draft is to change the language in order to enable streamed processing. Note that only very restricted transformations can be formulated through the stream primitives of XSLT 3.0. For XQuery there are several best-effort systems, e.g., Raindrop [13], GCX [5], and XTISP [3]. Note that XTISP is based on a more general transducers than dtops; they do not give memory guarantees, but show that their system performs on par with the state-of-the art XQuery streaming engine GCX.

2   Streams with References

An XML document is modeled as a *ranked tree*, in which the last child of a node represents the next sibling. For instance, if the content of an element consists of three lists, then the element is represented by a node of rank four. In this way, a ranked tree transducer can change the order of the content. See [6] where such encodings are called "DTD-based". For the rest of of the paper (except the Appendix) the details of the encoding are not relevant. A *ranked alphabet* is a finite set Σ of symbols each equipped with a non-negative integer called its *rank*. The rank of a symbol σ determines the number of children of nodes labeled by σ. We write $\sigma^{(k)}$ to denote that σ has rank k. Let Lab be an infinite set of *labels*. We require that the set Lab is equipped with a method new(), which, given a current active set of labels $L \subseteq$ Lab, returns a symbol in Lab $- L$. *Ref-trees* t (over Σ with forward references over Lab), *sequences* s *of definitions*, and *streams* p are defined by the following grammar:

$$t ::= a\,(t_1, \ldots, t_k) \mid \mathsf{ref}\ l$$
$$s ::= \epsilon \mid L\!:\!t\ s$$
$$p ::= t\ s$$

where $a \in \Sigma$ is of rank $k \geq 0$, $l \in$ Lab, $L \subseteq$ Lab with $L \neq \emptyset$, and, if ref l occurs in s then $L\!:\!t$ occurs in s to the right for some t and L with $l \in L$. The *size* of a ref-tree t, a sequence s, or a stream p counts the number of occurring symbols (ignoring the meta-symbols brackets, commas or colons; cf. Section 3) where a reference ref l counts for one and a label set L counts for the cardinality of L. The size function is denoted by $|\,.\,|$. A *tree* (*over* Σ) is a ref-tree t without occurrences of ref l and of $L : t\ s$. We denote by $\mathcal{S}_\Sigma$ the set of streams (with forward references over Lab) and by $\mathcal{T}_\Sigma$ the set of all trees (over Σ). Textually, label sets will be written as lists, i.e., without set delimiters. For $\Sigma = \{a^{(2)}, b^{(2)}\}$, the following

$$a(\mathsf{ref}\ 1, a(\mathsf{ref}\ 2, \mathsf{ref}\ 1))\ 1\!:\!b(\mathsf{ref}\ 2, \mathsf{ref}\ 2)\ 2\!:\!e$$

is a stream in $\mathcal{S}_\Sigma$. This stream represents the tree $a(b(e, e), a(e, b(e, e)))$. Formally, for a stream $p = t\ s$ we define $\mathsf{decode}(p) = \mathsf{dec\_t}(t, \mathsf{dec\_s}(s, \emptyset))$. The function $\mathsf{dec\_s}$ computes a partial function $E :$ Lab $\to \mathcal{T}_\Sigma$, while $\mathsf{dec\_t}$ computes a tree in $\mathcal{T}_\Sigma$:

$$
\begin{aligned}
\mathsf{dec\_t}(\mathsf{ref}\ l, E) \quad &= E(l)\\
\mathsf{dec\_t}(a(t_1, \ldots, t_k), E) &= a(\mathsf{dec\_t}(t_1, E), \ldots, \mathsf{dec\_t}(t_k, E))\\
\mathsf{dec\_s}(\epsilon, E) \quad &= E\\
\mathsf{dec\_s}(L\!:\!t\ s, E) \quad &= \mathsf{dec\_s}(s, E) \oplus \{l \mapsto \mathsf{dec\_t}(t, \mathsf{dec\_s}(s, E)) \mid l \in L\}
\end{aligned}
$$

where the operation $\oplus$ updates the partial function in the first argument according to the argument-value pairs provided in the second argument.

Left-Depth. The structure of an XML document corresponds to a well-balanced sequence of opening and closing tags. Checking well-balancedness of a sequence of tags can be done with memory proportional to the nesting-depth of the sequence: for each opening tag we push its name onto a stack, and for each closing tag we pop the matching name (or report an error). In our ranked tree encodings of XML structures, the last child represents the next sibling of the original node. Therefore, the *left-depth* corresponds to the nesting-depth of the XML structure. For a stream $p = t\,s$ we define $\mathsf{ldepth}(p) = \max\{\mathsf{ldepth\_t}(t), \mathsf{ldepth\_s}(s)\}$, where:

$$
\begin{aligned}
\mathsf{ldepth\_t}(\mathsf{ref}\ l) &= 1 \\
\mathsf{ldepth\_t}(a(t_1,\dots,t_k)) &= \max\{1+\mathsf{ldepth\_t}(t_1),\dots,1+\mathsf{ldepth\_t}(t_{k-1}), \mathsf{ldepth\_t}(t_k)\} \\
\mathsf{ldepth\_s}(\epsilon) &= 0 \\
\mathsf{ldepth\_s}(L\!:\!t\ s) &= \max\{\mathsf{ldepth\_t}(t), \mathsf{ldepth\_s}(s)\}.
\end{aligned}
$$

3 Top-Down Tree Transducer to Stream Processor Translation

Top-down tree transducers generalize (top-down) tree automata by producing output. They were invented in the 1970's as formal models for linguistics and compilers, and have recently been applied to XML [9,8,10,7]. Formally, a top-down tree transducer is a tuple $M = (Q, \Sigma, \Delta, q_0, R)$ where Q is a finite set of states, Σ and Δ are ranked alphabets of input and output symbols, $q_0 \in Q$ is the initial state, and R is a finite set of rules of the form $q(a(x_1,\dots,x_k)) \to t$, where a is an input symbol of rank k and t is generated by this grammar:

$$
t ::= q(x_i) \mid b(t_1,\dots,t_m)
$$

for $q \in Q$, $i \in \{1,2,\dots,k\}$, and $b \in \Delta$ of rank $m \geq 0$. If for each $q \in Q$ and $a \in \Sigma$ there is at most one rule in R of the form $q(a(x_1,\dots,x_k)) \to t$, then M is a *deterministic top-down tree transducer* (dtop); in this case t is denoted by $\mathsf{rhs}(q,a)$. Let M be a dtop, $q \in Q$, $a \in \Sigma$, and $t_1,\dots,t_k \in \mathcal{T}_\Sigma$. The q-translation $[\![q]\!]$ is the function from $\mathcal{T}_\Sigma$ to $\mathcal{T}_\Delta$ defined recursively as: $[\![q]\!](a(t_1,\dots,t_k)) = \mathsf{rhs}(q,a)[q'(x_i) \leftarrow [\![q']\!](t_i) \mid q' \in Q, i \in \{1,\dots,k\}]$. For leaf labels $d_1,\dots,d_n$ and trees $s_1,\dots,s_n$, $[d_j \leftarrow s_j \mid 1 \leq j \leq n]$ denotes the tree substitution of replacing each occurrence of d_j (with $1 \leq j \leq n$) by the tree s_j. The translation $[\![M]\!]$ of M is defined as $[\![q_0]\!]$.

As an example, let $\Sigma = \Delta = \{\mathsf{root}^{(2)}, a^{(2)}, e^{(0)}\}$. The dtop with states q_0, q and rules $q_0(\mathsf{root}(x_1,x_2)) \to \mathsf{root}(q(x_2),q(x_1))$ and $q(a(x_1,x_2)) \to a(q(x_1),q(x_2))$, and $q(e) \to e$ realizes the flip transformation of the Introduction: it translates $\mathsf{root}(t_1,t_2)$ into $\mathsf{root}(t_2,t_1)$, where t_1, t_2 are trees over $a^{(2)}$ and $e^{(0)}$. If we replace the first rule of this transducer by the rule $q_0(\mathsf{root}(x_1)) \to \mathsf{root}(q(x_1),q(x_1))$, then input trees $\mathsf{root}(t)$ are translated to $\mathsf{root}(t,t)$ for trees t over $a^{(2)}$ and $e^{(0)}$.

For giving correctness proofs of our constructions later, we now extend the partial functions $[\![q]\!]$ of the dtop from trees to ref-trees possibly containing references to labels $l \in \mathsf{Lab}$ by defining $[\![q]\!](\mathsf{ref}\ l) = \langle q, l \rangle$ for new output symbols $\langle q, l \rangle$. The following is a basic substitution lemma; it states that if a dtop translates a ref-tree with labels, then placing the translation $[\![q]\!](E(l))$ wherever l is visited by state q, yields the same tree as running the dtop over the decoded reference-less input tree. The lemma can be proved by induction on the size of t.

Lemma 1. *Let t be a ref-tree. Assume that L is the set of labels referenced in t, and that N is the set of all pairs $\langle q', l \rangle$ occurring in $[\![q]\!](t)$. Furthermore, assume that $E : L \to \mathcal{T}_{\Sigma}$. Then $[\![q]\!](\mathsf{dec\_t}(t, E))$ is defined iff $[\![q']\!](E(l))$ is defined for all $\langle q', l \rangle \in N$, where $[\![q]\!](\mathsf{dec\_t}(t, E)) = [\![q]\!](t)[\langle q', l \rangle \leftarrow [\![q']\!](E(l)) \mid \langle q', l \rangle \in N]$.* □

The Translation. Note that, due to the ranks of symbols $a \in \Sigma$, brackets and commas in streams are redundant. In the following, we therefore ignore these, when it comes to parsing or outputting ref-trees and streams. A stream processor S thus can be considered as a string transducer $S = (\mathsf{conf}, \bar{\Sigma}, \vdash, \mathsf{init}, \mathsf{Final})$ where conf is a (possibly infinite) set of *configurations*, init and Final are the initial configuration and the set of final configurations, $\bar{\Sigma}$ is the set of characters or tokens in the input or output, and $\vdash: \mathsf{conf} \times (\bar{\Sigma} \cup \{\epsilon\}) \to \mathsf{conf} \times \bar{\Sigma}^*$ is a partial function describing the one-step transitions of the processor. For the processor to operate deterministically, we demand that $\vdash (q, a)$ is undefined for every $a \in \bar{\Sigma}$ whenever $\vdash (q, \epsilon)$ is defined. As usual, the function $\vdash$ is written infix. The relation $\vdash$ is extended from symbols to strings and output strings as third component by defining $(q, aw, o) \vdash (q', w, o\ o')$ whenever $(q, a) \vdash (q', o')$ is a transition of M and o, w are strings. Now assume that we are given a dtop $M = (Q, \Sigma, \Delta, q_0, R)$. We construct the stream processor S_M which is meant to traverse an input stream p with $\mathsf{decode}(p) = t \in \mathcal{T}_{\Sigma}$, and produce an output stream p' with $\mathsf{decode}(p') = [\![M]\!](t)$. The set $\bar{\Sigma}$ of tokens of the stream processor S_M consists of the alphabets Σ and Δ extended by tokens ref l with $l \in \mathsf{Lab}$, and L: with L a finite subset of Lab. The configurations of the stream processor are of the following form:

$$\begin{aligned}
\mathsf{conf} &\subseteq (\mathsf{lmap} \times \mathsf{stack} \times \mathsf{Lab}) \cup \{\mathsf{init}\} \\
\mathsf{lmap} &\subseteq (Q \times \mathsf{Lab}) \to 2^{\mathsf{Lab}} \\
\mathsf{stack} &\subseteq \mathsf{local}^* \\
\mathsf{local} &\subseteq Q \to 2^{\mathsf{Lab}}.
\end{aligned}$$

Thus, a configuration of the stream processor either equals the initial configuration init, or consists of the following components:

- A mapping lmap which records for each state q of the dtop and label l in the input, the set of labels in the output stream which refer to a representation of the output for the corresponding subtree translated in state q of M. According to this usage, each label may occur at most once in any of the sets $\mathsf{lmap}(q, l)$.

– A stack of *local configurations* from local where each such local configuration ϕ provides for each state q of M the set of output labels whose definition is meant to be the output of the current part of the input for state q of M.

– A local configuration ϕ from local is a mapping from states of M to (possibly empty) sets of output labels;

– The next fresh label to be used in the output stream. For convenience in our implementation we assume labels to be natural numbers (starting from 0).

The transition relation of S_M is denoted $\vdash_M$. In the following, we use the convention that we only list nontrivial argument-value pairs of a mapping, i.e., where the value is different from $\emptyset$. The empty stack is denoted by $[]$. Initially (from configuration init), a first label $l_0 \in$ Lab is produced. According to our convention, $l_0 = 0$. Then a reference to 0 is outputted, and a configuration is created in which the initial state q_0 maps to 0: $(\text{init}, \epsilon) \vdash_M ((\emptyset, \{q_0 \mapsto \{0\}\}, 1), \text{ref } 0)$. For the next cases we consider a configuration $c = (\lambda, \sigma, l')$, where $\lambda : (Q \times \text{Lab}) \to 2^{\text{Lab}}$, $\sigma \in (Q \to 2^{\text{Lab}})^*$, and $l' \in$ Lab. Assume that the next symbol in the input stream is a constructor $a \in \Sigma$, and $\sigma = \phi :: \sigma'$ where $\phi = \{q_1 \mapsto L_1, \ldots, q_r \mapsto L_r\}$ for non-empty sets $L_j \subseteq$ Lab, and $r \geq 0$. Furthermore assume that $q_j(a(x_1, \ldots, x_k)) \to t_j \in R$ for $j = 1, \ldots, r$; note that if any such rule does not exists, then $\vdash$ is correctly undefined (because the dtop blocks and does not produce output). Then we consider a set L of fresh labels containing one distinct label $l_{q',i}$ for every $q'(x_i)$ occurring in any of the t_j. For $i = 1, \ldots, k$, let ϕ_i denote the mapping $\phi_i = \{q' \mapsto \{l_{q',i}\} \mid l_{q',i} \in L\}$. Let t'_j be the ref-tree obtained from t_j by replacing $q'(x_i)$ with ref $l_{q',i}$. Then

$$(c, a) \vdash_M (c', L_1 : t'_1 \ldots L_r : t'_r) \quad \text{where} \quad c' = (\lambda, \phi_1 :: \ldots :: \phi_k :: \sigma', l' + |L|).$$

Note that in the specific case where ϕ has been *empty*, i.e., the current ref-tree is not required for the prospected output, then $m = 0$ and the produced output string is ϵ where $L = \emptyset$ and $c' = (\lambda, \emptyset^k :: \sigma', l')$.

Now assume that the next token in the input is a reference ref l to some label l where $\phi = \{q_1 \mapsto L_1, \ldots, q_r \mapsto L_r\}$. Then we define a mapping λ' by $\lambda'(q', l) = \{l'_j\}$ for a new distinct label l'_j if $q' = q_j$ for some j with $\lambda(q_j, l) = \emptyset$, and otherwise $\lambda'(q', l) = \lambda(q', l)$. Thus, new labels are created for every pair (q_j, l) with $j = 1, \ldots, r$, which has not yet been defined in λ. Assume that the number of these new labels is m. Now let select denote a (partial) function which selects one label from each nonempty set of labels, and define $l'_j = \text{select}(\lambda'(q_j, l))$ for $j = 1, \ldots, r$. Then

$$(c, \text{ref } l) \vdash_M (c', L_1 : \text{ref } l'_1 \ldots L_r : \text{ref } l'_r) \quad \text{where} \quad c' = (\lambda', \sigma', l' + m).$$

This means we redirect the references as provided by ϕ to the new references as introduced in λ'. Again, if ϕ is the empty set, the empty output string is produced while $c' = (\lambda, \sigma', l')$.

Finally, assume that the next token in the input is a label set L followed by : indicating a definition. Then new mappings ϕ' and λ' are constructed by

$\phi'(q') = \bigcup\{\lambda(q',l) \mid l \in L\}$ and λ' is obtained from λ by setting the values for $(q',l), l \in L$, to $\emptyset$. With that, the transition is

$$(c, L :) \vdash_M (c', \epsilon) \quad \text{where} \quad c' = (\lambda', \phi' :: \sigma, l').$$

Note that if none of the $l \in L$ has ever been referred to, then $\lambda(q, l)$ is the empty set for every q. This implies that the mapping ϕ' is the empty mapping as well. Consequently, the stream processor will *ignore* the rest of the definition of the $l \in L$, i.e., the subsequent ref-tree in the stream. The stream processor terminates in configurations $(\emptyset, [], l')$.

Reference Reuse. The implementation so far referred to *fresh* labels whenever new labels are required. Labels therefore are unique throughout whole streams. This can cause a blow-up in the bit size of the output by a logarithmic factor. We reduce the number of distinct labels by *reference reuse*. If the same label l occurs several times, a reference ref l simply refers to the *next* occurrence of l to the right. In this way, only as many labels are required as have already been referred to but have not yet been defined. Technically, these consist of all labels which are mentioned either in the current stack σ or in one of the output sets of the global mapping λ of the current configuration. Let $L(\sigma, \lambda)$ denote this set. Then the next new label can be generated without referring to a counter which is passed around. Instead, it can be chosen as the smallest natural number not contained in $L(\sigma, \lambda)$. Fresh labels are possibly introduced when processing an input tag a or a reference ref l in the input. On the other hand, a label l' for the output goes *out of scope* as soon as some set L': is produced with $l' \in L'$. The label l' may be used as a fresh label already in the sub-stream succeeding the colon.

Theorem 1. *Every dtop M can be compiled into a stream processor S_M such that for every stream p and input document t with* $\mathsf{decode}(p) = t$ *the following holds:*

1. *If $[\![M]\!](t)$ is undefined then so is $S_M(p)$. If $[\![M]\!](t) = t'$ then $S_M(p) = p'$ for some stream p' with* $\mathsf{decode}(p') = t'$.
2. *The depth of the stack of S_M when processing the input stream p is bounded by $l = (k - 1) \cdot (\mathsf{ldepth}(p) + 1)$,*
3. *The size $|p'|$ of the output stream p' is bounded by $n \cdot d \cdot |p|$.*
4. *The maximal number of distinct labels used by S_M is bounded by $c \cdot n \cdot (l + r)$.*
5. *The cardinality of label sets L occurring in the output is at most $\max\{1, c\}$.*

Here k is the maximal rank of an input symbol occurring in p, n is the number of states of M, d is the maximal size of right-hand sides of the rules of M, r is the number of distinct labels in the input stream, and c is the maximal cardinality of label sets in the input.

Since each output token carries at most one set of labels, the total number of occurrences of labels in the output is bounded by $c \cdot n \cdot d \cdot |p|$, for an input stream p. Therefore, the total bit length of the output stream generated by a given stream processor is in $\mathcal{O}(m \cdot \log(m))$ if m is the bit length of the input stream

— even if the dtop is unboundedly copying and thus may produce output trees of size exponential in the input tree. This even holds for the stream processor *without* reuse of labels. In case of label reuse, the total number of labels to be stored in any configuration is bounded by the number of the labels stored in the stack plus the number of labels stored in the mapping λ, i.e., proportional to $c \cdot n \cdot (l + r)$. Therefore, the bit length of the output can more precisely be bounded by $\mathcal{O}(m \cdot \log(l + r))$ if sets of labels in the input stream (and thus also in the output stream) are assumed to have constant size. Likewise, the maximal memory consumption in bits used by the stream processor can be bounded by $\mathcal{O}((l + r) \cdot \log(l + r))$.

The correctness of the stream processor S_M in Theorem 1 with respect to the dtop M follows from Lemma 3. For simplicity, we only consider in these lemmas the vanilla version of S_M where always fresh labels are created. The proof of Lemma 3 uses Lemma 2 which can be proved by structural induction on t (using Lemma 1).

When we refer to input (resp. output) ref-trees, sequences of definitions, or streams, we mean ref-trees (etc) "over Σ (resp. Δ) with forward references over Lab".

Lemma 2. *Let* $\lambda \subseteq Q \times \mathsf{Lab} \to 2^{\mathsf{Lab}}$, $\phi = \{q_1 \mapsto L_1, \ldots, q_r \mapsto L_r\}$, $\sigma \subseteq (Q \to 2^{\mathsf{Lab}})^*$, l_1 *denote a label exceeding all labels in* λ, ϕ, σ, s *a sequence of definitions, and* o *an output string. Then the following two statements for every ref-tree* t *are equivalent:*

1. $[\![q_j]\!](t)$ *is defined for all* $j \in \{1, \ldots, r\}$;
2. $((\lambda, \phi::\sigma, l_1), t\,s, o) \vdash_M^* ((\lambda \cup \lambda_2, \sigma, l_2), s, o\,o_2)$ *for an output sequence of definitions* o', $l_2 \in \mathsf{Lab}$ *and some mapping* λ_2 *such that*
   (a) *All labels* l' *in the image of* λ_2 *are fresh, i.e.,* $l_1 \leq l' < l_2$;
   (b) *For all* $q' \in Q$, $l \in \mathsf{Lab}$, $\lambda_2(q', l)$ *has cardinality at most* 1;
   (c) $\lambda_2(q', l) \neq \emptyset$ *iff* $\lambda(q', l) = \emptyset$ *and* $\langle q', l \rangle$ *occurs in* $[\![q_j]\!](t)$ *for some* j.
   (d) *Let* $E(l') = \langle q', l \rangle$ *whenever* $l' \in \lambda_2(q', l)$. *Then for every* $j \in \{1, \ldots, r\}$ *and* $l \in L_j$, $\mathsf{dec\_t}_s(o', E)(l) = [\![q_j]\!](t)$.

Lemma 3. *Let* s *be an input sequence of definitions,* $E = \mathsf{dec\_t}_s(s, \emptyset)$, o *an output string,* $\lambda \subseteq Q \times \mathsf{Lab} \to 2^{\mathsf{Lab}}$, *and* l_1 *a label exceeding all lables occurring in the image of* λ. *Let* N *denote the set of pairs* $\langle q, l \rangle$ *where* $\lambda(q, l) \neq \emptyset$. *Then the following two statements are equivalent:*

1. $[\![q]\!](E(l))$ *is defined for all* $\langle q, l \rangle \in N$;
2. $((\lambda, [], l_1), s, o) \vdash_M^* ((\emptyset, [], l_2), \epsilon, o\,o')$ *for some output sequence of definitions* o' *and* $l_2 \in \mathsf{Lab}$ *where for all* $\langle q, l \rangle \in N$, $[\![q]\!](E(l)) = \mathsf{dec\_t}_s(o', \emptyset)(l')$ *for all* $l' \in \lambda(q, l)$.

Proof. Let $c = (\lambda, [], l_1)$. We proceed by structural induction on the sequence of definitions s. If $s = \epsilon$, then the assertion is vacuously true (because $N = \emptyset$). Now assume that $s = L{:}t\,s'$. Let $\bar{L}$ denote the set of labels l defined in s where $\lambda(q', l) \neq \emptyset$ for some $q' \in Q$. Let $\bar{L}_1 = \bar{L} \cap L$ where $\bar{L}'$ are the labels in $\bar{L}$ defined in s'.

If $\bar{L}_1 = \emptyset$, then $\lambda(q', l) = \emptyset$ for all $q' \in Q$ and $l \in L$. By the definition of $\vdash_M$, $(c, L : t\ s', o) \vdash_M (c', t\ s', o)$ where $c' = (\lambda, \emptyset :: [], l_1)$. According to the definition of $\vdash_M$, only the empty mapping $\emptyset$ is pushed and popped for each symbol of the ref-tree l until we arrive at s' with empty stack. Accordingly, $(c', t\ s', o) \vdash_M^*$ $((\lambda, [], l_1), s', o)$. Since by induction hypothesis $((\lambda, [], l_1), s', o) \vdash_M^* ((\emptyset, [], l_2), o\ o')$, the assertion follows.

Now assume that $\bar{L}_1 \neq \emptyset$, and consider the mapping $\phi(q') = \bigcup \{\lambda(q', l) \mid l \in L\}$, which we write as usual $\phi = \{q_1 \mapsto L_1, \ldots, q_r \mapsto L_r\}$. Assume that $[\![q]\!](E(l))$ is defined for all $\langle q, l \rangle \in N$, in particular those where $l \in L$.

Let λ_1 be obtained from λ by removing all entries for $l \in L$. Then by Lemma 2, there is a mapping λ_2, an output sequence o_1 of definitions, and a next label l_2 with the following properties:

1. $((\lambda_1, \phi :: [], l_1), ts', o) \vdash_M^* ((\lambda_1 \cup \lambda_2, [], l_2), s', o\ o_1)$
2. $\lambda_2(q', l) \neq \emptyset$ iff $\lambda_1(q', l) = \emptyset$ and $\langle q', l \rangle$ occurs in $[\![q_j]\!](t)$ for some j.
3. Let $E_1(l') = \langle q', l \rangle$ whenever $l' \in \lambda_2(q', l)$. Then for every j and $l \in L_j$, $\mathsf{dec\_t}_s(o_1, E_1)(l) = [\![q_j]\!](t)$.

Now let $\lambda' = \lambda_1 \cup \lambda_2$. By induction hypothesis for s' and λ', $((\lambda', [], l_2), s', o\ o_1) \vdash_M^* ((\emptyset, [], l_3), o\ o_1\ o_2)$. for some o_2 and l_3 where

$$[\![q']\!](E(l')) = \mathsf{dec\_t}_s(o_2, \emptyset)(l'')$$

for all $l'' \in \lambda'(q', l')$. Therefore, we have:

$$((\lambda, [], l_1), s, o) \vdash_M ((\lambda_1, \phi :: [], l_1), t\ s', o) \vdash_M^* ((\lambda', [], l_2), s', o\ o_1)$$
$$\vdash_M^* ((\emptyset, [], l_3, o\ o_1\ o_2)$$

where for $l \in \bar{L}_1$, and $l' \in \lambda(q, l)$,

$$[\![q]\!](E(l)) = [\![q]\!](t)[\langle q', l'' \rangle \leftarrow \mathsf{dec\_t}_s(o_2, \emptyset)(l_{q', l''})]$$
$$= \mathsf{decode}(o_1, \mathsf{dec\_t}_s(o_2, \emptyset))(l')$$
$$= \mathsf{decode}(o_1\ o_2, \emptyset)(l')$$

while for $l \in \bar{L}'$, the assertion follows by induction hypothesis for s'. This proves the first direction. The reverse direction follows analogously.   □

Altogether, we thus have shown that the stream processor S_M when applied to the stream representation of a tree t, produces a stream representation of the output tree returned by M for t.

Avoiding Reference Chains. The disadvantage of the construction so far is that it may abundantly generate references which themselves may point to references. In particular, this is the case if the dtop has erasing rules, i.e., rules whose right-hand sides do not produce any output nodes but consist of recursive calls only.

Example 1. Consider the dtop with the following rules

$$q_0(f(x_1, x_2)) \rightarrow q_1(x_1) \qquad q_1(f(x_1, x_2)) \rightarrow q_0(x_2)$$
$$q_0(\bot) \qquad\quad \rightarrow a \qquad\quad q_1(\bot) \qquad\quad \rightarrow b$$

where q_0 is the initial state. The corresponding stream processor translates the input stream $f(\text{ref } l_1, \bot) \; l_1 : f(\bot, \text{ref } l_2) \; l_2 : \bot$ into the stream $\text{ref } 0 \;\; 0 : \text{ref } 1 \;\; 1 : \text{ref } 2 \;\; 2 : a$. $\qquad\square$

In order to avoid such chains of references, we modify the rules for constructor applications and references as follows. Assume that $a \in \Sigma$ and $\sigma = \phi :: \sigma'$ with $\phi = \{q_1 \mapsto L_1, \ldots, q_r \mapsto L_r\}$. Furthermore, assume that for $j = 1, \ldots, r$, $q_j(a(x_1, \ldots, x_k)) \rightarrow \zeta_j$. Then we consider the set N_i the set of all states $q' \in Q$ with the following two properties:

- $q'(x_i)$ occurs as a proper subterm in any of the ζ_j;
- $q'(x_i)$ is different from any of the ζ_j.

For $i = 1, \ldots, k$, let ϕ_i denote the mapping defined by $\phi_i(q') = \{l_{q',i}\}$ if $q' \in N_i$, and otherwise $\phi_i(q') = \bigcup\{L_j \mid \zeta_j = q'(x_i)\}$, where the $l_{q',i}$ are fresh labels. Let $j_1, \ldots, j_m$ be the subsequence of the j' where ζ_j produces output nodes, i.e., is *not* just equal to some recursive call $q'(x_{i'})$. For $j' \in \{j_1, \ldots, j_m\}$, let $\zeta'_{j'}$ be obtained from $\zeta_{j'}$ by replacing $q'(x_i)$ with $\text{ref } l'_{q',i}$ for some $l'_{q',i} = \text{select}(\phi_i(q'))$. Then

$$o' = L_{j_1} : \zeta'_{j_1} \ldots L_{j_m} : \zeta'_{j_m} \quad \text{and} \quad c' = (\lambda, \phi_1 :: \ldots :: \phi_k :: \sigma').$$

A second modification takes place for processing references in the input. Let $\text{ref } l$ be a reference in the input and $\sigma = \phi :: \sigma'$ where $\phi = \{q_1 \mapsto L_1, \ldots, q_r \mapsto L_r\}$. We define a mapping λ' by $\lambda'(q_j, l) = \lambda(q_j, l) \cup L_j$ for $j = 1, \ldots, r$ and $\lambda'(q', l') = \lambda(q', l')$ otherwise. Thus, no new labels are created at all. Also, no output is produced, λ is updated to λ', and σ is popped to σ', i.e., $c' = (\lambda', \sigma')$.

Example 2. Consider again the dtop from Example 1 together with the input stream $f(\text{ref } l_1, \bot) \; l_1 : f(\bot, \text{ref } l_2) \; l_2 : \bot$. Then no new label is ever introduced. Instead, the output is given by $\text{ref } 0 \; 0 : a(\bot, \bot)$. $\qquad\square$

In the previous example, the output stream without reference chains is much simpler than the original output stream. In some cases, though, sets of labels in the output stream grow considerably.

Example 3. Consider the following dtop with states q, q':

$$q(a(x_1, x_2)) \rightarrow b(q(x_2), q'(x_2)) \qquad q'(a(x_1, x_2)) \rightarrow q'(x_2)$$
$$q(\bot) \qquad\quad \rightarrow c(\bot, \bot) \qquad\qquad q'(\bot) \qquad\quad \rightarrow d(\bot, \bot)$$

Assume that q is the initial state. The input stream $a(\bot, a(\bot, a(\bot, \bot)))$ is translated to $\text{ref } 0 \;\; 0 : b(\text{ref } 1, \text{ref } 2) \; 1 : b(\text{ref } 3, \text{ref } 4) \; 3 : b(\text{ref } 5, \text{ref } 6) \; 5 : c(\bot\bot) \; 2, 4, 6 : d(\bot, \bot)$ Thus, the labels for the erasing calls of state q' all are collected into one label set. $\qquad\square$

A second source for the growing of label sets may be sharing present in the input stream.

Example 4. Consider the following dtop with the single state q implementing the identity by these two rules: $q(a(x_1, x_2)) \rightarrow a(q(x_1), q(x_2))$ and $q(\perp) \rightarrow \perp$. Consider the input stream $a(\text{ref } l_1, a(\text{ref } l_1, a(\text{ref } l_2, \perp)))$ l_2 : ref l_1 l_1 : $\perp$ containing four occurrences of references to the same node. This input stream is translated into the output stream

$$\text{ref } 0 \quad 0 : a(\text{ref } 1, \text{ref } 2) \ 2 : a(\text{ref } 3, \text{ref } 4) \ 4 : a(\text{ref } 5, \text{ref } 6) \ 6 : \perp 1, 3, 5 : \perp$$

This means that all labels introduced for the occurrences of the references ref l_1, ref l_2 are collected into one label set.   □

In fact, these two mechanisms are the only reasons how large sets of labels in the output set may be accumulated. We have:

Theorem 2. *For a dtop M, a streaming processor S'_M can be constructed with the following properties: (1) The output stream does not contain chains of references even if such subexpressions occur in the input stream. (2) The cardinality of any label set in the output is bounded by $a \cdot b$, where a is the maximal sharing in the input, and b is the number of visits of M to nodes in the tree unfolding of the input.*

Hereby, the *maximal sharing* of a stream p is the maximal number of occurrences of references ref l all of which directly or indirectly point to the same subtree. The maximal number of visits to subtrees of the input, on the other hand, is the maximal number of occurrences of leaves $\langle q, l \rangle$ in the output produced by M for any input tree which contains a single reference ref l.

Example 5. Consider the dtop implementing the identity from Example 4, now together with the input stream $a(\text{ref } l_1, a(\text{ref } l_1, a(\text{ref } l_1, \perp)))$ $l_1 : \perp$ which contains three occurrences of references to the same leaf with label l_1. This input stream is translated into ref 0 $0 : a(\text{ref } 1, \text{ref } 2)$ $2 : a(\text{ref } 3, \text{ref } 4)$ $4 : a(\text{ref } 5, \text{ref } 6)$ $6 : \perp 1, 3, 5 : \perp$. This means that all labels introduced for the occurrences of the reference ref l_1 are collected into one label set.   □

Experiments. We implemented a prototype of our stream processor in C++, using the Xerces-C++ parser version 3.1.1. One experiment is to filter out certain subtrees (e.g., all article- or all book-subtrees) from a Dblp bibliography XML file. This is done in different ways: (a) produce only the article-subtrees, (b) produce the original document together with the list of all article-subtrees, and (c) produce with each article a list of all following articles. Here inlining makes a huge difference, essentially halving the output size for (a) and (b). Since (c) is of quadratic size increase, the output documents with references are dramatically smaller than output documents without references.

Another experiment recursively flips first and second subtree; here we observe that the space overhead caused by references is a factor between two and three, and, if references are reused and inlining is applied, this factor goes down to between 1.2 and 1.7. The importance of reference chain removal can be seen when applying the flip transformation several times in sequence: no extra overhead is incurred, while without chain removal an increasing amounts of overhead is produced.

For all experiments we ran, even when running many transducers in sequence, we have never seen a larger space overhead than a factor three. The memory consumption of the our processor is minuscule throughout all experiments, at about 5% of the memory usage used by the XML parser.

4   Conclusion

Through the enhancement of XML streams by forward references, the large and natural class of deterministic top-down tree transformations can be realized with bounded memory. The memory only depends on the depth of the XML document (corresponding to the ldepth of the tree encoding), the number of distinct references in the input, and the transducer.

The choice of dtops as class of tree transformations, and forward references as way of achieving LDBM, may seem rather ad hoc. Let us explain why these two choices are not only natural, but also (in a certain sense) maximal. Our decisions are directed by the two desires (1) never to produce *garbage* (label definitions that are not referenced), and (2) to be compositional, i.e., the output of one processor can serve as input to another processor. A priori, we do not have a preference with respect to forward or backward references. In fact, even allowing both types of references is a viable choice. It is not difficult to see that for realizing the translation of a dtop, the output either may be represented by means of forward references as suggested so far, or alternatively by means of backward references. For instance, for the flip of two lists we can as well produce this stream with backward references, $1 : l_1$ $2 : l_2$ $0 : \mathsf{root}(\mathsf{ref}\ 2, \mathsf{ref}\ 1)$ where the label 0 represents the root of the output tree. But with such a representation, are we compositional? Consider a second dtop which translates $\mathsf{root}(l_1, l_2)$ into the tree $\mathsf{root}(l_1', l_2)$ where l_1' contains only every odd element of the list l_1. Let τ_{odd} denote this translation. Taking as input the stream displayed above, how can this transformation be realized? This *cannot* be done with LDBM, because each element of l_1 must be kept in memory (until we know where in the input relative to the root it appears). If on the other hand, garbage is allowed, then LDBM is possible by producing both translations l_1 and l_1', and later inserting a reference to the correct one (leaving the other as garbage).

Proposition 1. *The dtop translation* τ_{odd} *cannot be realized in* LDBM *over streams with backward references (unless garbage is allowed).*

We conclude that backward references in the input must be ruled out, if we want to handle arbitrary dtops. On the other hand, can we handle larger classes of transformations than dtops, by using forward references only (and no garbage)? Clearly, the addition of regular look-ahead is not possible: in some cases, it would require the generation of garbage for similar reasons as above.

Let us review the two other important generalizations of dtops: (1) deterministic top-down tree-to-string (dts) transducers, and (2) left-to-right attribute grammars (LR-AG), seen as tree transducers. LR-AGs have been used in the context of XML streaming [4] and already Bochman [12] showed that all attribute values of an LR-AG can be computed in one left-to-right pass through the input tree. Accordingly, LR-AGs seem a natural candidate in our context. Is it always possible for an LR-AG to produce with LDBM output streams with forward references? As an example, consider the transformation which reverses a list, given as binary tree, i.e., which translates trees $r(a_1(\bot, a_2(\bot \ldots a_n(\bot, \bot) \ldots)), \bot)$ into $r(a_n(\bot, a_{n-1}(\bot \ldots a_1(\bot, \bot) \ldots)), \bot)$. This translation τ_{reverse} can be realized by an LR-AG: it uses one inherited attribute to compute the reverse of the a_i-path. This translation can also be realized by a simple dts transducer using two states. It should be clear that this translation *cannot* be realized in LDBM using forward references alone. Therefore, we have:

Proposition 2. τ_{reverse} *cannot be realized in* LDBM *on streams with forward references.*

Propositions 1 and 2 imply that LDBM streaming can neither be extended from dtops to dts transducers nor to LR-AGs without introducing backward references and thus loosing composability with dtops.

References

1. Buneman, P., Grohe, M., Koch, C.: Path queries on compressed XML. In: VLDB, pp. 141–152 (2003)
2. Filiot, E., Gauwin, O., Reynier, P.-A., Servais, F.: Streamability of nested word transductions. In: FSTTCS, pp. 312–324 (2011)
3. Hakuta, S., Maneth, S., Nakano, K., Iwasaki, H.: XQuery streaming by forest transducers. In: ICDE, pp. 952–963 (2014)
4. Koch, C., Scherzinger, S.: Attribute grammars for scalable query processing on XML streams. VLDB J. **16**(3), 317–342 (2007)
5. Koch, C., Scherzinger, S., Schmidt, M.: The GCX system: Dynamic buffer minimization in streaming XQuery evaluation. In: VLDB, pp. 1378–1381 (2007)
6. Lemay, A., Maneth, S., Niehren, J.: A learning algorithm for top-down XML transformations. In: PODS, pp. 285–296 (2010)
7. Maneth, S., Neven, F.: Structured Document Transformations Based on XSL. In: Connor, R.C.H., Mendelzon, A.O. (eds.) DBPL 1999. LNCS, vol. 1949, pp. 80–98. Springer, Heidelberg (2000)
8. Martens, W., Neven, F.: On the complexity of typechecking top-down XML transformations. Theor. Comput. Sci. **336**(1), 153–180 (2005)
9. Martens, W., Neven, F., Gyssens, M.: Typechecking top-down XML transformations: Fixed input or output schemas. Inf. Comput. **206**(7), 806–827 (2008)
10. Perst, T., Seidl, H.: Macro forest transducers. Inf. Process. Lett. **89**(3), 141–149 (2004)
11. Segoufin, L., Vianu, V.: Validating streaming XML documents. In: PODS, pp. 53–64 (2002)
12. von Bochmann, G.: Semantic evaluation from left to right. CACM **19**(2), 55–62 (1976)
13. Wei, M., Rundensteiner, E.A., Mani, M., Li, M.: Processing recursive XQuery over XML streams: The Raindrop approach. Data Knowl. Eng. **65**(2), 243–265 (2008)

Efficient Term Set Prediction
Using the Bell-Wigner Inequality

Massimo Melucci[✉]

Department of Information Engineering, University of Padua, Padua, Italy
massimo.melucci@unipd.it

Abstract. The task of measuring the dependence between terms is computationally expensive for IR systems which have to deal with large and sparse datasets. The current approaches to mining frequent term sets are based on the enumeration of the term sets found in a set of documents and on monotonicity, the latter being the property that a term set is frequent only if all its subsets are frequent as implemented by Apriori. However, the computational time can be very large. An alternative approach is to store the dataset in a FPT and to visit and prune the tree in a recursive way as implemented by FPGrowth. However, the storage space can still be very large. We introduce the BWI as a conceptual enhancement of monotonicity to predict with certainty when an itemset is frequent and when it is infrequent. We describe the empirical validation that the BWI can significantly reduce both the computational time of Apriori and the storage space of pattern tree-based algorithms such as FPGrowth. The empirical validation has been performed using some runs produced by IR systems from the TIPSTER test collection.

1 Introduction

In Information Retrieval (IR), term dependence has been difficult to implement because of the high computational cost needed for finding the term correlation based on the co-occurrence within query logs and collections of documents or other textual units such as passages or queries. When term dependence is needed, the standard approach to detecting frequent term sets is based on the Frequent Itemset (FI) mining algorithms. In FI mining, a database of transactions is scanned and the most frequent subsets of items (i.e. itemsets) are mined; a transaction is a set of items. FI mining is a major task of association analysis since high frequency is a signal that an itemset is not occurring by chance [1]. The application of FI mining to IR can be straightforward since documents are transactions, words are items and term sets are itemsets.

The main drawback of this approach to detecting frequent term sets is the exponential size of the number of candidate term sets which have to be extracted from a set of documents. In IR the number of items can be very large, the datasets can be very sparse (i.e. there are many more items than transactions and transactions tend to include different items) and then finding discriminative terms by mining FI may be affected by high computational cost which may not

© Springer International Publishing Switzerland 2015
C. Iliopoulos et al. (Eds.): SPIRE 2015, LNCS 9309, pp. 46–53, 2015.
DOI: 10.1007/978-3-319-23826-5_5

acceptable especially when the system has to react in very short time or the space for storing final or intermediate results is limited.

Many approaches to FI mining are based on monotonicity: an itemset is frequent only if each subset of the itemset is frequent. The standard algorithm that implements monotonicity is Apriori [2]. Despite the significant contribution of monotonicity to the reduction of the computational time, the latter can still be high especially. One of the fastest algorithms for dense datasets is based on Frequent Pattern (FP) Trees (FPTs) and is known as FPGrowth [3]. Apriori is slower than FPGrowth which however consumes more storage space [1].

We have addressed the problem of FI mining using the Bell-Wigner Inequality (BWI) (see [4,5]). This inequality provides additional bounds to the principle of monotonicity which state that the frequency of a term set belongs to a range such that it is possible to certainly state that an itemset is frequent if the minimum of the range is above a minimum support threshold or that an itemset cannot be frequent if the maximum of the range is below the threshold.

2   Methodology

This work enhances monotonicity by using the BWI defined as follows. Let $\Omega = \{\omega_1, \ldots, \omega_N\}$ be a transaction database. Let X be an itemset which determines the set, Ω_X, of transactions containing X. The function $p_{1\ldots m}$ gives the probability that a randomly selected transaction contains $X_1 \cup \cdots \cup X_m$.

$$p_{123} \geq l(w) = \max\{0, -p_1 + p_{12} + p_{13}, -p_2 + p_{12} + p_{23}, -p_3 + p_{13} + p_{23}\} \quad (1)$$
$$p_{123} \leq u(w) = \min\{p_{12}, p_{13}, p_{23}, 1 - (p_1 + p_2 + p_3 - p_{12} - p_{13} - p_{23})\} \quad (2)$$

The bounds (1) and (2) of the BWI are exploited in this work to decide whether an itemset is definitely frequent, definitely infrequent or possibly frequent.

- An itemset X is defined as *definitely frequent* when it is certain that the number of transactions which include X is not below a given minimum support threshold without passing the transaction database.
- An itemset X is defined as *definitely infrequent* when it is certain that the number of transactions which include X is below a given minimum support threshold without passing the transaction database and despite the fact that every its subset is frequent.
- An itemset X is *possibly frequent* in the sense of Apriori, that is, every its subset is frequent, but it can be stated as definitely frequent or infrequent only if the transaction database is passed.

The BWI can be used to predict with certainty whether p_{123} will be below a given minimum support μ, thus without scanning the transaction database. In this way, it is possible to predict with certainty whether $X_1 \cup X_2 \cup X_3$ is either frequent or infrequent, or if it might be frequent, the latter being the only case that needs additional computation using Apriori.

3   Validation

Predicting Terms. In this section we report on the use of BWI-Apriori for mining frequent terms from a collection of documents. We concentrate on itemsets consisting of three words since the terms of no more than three words are very common in IR.

We found that a large proportion of itemsets (i.e. three-word terms) can be definitely declared as either frequent or infrequent and can be removed from the set of possible FIs, thus significantly improving the efficiency of Apriori.

Suppose the event $X_1 \cup X_2 \cup X_3$ is under scrutiny to decide whether it is worth consideration where X_i is the i-th term. The difference between the lower bound (1) and the upper bound (2) is a measure of the prediction error of p_{123} because the smaller the difference the better BWI-Apriori is able to decide whether an itemset is definitely frequent or definitely infrequent. The smaller the difference the higher the accuracy. Indeed, if the difference were null, p_{123} would necessarily coincide with the bounds. We have thus defined the error of prediction of p_{123} for a given term w as $u(w) - l(w)$. Note that this error measure is between 0 and 1 because the bounds are probabilities.

The evaluation was performed on the TIPSTER document collection (discs 4 and 5) and the TREC 6, 7, and 8 topic sets. The set of documents that have been used to mine terms are the most successful runs submitted to three editions of TREC.[1] It was therefore assumed that the user was interacting with the "best" search engines available at that time. The adoption of these runs allowed the experimental results of this paper to be replicated and considered quite general and "independent" of the ranking function adopted by the IR system behind each run.

For each run, the 1,000 most frequent keywords and the 1,000 most frequent word pairs were computed; this selection was based on a heuristic yet reasonable assumption, that is, the content of the document set was best represented by the most frequent keywords, thus speeding up data processing. The three-word terms that are predicted by BWI have been ranked by increasing error; for example, the probability p_{123} of term "kong public security" mined from the topic set 301-350 has error 0.003 since u(kong public security) $- l$(kong public security) $= 0.003$. In the case of topic 301 and of run "anu6alo1", 20 terms have an error not greater than 0.001 and there are 10 terms which are predicted without error.

As for the distribution of error values computed for all the topics and runs, the minimum was zero, the first quartile was 0.009, the median was 0.013, the mean was 0.017, the third quartile was 0.020 and the maximum was 0.342. The results suggest that 25% of the set of 3-word terms have been predicted with an error not greater than 0.009, half of the 3-word terms have been predicted with an error not greater than 0.013 and that only 25% of these terms have been predicted with an

---

[1] The run tags were T3D1N0, acsys8aln2, anu6alo1, att98atdc, att99atde, bbn1, Brkly21, Brkly22, Brkly26, city6al, CL99SDopt2, Cor6A3cll, Cor7A3rrf, Flab8as, Flab8atd2, Flab8atdn, Flab8ax, fub99td, ibmg97b, ibms99a, INQ502, mds602, mds98td, Mercure1, MITSLStd, ok7ax, ok8amxc, pir9Attd, pirc7Aa, pirc8Aa2, tno7exp1, tno8d3.

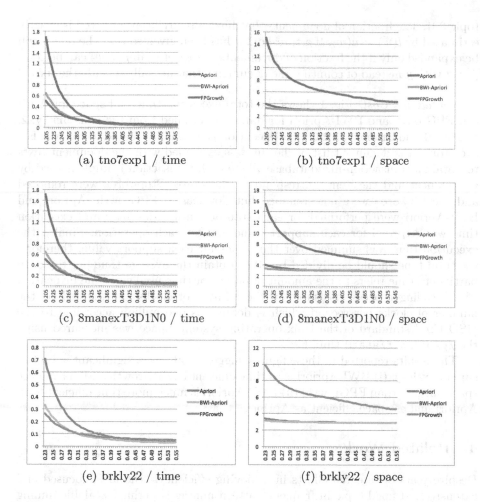

Fig. 1. Execution time and space of Apriori, BWI-Apriori and FPGrowth. The values of the x-axis are support values.

error greater than 0.020. Similar results were found for individual topic sets and the related runs, but cannot be reported for space reasons.

Sometimes the generated terms do not conform to grammatical rules. However, short queries are often bags-of-words that cannot be arranged in correct grammatical forms because of their small size or inscrutable user intents. Moreover, the generated terms look strange and do not match our expectations. However, it was found that these strange terms discriminate relevant documents better than grammatically correct terms [6].

It would then be possible to exploit $l(w)$ or $u(w)$ to predict the probability of w's without computing p_{123}. Suppose it has to be decided whether $w =$ "alternative development government" should be added to the query originated from topic 301. The fraction of documents that have already been retrieved is exactly estimated by $l(w) = u(w)$. If a threshold μ has been given so that the query will be expanded only if the fraction of documents is greater than μ, $l(w)$ can be compared with μ instead of counting the documents that actually include the term.

Computational Cost. Fig. 1 shows the total execution time and space for Apriori, FPGrowth and BWI-Apriori for datasets obtained from the runs `brkly22`, `tno7exp1` and `8manexT3D1N0`;[2] the execution time was calculated as follows. For each topic, the document list of the run was expanded with the document texts to obtain a transaction-like database where each transaction is implemented by a document and the items correspond to the words. Stopwords were removed and the keywords were not stemmed but downcased. FPGrowth, Apriori and BWI-Apriori were performed for each transaction-like database and execution time was recorded for each support value and for each operation. Finally, the execution time was summed over the topics for each support value; therefore, the total time should be divided by 50 to obtain the average time per topic. In particular, time refers to the amount of processor time used since the invocation of the calling process, measured in the number of clocks and divided by the number of clocks per second which is defined as one million according to the "ISO C90" standard of the Unix operating system. Space was measured using the `/proc/self/statm` Unix file.

The results reported in these figures suggest that, for a significant subset of support values: (i) BWI-Apriori is as time efficient as FPGrowth and much more space efficient than FPGrowth; (ii) BWI-Apriori is much more time efficient than Apriori and as space efficient as Apriori.

4   Related Work

Despite years of research efforts in achieving efficiency, "the most focused and extensively studied topic in frequent pattern mining is perhaps scalable mining methods" [7]. The foundational issue of the problem of mining all FIs is the intrinsic exponential computational cost. This issue is independent of the algorithm since any algorithm aiming at mining itemsets is likely to incur high costs due to the exponential number of candidate FIs to be examined [8]. Indeed, this paper aims to address this issue. Improvements of Apriori were proposed in [9], [10], and [11]. In contrast, this paper aims to address the problem at the conceptual level by defining additional, more precise bounds. Efficient data structures such as the FPT [3] can also be used when the transaction database is dense. When the database is sparse, an FPT may explode and exhausts memory. FPGrowth performs better on artificial datasets whereas Apriori performs sufficiently well on all datasets [12], and in particular on sparse datasets. To deal with these issues, some enhancements of FPGrowth have been proposed in [13], [14], [15] and [16]. In IR, FI mining has been applied to indexing and retrieving documents[17]. Variations on the theme were discussed in [18] and [19].

---

[2] We cannot include further plots for the space reasons.

Another use of FI mining in IR was investigated in query expansion and suggestion and reported in [20], [21], [22]. The BWI can be utilised to enhance the previous approaches to finding term sets. Other approaches to dealing with terms that discriminate the relevant documents are based on recording word position information in documents. However, processing (or even storage) lists including word position should be avoided whenever possible [23] since position information has been shown to have little effect on retrieval accuracy on average [24]. The use of the BWI in this paper indeed allows us to avoid the recording of word position.

5 Concluding Remarks

In this paper we have validated the hypothesis that the BWI can successfully be utilised to reduce the computational time of Apriori, thus making the computation of term sets for IR purposes less expensive and more feasible. We focussed on efficiency of term set mining since effectiveness was addressed in the previous, related work. BWI-Apriori is still competitive with FPGrowth in time and space when the datasets are derived from the runs of an IR system. While FPGrowth demands much more space than BWI-Apriori, BWI-Apriori is only slightly slower when the minimum support is not very low. This result opens up the use of FI and makes some inefficient algorithms designed for indexing, retrieval and query expansion and suggestion useable by IR systems.

The BWI is not an ad-hoc data structure, a programming solution, or a code optimisation. It is on the contrary a conceptual enhancement of Apriori which would require an efficient implementation to manifest its own potential more than manifested in this paper. Some research work was done in order to achieve efficiency in mining FIs from large transaction databases by using ad-hoc data structures and specific programming solutions. These data structures and programming solutions could considerably reduce the execution time of FI mining implementations to an extent that it is possible to find efficient implementations of naive mining algorithms that can run faster than inefficient implementations of BWI-Apriori.

In this work we concentrated on three words since this is the most frequent case other than the case of two words [25], [26], [27], [28], [29], [30], [31], [32]. The extension of the validation to the event of term sets of more than three words would require the use of inequalities defined for four or more events. Some inequalities similar to the BWI were defined, however, the number of these inequalities is of exponential order and deciding whether an event satisfies all the inequalities of any dimension is NP-complete [4].

The empirical validation has shown that the number of itemsets that can certainly be declared either as frequent or as infrequent is high, thus reducing the computational time by 50%-70% or higher in some cases. What this empirical validation did not tell is the reason why this number is high. In fact, the main aim of this paper has been to show that the BWI is an efficient means for reducing the computational time of Apriori which is notoriously very high. The theoretical explanation is left to future work. What we may at present suggest is that the efficiency of the BWI and in particular of the frequency of the operations might be due to the distributional properties of the transaction databases and that these properties are quite general since the reduction of the computational time of Apriori has been strong with different datasets.

FPGrowth has been an approach to FI mining alternative to Apriori and it is currently utilized whenever datasets are dense; in this event, the resulting FPT is a compact representation of the transaction database and memory swapping is reduced if not eliminated, thus decreasing computational time by orders of magnitude. In fact, a compact FPT can be entirely stored in the main memory and FI mining is fast until the main memory becomes scarse and swapping becomes a necessity. Unfortunately, FPGrowth suffers from sparse datasets such as those used in IR, the resulting FPT may be huge and the algorithm can require more storage space than Apriori and than BWI-Apriori. The incorporation of the BWI in FPGrowth may be an interesting venue for the next research works.

References

1. Tan, P.N., Steinbach, M., Kumar, V.: Introduction to Data Mining. Pearson International Edition (2006)
2. Agrawal, R., Imielinski, T., Swami, A.: Mining association rules between sets of items in large databases. In: Proceedings of SIGMOD, Washington, D.C., pp. 207–216 (1993)
3. Han, J., Pei, J., Yin, Y.: Mining frequent patterns without candidate generation. In: Proceedings of SIGMOD, pp. 1–12 (2000)
4. Pitowsky, I.: Correlation polytopes: Their geometry and complexity. Mathematical Programming **50**, 395–414 (1991)
5. Pitowsky, I.: Quantum Probability - Quantum Logic. Springer (1989)
6. Blanco, R., Boldi, P.: Extending BM25 with multiple query operators. In: Proceedings of SIGIR, pp. 921–930 (2012)
7. Han, J., Cheng, H., Xin, D., Yan, X.: Frequent pattern mining: current status and future directions. Data Mining and Knowledge Discovery **15**, 55–86 (2007)
8. Kirsch, A., Mitzenmacher, M., Pietracaprina, A., Pucci, G., Upfal, E., Vandin, F.: An efficient rigorous approach for identifying statistically significant frequent itemsets. Journal of the ACM 59(3) (2012)
9. Wang, K., He, Y., Han, J.: Mining frequent itemsets using support constraints. In: Proceedings of VLDB (2000)
10. Burdick, D., Calimlim, M., Flannick, J., Gehrke, J., Yiu, T.: MAFIA: A maximal frequent itemset algorithm. IEEE Transactions on Knowledge and Data Engineering **11**, 1490–1504 (2005)
11. Gouda, K., Zaki, M.J.: Efficiently mining maximal frequent itemsets. In: Proceedings of ICDM (2001)
12. Zheng, Z., Kohavi, R., Mason, L.: Real world performance of association rule algorithms. In: Proceedings of KDD, pp. 401–406. ACM New York (2001)
13. Liu, J., Pan, Y., Wang, K., Han, J.: Mining frequent item sets by opportunistic projection. In: Proceedings of KDD, pp. 229–238. ACM, New York (2002)
14. Pei, J., Han, J., Lu, H., Nishio, S., Tang, S., Yang, D.: H-mine: Hyperstructure mining of frequent patterns in large databases. In: Proceedings of ICDM, pp. 441–448. IEEE Computer Society, Washington, DC (2001)

15. Pietracaprina, A., Zandolin, D.: Mining frequent itemsets using patricia tries. In: Goethals, B., Zaki, M.J. (eds.) FIMI. CEUR Workshop Proceedings, vol. 90. CEUR-WS.org (2003)
16. Schlegel, B., Gemulla, R., Lehner, W.L.W.: Memory-efficient frequent-itemset mining. In: Proceedings of EDBT, pp. 461–472 (2011)
17. Pôssas, B., Ziviani, N., Meira Jr, W., Ribeiro-Neto, B.: Set-based vector model: An efficient approach for correlation-based ranking. ACM Trans. Inf. Syst. **23**(4), 397–429 (2005)
18. Amir, A., Aumann, Y., Feldman, R., Fresko, M.: Maximal association rules: A tool for mining associations in text. J. Intell. Inf. Syst. **25**(3), 333–345 (2005)
19. Sarawagi, S., Thomas, S., Agrawal, R.: Integrating association rule mining with relational database systems: Alternatives and implications. Data Min. Knowl. Discov. **4**(2–3), 89–125 (2000)
20. Fonseca, B.M., Golgher, P., Pôssas, B., Ribeiro-Neto, B., Ziviani, N.: Concept-based interactive query expansion. In: Proceedings of CIKM, CIKM 2005, pp. 696–703. ACM, New York (2005)
21. Fonseca, B.M., Golgher, P.B., De Moura, E.S., Pôssas, B., Ziviani, N.: Discovering search engine related queries using association rules. J. Web Eng. **2**(4), 215–227 (2003)
22. Song, D., Huang, Q., Rüger, S.M., Bruza, P.D.: Facilitating Query Decomposition in Query Language Modeling by Association Rule Mining Using Multiple Sliding Windows. In: Macdonald, C., Ounis, I., Plachouras, V., Ruthven, I., White, R.W. (eds.) ECIR 2008. LNCS, vol. 4956, pp. 334–345. Springer, Heidelberg (2008)
23. Zobel, J., Moffat, A.: Inverted files for text search engines. ACM Comput. Surv. **38**(2) (2006)
24. Robertson, S., Zaragoza, H.: The probabilistic relevance framework: BM25 and beyond. Foundations and Trends in Information Retrieval **3**(4), 333–389 (2009)
25. Keyword Discovery. http://www.keyworddiscovery.com/keyword-stats.html (visited on April 2014)
26. Bendersky, M., Croft, W.B.: Analysis of long queries in a large scale search log. In: Proceedings of the Workshop on Web Search Click Data, WSCD 2009, pp. 8–14. ACM, New York (2009)
27. Gan, Q., Attenberg, J., Markowetz, A., Suel, T.: Analysis of geographic queries in a search engine log. In: Proceedings of the International Workshop on Location and the Web, LOCWEB 2008, pp. 49–56. ACM New York (2008)
28. Jansen, B.J., Spink, A.: How are we searching the world wide web?: a comparison of nine search engine transaction logs. Inf. Process. Manage. **42**, 248–263 (2006)
29. Jansen, B.J., Booth, D.L., Spink, A.: Determining the user intent of Web search engine queries. In: Proceedings of WWW, pp. 1149–1150. ACM, New York (2007)
30. Jansen, B.J., Booth, D.L., Spink, A.: Determining the informational, navigational, and transactional intent of Web queries. Inf. Process. Manage. **44**, 1251–1266 (2008)
31. Jansen, B.J., Booth, D.L., Spink, A.: Patterns of query reformulation during Web searching. Journal of the American Society for Information Science and Technology **60**, 1358–1371 (2009)
32. Huston, S., Croft, W.B.: Evaluating verbose query processing techniques. In: Proceedings of SIGIR, pp. 291–298. ACM, New York (2010)

On Prefix/Suffix-Square Free Words

Marius Dumitran[1], Florin Manea[2]([✉]), and Dirk Nowotka[2]

[1] Faculty of Mathematics and Computer Science, University of Bucharest,
Str. Academiei 14, RO-010014 Bucharest, Romania
`marius.dumitran@fmi.unibuc.ro`
[2] Department of Computer Science, Christian-Albrechts University of Kiel,
Christian-Albrechts-Platz 4, 24118 Kiel, Germany
`{flm,dn}@informatik.uni-kiel.de`

Abstract. We present a series of algorithms identifying efficiently the factors of a word that neither start nor end with squares (called, accordingly, prefix-suffix-square free factors). A series of closely related algorithmic problems are discussed.

1 Introduction

The *prefix duplication* (for short, *PD*), *suffix duplication* (*SD*), and *prefix-suffix duplication* (*PSD*) are formal operations on words that were introduced in [1] following a biological motivation. Essentially, by suffix duplication, from a word w one can derive any word wx where x is a suffix of w; that is, suffixes of w are duplicated. Prefix duplication is the symmetric operation: in this case, one duplicates prefixes of w. Finally, the prefix-suffix duplication is a combination of the previous two operations: by it, one can duplicate either a suffix or a prefix of the initial word. As it is the case in the study of many word operations, the authors of [1] were mainly interested in the properties of the words generated by iteratively applying these operations to an initial word or set of words.

The basic motivation of introducing these operations was to mathematically model a special type of duplications in DNA sequences, that appear only at the ends of these sequences, also known as telomeres. Generally, telomeres consist of tandem repeats of a small number of nucleotides and their role is to stabilise the linear chromosomal DNA molecule [2,3]; the loss of telomeric repeat sequences may result in chromosome fusion and lead to chromosome instability [4]. Another motivation would be to model the process of generating long terminal repeats (LTRs): identical sequences of DNA that repeat hundreds or thousands of times at either end of some specific DNA sequence. Such sequences are used, for instance, by viruses to insert their genetic material into the host genomes.

According to this motivation, it seems natural to investigate the problem of efficiently detecting the existence of repetitive structures occurring at both ends of some sequence. For instance, words that do not end or start with repetitions may model DNA sequences that went through some degenerative process that destroyed the terminal repeats, affecting their stability or functionalities. However, while our paper draws its motivation from biology, it essentially approaches

© Springer International Publishing Switzerland 2015
C. Iliopoulos et al. (Eds.): SPIRE 2015, LNCS 9309, pp. 54–66, 2015.
DOI: 10.1007/978-3-319-23826-5_6

these concepts from a purely (and somehow oversimplified) mathematical point of view. So, moving away from the biological motivation, words that do not start nor end with repetitions seem to be interesting from a combinatorial point of view, as well. Indeed, repetitions-free words (i.e., words that do not contain consecutive occurrences of the same factors) are central in combinatorics on words, stringology, and their applications (see, e.g., [5,6]); words that do not have repetitive prefixes or suffixes model a weaker, but strongly related, notion.

In this paper we address a series or questions related to the simplest repetitive structure that may occur in a word: squares. We say that a word is suffix-square free (respectively, prefix-square free or prefix-suffix-square free) if that word does not end (respectively, start, or both start and end) with two consecutive occurrences of a factor. Alternatively, these are the words that cannot be obtained from shorter factors by the SD−operation (respectively, PD, or PSD).

We show how the following tasks can be performed efficiently, for an input word w of length n. First, we show how w can be processed in linear time so that we can answer in constant time queries asking whether the factors of w are suffix-square free, prefix-square free, or prefix-suffix-square free. Then, we give an algorithm that outputs in $\mathcal{O}(n + |S|)$ the set S of all factors of w which are prefix-suffix-square free (respectively, suffix-square free or prefix-square free); computing the size of S (without enumerating all its elements) can be done in $\mathcal{O}(n \log n)$. Note that, for square-free words (see [5]) of length n we have $|S| \in \Theta(n^2)$, so there are indeed cases in which we can compute $|S|$ faster than just going through its elements. Finally, the longest prefix-suffix square free factor of a word can be obtained in $\mathcal{O}(n)$ time.

In [1] a series of language theoretic and algorithmic results related to the iterated PSD and SD were given; clearly, all the results stated for SD hold canonically for PD. Interesting to us, in [1] the authors gave algorithms deciding in $\mathcal{O}(n \log n)$ and, respectively, $\mathcal{O}(n^2 \log n)$ time, for two given words, whether the longer one (whose length was n) can be generated from the shorter one by iterated SD, respectively, PSD. These algorithms can be adapted to compute within the same time complexity bounds, respectively, all the SD-irreducible roots of a word or all its PSD-irreducible roots. These are suffix-square free (respectively, prefix-suffix-square free) factors of a word that can generate it by iterated SD, respectively, PSD operations. Intuitively, these are the factors of a word containing its core information: on one hand, the rest of the word consists only in repetitions of parts of these factors, and on the other hand, they cannot be further reduced by eliminating repeated information from their ends, as they do not end with any repetition. It is not trivial to see how many irreducible roots a word may have, but we give examples of words of length n having $\Theta(n)$ SD-irreducible roots and, respectively, words with $\Theta(n^2)$ PSD-irreducible roots; so our algorithms finding all these roots are only a $\log n$-factor slower than what one could expect in the worst case. However, finding only one (not all) SD or one PSD−irreducible root of a given word takes linear time.

We conclude this paper by addressing a series of related problems. Essentially, detecting a SD-irreducible root of some word w is equivalent to asking whether there is a (non-trivial) factorisation of that word into $k > 1$ factors $w = s_1 \cdots s_k$ such that s_i is a suffix of $s_1 \cdots s_{i-1}$ for $i > 1$; that is, for all $i > 1$, we have the square s_i^2 centred on the position that follows immediately after $s_1 \cdots s_{i-1}$, and s_1 is suffix-square free. This factorisation, as well as the similar factorisations derived from the detection of PD or PSD−roots of a word, seems to be strongly connected to the square-structure of w. So, we further discuss other factorisations also strongly related to the way squares and, more general, repetitions occur in the factored words. Namely, we address the problem of factoring a word into squares, or into periodicities of exponent at least 2, or in such a way that the number of square factors of the factorisation is maximum, compared to any other possible factorisation of that word.

It seems important to us to stress that, in fact, all our results are about understanding the way squares occur in a word. Either we discuss about prefix or suffix-square free factors of a word or about various factorisations of a word, in fact we gather information on the occurrences of squares inside the given word: sometimes we are interested in avoiding these occurrences (i.e., finding words that do not start or end with such an occurrence), while sometimes we want to use them to factor that word. Thus, it is no surprise that our main tool is a lemma showing how one can compute in linear time the shortest square starting (respectively, ending) at each position of the input word. This result was settled in [7], but only for inputs over constant size alphabets. In [8] the result was shown for integer alphabets, using a proof based on a data-structure (see [9]) designed to maintain efficiently a family of disjoint sets under union and find operations (however, the proof was not published in the conference version of the paper [8]). Here, we give a new proof of this result based on a Lempel-Ziv-like factorisation of the input word and a series of combinatorial remarks on the structure of this factorisation. A similar approach was used in [10,11] to find efficiently all the periodicities of a word (easily leading to an algorithm testing in linear time the existence of squares in words), and in [12] to compute, also in linear time, all local periods of a word; the details of our algorithm are, however, quite different.

2   Preliminaries

§ **Basic Facts.** Let Σ be a finite alphabet; Σ^* denotes the set of all finite words over Σ. The *length* of a word $w \in \Sigma^*$ is denoted by $|w|$. The *empty word* is denoted by ϵ. A word $u \in \Sigma^*$ is a *factor* of $v \in \Sigma^*$ if $v = xuy$, for some $x, y \in \Sigma^*$; we say that u is a *prefix* of v, if $x = \epsilon$, and a *suffix* of v, if $y = \epsilon$. We denote by $w[i]$ the symbol occurring at position i in w, and by $w[i..j]$ the factor of w starting at position i and ending at position j, consisting of the catenation of the symbols $w[i], w[i+1], \ldots, w[j]$, where $1 \le i \le j \le n$; we define $w[i..j] = \epsilon$ if $i > j$. A range $[i, j]$ in a word w is the set of positions $\{i, i+1, \ldots, j\}$ of w.

The powers of a word w are defined recursively by $w^0 = \epsilon$ and $w^n = ww^{n-1}$ for $n \ge 1$. A repetition w^2 is called square; the centre of the square w^2 is the first position of the second factor w of the square. If w cannot be expressed as a nontrivial power (i.e., w is not a repetition) of another word, then w is *primitive*.

A *period* of a word w over Σ is a positive integer p such that $w[i] = w[j]$ for all i and j with $i \equiv j \pmod{p}$. Let $per(w)$ be the smallest period of w. A run is a word w with $per(w) \leq \frac{|w|}{2}$ which cannot be extended to the left or right to get a word with the same period p, i.e., $i = 1$ or $w[i-1] \neq w[i+p-1]$, and $j = |w|$ or $w[j+1] \neq w[j-p+1]$.

Let $w \in \Sigma^*$ be a word. The s-factorisation of w is defined as follows (see [10]). We factorise $w = u_1 \cdots u_r$ if the following hold for all $i \geq 1$:

- If a letter a occurs in w immediately after $u_1 \cdots u_{i-1}$ and a did not appear in $u_1 \cdots u_{i-1}$ then $u_i = a$
- Otherwise, u_i is the longest word such that $u_1 \cdots u_{i-1}u_i$ is a prefix of w and u_i occurs at least once in $u_1 \cdots u_{i-1}$.

§ **Square Duplication Operations.** The prefix, suffix, and prefix-suffix duplication operations were defined in [1]. Given a word $w \in \Sigma^*$, we have:

- *prefix duplication:* $PD(w) = \{xw \mid w = xw' \text{ for some } x \in \Sigma^+\}$.
- *suffix duplication:* $SD(w) = \{wx \mid w = w'x \text{ for some } x \in \Sigma^+\}$.
- *prefix-suffix duplication:* $PSD(w) = PD(w) \cup SD(w)$.

A word w is called SD-irreducible (respectively, PD-irreducible) if $w \notin SD(w')$ (respectively, $w \notin PD(w')$) for any factor w' of w; a word w is PSD-irreducible if it is both SD- and PD-irreducible. Alternatively, an SD-irreducible (respectively, PD-irreducible) word is called *suffix-square free* word (respectively, *prefix-square free*). PSD-irreducible words are called *prefix-suffix-square free*.

We further define, for $\Theta \in \{PD, SD, PSD\}$, its iteration:

$$\Theta_k^0(x) = \{x\}, \Theta_k^{n+1}(x) = \Theta_k^n(x) \cup \Theta_k(\Theta_k^n(x)), \text{ for } n \geq 0, \ \Theta_k^*(x) = \bigcup_{n \geq 0} \Theta_k^n(x).$$

For a word w and $\Theta \in \{PD, SD, PSD\}$, its factor $w[i..j]$ is a Θ-irreducible root of w if $w \in \Theta^*(w[i..j])$ and $w[i..j]$ is Θ-irreducible.

§ **Algorithmic Prerequisites.** The computational model we use to design and analyse our algorithms is the standard unit-cost RAM (Random Access Machine) with logarithmic word size, which is generally used in the analysis of algorithms. In the upcoming algorithmic problems, we assume that the words we process are sequences of integers (called letters, for simplicity). In general, if the input word has length n, we assume its letters are in $\{1, \ldots, n\}$, so each letter fits in a single memory-word. This is a common assumption in stringology (see, e.g., [13]).

For a word u, $|u| = n$, over $\Sigma \subseteq \{1, \ldots, n\}$ we build in $\mathcal{O}(n)$ time its suffix array, as well as LCP-data structures, allowing us to retrieve in constant time the length of the longest common prefix of any two suffixes $u[i..n]$ and $u[j..n]$ of u, denoted $LCP_u(i, j)$ (the subscript u is omitted when there is no danger of confusion). See, e.g., [6,13], and the references therein.

We also use the fact that the number of runs of a word is linear and their list (with a run $w[i..j]$ represented as the triple $(i, j, per(w[i..j]))$) can be computed in linear time in the RAM with logarithmic word size model (see [11]). The exponent of a run $w[i..j]$ occurring in w is defined as $\frac{j-i+1}{per(w[i..j])}$; in [11] it is shown that the sum of the exponents of runs in a word of length n is $\mathcal{O}(n)$.

Note also that the s-factorisation of a word w can be computed in linear time when w is from an integer alphabet (see, e.g., [14]).

We also recall the *range minimum query* data structure (see [15]). Given an array of n integers $T[\cdot]$, we can produce in $\mathcal{O}(n)$ time several data-structures for this array, allowing us to answer in constant time to queries $RMQ(i,j)$ and $posRMQ(i,j)$, asking for the minimum value and, respectively, its position among $T[i], T[i+1], \ldots, T[j]$.

§ **Main Technical Tool.** We conclude the Preliminaries section with our main technical lemma, used throughout the paper. For a word w of length n we define the array $left[\cdot]$ as follows: for $1 \le i \le n$, $left[i] = \max\{j \mid w[j..i]$ is a square$\}$.

Lemma 1. *Given a word w we can compute in linear time the array $left[\cdot]$.*

Proof. In fact, on $left[j]$ starts the shortest square ending at position j of w.

Before describing our algorithm, we note that if x^2 and y^2 are two squares ending on position j of w, and $\frac{3\ell}{2} \ge |y| > |x| \ge \ell$ for some $\ell > 0$, then there exists a square u^2 with $|u| \le \frac{\ell}{2}$ also ending at position j. Indeed, $y = ux$ for some word u with $|u| \le \frac{\ell}{2}$. As xx is a suffix of yy and $2|x| \ge |y|$, we get that u is a suffix of (the first) x. It follows that u is also a suffix of y. Consequently, $uy = uux$ is a suffix of y^2. However, as $|u| \le \frac{\ell}{2}$, we get that x^2 is longer than $uy = uux$, so uu is also a suffix of (both occurrences of) x. This shows our claim.

Moving now to the algorithm, we first compute the s-factorisation of the input word $w = u_1 \cdots u_k$, using the tools from [14]. During this computation of the s-factorisation we also get for each $i \le k$ the position ℓ_i where u_i occurs in $u_1 \cdots u_{i-1}$; let also $k_i = |u_1 \cdots u_{i-1}| + 1$ be the starting position of u_i, for all i.

We compute separately, for each i from 1 to k, considered in increasing order, the values $left[j]$ for each position j of the factor u_i. In the following, we explain how these values are computed for some fixed i; our approach ensures that when considering u_i we already know $left[j']$ for every position j' of $u_1 \cdots u_{i-1}$.

First, note that if x^2 is a square ending on position j of u_i then the centre of this square occurs inside $u_{i-1} u_i$. Otherwise, u_{i-1} was not correctly chosen: in that case, a longer factor starting on position k_{i-1} would be $w[k_{i-1}..j]$, a suffix of the second x factor of the square. Hence, $|x| \le |u_{i-1} u_i|$ and there are three cases to be analysed: the shortest square ending on position j might be completely contained in u_i, centred in u_i but starting in u_{i-1}, or centred in u_{i-1}.

We begin with the simplest case: if the shortest square ending on position j of u_i is completely contained in u_i then it should be equal to the shortest square ending at position $\ell_i + (j - k_i)$ (that is, the shortest factor ending on the position corresponding to j from the previous occurrence of u_i inside $u_1 \cdots u_{i-1}$). So, in the first step of our algorithm, we just check for every position j of u_i if the shortest square ending on $\ell_i + (j - k_i)$ is short enough to be contained in u_i, and, if yes, we decide that the respective square occurs again as the shortest square ending at j.

Secondly, we detect for each position j of u_i the shortest square x^2 ending on j, starting in u_{i-1} and whose centre is in u_i; clearly, $|x| \le |u_i|$. Following the strategy of [10], we detect for each possible length ℓ of x a range of u_i where the centre of x^2 may occur. Basically, for each ℓ, we compute the longest common prefix $u_i[1..b_\ell]$ of $u_i[1..\ell]$ and $u_i[\ell+1..|u_i|]$ and the longest common suffix $u_i[a_\ell..\ell]$ of u_{i-1} and $u_i[1..\ell]$; the range we look for is $[a_\ell, b_\ell]$. Now, the range where a square x^2 as above may end is obtained by intersecting $[a_\ell + \ell - 1, b_\ell + \ell - 1]$ with u_i; in this way, we obtain a new range $[c_\ell, d_\ell]$ where the squares of length ℓ may end. As $b_\ell \leq \ell$, we have $d_\ell \leq b_\ell + \ell - 1 < 2\ell$.

So, for each possible length ℓ of x (i.e., $1 \leq \ell \leq |x_i|$) we now have a range $[c_\ell, d_\ell]$ where a square x^2 may end inside u_i; each range was computed using a constant number of longest common prefix queries, so in constant time. Now we just have to report for each position j of u_i which is the minimum ℓ such that j is in the range $[c_\ell, d_\ell]$. To do this, we report for each $k \geq 0$ the positions where a square x^2 with $(3/2)^k \leq |x| \leq (3/2)^{k+1}$ ends, and we did not find so far a shorter square ending at that position. Clearly, for ℓ with $(3/2)^k \leq \ell \leq (3/2)^{k+1}$ we have $d_\ell \leq k_i + 2 \cdot (3/2)^{k+1}$. So, we can sort the ends of the ranges $[c_\ell, d_\ell]$ with $(3/2)^k \leq \ell \leq (3/2)^{k+1}$ in $\mathcal{O}((3/2)^{k+1})$ time. This allows us to find in $\mathcal{O}((3/2)^{k+1})$ time the positions of u_i contained in exactly one range $[c_\ell, d_\ell]$ (all these positions are in the prefix of length $2 \cdot (3/2)^{k+1}$ of u_i). For each such position j we store the value ℓ such that $j \in [c_\ell, d_\ell]$; we conclude that a square xx with $|x| = \ell$ ends on j, and, if we did not already find a shorter square ending on that position, we store this square as the shortest one we found so far that ends on j. By the claim shown at the beginning of this proof, if a position is contained in at least two ranges $[c_\ell, d_\ell]$ and $[c_{\ell'}, d_{\ell'}]$ with $(3/2)^k \leq \ell \leq \ell' \leq (3/2)^{k+1}$, then there exists a square with root of length at most $\frac{(3/2)^k}{2}$ ending at that position, so we do not have to worry about it: we should have already found the shorter square occurring at there (note that this square might be completely contained in u_i).

Consequently, processing all the ranges $[c_\ell, d_\ell]$ with $(3/2)^k \leq \ell \leq (3/2)^{k+1}$ takes $\mathcal{O}((3/2)^{k+1})$ time. Now, we iterate this process for all $k \leq \log_{3/2}|u_i|$, and, alongside the analysis of the first case, obtain for each position j the shortest square ending there that either starts in u_{i+1} but it is centred in u_i or is completely contained in u_i. The total time is $\mathcal{O}\left(\sum_{k \leq \log_{3/2}|u_i|}(3/2)^{k+1}\right) = \mathcal{O}(|u_i|)$.

Finally, we look for the shortest squares ending in u_i that are centred in u_{i-1}; the length of such squares may go up to $|u_{i-1}u_i|$. The analysis is very similar to the one of the second case, when we searched for squares centred in u_i, starting in u_{i-1}. In $\mathcal{O}(|u_{i-1}u_i|)$ time we find, for all $\ell \leq |u_{i-1}u_i|$ the ranges where the centres of squares x^2 ending in u_i, with $|x| = \ell$, occur in u_{i-1}. This gives the ranges of u_i where such squares end. Now, using the same ideas as above, we detect, in $\mathcal{O}(|u_{i-1}| + |u_i|)$ time, the shortest square ending at each position where we did not already find a shortest square. Putting together the results of the three analysed cases, we get for each position j of u_i the value $left[j]$.

In conclusion, the time needed to compute for all i from 1 to k the values $left[j]$ for every position j of the factors u_i is $\mathcal{O}(\sum_{1 \leq i \leq k} |u_{i-1}u_i|) = \mathcal{O}(n)$. $\square$

Clearly, the same algorithm can be used on the mirror image of w to obtain $right[\cdot]$ defined as follows: for $1 \leq i \leq n$, $right[i] = \min\{j \mid w[i..j]$ is a square$\}$.

3   Prefix/Suffix-Square Free Words

The previous lemma allows us to answer efficiently queries asking whether the factors of a given word are or not prefix-square free, or suffix-square free, or prefix-suffix-square free.

Theorem 1. *Given a word w of length n, we can construct in $\mathcal{O}(n)$ data structures allowing us to answer in constant time the following three types of queries (for all $1 \leq i \leq j \leq n$): $q_p(i,j)$: "is $w[i..j]$ prefix-square free?"; $q_s(i,j)$: "is $w[i..j]$ suffix-square free?"; $q_{ps}(i,j)$: "is $w[i..j]$ prefix-suffix-square free?".*

Proof. We construct the arrays *left* and *right*. Then we can answer a query $q_p(i,j)$ positively if $right[i] > j$ (the shortest square that starts at i ends after j); otherwise, we answer it negatively. Similarly, a query $q_s(i,j)$ returns true if and only if $left[j] < i$ (the shortest square that ends at j starts before i). Finally, $q_{ps}(i,j)$ is answered positively if and only if both $q_s(i,j)$ and $q_p(i,j)$ are true.   □

However, this lemma does not enable us to enumerate the prefix-suffix-square free factors efficiently. The following theorem shows how this can be done.

Theorem 2. *Given a word w of length n, we can find the set S of prefix-suffix-square free factors of w in $\mathcal{O}(n + |S|)$ time.*

Proof. The idea is to find separately, for each $i \leq n$, all the prefix-suffix-square free factors $w[i..j]$. This can be done as follows. We first construct the arrays $left[\cdot]$ and $right[\cdot]$ for the word w, and define data-structures allowing us to answer in constant time range minimum queries for $left[\cdot]$. Let $RMQ(j_1, j_2)$ denote the minimum value occurring in the range between j_1 and j_2; this is, in fact, the rightmost starting position of a square x^2 that ends on a position between j_1 and j_2. Let $posRMQ(j_1, j_2)$ denote the position where $RMQ(j_1, j_2)$ occurs (in case of equality, we take the rightmost such position); this position is the ending position of the aforementioned square x^2.

Let us now fix some $i \leq n$. We note that if $w[i..j]$ is prefix-square free then $j < right[i]$ (or $w[i..right[i]]$ would be a square prefix). So, we only consider the values between $i + 1$ and $right[i] - 1$ as possible ending positions of prefix-suffix-square free words starting on i. The procedure detecting these words works recursively. In a call of the procedure, we find the prefix-suffix-square free words starting on i and ending somewhere between two positions j_1 and j_2; initially, $j_1 = i + 1$ and $j_2 = right[i] - 1$, and j_1 and j_2 are always between $i + 1$ and $right[i] - 1$.

So, let us explain how our search is conducted for a pair of positions (j_1, j_2), where $j_1 < j_2$. We compute $RMQ(j_1, j_2)$ and $posRMQ(j_1, j_2)$. If $RMQ(j_1, j_2) < i$ and $posRMQ(j_1, j_2) = j$ then clearly the shortest square ending on j starts before i, and the shortest square starting on i ends after j, so $w[i..j]$ is prefix-suffix-square free; in this case we run the procedure for the two new pairs $(j_1, posRMQ(j_1, j_2) - 1)$ and $(posRMQ(j_1, j_2) + 1, j_2)$. The other case is when $RMQ(j_1, j_2) \geq i$. Then the shortest square ending on any position j between j_1 and j_2 ends after i, so none of the words $w[i..j]$ is suffix-square free; thus, we stop the procedure. A call of the recursive procedure for a pair (j_1, j_2) where j_1 is not strictly smaller than j_2 does not do anything.

It is straightforward that, for some $i \leq n$, we obtain in this manner all the positions $j < right[i]$ such that $w[i..j]$ is suffix-square free. But these are, in fact, all the positions $j \leq n$ such that $w[i..j]$ is prefix-suffix-square free. We iterate this algorithm for all i, and obtain all prefix-suffix-square free factors of w.

To evaluate the total time complexity of this algorithm, let us again fix i and see how much time we spend to detect the prefix-suffix-square free words starting on i. Each call of the recursive procedure either returns a valid position j and calls the procedure for two new pairs (defining disjoint ranges), or stops the search in the range defined by the pairs for which it was called. So, the calls of this procedure can be pictured as a binary tree with $|S_i|$ internal nodes, where $S_i = \{j \mid w[i..j]$ is prefix-suffix-square free$\}$. Thus, we have, in total $\mathcal{O}(|S_i|)$ calls. With the help of the data structures we constructed, each call of the procedure can be executed in constant time. Thus, the time needed to compute S_i is $\mathcal{O}(|S_i|)$. Now, adding this up for all possible i, we get that the set S of prefix-suffix-square free factors of w can be computed in $\mathcal{O}(n + |S|)$ time (including the time used to construct $left$, $right$ and RMQ-structures for $left$). $\qquad\square$

Remark 1. The suffix-square-free (resp., prefix-square-free) factors of a word can be detected easier. That is, $w[i..j]$ is suffix-square (resp., prefix-square) free iff $i > left[j]$ (resp., $j < right[i]$). So, we can output the suffix-square-free (resp., prefix-square-free) factors of w in $\mathcal{O}(n + |S|)$ time, where S is the solution set.

It is not hard to construct words with a high number of prefix-suffix-square-free factors. E.g., take any finite prefix w of a right-infinite square-free word $\mathbf{w}$ (see [5]). Then w does not contain squares, so all its factors are prefix-suffix-square free and the size of the set of prefix-suffix-square free factors of w is $\Theta(n^2)$. However, the size of S can be computed without enumerating all its elements. The main idea is that, for a position j of the input word w, we need to report as many squares $w[i..j]$ as there are positions i greater than $left[j]$ and less or equal to j such that $right[i] > j$. This can be done efficiently using segment trees [16].

Theorem 3. *Given a word w of length n, we can compute the number of prefix-suffix-square free factors of w in $\mathcal{O}(n \log n)$ time.*

We conclude this section by showing that the longest prefix-suffix-square free factor $w[i..j]$ (i.e., the longest element of the set S from Theorem 2) can be computed in linear time, so without enumerating all the elements of S.

Theorem 4. *Given a word w of length n, we can find its longest prefix-suffix-square free factor in $\mathcal{O}(n)$ time.*

Proof. We first construct the arrays $left[\cdot]$ and $right[\cdot]$ for the input word w. Moreover, this time we produce a range minimum query structure for the array $left[\cdot]$. Let $RMQ(j_1, j_2)$ denote the minimum value occurring in the range between j_1 and j_2 and $posRMQ(j_1, j_2)$ denote the position where $RMQ(j_1, j_2)$ occurs (in case of equality, we take the rightmost such position).

The main idea of our algorithm is the following. We go through the positions of the word w, and try to maintain the rightmost position j where a prefix-suffix-square free factor starting on one of the positions already considered may end. Initially, before considering the first position of the word, we have $j = 0$. Let us assume that we reached the point where we consider position i, and the factor $w[i'..j']$ with $i' < i$ is the prefix-suffix-square free factor that ends on the rightmost position among all the prefix-suffix-square free factors that start on positions less than i. We want to see whether we can construct a factor $w[i..j]$ with $j > j'$ and $j < right[i]$ (so that the factor $w[i..j]$ is prefix-square free); if yes, we want to find the largest such j. First, such a position j exists if and only if $RMQ(j' + 1, right[i] - 1) < i$; then $j = posRMQ(j' + 1, right[i] - 1)$. Indeed, if $RMQ(j' + 1, right[i] - 1) < i$ then there is some position j such that $j' < j < right[i]$ and the shortest square ending on j starts before i; this means that $w[i..j]$ is both prefix and suffix-square free, and it also ends after j', so it is the kind of factor we were looking for. Now, we try to maximise j. For this we set $j' = j$ and repeat the procedure above: check if $RMQ(j'+1, right[i]-1) < i$ and, if yes, set $j = posRMQ(j'+1, right[i]-1)$. We keep repeating this procedure until if $RMQ(j' + 1, right[i] - 1) \geq i$ or $j' + 1 = right[i]$. After the procedure cannot be repeated anymore, we have the rightmost position j where a prefix-suffix-square free factor $w[i'..j]$ with $i' \leq i$ may end. Clearly, before considering the next position $i+1$ and repeating the entire process above, we have also obtained the longest prefix-suffix-square free factor starting on one of the positions less or equal to i: if this factor starts on i' and is $w[i'..j']$ then it was exactly the factor ending on the rightmost position produced by our algorithm when the position i' was considered as a potential starting position of the prefix-suffix-square free factor with the rightmost ending position. We then repeat the whole process for $i + 1$, and so on, until each position of the word was considered.

By the explanations given above, it is clear that this process correctly identifies the longest prefix-suffix-square free factor of w. The time complexity is $\mathcal{O}(n)$. Indeed, the time used to process each i is proportional to the number of RMQ and $posRMQ$ queries we perform during this processing. However, each position of w can be the answer to at most one such $posRMQ$-query (when some new RMQ and $posRMQ$ queries are asked, they are asked for a range strictly to the right of the position that was the answer of the previous $posRMQ$-query). So, in total (for all i) we only ask $\mathcal{O}(n)$ queries, and each is answered in constant time. Consequently, the total running time of our algorithm is linear.   □

4   Duplication and Related Factorisation Problems

The previous section explains how PD-, SD-, PSD-irreducible factors of a word can be efficiently identified. However, it seems interesting to us whether this helps us identify the PD, SD, or PSD-irreducible roots of w.

Let us discuss first the case of the PSD operation. First, the ideas of [1] can be used to identify in $\mathcal{O}(n^2 \log n)$ all the factors $w[i..j]$ of a given word w of length n

such that $w \in PSD^*(w[i..j])$. Indeed, if $w \in PSD^*(w[i..j])$ and t^2 is a primitively rooted square starting on i (respectively, ending on j) and $|t^2| \leq j - i + 1$, then $w \in PSD^*(w[i + |t|..j])$ (respectively, $w \in PSD^*(w[i..j - |t|])$). So, we can use a dynamic programming approach to find the roots of w, going from the longer to the shorter ones. This approach can be implemented in $\mathcal{O}(n^2 \log n)$ time as for each position i of w there are $\mathcal{O}(\log n)$ primitively rooted squares starting and ending at i, and we can produce the lists of such squares for all positions in $\mathcal{O}(n \log n)$ time [17]. This strategy also provides the irreducible roots: from the factors $w[i..j]$ we just computed, so with $w \in PSD^*(w[i..j])$, we keep those for which $q_{ps}(i, j)$ returns true (i.e., $left[j] < i$ and $right[i] > j$).

However, it is clear that the most time consuming step in the above strategy is finding all the factors from which w is generated. If that step could be implemented more efficiently than $\mathcal{O}(n^2 \log n)$, then the whole process would take less time (as with $\mathcal{O}(n)$ preprocessing, we can test in $\mathcal{O}(1)$ time each root of w to see whether it is irreducible).

Similarly, detecting all the PD or SD-irreducible roots of w takes $\mathcal{O}(n \log n)$ time (we can use the same strategy, but this time we only need to analyse suffixes or, respectively, prefixes of the word). Again, the most time consuming step is finding the roots of w, not testing which of them are irreducible.

On the other hand, finding only one PD, SD, or PSD-irreducible root can be done in linear time. We only give the proof for PSD.

Theorem 5. *Given a word w of length n, we can find one PSD-irreducible root of w in $\mathcal{O}(n)$ time.*

Proof. Let $w_0 = w$. We construct a sequence $(w_i)_{i \geq 0}$ such that $w_i \in PSD(w_{i+1})$. Assume that w_i has a square suffix and let t^2 be the shortest square suffix of w_i. Then let $w_{i+1} = w_i[1..|w_i| - |t|]$. If w_i does not end with a square, but it has a square prefix, let t^2 the shortest square prefix of w_i. Then let $w_{i+1} = w_i[|t| + 1..|w_i|]$. If w_i is prefix-suffix-square free, then we stop: we reached an irreducible root of w. Using the arrays *left* and *right* defined in the last section, we can implement this strategy in linear time.   □

Remark 2. Given a word w of length n, we can find an SD-irreducible (respectively, PD-irreducible) root of w in $\mathcal{O}(n)$ time. We just use the same procedure, but only extend our sequence as long as the current word is not suffix-square free (respectively, prefix-square free).

Example 1. We define a family of words $(w_n)_{n \in \mathbb{N}}$ such that w_n has $\Theta(|w_n|)$ SD-irreducible roots. Let $w_1 = aabbab$; this word is SD-irreducible. Then define $w_i = w_{i-1}w_{i-1}bb$ for $i \geq 2$. The length of w_i is $2^{i+2} - 2$ for all $i \geq 1$. Note that w_2 has the SD-irreducible root w_1 and let $R_2 = \{w_1\}$. Now, for $i \geq 3$, w_i has at least the SD-irreducible roots $R_i = R_{i-1} \cup \{w_{i-1}r \mid r \in R_{i-1}\}$. So w_i has, indeed, at least $|R_i| \geq \frac{|w_i|}{16}$ irreducible roots. Further, it is not hard to see that $w_i'^R w_i$, where w_i' is obtained from w_i by replacing a by c and b by d, has $\Theta(|w_i'^R w_i|^2)$ PSD-irreducible roots. Indeed, if r_1 and r_2 are SD-irreducible roots of w_i, then $r_1'^R r_2$ is a PSD-irreducible root of $w_i'^R w_i$ (where r_1' is obtained from

r_1 by replacing a by c and b by d). So the words of the family $(w_n'^R w_n)_{n \in \mathbb{N}}$ are such that $w_n'^R w_n$ has $\Theta(|w_n'^R w_n|^2)$ *PSD*-irreducible roots.

Essentially, deciding for a word w the existence of a *SD*-irreducible root is equivalent to asking whether we can factor w in $k > 1$ factors $w = s_1 \cdots s_k$ such that s_i is a suffix of $s_1 \ldots s_{i-1}$, for $1 < i \leq k$, and s_1 is suffix-square free. It is not hard to see that such a factorisation is deeply related to the square-structure of w: the square s_i^2 occurs centred at the border between the factors s_{i-1} and s_i of the factorisation, for all $i > 1$. So, it seems natural to us to also investigate other factorisations that can be easily linked to the square-structure, or, more generally, to the repetitions-structure, of the word.

To this end, the first question we ask is how to test whether a word can be factored into squares. If such a factorisation exists, then one where every factor is a primitively rooted square also exists. As the list of $\mathcal{O}(\log n)$ primitively rooted squares ending at each position of a word of length n can be obtained in total time $\mathcal{O}(n \log n)$, we can easily obtain within the same complexity a factorisation of w in primitively rooted squares by dynamic programming.

Theorem 6. *Given a word w of length n, we can find (if it exists) in $\mathcal{O}(n \log n)$ time a factorisation $w = s_1 \cdots s_k$ of w, such that s_i is a square for all $1 \leq i \leq k$.*

While we were not able to find in linear time a factorisation of a word into squares, we can find in linear time a factorisation of w that contains as many squares as possible. In fact, there exists such a factorisation where all the square factors are the shortest squares occurring at their respective position. The factorisation can be obtained in linear time by dynamic programming, using the $left[\cdot]$ array constructed in the previous section.

Theorem 7. *Given a word w, we can find (if it exists) in linear time a factorisation $w = s_1 \cdots s_k$ of w, such that, for any other factorisation $= s_1' \cdots s_p'$ we have $|\{i \mid 1 \leq i \leq k, s_i \text{ is a square}\}| \geq |\{i \mid 1 \leq i \leq p, s_i' \text{ is a square}\}|$.*

Another extension of the factorisation into squares of a word, that can still be solved in linear time, is to decide whether a given word can be factored into k factors $w = s_1 \cdots s_k$, such that $per(s_i) \leq \frac{|s_i|}{2}$ for all i; in such a factorisation, each s_i is a run (not necessarily maximal). The main idea in finding such a factorisation is that if $w[1..i]$ is a prefix of w that can be factored in runs, and $w[i'..j']$ is a maximal run of period p containing $i + 1$ and with $i + 2p \leq j'$, then all the factors $w[1..k]$ with $i + 2p \leq k \leq j'$ can also be factored in runs. Using an interval union-find data-structure ([9]) to maintain the positions i incrementally discovered during the execution of our algorithm such that $w[1..i]$ can be factored into runs we obtain in linear time a factorisation of w in runs.

Theorem 8. *Given a word w, we can find (if it exists) in linear time a factorisation $w = s_1 \cdots s_k$ of w, such that $per(s_i) \leq \frac{|s_i|}{2}$ for all $1 \leq i \leq k$.*

5   Conclusions

This paper introduced the concept of prefix-suffix-square free word, and several initial results on this concept were obtained. While the algorithms identifying all prefix-suffix-square free factors of a word or the longest such factor run in linear (optimal) time, it seems interesting to see whether counting these factors can also be done in linear time; our solution runs in $\mathcal{O}(n \log n)$ time.

The notion of prefix-suffix-square free word is strongly related to the notion of irreducible root of a word w.r.t. the operations of prefix/suffix-duplication. It seems interesting to us to see whether we can detect the SD-irreducible roots (resp., PSD-irreducible roots) of a word faster than $\mathcal{O}(n \log n)$ (resp., $\mathcal{O}(n^2 \log n)$) time. We showed here how to find in linear time one (non-specific) PSD-irreducible root of a word; how about finding the longest or shortest root?

Several questions related to possible factorisations of a word were addressed at the end of our paper. To this end, an interesting open problem is whether one can factorise a word in square factors in linear time.

References

1. García-López, J., Manea, F., Mitrana, V.: Prefix-suffix duplication. J. Comput. Syst. Sci. **80**, 1254–1265 (2014)
2. Chan, S., Blackburn, E.: Telomeres and telomerase. Philos. Trans. R. Soc. Lond. B. Biol. Sci. **359**, 109–121 (2004)
3. Preston, R.: Telomeres, telomerase and chromosome stability. Radiat. Res. **147**, 529–534 (1997)
4. Murnane, J.: Telomere dysfunction and chromosome instability. Mutat. Res. **730**, 28–36 (2012)
5. Lothaire, M.: Combinatorics on Words. Cambridge University Press (1997)
6. Gusfield, D.: Algorithms on strings, trees, and sequences: computer science and computational biology. Cambridge University Press, New York (1997)
7. Xu, Z.: A Minimal Periods Algorithm with Applications. In: Amir, A., Parida, L. (eds.) CPM 2010. LNCS, vol. 6129, pp. 51–62. Springer, Heidelberg (2010)
8. Gawrychowski, P., Manea, F., Nowotka, D.: Testing generalised freeness of words. In: Proc. STACS 2014. LIPIcs, vol. 25, pp. 337–349 (2014)
9. Gabow, H.N., Tarjan, R.E.: A linear-time algorithm for a special case of disjoint set union. In: Proc. STOC, pp. 246–251 (1983)
10. Main, M.G.: Detecting leftmost maximal periodicities. Discrete Appl. Math. **25**, 145–153 (1989)
11. Kolpakov, R., Kucherov, G.: Finding maximal repetitions in a word in linear time. In: Proc. FOCS, pp. 596–604 (1999)
12. Duval, J.-P., Kolpakov, R., Kucherov, G., Lecroq, T., Lefebvre, A.: Linear-Time Computation of Local Periods. In: Rovan, B., Vojtáš, P. (eds.) MFCS 2003. LNCS, vol. 2747, pp. 388–397. Springer, Heidelberg (2003)
13. Kärkkäinen, J., Sanders, P., Burkhardt, S.: Linear work suffix array construction. J. ACM **53**, 918–936 (2006)

14. Crochemore, M., Iliopoulos, C.S., Kubica, M., Rytter, W., Waleń, T.: Efficient Algorithms for Two Extensions of LPF Table: The Power of Suffix Arrays. In: van Leeuwen, J., Muscholl, A., Peleg, D., Pokorný, J., Rumpe, B. (eds.) SOFSEM 2010. LNCS, vol. 5901, pp. 296–307. Springer, Heidelberg (2010)
15. Bender, M.A., Farach-Colton, M.: The LCA problem revisited. In: Gonnet, G., Panario, D., Viola, A., (eds.) LATIN 2000. LNCS, vol. 1776, pp. 88–94. Springer, Heidelberg (2000)
16. Bentley, J.: Decomposable searching problems. Inform. Proc. Letters **8**, 244–251 (1979)
17. Crochemore, M.: An optimal algorithm for computing the repetitions in a word. Inform. Proc. Letters **12**, 244–250 (1981)

Temporal Analysis of CHAVE Collection

Olga Craveiro[1,2]($\boxtimes$), Joaquim Macedo[3], and Henrique Madeira[1]

[1] CISUC/Department of Informatics Engineering, University of Coimbra,
Coimbra, Portugal
{marine,henrique}@dei.uc.pt
[2] ESTG, Polytechnic Institute of Leiria, Leiria, Portugal
[3] Centro Algoritmi/Department of Informatics, University of Minho, Braga, Portugal
macedo@di.uminho.pt

Abstract. The importance of temporal information (TI) is increasing in several Information Retrieval (IR) tasks. CHAVE, available from Linguateca's site, is the only ad hoc IR test collection with Portuguese texts. So, the research question of this work is whether this collection is sufficiently rich to be used in Temporal IR evaluation. The obtained answer was yes. By the analysis of the CHAVE collection, we verified that 22% of the topics and 86% of the documents have at least one *chronon*. 49% of topics are time-sensitive. Analyzing the relation of topics with documents, relevant documents of time-sensitive topics converge to a specific date(s), while the non-relevant ones are dispersed along the timeline. Finally, we used a peak dates strategy as a time-aware query expansion (QE) process. Experiments showed effectiveness improvements for time-sensitive queries.

1 Introduction

The recency of information contained in a document, or the time periods of the events reported is a significant aspect in assessment of its quality, importance and relevance. Several IR tasks can benefit by incorporating the TI.

The focus of this work is the use of TI in ad hoc IR tasks. In particular, we are interested in the usability of a given test collection for the evaluation of a time-aware IR system. Such collection includes three components: documents, topics and relevance judgements (RJs). As our research work uses Portuguese texts, and CHAVE[1] is the only test collection available, the result of such evaluation is very important for our experimental work. TimeBankPT[1] is a temporal annotated Portuguese Collection, but intended for natural language processing.

Supposing that Chave is suitable for temporal ad hoc IR research, we must answer the following questions: 1) Are there a significant percentage of time-sensitive topics? 2) Are there enough temporal information in the content of the majority of documents? 3) Does the usage of time-aware retrieval strategies improve the effectiveness results for time-sensitive topics?

---

[1] Accessible in Linguateca's site http://www.linguateca.pt/

© Springer International Publishing Switzerland 2015
C. Iliopoulos et al. (Eds.): SPIRE 2015, LNCS 9309, pp. 67–74, 2015.
DOI: 10.1007/978-3-319-23826-5_7

The structure of the document is the following: the Chave test collection and its overall temporal statistics (extracted using temporal preprocessing described in the same section) are presented in Section 2; TI of documents and segments is summarized in Section 3; Section 4 synthesizes the TI of topics and establishes the relationship with their RDs and non-RDs; Section 5 compares the effectiveness results of a time-aware strategy (peak dates for QE) and two baselines (one with QE and another without); Section 6 discusses the overall results and some further work directions.

2   The CHAVE Collection

The CHAVE collection, created by Linguateca[2], is composed of full-text articles from two major daily Portuguese and Brazilian newspapers, namely the PUBLICO and the Folha de Sao Paulo, with complete editions of 1994 and 1995. The texts are written in two variants of Portuguese language, from Portugal and Brazil. This collection has a total of 574768 documents, 4682363 sentences, and 90646837 words from which 866702 are distinct.

For the purpose of this work, the suitable resources are the ones available for the ad hoc IR track. All the resources are in TREC [3] format. Documents and topics are annotated in SGML. All the CHAVE topics were judged. 4 topics (C216, C220,C227, and C240) do not have any RD. On average, each topic has 41.7 RDs for 378.3 non-RDs.

Collection Temporal Preprocessing
The CHAVE collection was preprocessed to obtain the statistics required for this temporal analysis. A key requirement for our research work is the temporal segmentation of texts, to establish temporal relationships between words which are used for automatic QE [4,5]. We developed a testbed software for Portuguese text processing, composed for Annotator, Resolver and Temporal Segmentator. The testbed is fully described in [6].

Our temporal approach considers not only the publication date of each news, but also all the temporal expressions included in its content. However, the most important are the ones that can be mapped into *chronons*, once they can be used later in IR tasks. Some temporal expressions are too vague and can not be transformed into a *chronon*. *Chronons* are normalized dates with a certain granularity (year, month, day or hour) and a mark on a timeline [7].

The source documents are processed by the Annotator which identifies the temporal expressions and assigns to them a classification. Then, the Resolver normalizes the annotated temporal expressions transforming them into *chronons*. Finally, the documents are temporally segmented. For this purpose, the documents are modeled as a chronologic narrative of events. The objective of segmentation is to detect temporal discontinuities in the text and split it into temporally coherent pieces. Each segment is labeled with the *chronons* that occur on its contents.

Note that the results presented here are subject to an error for two main reasons: first, our testbed system does not have an effectiveness of 100%, as we explained in [6]; second, the original documents have also some typo errors[2].

Temporal Statistics of Collection

The temporal scope of CHAVE is from year 94AD to 2577, though their documents are dated in [1994-01-01; 1995-12-31]. TI was found in the content of about 90% of documents, but as this information is not all resolved, the percentage of documents with at least one *chronon* is lower (about 86%). A total of 869051 *chronons* (only 12796 distinct), were obtained from the 1022846 temporal expressions that were annotated. Note that these two values do not have a direct correspondence. There are some temporal expressions without any *chronon*, and others with two *chronons* ("1992-93" has two *chronons*: 1992-XX-XX and 1993-XX-XX).

DATE was the classification given to about 90% of the temporal expressions. For example, 1995-09-XX is associated to the timeline with month granularity. We observed that the *chronons* are distributed by four timelines: hour (7.74%), day (49.3%), month (13.89%), and year (29.07%). So, the temporal specificity[3] of CHAVE is day, as expected, since documents are news of daily newspapers.

3   Documents and Segments

The results show that about 54% of the documents have at least 1 *chronon* and a maximum of 4 *chronons*. The number of *chronons* variates from 0 to 228. 14.16% has 0 *chronons*; 17.46% has 1 chronon; 15.,5% has 2 *chronons*; 11.88% has 3 *chronons*; 9.27% has 4 *chronons*; 12.49% has 5-6 *chronons*; 9.80% has 7-9 *chronons*; 4.47% has 10-12 *chronons*; and 4.99% has 13-228 *chronons*.

The higher temporal richness is shared by two documents, each one with 228 *chronons*, the maximum number of *chronons* per document. Although day is the predominant temporal specificity (about 47% of the documents), there is also a considerable number of documents (about 40%) with year as the most representative time granularity. Month and hour are also temporal specificity of 10.26% and 2.62% documents. The 6 *chronons* that have an occurrence in more documents are very close to the publication date of the documents: 1990-XX-XX with 7220 documents; 1991-XX-XX, 8119; 1992-XX-XX, 10566; 1993-XX-XX, 16185; 1994-XX-XX, 21515; and 1995-XX-XX, 14662.

As temporal segmentation is important for our work, some related statistics were included here. The number of temporal segments per document closely follows the number of *chronons* of a document, as foreseen, since the *chronons* give an indication about the number of temporal discontinuities per document. Thus, the average number of temporal segments per document is about 6 with a median value of 5 and a range from 1 to 388.

---

[2] For instance, 'March of 9160' instead of 'March of 1960'.

[3] Most frequent timeline[7].

It was not possible to carry out any temporal segmentation in 29847 documents (about 14%), as these documents do not have any *chronon*. Such documents have a single temporal segment. The number of temporal segments in the collection is 1255416, with an average of about 69 words and 1 *chronons* per segment. The number of words and distinct words is very close, which means that in most segments, there are a few repeated words.

Summarizing, the segments has a maximum of 29 *chronons*, 4202 words and 1305 distinct words; in terms of minimum they have 0 *crhonon*, 1 word and 1 distinct word; the average are 1 *chronon*, 68.64 words and 53.67 words; and for median 1 *chronon*, 34 words and 33 distinct words.

4 Topics and Their Relevant Documents

The topics of CHAVE are divided into three groups: TG1(C201-C250), TG2(C251-C300), and TG3(C301-C350).Almost all the temporal expressions in the text of the topics could be mapped into *chronons*, since these expressions were classified as DATE, having explicit temporal references. TG1 has 24 temporal expressions and 22 *chronons*; its temporal specificity is year and it has explicit TI in 14 topics. TG2 has only one topic with TI but there is not any *chronon*, since the only temporal expression *Nos dias de hoje*[4]. has a classification of GENERIC, which means that this TI cannot be used as explicit TI. TG3 has 25 temporal expressions and 26 *chronons*; its temporal specificity is year and it has 18 topics with explicit TI.

We verified that only 32 of the 150 topics (21.3%) have explicit TI. Almost all these topics aim to obtain documents referencing to events that occur during the time period of the collection (1994-1995), only considering the publication date of the documents. More precisely, the TI of 13 topics of TG1 and 17 topics of TG3 reference dates in that time period. References outside that time period exist only in C339 (December 1993) and C221 (year of 2002).

We also analyzed the relationship between the *chronons* of the topic and the ones found in their RDs. For each topic, we computed the number of distinct *chronons* of their RDs that are equal when they are mapped into the same timeline of the topic's *chronons*. The number of distinct *chronons* of the RDs which represents a date before or after the date of the topic is also computed. Considering only topics with explicit TI, it was selected a set with 7 with month granularity and 13 with year granularity. For each date of the topic, RDs with dates before, equal and after were counted. The result of counting was: 10 of the topics (50 %) have a higher number of RD with equal dates; 4 (20 %) have higher number of RD with later dates; and 6 topics (30 %) have a higher number of RD with before dates.

We also computed the maximum, minimum, average and median values of *chronons* (distinct *chronons*) in RDs per topic in the three topic groups. TG1 has maximum 842 (243), minimum 1(1), average 84.67 (39,11) and median 37.50

---

[4] *Nowadays, in English*

(20.00). TG2 has maximum 1863(612), minimum 12(8), average 350.02 (154.46) and median 246.50 (145.50). TG3 has maximum 2226(543), minimum 8(8), average 372.13 (130.20) and median 241.50 (104.50).

In general, the topics of TG1 have less RDs than the topics of other groups. So, it is normal that the topics of this group also have less *chronons*. However, the three groups of topics have approximately the same percentage of distinct *chronons*. We can notice that the distinct *chronons* are around 30% of all *chronons* for each group of topics. TG1 has 35%, TG2 has 33% and TG3 has 27% of distinct *chronons*.

The *chronons* extracted from topics with explicit TI can be placed in the year timeline (Ty) and month timeline (Tm), namely, 1995-XX-XX, 1994-XX-XX, and 2002-XX-XX in Ty, and 1993-12-XX, 1994-02-XX, 1994-09-XX, 1994-11-XX, 1995-05-XX, 1995-10-XX, and 1995-11-XX in Tm. In conclusion, only 8 of 322 *chronons*[5] have representation in timeline Tm; the others are not so specific, having only the year.

We verified that there are also temporal references of about 60% of topics in the RDs of the topic. Note that the content of the RDs has more temporal references to present or past events than future events. In general, the temporal references, both in topic text and content of RDs, are in the interval of date publication of the CHAVE documents.

Analysis Per Topic

In order to know the temporal scope of each topic, we determined the frequency per *chronon* of RDs and non-RDs. For simplicity, the *chronons* represented only in a timeline with hours granularity (Th) were not considered in this analysis.

The temporal sensitivity of topics was also evaluated by the analysis of the TI in RDs and non-RDs. A topic is time-sensitive when the temporal references, represented as *chronons*, of their RDs converge to a date or a time period. TG1 has 16 time-insensitive topics and 34 time-sensitive topics, from which 1 have explicit TI. TG2 has 15 with implicit TI and 35 time-insensitive topics. TG3 has 18 topics with explicit TI, 7 with implicit TI and 25 time-insensitive topics. In conclusion, the time-sensitive topics are well represented in two of the topics groups, namely 68% and 50% of topics of TG1 and TG3, respectively. TG2 only contains 30% of time-sensitive topics. Due to space constraints, we only present the analysis of three topics, each one with a different classification: C222 (explicit TI), C254 (implicit TI) and C310 (time-insensitive).

The topic C222 has explicit TI namely 1995-05-XX. In RD set, we can observe that almost all *chronons* are in the period of time defined by the topic. The *chronons* with higher frequency of RDs are 1995-05-08 and 1995-05-09. Considering only the 10 *chronons* with higher frequency for non-RD, 1995-XX-XX is the one with more occurrences. However, these documents are scattered along the timeline. We verified that 100% of the RDs have at least one reference to May 1995, while only about 25% of the non-RDs have the same reference.

---

[5] The *chronon* 1955-05-XX occurs twice.

Although topic C254 does not have explicit TI, the *chronons* of RDs converge to a date. The temporal references of both RDs and non-RDs present the same tendency of the topic C222. In other words, the RDs converge to a date unlike non-RDs. In topic C254, about 85% of the 129 RDs have a reference to January 1995, whilst only 25% of the 181 non-RDs have a reference to this date and their *chronons* are scattered along the timeline, such as topic C222.

A similar analysis was done for topic C310. The *chronons* in both document sets (RDs and non-RDs) are distributed along the all timeline. Since there is not a concentration of RDs around a date, we conclude that this topic is not time-sensitive. The *chronons* which occur in most documents are 1994-XX-XX (in 20% of non-RDs) and 1993-XX-XX (in 10% of RDs).

5   Experiments with Temporal Query Expansion

To answer the last question (see third paragraph of Section 1), we used a strategy based on peak dates to modify the QE algorithm. Documents without an occurrence in a peak date are removed from the set of the pseudo-relevant documents (set R) to the original query q0. If there are no peaks, R remains unchanged. The date of publication and *chronons* extracted from the documents content of the set R are considered for computing of peak dates. First, the number of documents for each date is determined. It is assumed a normal distribution of documents across dates. Subsequently, the peaks are selected based on an outlier detection criterion, as the one defined by Chauvenet [8].

This evaluation used 100 CHAVE topics, C251-C350, considering their title and their description for automatic query generation. The topics were manually divided according to their time sensitivity. Therefore, we performed experiments using 3 topic sets, namely 60 time-insensitive queries, 40 time-sensitive queries (with 22 implicit and 18 explicit temporal information) and 18 temporal explicit queries. With this last set, we performed 2 experiment: one that does not consider the TI expressed in the text of the query and another considering it. In the last case, all dates not contained in the query date are not considered for computing peak dates.

For the implementation, we modified the Terrier search-engine software package [9] to include temporal pre-processing of the documents and topics and we used the peak dates strategy on the QE procedures.

The retrieval was performed using TF-IDF, considering only the top 1000 documents of the results set. The results obtained were used as the baseline noQE. In all the experiments, the queries were expanded with 10 terms of the top 3 retrieved documents. The terms weight was computed by the Bo1 model's formula[10]. We also report a run with the Bo1 model as the stronger baseline, named as Bo1QE. For each run we present Precision at top 10 and 15 documents (P@10 and P@15), MAP, Robustness[11] compared with noQE (Rob_1), and with Bo1QE(Rob_2). I_i and Q_i are the number of queries for which the effectiveness increases or decreases, respectively. Table 1 shows the results comparison between the strategy based on peak dates (PeakDates) and baselines (noQE, Bo1QE). We verified that the evaluated strategy achieved a better effectiveness

Table 1. Results using CHAVE for time sensitive and insensitive queries

	MAP	P@10	P@15	Rob$_1$	I$_1$;Q$_1$	Rob$_2$	I$_2$;Q$_2$
40 time-sensitive queries							
noQE40	0.358	0.458	0.410				
BolQE40	0.387	0.473	0.430	0.500	29 ;9		
PeakDates	**0.402**	**0.493**	**0.455**	0.500	29;9	0.075	18 ; 15
18 queries with explicit temp. information without query dates							
noQE18	0.431	0.444	0.411				
BolQE18	0.441	0.472	0.430	0.333	11;5		
PeakDates	**0.463**	**0.494**	**0.448**	0.333	11;5	0.333	9;3
18 queries with explicit temp. information with query dates							
noQE18	0.431	0.444	0.411				
BolQE18	0.441	0.472	0.430	0.333	11;15		
PeakDates	**0.451**	**0.494**	**0.452**	0.389	12;5	0.000	7;7
60 time-insensitive queries							
noQE60	0.302	0.502	0.457				
BolQE60	**0.344**	**0.515**	**0.498**	0.333	40;20		
PeakDates	0.341	0.513	0.488	**0.483**	44;15	-0.067	25;29

than the baselines when using time-sensitive queries. MAP, Precision@10 and Precision@15 were increased from 2% to 6%, when compared with the baseline BolQE. Robustness is also better in time-insensitive queries (more 4 queries than Bo1QE).

6 Conclusions

This paper uses some statistical measures for temporal characterization of CHAVE, a Portuguese text collection. About 22% of the topics and 86% of the documents have at least one *chronon*. The temporal references of documents and topics are around the time period of their publication date [1994-1995]. Three groups of topics were identified: time-sensitive with explicit temporal TI, time-sensitive without explicit TI and time-insensitive. Almost half of the topics are time-sensitive. For each one of these types, it was analyzed the distribution of the dates in RDs and non-RDs along timelines. The results obtained show that the RDs of the time-sensitive topics converge to a specific date, as opposed to the non-RDs which are dispersed along the timeline. These results are similar to those obtained by previous studies [12]. The set of obtained results confirms the usability of the CHAVE test collection for temporal ad hoc IR research. Furthermore a peak dates strategy was used on QE to increase the retrieval effectiveness, presenting auspicious results. With temporal segmentation of documents, the goal is to establish a direct relationship between words and time. Due to excessive processing costs, this issue is a topic for further work. Equally important is the comparison of several outlier detection criteria for peak dates.

Acknowledgments. This work is partially supported by Algoritmi and CISUC, financed by Fundação para a Ciência e Tecnologia, within the project scope UID/CEC/00319/2013 and co-supported by iCIS project (CENTRO-07-ST24-FEDER-002003) is co-financed by QREN, in the scope of the Mais Centro Program and European Union's FEDER.

References

1. Costa, F., Branco, A.: TimeBankPT: a TimeML annotated corpus of Portuguese. In: Calzolari, N., Choukri, K., Declerck, T., Dogan, M.U., Maegaard, B., Mariani, J., Odijk, J., Piperidis, S. (eds.) Proceedings of the 8th International Conference on Language Resources and Evaluation (LREC 2012), pp. 3727–3734. ELRA, Istanbul (2012)
2. Santos, D., Rocha, P.: The key to the first CLEF with Portuguese: topics, questions and answers in CHAVE. In: Peters, C., Clough, P., Gonzalo, J., Jones, G.J.F., Kluck, M., Magnini, B. (eds.) CLEF 2004. LNCS, vol. 3491, pp. 821–832. Springer, Heidelberg (2005)
3. Voorhees, E.M., Harman, D.K.: TREC: Experiment and Evaluation in Information Retrieval (Digital Libraries and Electronic Publishing). The MIT Press (2005)
4. Craveiro, O., Macedo, J., Madeira, H.: Query expansion with temporal segmented texts. In: de Rijke, M., Kenter, T., de Vries, A.P., Zhai, C.X., de Jong, F., Radinsky, K., Hofmann, K. (eds.) ECIR 2014. LNCS, vol. 8416, pp. 612–617. Springer, Heidelberg (2014)
5. Craveiro, O., Macedo, J., Madeira, H.: Words temporality for improving query expansion. In: Baptista, J., Mamede, N., Candeias, S., Paraboni, I., Pardo, T.A.S., Volpe Nunes, M.G. (eds.) PROPOR 2014. LNCS, vol. 8775, pp. 262–267. Springer, Heidelberg (2014)
6. Craveiro, O., Macedo, J., Madeira, H.: It Is the time for portuguese texts!. In: Caseli, H., Villavicencio, A., Teixeira, A., Perdigão, F. (eds.) PROPOR 2012. LNCS, vol. 7243, pp. 106–112. Springer, Heidelberg (2012)
7. Alonso, O., Gertz, M., Baeza-Yates, R.: Temporal analysis of document collections: framework and applications. In: Chavez, E., Lonardi, S. (eds.) SPIRE 2010. LNCS, vol. 6393, pp. 290–296. Springer, Heidelberg (2010)
8. Grubbs, F.E.: Procedures for detecting outlying observations in samples. Technometrics **11**(1), 1–21 (1969)
9. Ounis, I., Amati, G., Plachouras, V., He, B., Macdonald, C., Johnson, D.: Terrier information retrieval platform. In: Losada, D.E., Fernández-Luna, J.M. (eds.) ECIR 2005. LNCS, vol. 3408, pp. 517–519. Springer, Heidelberg (2005)
10. Amati, G.: Probability models for information retrieval based on divergence from randomness, Ph.D. thesis, University of Glasgow (2003)
11. Carpineto, C., Romano, G.: A survey of automatic query expansion in information retrieval. ACM Comput. Surv. **44**(1), 1:1–1:50 (2012)
12. Jones, R., Diaz, F.: Temporal profiles of queries. ACM Trans. Inf. Syst. **25**(3)

DeShaTo: Describing the Shape of Cumulative Topic Distributions to Rank Retrieval Systems Without Relevance Judgments

Radu Tudor Ionescu[1]([⊠]) , Adrian-Gabriel Chifu[2], and Josiane Mothe[3]

[1] Faculty of Mathematics and Computer Science,
University of Bucharest, Bucharest, Romania
raducu.ionescu@gmail.com
[2] IRIT UMR5505, CNRS, Université de Toulouse,
Université Paul Sabatier, Toulouse, France
adrian.chifu@irit.fr
[3] IRIT UMR5505, CNRS, Université de Toulouse, ESPE, Toulouse, France
josiane.mothe@irit.fr

Abstract. This paper investigates an approach for estimating the effectiveness of any IR system. The approach is based on the idea that a set of documents retrieved for a specific query is highly relevant if there are only a small number of predominant topics in the retrieved documents. The proposed approach is to determine the topic probability distribution of each document offline, using Latent Dirichlet Allocation. Then, for a retrieved set of documents, a set of probability distribution shape descriptors, namely the skewness and the kurtosis, are used to compute a score based on the shape of the cumulative topic distribution of the respective set of documents. The proposed model is termed *DeShaTo*, which is short for *De*scribing the *Sha*pe of cumulative *To*pic distributions. In this work, DeShaTo is used to rank retrieval systems without relevance judgments. In most cases, the empirical results are better than the state of the art approach. Compared to other approaches, DeShaTo works independently for each system. Therefore, it remains reliable even when there are less systems to be ranked by relevance.

Keywords: Information retrieval · Topic modeling · LDA · Document topic distribution · Skewness · Kurtosis · Ranking retrieval systems

1 Introduction

Automatically estimating the effectiveness of any information retrieval system is one of the most important tasks in information retrieval (IR). An approach that could solve this task with a high degree of accuracy would have a broad range of applications including selective IR, selective query expansion [4,15], ranking retrieval systems without relevance judgments [9,12,13], query difficulty prediction [3,11], to name only a few. Being able to understand and distinguish

© Springer International Publishing Switzerland 2015
C. Iliopoulos et al. (Eds.): SPIRE 2015, LNCS 9309, pp. 75–82, 2015.
DOI: 10.1007/978-3-319-23826-5_8

the behavior of a highly effective IR system from a poorly effective one (on a per query basis) is the key in solving the task of estimating IR effectiveness. Intuitively, a highly effective system should return a set of documents in which there are only one or a few predominant topics[1] (related to the query), while an average or poorly performing system will return documents from various topics, since not all the documents will be relevant for the given query. In other words, more topics indicate that the given query is more ambiguous from the point of view of an IR system. Interestingly, this hypothesis represents the cornerstone of the clarity score [3], but there are many other aspects of relevance that are ignored by this supposition. Nevertheless, the same hypothesis is explored into a different direction in this work. More precisely, the current work proposes an approach that can potentially be used for estimating the relevance level of any IR system. Latent Dirichlet Allocation (LDA) [2] is employed to model the topics within a document collection offline. Then, by describing the shape of the cumulative topic distribution generated by the top documents retrieved by an IR system for a given query, it can be easily determined if the behavior of the respective system resembles the behavior of a highly or rather poorly effective system. Therefore, the proposed approach is termed *DeShaTo*, which is short for *De*scribing the *Sha*pe of cumulative *To*pic distributions. Finally, the proposed approach computes a score based on a combination of two probability distribution shape descriptors, namely skewness and kurtosis [5], which are computed on the cumulative topic distribution of the retrieved documents.

A series of experiments are conducted to validate the underlying hypothesis of the DeShaTo approach. Relevant document sets are tested against non-relevant document sets for all the queries available in the TREC Robust Track, Web Track 2013 and Web Track 2014 collections. In almost 95% of the cases, the DeShaTo approach is able to identify which set of documents is relevant, proving that the underlying hypothesis holds in most cases. Next, the DeShaTo score is averaged on the queries of each data set to produce rankings of the retrieval systems submitted for the respective TREC tracks. The DeShaTo approach is compared with a state of the art approach for the task of ranking retrieval systems without relevance judgments, namely *nruns* [9], using the Kendall Tau correlation measure. The results presented in [9] indicate that nruns is more accurate compared to the previous works [1,12,13]. Therefore, DeShaTo is only compared with nruns in the experiments. The overall empirical results presented in this work indicate that the DeShaTo approach is able to obtain a higher correlation with the true Average Precision (AP) scores.

Unlike most approaches for ranking retrieval systems [10,13], including the state of the art approach [9], DeShaTo does not require information about other retrieval systems when dealing with one system. Indeed, nruns [9] is based on

---

[1] *Topic* represents here the theme of a text, as in topic modeling. In IR evaluation programs such as TREC, a *topic* refers to the information need. To avoid any confusion with the LDA topics, TREC topics are referred to as *queries* throughout this paper, therefore *query* can mean either the information need or to the text submitted to the search engine.

sharing information among systems to produce a set of pseudo-relevant documents for each query, while DeShaTo works independently for each system and thus, it can produce accurate results when there are less systems to be ranked. Some approaches, such as [8], require human assessments, while DeShaTo does not involve human effort. Another distinctive trait of DeShaTo is the employment of topic modeling for ranking retrieval systems. Moreover, DeShaTo is a general approach with high potential for other applications such as query difficulty prediction, selective query expansion and selective IR.

The rest of the paper is organized as follows. The DeShaTo approach is presented in Section 2. The validation and the experiments are described in Section 3. The final remarks are drawn in Section 4.

2   Describing the Shape of Cumulative Topic Distributions

The DeShaTo approach is based on the hypothesis that if there are more predominant topics that emerge in the set of documents retrieved for a given query, then the system effectiveness for the respective query is lower. On the other hand, if there are less predominant topics in the set of documents, then it means that the system is highly effective. Naturally, the topic distributions of the retrieved documents have to be computed in order to determine the effectiveness level of the IR system. In this work, Latent Dirichlet Allocation based on Gibbs sampling [2] is employed to compute the topic distributions of the documents, but other topic modeling approaches could possibly work equally well or even better [6]. Nevertheless, it is worth mentioning that LDA has successfully been used in different contexts in IR [7,14].

Although DeShaTo is a post-retrieval approach, the topic distribution can be computed offline, right after indexing the documents, in order to reduce the online processing time. If LDA is carried out offline, the topic distributions can be immediately retrieved along with the documents when necessary.

In order to obtain a unique representation from all the topic distributions associated to the top retrieved documents for a query, the distributions have to be somehow combined into a single distribution. Instead of choosing only one way of cumulating the topic distributions, three alternative ways are simultaneously employed, namely, the component-wise sum, the component-wise minimum, and the component-wise product, respectively. It is important to note that the three cumulative distributions have to be normalized, such that they all remain probability distributions (the sum of all the components is 1). The sum, the minimum and the product produce slightly different cumulative distributions and using them all together provides useful information for the next step. The relevant documents set contains only one or two predominant topics, while the non-relevant documents produce a mixture of predominant topics. What remains to be done from this point on is to find a way of comprising this difference in a measure or score. More formally, the next step is to find a robust approach to describe the shape of the cumulative topic distributions. The proposed approach uses two probability distribution shape descriptors, namely skewness and kurtosis [5]. More common statistics such as the mean or the standard deviation have also been tested out, but they have been found to be less informative. In probability theory and statistics, *skewness* is a measure of the asymmetry of the probability distribution of a real-valued random variable about its mean. The skewness can be computed as the third central moment of the input probability distribution, divided by the cube of its standard deviation. In a similar way to the skewness, *kurtosis* is a descriptor of the shape of a probability distribution. More precisely, the kurtosis quantifies the peakness (width of peak) of the probability distribution of a real-valued random variable. The kurtosis can be computed as the fourth central moment of the probability distribution, divided by fourth power of its standard deviation.

High values of skewness and kurtosis indicate that the topic distribution is characterized by a small number of predominant topics, while low values of these statistics indicate that there are more (or even no) predominant topics. Therefore, the two statistics reflect exactly what is required to determine if the system behavior is good or poor with respect to a given query. Finally, the skewness and the kurtosis are embedded in the DeShaTo score that can be computed using the following closed form equation:

$$score = k(S) + s(S) + k(M) + s(M) + k(P) + s(P), \qquad (1)$$

where k and s are two functions that return the kurtosis and the skewness of a probability distribution given as parameter, and S, M and P are the cumulative topic distributions obtained by computing the sum, the minimum and the product of the topic distributions corresponding to the retrieved documents. The probability distribution shape descriptors are combined in a very natural straightforward manner in Equation (1). By trying various combination schemes, a more efficient way of combining these descriptors can supposedly be found. For instance, a weighted sum could probably work better in practice, if the weights are learned on some training data. However, adding more parameters to DeShaTo is not necessarily desirable. Proposing alternative combination schemes will be properly addressed in future work.

3   Experiments

3.1   Data Sets Description

The TREC collections[2] that are being used in the experiments are presented next. They contain a set of information need statements, the document set and the relevance judgments for each query. The experiments are conducted on precisely three data sets, namely Robust, TREC Web Track 2013 and TREC Web Track 2014.

The results provided by participants are termed runs. TREC evaluates the runs using various effectiveness measures. The participant runs can thus be ranked according to one of these measures, such as the Mean Average Precision (MAP) over queries. In the experiments, all the queries along with all the

---

[2] http://mitpress.mit.edu/catalog/item/default.asp?ttype=2\&tid=10667

Table 1. A summary of the data sets used for the task of ranking retrieval systems without relevance judgments.

Data Set	Query IDs	# Queries	# Systems
Robust	$301 - 450, 601 - 700$	249	110
Web Track 2013	$201 - 250$	50	61
Web Track 2014	$251 - 300$	50	30

Table 2. Accuracy of the DeShaTo score for correctly identifying the set of relevant documents tested against a set of non-relevant documents. The number of LDA topics is 100.

Data Set	# Queries	# Correct Predictions	Accuracy
Robust	249	231	92.77%
Web Track 2013	50	50	100%
Web Track 2014	50	50	100%
Overall	349	331	94.84%

participant runs from Robust, Web Track 2013 and Web Track 2014 are used to rank retrieval systems without relevance judgments. A summary of the data used in the experiments is given in Table 1.

The DeShaTo approach is compared with a state of the art approach, namely nruns [9], using the Kendall Tau correlation measure, as in [9]. To reduce the offline processing time of DeShaTo, only the documents retrieved in the participant runs where included in the topic modeling process.

3.2   Empirical Validation of the Hypothesis

To validate the underlying hypothesis of DeShaTo, a simple procedure has been designed as described next. For each query in the three collections, a set of 30 relevant documents is produced by randomly sampling the documents. Likewise, another set of 30 non-relevant documents is produced for each query. For each query, 20 draws were made to randomly select the documents within the relevant and the non-relevant sets, in order to reduce the result of chance. Remarkably, the results of different trials are consistent with each other.

The cornerstone hypothesis of this work can be validated if it can identify, with a high degree of accuracy, which is the set of relevant documents only by using the DeShaTo score proposed in Equation (1). As such, the DeShaTo score was put to the test and the results are presented in Table 2. The DeShaTo score seems to be able to make good predictions in most of the cases. Indeed, there are only 18 queries from the Robust collection for which the score associated to the non-relevant document set is higher than the score associated to the relevant document set. The accuracy goes up to 100% for the Web Track 2013 and 2014 data sets. Overall, the accuracy of the DeShaTo approach is 94.84%. Although not perfect, this result offers some empirical proof that the underlying hypothesis of DeShaTo works well enough in practice.

Table 3. Kendall Tau correlation between the ground truth ranking according to the $MAP@30$ measure and the systems ranking determined by the DeShaTo score. The best correlation per data set is highlighted in bold.

Data Set	50 Topics	100 Topics	250 Topics
Robust	**0.4286**	0.3524	0.3143
Web Track 2014	0.1190	0.1905	**0.3048**

3.3 Parameter Tuning

The DeShaTo approach is based on the topics modeled by LDA, but the number of topics could influence the accuracy of the proposed approach. Wei and Croft also observed that the number of topics affects the retrieval performance [14]. Therefore, the number of topics is tuned on the Robust and the Web Track 2014 data sets through a validation procedure. Since the documents from the Web Track 2013 and 2014 data sets are from ClueWeb12, the number of topics validated on Web Track 2014 is also used on the Web Track 2013 collection. From each data set, 10% of the queries and 15 systems are chosen at regular interval. More precisely, one in every 10 queries are used for validation. One in every 7 systems are used for the Robust data set, while for the Web Track 2014, one in every two systems are used for validation. The amount of observations ($\#queries \times \#systems$) used for validation is deliberately chosen such that it is significantly smaller than the total amount of data, in order to prevent any kind of overfitting. Only 1.37% of the Robust data is used for validation. In a similar manner, 1.65% of the Web Track 2013 and 2014 data is used for tuning the number of topics.

The validation procedure aims to choose between using 50, 100 or 250 topics by evaluating the Kendall Tau correlation between the systems ranking determined by the DeShaTo score and the ground truth ranking determined by the Average Precision of the top 30 retrieved documents per run, namely $AP@30$. Actually, to produce the rankings, the score of each system has to be averaged over all the validation queries. Thus, the ground truth rankings are given by the Mean Average Precision of the top 30 documents, namely $MAP@30$. An interesting remark is that very similar results are obtained using the top 10 or the top 100 retrieved documents, but since nruns [9] was evaluated using the top 30 documents, the results presented in Table 3 and throughout this paper are also based on the top 30 documents per run. According to the best Kendall Tau correlations reported in Table 3, 50 topics will be used when LDA is carried out on the Robust documents. On the other hand, 250 topics will be used when LDA is carried out on the Web Track 2013 and 2014 documents. This difference can probably be explained by the type of documents that constitute the collections. The documents within the Robust collection are quite homogeneous since they are extracted from newspapers, while the documents within ClueWeb12 are web documents. In the latter collection, documents are much more heterogeneous and may contain topics that are not related to the document content such as links to the home page, menu buttons and so on.

Table 4. Kendall Tau correlation between the ground truth ranking according to the $MAP@30$ measure and the systems ranking determined by the DeShaTo score, on one hand, and the systems ranking produced by nruns, on the other. The best correlation per data set is highlighted in bold.

Data Set	nruns [9]	DeShaTo
Robust	**0.6195**	0.6112
Web Track 2013	0.1005	**0.2306**
Web Track 2014	0.4529	**0.4966**

3.4   Ranking Retrieval Systems Results

The DeShaTo score is compared with nruns [9] for the task of ranking retrieval systems without relevant judgements and the results are presented in Table 4. The values given in Table 4 represent the Kendall Tau correlations between the ground truth ranking given by the $MAP@30$ values and the systems ranking determined by the DeShaTo score, on one hand, and by nruns, on the other hand. It is important to mention that the correlation reported for nruns in this paper (0.6195) is lower than the correlation reported in [9] (0.640). The difference comes from the fact that here the correlation is based on 249 queries from 2004 and 2005, instead of only the 150 queries from 2004. Furthermore, the correlation is here computed with respect to the $MAP@30$ score instead of the MAP score as in [9], which is actually more fair, since the predictions are made on the top 30 documents per query. Compared to nruns, the DeShaTo score gives a higher correlation for the Web Track 2013 and 2014 participant runs, while it produces only a slightly lower correlation for the Robust runs. This could be explained by the fact that nruns becomes unreliable for a small set of runs, because it leverages the information from multiple systems to produce a good set of pseudo-relevant documents. Web Track 2013 and 2014 have considerably less participants than Robust, and the nruns approach is less accurate on the Web Track 2013 and 2014 data sets. Unlike nruns, the DeShaTo approach relies solely on the results of a system to compute its score, which seems to be an advantage for the newer TREC collections. The overall results seem to indicate that DeShaTo is better than nruns.

4   Conclusion

This paper presented an approach that is able to distinguish between a highly effective IR system and a less effective IR system for some queries. The proposed approach is based on *De*scribing the *Sha*pe of cumulative *To*pic distributions modeled by LDA, hence the name DeShaTo. A set of experiments have been conducted in order to validate the underlying hypothesis of DeShaTo in practice. Moreover, another set of experiments have been conducted to compare DeShaTo with nruns [9] for the task of ranking retrieval systems without relevance judgments. The results indicate that DeShaTo gives a higher correlation with the $MAP@30$ score in most cases, most likely because its accuracy does not depend on the number of systems used. The described approach does not take into account the query text itself, but this will be covered in future work by analyzing the topic distribution of the query in relation to the document distributions.

References

1. Aslam, J.A., Savell, R.: On the effectiveness of evaluating retrieval systems in the absence of relevance judgments. In: Proceedings of the 26th Annual International ACM SIGIR Conference on Research and Development in Informaion Retrieval, pp. 361–362 (2003)
2. Blei, D.M., Ng, A.Y., Jordan, M.I.: Latent Dirichlet Allocation. Journal of Machine Learning Research **3**, 993–1022 (2003)
3. Cronen-Townsend, S., Zhou, Y., Croft, W.B.: Predicting query performance. In: Proceedings of the 25th Annual International ACM SIGIR Conference on Research and Development in Information Retrieval, pp. 299–306 (2002)
4. Cronen-Townsend, S., Zhou, Y., Croft, W.B.: A framework for selective query expansion. In: Proceedings of the Thirteenth ACM International Conference on Information and Knowledge Management, pp. 236–237 (2004)
5. Groeneveld, R.A., Meeden, G.: Measuring Skewness and Kurtosis. Journal of the Royal Statistical Society. Series D (The Statistician) **33**(4), 391–399 (1984)
6. Mimno, D.M., McCallum, A.: Topic models conditioned on arbitrary features with Dirichlet-multinomial regression. In: Proceedings of the 24th Conference in Uncertainty in Artificial Intelligence, pp. 411–418 (2008)
7. Park, L.A.F., Ramamohanarao, K.: The sensitivity of latent Dirichlet allocation for information retrieval. In: Buntine, W., Grobelnik, M., Mladenić, D., Shawe-Taylor, J. (eds.) ECML PKDD 2009, Part II. LNCS, vol. 5782, pp. 176–188. Springer, Heidelberg (2009)
8. Pavlu, V., Rajput, S., Golbus, P.B., Aslam, J.A.: IR system evaluation using nugget-based test collections. In: Proceedings of the Fifth ACM International Conference on Web Search and Data Mining, pp. 393–402 (2012)
9. Sakai, T., Lin, C.Y.: Ranking retrieval systems without relevance assessments – revisited. In: The 3rd International Workshop on Evaluating Information Access (EVIA) - A Satellite Workshop of NTCIR-8, pp. 25–33 (2010)
10. Shi, Z., Wang, B., Li, P., Shi, Z.: Using global statistics to rank retrieval systems without relevance judgments. In: Shi, Z., Vadera, S., Aamodt, A., Leake, D. (eds.) IIP 2010. IFIP AICT, vol. 340, pp. 183–192. Springer, Heidelberg (2010)
11. Shtok, A., Kurland, O., Carmel, D., Raiber, F., Markovits, G.: Predicting Query Performance by Query-Drift Estimation. ACM Transactions on Information Systems **30**(2), 11:1–11:35 (2012)
12. Soboroff, I., Nicholas, C., Cahan, P.: Ranking retrieval systems without relevance judgments. In: Proceedings of the 24th Annual International ACM SIGIR Conference on Research and Development in Information Retrieval, pp. 66–73 (2001)
13. Spoerri, A.: Using the structure of overlap between search results to rank retrieval systems without relevance judgments. Information Processing & Management **43**(4), 1059–1070 (2007)
14. Wei, X., Croft, W.B.: LDA-based document models for Ad-hoc retrieval. In: Proceedings of the 29th Annual International ACM SIGIR Conference on Research and Development in Information Retrieval, pp. 178–185 (2006)
15. Winaver, M., Kurland, O., Domshlak, C.: Towards robust query expansion: model selection in the language modeling framework. In: Proceedings of the 30th Annual International ACM SIGIR Conference on Research and Development in Information Retrieval, pp. 729–730 (2007)

Induced Sorting Suffixes in External Memory with Better Design and Less Space

Wei Jun Liu[1,2], Ge Nong[1,3]($\boxtimes$), Wai Hong Chan[4]($\boxtimes$), and Yi Wu[1]

[1] Sun Yat-sen University, Guangzhou, China
issng@mail.sysu.edu.cn
[2] Gannan Normal University, Ganzhou, China
[3] SYSU-CMU Shunde International Joint Research Institute, Guangzhou, China
[4] The Hong Kong Institute of Education, Tai Po, Hong Kong
waihchan@ied.edu.hk

Abstract. Recently, several attempts have been made to extend the internal memory suffix array (SA) construction algorithm SA-IS to the external memory model, e.g., eSAIS, EM-SA-DS and DSA-IS. While the developed programs for these algorithms achieve remarkable performance in terms of I/O complexity and speed, their designs are quite complex and their disk requirements remain rather heavy. Currently, the core algorithmic part of each of these programs consists of thousands of lines in C++, and the average peak disk requirement is over $20n$ bytes for an input string of size $n < 2^{40}$. We re-investigate the problem of induced sorting suffixes in external memory and propose a new algorithm SAIS-PQ (SAIS with Priority Queue) and its enhanced alternative SAIS-PQ+. Using the library STXXL, the core algorithmic parts of SAIS-PQ and SAIS-PQ+ are coded in around 800 and 1600 lines in C++, respectively. The time and space performance of these two programs are evaluated in comparison with eSAIS that is also implemented using STXXL. In our experiment, eSAIS runs the fastest for the input strings not larger than 16 GiB, but it is slower than SAIS-PQ+ for the only two input strings of 32 and 48.44 GiB. For the average peak disk requirements, eSAIS and SAIS-PQ+ are around $23n$ and $15n$ bytes, respectively.

Keywords: Suffix array · Sorting algorithm · External memory

1 Introduction

During the past two decades, a plethora of suffix array construction algorithms (SACAs) have been proposed [11]. Most of SACAs assume that the input string T is completely stored in the internal memory, so that the characters of T can be

Corresponding authors: Ge Nong (issng@mail.sysu.edu.cn) and Wai Hong Chan (waihchan@ied.edu.hk). Nong is supported by the Project of DEGP (2012KJCX0001). Chan is partially supported by the General Research Fund (810012), The Research Grant Council, Hong Kong SAR.

C. Iliopoulos et al. (Eds.): SPIRE 2015, LNCS 9309, pp. 83–94, 2015.
DOI: 10.1007/978-3-319-23826-5_9

randomly accessed for constructing the suffix array (SA). To meet the demand for building a big SA, a number of external memory SACAs have been proposed, such as DC3 [2], bwt-disk [5], SAscan [6], eSAIS [1], EM-SA-DS [9] and DSA-IS [8]. Among them, the last 3 algorithms commonly adopt the induced sorting principle of the internal memory algorithm SA-IS [10], and achieve better time and space results than the others in the conducted experimental studies.

A key operation in the process of induced sorting suffixes is to access the preceding character of a sorted suffix. This can be easily done if T is completely stored in the internal memory, but becomes difficult when T is stored in the external memory. In the latter case, random accesses to T must be avoided as most as possible for high I/O efficiency, which recently has been attempted by different methods in eSAIS [1], EM-SA-DS [9] and DSA-IS [8]. Specifically, the former two split variable-size LMS-substrings (defined in Section 2) into fixed size tuples, and restore the characters of an LMS-substring on the fly, by retrieving its tuples one by one that is guaranteed to be available in the internal memory when it is needed. However, the latter divides T into blocks, compute the SA of each block in the internal memory one by one, then merge all of them in the external memory to produce the final SA.

While eSAIS, EM-SA-DS and DSA-IS have already achieved remarkable performance in terms of speed and I/O complexity, they are still facing of at least two drawbacks: (1) In their current implementations in C++, the core algorithmic part of each is coded in thousands of lines, this makes it challenging to be revised for a specific application. (2) The space requirement remains rather heavy, e.g., the average peak disk requirement is over $20n$ bytes for $n < 2^{40}$. To overcome these drawbacks, we re-investigate the problem of induced sorting suffixes in external memory and propose a new algorithm SAIS-PQ (SAIS with Priority Queue) and its enhanced alternative SAIS-PQ+. Using the library STXXL [3,4], the core algorithmic parts of SAIS-PQ and SAIS-PQ+ are coded in around 800 and 1600 lines in C++, respectively. The time and space performance of these two programs are evaluated in comparison with eSAIS that is also implemented using STXXL. In our experiment, eSAIS runs the fastest for the input strings not larger than 16 GiB, but it is slower than SAIS-PQ+ for the only two input strings of 32 and 48.44 GiB. For the average peak disk requirements, eSAIS and SAIS-PQ+ are around $23n$ and $15n$ bytes, respectively.

The other sections are organized as follows. Section 2 gives the preliminaries, Section 3 the algorithms, Section 4 the performance evaluation experiments and Section 5 the conclusions.

2   Preliminaries

Consider an input string $T = T[0]T[1]\ldots T[n-1]$ of n symbols over an ordered alphabet $\Sigma \bigcup \{\$\}$. The first $n-1$ symbols are from Σ and $T[n-1] = \$$ is the sentinel which is lexicographically smaller than any character in Σ. Let $T[i,j]$ denote the substring of T running from $T[i]$ to $T[j]$, a suffix $\mathsf{suf}(T,i) = T[i,n-1]$ is a substring that starts at position i and ends at position $n-1$. The following definitions and notations are used for presentation convenience.

- Suffix array (SA) and inverse suffix array (ISA). The suffix array $SA[0, n-1]$ of T is a permutation of integers in $[0, n)$ such that $\mathsf{suf}(T, SA[0]) < \mathsf{suf}(T, SA[1]) \ldots < \mathsf{suf}(T, SA[n-1])$ are in increasing lexicographical order. The inverse suffix array of T is ISA, where $ISA[SA[i]] = i$ for all $i \in [0, n)$.
- S-type, L-type, LMS character. For $i \in [0, n)$, $T[i]$ is S-type, if (1) $i = n-1$, or (2) $T[i] < T[i+1]$, or (3) $T[i] = T[i+1]$ and $T[i+1]$ is S-type; otherwise, $T[i]$ is L-type. Moreover, $T[i]$ is an LMS (left most S-type) character, if $T[i]$ is S-type and $T[i-1]$ is L-type. The first character is never an LMS character.
- S-type, L-type, LMS suffix/substring. For $i \in [0, n)$, $\mathsf{suf}(T, i)$ is an S-type, L-type or LMS suffix, if its heading character $T[i]$ is S-type, L-type or LMS, respectively. For $0 \le i \le j < n$, $T[i, j]$ is an LMS substring, if $T[i]$ and $T[j]$ are LMS and there is no other LMS character in $T[i, j]$. Given $T[i, j]$ is an LMS substring and $k \in [i, j)$, $T[k, j]$ is an S- or L-type substring if $T[k]$ is S- or L-type, respectively.
- Bucket in SA. The suffixes of an identical heading character occupy a contiguous interval in SA, where L- and S-type suffixes are gathered to the left and right sides of the interval, respectively. We split SA into one or multiple buckets. For example, for one character $c \in \Sigma \bigcup \{\$\}$, $\mathsf{bucket}(c)$ contains the suffixes starting with character c.
- Preceding character. For $i \in [1, n-1]$, the preceding character $T[i-1]$ of $T[i]$ or $\mathsf{suf}(T, i)$ is denoted by $\mathsf{prec}(T, i)$.

2.1 Preceding Cache Item and Preceding Cache Array

The difficulty for induced sorting suffixes in external memory is to obtain the preceding character of a sorted suffix without a random access to T in the external memory. To overcome this difficulty, we introduce to the design of our algorithm the following two data structures:

- Preceding cache item (PCI). For $p \in [1, n-1]$, if $p = n-1$ or $T[p] \neq T[p+1]$, then the PCI of $T[p]$ is a tuple $\langle c_0, p, t, c_1, h \rangle$, where $c_0 = T[p]$, $c_1 = T[p-h-1]$, t is the L/S-type of $T[p]$ and $h \ge 0$ is the maximum integer such that the character(s) in $T[p-h, p]$ are identical and different from $T[p-h-1]$. The PCI of $T[p]$ is also referred as the PCI of $\mathsf{suf}(T, p)$.
- Preceding cache array (PCA). Stably sort all PCI $\langle c_0, p, t, c_1, h \rangle$ by $\langle c_0, p \rangle$, the sorting result is an array called the preceding cache array (PCA) of T, denoted by $\mathsf{PCA}(T)$.

It should be mentioned that, the PCI tuple is different from the tuples defined in eSAIS. In particular, h is essentially different from the "repetition count" in eSAIS. Using PCA, the method for retrieving the preceding character of a sorted suffix is straightforward. In principle, for each sorted suffix $\mathsf{suf}(T, p)$, we know its $\langle c_0, p \rangle$ and use it as the sorting key to find in $\mathsf{PCA}(T)$ the PCI of the same key for sorting the preceding suffix $\mathsf{suf}(T, p-1)$. There is no need to divide an LMS-substring into a number of fixed-size tuples as required for eSAIS, yielding the simple design of our algorithm.

2.2   Priority Queue in STXXL

STXXL is a C++ template library that implements a set of containers and algorithms capable of processing massive data in external memory. The priority queue (PQ) in STXXL has been successfully applied to implement the algorithm eSAIS for inducing the suffix and LCP (longest common prefix) arrays in external memory. The PQ can function as a minimum or maximum heap, where the top element has the minimum or maximum priority, respectively. Similar to eSAIS, our algorithm also uses a number of PQs as the underlying data structures for induced sorting both suffixes and LMS-substrings. For this sake, we call it SAIS-PQ (SAIS with PQ). In SAIS-PQ, by using PCA and PQ, the same induced sorting method is applied to sort both substrings and suffixes for reducing problem and inducing solution. This is similar to the internal memory algorithm SA-IS. As a result, the design of SAIS-PQ is simple and natural to be implemented. Moreover, the I/O, time and space complexities of SAIS-PQ are amortized by using a PQ to sort $O(n)$ integers.

3   The Algorithms

3.1   Algorithmic Framework

The algorithmic framework of SAIS-PQ is shown in Algorithm 1, and each step is explained below.

PQ_L or PQ_S are two PQs with each element being a tuple $\langle c_0, t, r, p \rangle$, where c_0 is the head character of $\mathsf{suf}(T, p)$, t the type of $\mathsf{suf}(T, p)$: 0 for L-type and 1 for S-type, r the rank of $\mathsf{suf}(T, p)$.

The reduction phase of lines 4-7 computes the reduced string T_1 from T. First, it puts all the LMS suffixes into PQ_S with the sorting key as $\langle c_0, 1, 0, p \rangle$. Then, it scans the elements of PQ_S in their ascending order to sort all the L-substrings into PQ_L, and scans PQ_L to sort all the LMS-substrings into PQ_S. In this step, when an LMS-substring is sorted, its name is also computed by our new naming algorithm given in Section 3.3. The names of all the sorted LMS-substrings are then sorted back to their positional order in T to produce the reduced string T_1.

The recursion phase of lines 10-14 calls recursion if each character in T_1 is not yet unique, or else directly computes SA_1 by sorting each suffix in T_1 by its head character. Given SA_1, line 15 computes ISA_1 by sorting each i by $SA_1[i]$, this can be easily done by an integer sorter in STXXL.

Given ISA_1, the induction phase of lines 18-20 computes SA from SA_1. First, each LMS-suffix of T is put into PQ_S with r in $\langle c_0, t, r, p \rangle$ being its rank given by ISA_1. Then, the elements of PQ_S are scanned in their ascending order to sort all the L-suffixes into PQ_L, and scans PQ_L to sort all the S-suffixes into PQ_S. Finally, SA is generated by scanning PQ_L and PQ_S once.

3.2 Retrieving a Preceding Character

Given $PCA(T)$, it is simply an integer sorting task for retrieving the preceding character of a sorted suffix. Due to the similarity between the processes of sorting L- and S-suffixes, we explain the method for sorting L-suffixes only. When scanning PQ_S to sort all the L-suffixes, we scan the suffixes in PQ_S bucket by bucket, i.e. each L-suffix in a bucket is popped up and pushed into a temporary PQ tmp with the sorting key as the position index of the suffix in T. As a result, the suffixes in tmp are sorted in the same order as their PCIs in $PCA(T)$, and we can sequentially retrieve from $PCA(T)$ the PCI for each suffix in tmp.

Algorithm 1. The algorithmic framework of SAIS-PQ

---

1 SAIS-PQ(T)
2 **begin**
3   /* Reduction */
4   create tuple $\langle c_0, 1, 0, p \rangle$ for each LMS-suffix and put it into PQ_S;
5   scan PQ_S to sort and name each L-substring into PQ_L;
6   scan PQ_L to sort and name each S-substring;
7   sort all the LMS-substring's names by the substring positions in T to get a reduced string T_1;
8
9   /* Recursion */
10   **if** *each character in T_1 is unique* **then**
11 | directly compute SA_1 from T_1;
12   **else**
13 | compute SA_1 by SAIS-PQ(T_1);
14   **end**
15   compute ISA_1 from SA_1;
16
17   /* Induction */
18   create tuple $\langle c_0, 1, r, p \rangle$ for each LMS-suffix and put it into PQ_S;
19   scan PQ_S to sort each L-suffix into PQ_L;
20   scan PQ_L to get SA;
21   return SA;
22 **end**

3.3 Naming Substrings

In order to reduce T to T_1, naming all the sorted LMS-substrings is required. In the existing algorithms such as eSAIS, EM-SA-DS and DSA-IS, this job is done after all the LMS-substrings have been sorted, by retrieving and comparing the substrings in their lexicographical order in different ways. However, SAIS-PQ integrates this job into the process of induced sorting substrings, i.e. sorting and naming of a substring are done simultaneously. This will not only improve the algorithm's performance, but also simplify the algorithm's design.

An LMS prefix [10], denoted by $\mathsf{pref}(T,i)$, is defined as $T[i,j]$, $i < j$, satisfying that $T[j]$ is LMS and none in $T[i+1,j-1]$ are. Furthermore, the LMS prefix is called L- or S-prefix if its first character is L- or S-type, respectively.

As what is done in the internal memory algorithm SA-IS, we sort all the LMS-substrings of T by reusing the space of SA to sort all the LMS prefixes. The computation of the name for each LMS-substring is coupled into the process for induced sorting all the substrings. For this purpose, we introduce a naming tuple $\langle c_0, r_0, r_1 \rangle$ to each LMS prefix in $SA[i]$, where $c_0 = T[SA[i]]$, and r_0 and r_1 are used to record the names of the prefix $\mathsf{pref}(T, SA[i])$ and its succeeding prefix $\mathsf{pref}(T, SA[i]+1)$, respectively.

Our naming algorithm given in Algorithm 2 is designed from the following observation:

Observation 1. *For $i \neq j$, $\mathsf{pref}(T,i) = \mathsf{pref}(T,j)$ if and only if $T[i] = T[j]$ and $\mathsf{pref}(T,i+1) = \mathsf{pref}(T,j+1)$.*

Algorithm 2. Algorithm for naming the LMS prefixes

---

(1) Set $r = 0$. Increase i from 0 to $n-1$ to scan SA from left to right to sort all the L-prefixes, and do the following for each scanned prefix in $SA[i]$:

   (a) If $\mathsf{pref}(T, SA[i])$ is L- or LMS-type, then: (1) set $r = i$ if the tuple $\langle c_0, r_1 \rangle$ for the prefix in $SA[i]$ is not equal to that in $SA[i-1]$; (2) set r_0 for the prefix in $SA[i]$ as r.

   (b) If $\mathsf{pref}(T, SA[i]) - 1$ is L-type and sorted into $SA[i_1]$, $i_1 > i$, then set r_1 for the prefix in $SA[i_1]$ as r_0 for the prefix in $SA[i]$.

(2) Set $r = n - 1$. Decrease i from $n-1$ to 0 to scan SA from right to left to sort all the S-prefixes, and do the following for each scanned prefix in $SA[i]$:

   (a) If $\mathsf{pref}(T, SA[i])$ is S-type, then: (1) set $r = i$ if the tuple $\langle c_0, r_1 \rangle$ for the prefix in $SA[i]$ is not equal to that in $SA[i+1]$; (2) set r_0 for the prefix in $SA[i]$ as r.

   (b) If $\mathsf{pref}(T, SA[i]) - 1$ is S-type and sorted into $SA[i_1]$, $i_1 < i$, then set r_1 for the prefix in $SA[i_1]$ as r_0 for the prefix in $SA[i]$.

The overall process of induced sorting LMS-substrings consists of two processes for sorting L- or S-prefixes, respectively. The naming methods used in these processes are given by Steps (1) and (2), respectively. In the processes for induced sorting L- or S-prefixes, when we are scanning a sorted prefix in $SA[i]$, the name of this prefix is computed by comparing $\langle c_0, r_1 \rangle$ in its naming tuple $\langle c_0, r_0, r_1 \rangle$ with that of the prefixes in $SA[i-1]$ or $SA[i+1]$ for induced sorting L- or S-prefixes, respectively; then the name is stored in r_0 of the naming tuple of this prefix, and further recorded in r_1 of the naming tuple of the preceding prefix $\mathsf{pref}(T, SA[i] - 1)$ for later use.

The correctness of this naming algorithm is established on Observation 1. In both steps, r_1 in the naming tuple of $\mathsf{pref}(T, SA[i] - 1)$ is used to record r_0 in the naming tuple of $\mathsf{pref}(T, SA[i])$, i.e. the name of $\mathsf{pref}(T, SA[i])$. In step (1),

r is used to record the name of the currently largest scanned L-prefix, which is initialized as 0 and increasingly set as i when $SA[i]$ stores a larger L-prefix than all the scanned L-prefixes by so far. Analogously, in step (2), r is used to record the name of the currently smallest scanned S-prefix, which is initialized as $n-1$ and decreasingly set as i when $SA[i]$ stores a smaller S-prefix than all the scanned S-prefixes by so far.

3.4 Further Improvements

The implementation of SAIS-PQ using STXXL is natural and simple, but its time and space performance needs to be further enhanced. We improve SAIS-PQ to be SAIS-PQ+ mainly by the following trick. In a PCI for $\mathsf{suf}(T,p)$, if $T[p] \neq T[p-1]$ and $T[p-1] \neq T[p-2]$, then h is reused to store $T[p-2]$. Hence, the PCI for $\mathsf{suf}(T,p-1)$ is not needed to be included in $\mathsf{PCA}(T)$ and its space is saved.

Table 1. Input files, n in Gi, 1 byte per character.

Name	n	$\|\|\Sigma\|\|$	Description
proteins	1.10	27	Swissprot database, at http://pizzachili.dcc.uchile.cl/texts.html
uniprot	2.42	96	UniProt Knowledgebase release 4.0, at http://www.uniprot.org/news/2005/02/01/release.
genome	2.86	6	Human genome data used in [2], at http://algo2.iti.kit.edu/dementiev/esuffix/instances.shtml.
guten	3.05	256	Gutenberg collection used in [2], at http://algo2.iti.kit.edu/dementiev/esuffix/instances.shtml.
random2	4.00	256	A concatenation of two copies of a string with characters randomly selected from [0, 255], with a maximum LCP of 2.0 GiB. The exact size of this file is $2^{32} - 2$ bytes.
genome2	5.72	6	A concatenation of two copies of the corpus "genome," with a maximum LCP of 2.86 GiB.
enwiki8g	8	256	The 8GiB prefix of enwiki1503.
guten1209	22.44	256	Gutenberg collection used in [1], at http://algo2.iti.kit.edu/bingmann/esais-corpus/gutenberg-201209.24090588160.xz.
enwiki1503	48.44	256	A dump of "enwiki-20150304-pages-articles-xml.bz2" for the English Wikipedia, at http://meta.wikimedia.org/wiki/Data_dump_torrents#enwiki.

4 Performance Evaluation

An experimental study has been conducted to evaluate the time and space performance of SAIS-PQ, its enhanced alternative SAIS-PQ+ and eSAIS [1], where

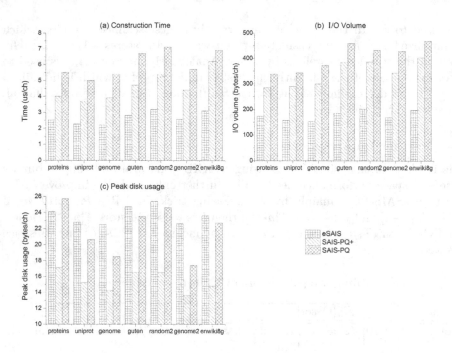

Fig. 1. Construction times, I/O volumes and peak disk usages for the first 7 input data in Table 1.

the files listed in Table 1 of different sizes and alphabets are used as the input strings. The program for eSAIS is downloaded from its project website[1], and that for SAIS-PQ and SAIS-PQ+ were codes by ourselves. The machine in use has the configuration as: 1 CPU (Intel(R) Core(TM) i3 3.07 GHz), 4 GiB RAM (1333 MHz DDR3), 1 Disk (1.8 TiB, 7200 rpm, SATA2) and Linux (Ubuntu 11.04). In this disk, around 170 GiB is already occupied, i.e. the free space is around 1.6 TiB. Each program is set to use at most 3 GiB RAM and 40-bit integers. The performance metrics are the mean running time in microseconds per character (μs/ch), mean I/O volume and peak disk usage in bytes per character (bytes/ch). The time of each program on an input file is a mean of 2 runs to absorb fluctuations. All these 3 performance metrics were collected using the statistics functions provided by STXXL, in a way similar to that in the program for eSAIS. To be consistent with the results reported by the program of eSAIS, the peak disk usage does not include the input string. To take into account the input string, n bytes should be added to the maximum disk allocation recorded by STXXL, this translates to adding 1 byte to the peak disk usage results reported here.

To see the effects of alphabet and LCP, Fig. 1 shows the experiment results for the first 7 corpora. For each corpus, both the construction times and the I/O

---

[1] http://panthema.net/2012/1119-eSAIS-Inducing-Suffix-and-LCP-Arrays-in-External-Memory/.

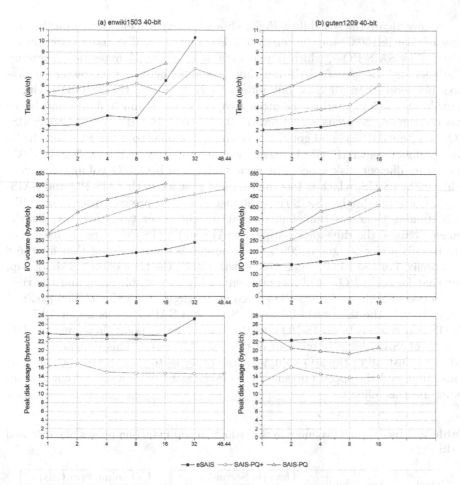

Fig. 2. Construction times, I/O volumes and peak disk usages for the prefixes in GiB of "enwiki1503" and "guten1209".

volumes of SAIS-PQ and SAIS-PQ+ is about twice as that for eSAIS. However, SAIS-PQ+ is more space-efficient. The peak disk usage of SAIS-PQ+ is only about 15 bytes/ch, which is around 2/3 (i.e. 15/23=0.65) of that for eSAIS. For each performance metrics, all the 3 programs vary in a similar trend. The alphabet size has positive influences on the time and I/O volume results, e.g., the construction times for "guten", "random2" and "enwiki8g" are larger than that for the bioinformatics datasets of smaller alphabets. The LCP causes no significant variations on the times and I/O volumes, as observed from the results on "genome" and "genome2". The peak disk usages randomly vary for different datasets without a predictable trend, indicating that it is independent of alphabet and LCP.

To investigate the scalability, Fig. 2 shows the experiment results on the prefixes of "enwiki1503" and "guten1209". The best peak disk usages are always achieved by SAIS-PQ+, which remains stable as about 15 bytes per character. The peak disk usages of eSAIS remains around 23 bytes per character for prefixes not more than 16 GiB, but jumps to 27 bytes for the 32 GiB prefix of "enwiki1503". Compared to the other two, SAIS-PQ is much slower and hence was not run on the prefixes of "enwiki1503" beyond 16 GiB. We also ran SAIS-PQ+ on "enwiki1503" and got the performance similar to SAIS-PQ+ on the 32 GiB prefix of "enwiki1503". However, we failed to run eSAIS on "enwiki1503" due to insufficient disk space. For each prefix, the best I/O volume is always achieved by eSAIS, which is less than half of that for both SAIS-PQ and SAIS-PQ+. Correspondingly, eSAIS runs the fastest for the prefixes from 1 to 8 GiB, in a running time less than half of that for the other two. However, for the longer prefixes, the time gaps between eSAIS and SAIS-PQ+ are shrinked. For the 16 GiB prefix of "guten1209", the time gap is around 2/3 of that for the 8 GiB prefix. For the 16 GiB prefix of "enwiki1503", the time of eSAIS shoots up, but that for SAIS-PQ+ still increases smoothly at a stable rate, and the time gap between eSAIS and SAIS-PQ+ becomes negligible. For the 32 GiB prefix of "enwiki1503", the time ratio between eSAIS and SAIS-PQ+ is about 10:7, i.e. eSAIS is overtaken by SAIS-PQ+ in this case. At this point, however, the I/O volume of eSAIS keeps about half of that of SAIS-PQ+. This indicates that in this case, instead of the I/O volume, some other factors in eSAIS become time consuming and the speed bottleneck. To investigate the reason, we conduct an experiment as follows.

Table 2. Times and I/O volumes in the reduction and induction phases of eSAIS and SAIS-PQ+.

Corpus	Algorithm	Time (in Seconds)			I/O volume (in GiB)		
		Reduction	Induction	Total	Reduction	Induction	Total
enwiki4g	eSAIS	3458	9168	12626	139	588	727
	SAIS-PQ+	12086	15605	27691	753	880	1633
enwiki32g	eSAIS	29038	325418	354456	1081	6651	7732
	SAIS-PQ+	127713	113870	241583	7538	7115	14653

Table 2 shows the times and I/O volumes for the reduction and induction phases of eSAIS and SAIS-PQ+ on the 4 and 32 GiB prefixes of "enwiki1503", i.e. "enwiki4g" and "enwiki32g". In the reduction phase, for both input files, the times and I/O volumes of eSAIS are far better than that of SAIS-PQ+, i.e., the time and I/O volume ratios between two algorithms are about 1:4. This is due to the different approaches for sorting LMS-substrings in the reduction phase. Specifically, eSAIS calls the in-place radix sort [7] for ASCII strings and gcc-4.4 STL's version of introsort for larger data types (i.e. 40-bit integers), while SAIS-PQ+ employs the induced sorting method. In the induction phase, for both input files, the I/O volumes for both algorithms are quite close while eSAIS is better, because both algorithms sort the suffixes using the induced sorting principle in two different ways. However, the better I/O volumes of eSAIS do not guarantee faster speeds. For example, eSAIS runs faster than SAIS-PQ+ for "enwiki4g", but slower than SAIS-PQ+ for "enwiki32g". This indicates that the time performance of the induced sorting method used by SAIS-PQ+ might be more stable for our experimental platform.

5 Conclusions

We present in this paper an algorithm SAIS-PQ and its enhanced alternative SAIS-PQ+ to extend the internal memory algorithm SA-IS to the external memory model. This algorithm is natural and simple to be implemented using the STXXL library, and its I/O, time and space complexities are amortized by using a PQ to sort $O(n)$ integers. In our programs for SAIS-PQ and SAIS-PQ+ used in the experiments of this paper, the core algorithmic parts are composed of around 800 and 1600 lines in C++, respectively. Such programs are feasible and flexible to be revised for a specific application in practice. Another distinct advantage of the proposed algorithm is that it is more space efficient compared to the existing external algorithms using the induced sorting principle, i.e. the peak disk usage for SAIS-PQ+ is only about 15 bytes per character for $n < 2^{40}$. This makes it a competitive candidate for applications where the disk space is a main concern. The I/O volumes of both SAIS-PQ and SAIS-PQ+ are quite large and constitute the speed bottlenecks. We are currently seeking ways to overcome this problem. Meanwhile, we are also employing the techniques developed in this work to optimize the design and implementation of the DSA-IS algorithm [8] for more promising time and space performance.

References

1. Bingmann, T., Fischer, J., Osipov, V.: Inducing suffix and LCP arrays in external memory. In: Proceedings of ALENEX, pp. 88–102 (2013)
2. Dementiev, R., Kärkkäinen, J., Mehnert, J., Sanders, P.: Better external memory suffix array construction. ACM Journal of Experimental Algorithmics **12**, 3.4:1–3.4:24 (2008). http://dx.doi.org/10.1145/1227161.1402296
3. Dementiev, R., Kettner, L., Sanders, P.: Stxxl: standard template library for xxl data sets. Software: Practice and Experience **38**(6), 589–637 (2008). http://dx.doi.org/10.1002/spe.844
4. Dementiev, R., Kettner, L., Sanders, P.: Stxxl: standard template library for XXL data sets. In: Brodal, G.S., Leonardi, S. (eds.) ESA 2005. LNCS, vol. 3669, pp. 640–651. Springer, Heidelberg (2005)
5. Ferragina, P., Gagie, T., Manzini, G.: Lightweight data indexing and compression in external memory. Algorithmica **63**, 707–730 (2012). http://dx.doi.org/10.1007/s00453-011-9535-0
6. Kärkkäinen, J., Kempa, D.: Engineering a lightweight external memory suffix array construction algorithm. In: Proceedings of the 2nd International Conference on Algorithms for Big Data, pp. 53–60 (2014)

7. Kärkkäinen, J., Rantala, T.: Engineering radix sort for strings. In: Amir, A., Turpin, A., Moffat, A. (eds.) SPIRE 2008. LNCS, vol. 5280, pp. 3–14. Springer, Heidelberg (2008). http://dx.doi.org/10.1007/978-3-540-89097-3_3

8. Nong, G., Chan, W.H., Hu, S.Q., Wu, Y.: Induced sorting suffixes in external memory. ACM Transactions on Information Systems **33**(3), 12:1–12:15 (2015). http://dx.doi.org/10.1145/2699665

9. Nong, G., Chan, W.H., Zhang, S., Guan, X.F.: Suffix array construction in external memory using d-critical substrings. ACM Transactions on Information Systems **32**(1), 1:1–1:15 (2014). http://doi.acm.org/10.1145/2518175

10. Nong, G., Zhang, S., Chan, W.H.: Two efficient algorithms for linear time suffix array construction. IEEE Transactions on Computers **60**(10), 1471–1484 (2011). http://dx.doi.org/10.1109/TC.2010.188

11. Puglisi, S.J., Smyth, W.F., Turpin, A.H.: A taxonomy of suffix array construction algorithms. ACM Comput. Surv. **39**(2), 1–31 (2007). http://doi.acm.org/10.1145/1242471.1242472

Efficient Algorithms for Longest Closed Factor Array

Hideo Bannai[1], Shunsuke Inenaga[1], Tomasz Kociumaka[2],
Arnaud Lefebvre[3], Jakub Radoszewski[2(✉)], Wojciech Rytter[2],
Shiho Sugimoto[1], and Tomasz Waleń[2]

[1] Department of Informatics, Graduate School of Information Science
and Electrical Engineering, Kyushu University, Fukuoka, Japan
{bannai,inenaga,shiho.sugimoto}@inf.kyushu-u.ac.jp
[2] Faculty of Mathematics, Informatics and Mechanics,
University of Warsaw, Warsaw, Poland
{kociumaka,jrad,rytter,walen}@mimuw.edu.pl
[3] Normandie Université, LITIS EA4108, NormaStic CNRS FR 3638,
IRIB, Université de Rouen, 76821 Mont-saint-aignan Cedex, France
arnaud.lefebvre@univ-rouen.fr

Abstract. We consider a family of strings called closed strings and a
related array of Longest Closed Factors (LCF). We show that the reconstruction of a string from its LCF array is easier than the construction and verification of this array. Moreover, the reconstructed string is
unique. We improve also the time of construction/verification, reducing it
from $\mathcal{O}(n \log n / \log \log n)$ (the best previously known) to $\mathcal{O}(n \sqrt{\log n})$. We
use connections between the LCF array and the longest previous/next
factor arrays.

1 Introduction

A *closed string* is a string with a proper (possibly empty) factor that occurs
in the string as a prefix and as a suffix, but not elsewhere. For example, a,
abaab, and ababababa are closed strings (with the corresponding factors ε, ab,
and abababa), whereas abaca and abc are not. Closed strings were first defined
by Fici in [8] and since then have found applications, mostly in the field of
combinatorics on words. Closed prefixes of Sturmian words were studied in [10].
A relation between closed factors and palindromic factors of a string was studied

T. Kociumaka—Supported by Polish budget funds for science in 2013-2017 as a
research project under the 'Diamond Grant' program.
J. Radoszewski and T. Waleń—Supported by the Polish Ministry of Science
and Higher Education under the 'Iuventus Plus' program in 2015-2016 grant no
0392/IP3/2015/73.
J. Radoszewski—Receives financial support of Foundation for Polish Science.
W. Rytter—Supported by the Polish National Science Center, grant no 2014/13/
B/ST6/00770.

C. Iliopoulos et al. (Eds.): SPIRE 2015, LNCS 9309, pp. 95–102, 2015.
DOI: 10.1007/978-3-319-23826-5_10

in [3]. The first algorithmic study of closed factors and closed factorizations was presented in [2].

The *longest closed factor array* (LCF array) of a string X stores for every suffix of X the length of its longest closed prefix. It was introduced in [2] in connection with closed factorizations of a string. In [2] an $\mathcal{O}(n \log n / \log \log n)$-time algorithm for computing this array for a string of length n was presented. Here we consider the problem of reconstructing a string from its LCF array.

We show that a (correct) LCF array corresponds to exactly one string, up to a permutation of the alphabet. We present an $\mathcal{O}(n)$-time randomized and an $\mathcal{O}(n \min(\log |\Sigma|, \log \log n + \frac{\log^2 \log |\Sigma|}{\log \log \log |\Sigma|}))$-time deterministic algorithm for reconstructing such a string if it exists. Here Σ is the alphabet of the string. Finally we present an $\mathcal{O}(n\sqrt{\log n})$-time construction algorithm for LCF array which improves the algorithm of [2]. We use it for verification of the LCF array, that is, for checking if it corresponds to any string. As a by-product we obtain $\mathcal{O}(n\sqrt{\log n})$-time computation of the so-called closest (rightmost) Longest Previous Factor array. The complexity of the LCF construction algorithm depends on the assumption of a linearly sortable alphabet of the input string.

2 Preliminaries

Let X be a string of length n composed of characters $X[1], \ldots, X[n]$. We denote $|X| = n$. By $X[i..j]$ we denote a factor of X consisting of the letters $X[i], \ldots, X[j]$. A factor is called a prefix (suffix) of X if $i = 1$ ($j = n$ respectively). A factor is called proper if $i > 1$ or $j < n$. If $i > j$ then the factor is assumed to be the empty string ε. A border of X is a factor of X that occurs as a prefix and as a suffix of X. By $border(X)$ we denote the length of the longest proper border of X.

The string X is called *closed* if it has a proper border that does not occur elsewhere in X. In particular, every single-letter string is closed. It is easy to see that a string is closed if and only if its longest proper border does not occur elsewhere in X. The *longest closed factor array* of X is an array $LCF[1..n]$ such that $LCF[i]$ is the length of the longest prefix of $X[i..n]$ that is closed. We denote by $lcf[i]$ the factor $X[i..i + LCF[i] - 1]$.

The *longest next factor array* of X is an array $LNF[1..n]$ such that $LNF[i]$ is the length of the longest prefix of $X[i..n]$ that is a factor of $X[i + 1..n]$. We denote by $lnf[i]$ the factor $X[i..i + LNF[i] - 1]$.

Example 1. $X = \texttt{abaabababbabbb}$ has the following LCF and LNF arrays:

position i	1	2	3	4	5	6	7	8	9	10	11	12	13	14
$X[i]$	a	b	a	a	b	a	b	a	b	b	a	b	b	b
$LCF[i]$	6	5	2	6	5	4	7	6	5	3	1	3	2	1
$LNF[i]$	3	2	1	4	3	2	4	3	2	1	0	2	1	0

Here the $lcf[]$ array is as follows: [abaaba, baaba, aa, ababab, babab, abab, babbabb, abbabb, bbabb, bab, a, bbb, bb, b].

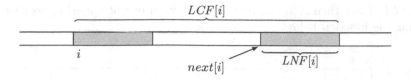

Fig. 1. Illustration of LCF, LNF and $next$ arrays. Here $LCF[i] = |lcf[i]|$ is the length of the longest closed factor starting at position i and $LNF[i] = |lnf[i]|$ is the length of the longest factor starting at position i that occurs to the right of position i.

3 Reconstruction Algorithm

In this section we show efficient reconstruction of a string from its LCF array. It is based on uniqueness of the output.

3.1 Uniqueness of Reconstruction

The following fact shows a correspondence between longest closed factors and longest next factors. We denote $next[i] = i + LCF[i] - LNF[i]$; see Fig. 1.

Lemma 2. *The longest proper border of $lcf[i]$ is $lnf[i]$.*

Proof. Let $\ell = border(lcf[i])$ and $t = i + LCF[i] - \ell$ ($t > i$). $X[i..i + \ell - 1]$ occurs also at the position t in X. Therefore, by the definition of the longest next factor, $\ell \leq LNF[i]$. We will show that $\ell = LNF[i]$. Assume to the contrary that $\ell < LNF[i]$. Let j be the first position greater than i where $lnf[i]$ occurs in X. Consider the factor $Y = X[i..j + LNF[i] - 1]$.

Claim. Y is closed.

Proof. Y has a border $lnf[i]$. Moreover, it is its longest border, as any longer border would imply a next factor longer than $lnf[i]$. By the definition of j, the string $lnf[i]$ occurs exactly twice in Y. Hence, Y is a closed factor of X.   □

Claim. $|Y| > LCF[i]$.

Proof. We have $|Y| = j + LNF[i] - i$ and $LCF[i] = t + \ell - i$. By the assumption, $LNF[i] > \ell$. The longest proper border of $lcf[i]$ is a prefix of $lnf[i]$ and so occurs at position j. If $j < t$, then this would mean the third occurrence within $lcf[i]$, which is impossible. Hence, $j \geq t$. Consequently, $|Y| > LCF[i]$.   □

We conclude that Y is a closed factor longer than $lcf[i]$, a contradiction.   □

We proceed with the first algorithm for recovering a string X from its LCF array. The algorithm is simple for positions i where $LCF[i] = 1$.

Fact 3. *The number of distinct letters of X equals the number of 1-entries in its LCF array. Moreover, each i such that $LCF[i] = 1$ corresponds to the rightmost occurrence of one of the letters in X.*

If $LCF[i] > 1$ then $X[i]$ can be recovered from one of following positions in X using the function below.

Algorithm *ComputeSingleSymbol(i)*
Input: $X[i+1..n]$, $LCF[i..n]$
Output: $X[i]$
$bord := border(X[i+1..i+LCF[i]-1])$;
$X[i] := X[i+LCF[i]-1-bord]$;

In the correctness proof of the function we use the following auxiliary lemma.

Lemma 4. *If* $LCF[i] > 1$ *then* $border(X[i+1..i+LCF[i]-1]) = LNF[i]-1$.

Proof. Let $S = X[i+1..i+LNF[i]-1]$ and $T = X[i+1..i+LCF[i]-1]$. By Lemma 2, $lnf[i]$ is a border of $lcf[i]$. Hence, S is a border of T. Assume to the contrary that T has a longer proper border S'. Then $lnf[i] = X[i]S$ is a suffix of S', as it is a suffix of T. Hence, $lnf[i]$ occurs also at position $i+1+|S'|-LNF[i] < i+LCF[i]-LNF[i] = next[i]$, which contradicts the fact that $lcf[i]$ is closed. □

Lemma 5. *Assume* $LCF[i] > 1$. *Then the function ComputeSingleSymbol correctly computes* $X[i]$.

Proof. By Lemma 4, we have $bord = LNF[i]-1$. Consequently, $next[i] = i + LCF[i]-1-bord$, which yields the correctness of the function. □

Fact 3 and Lemma 5 show that a string can be restored from its LCF array.

Theorem 6. *If there exists a string with the given* LCF *array then it is uniquely determined up to permutation of the corresponding alphabet.*

Proof. For each position i such that $LCF[i] = 1$ we introduce a unique letter $X[i]$. For each of the remaining positions, $X[i]$ can be determined from the following letters using function *ComputeSingleSymbol*. □

The algorithm of Theorem 6 is not efficient yet. In the following section we introduce an additional combinatorial fact and algorithmic tools that make the solution efficient.

3.2 Efficient Reconstruction

In the reconstruction algorithm we compute the string X together with the corresponding LNF array. We use the following crucial fact.

Lemma 7. *For every* $1 \le i < n$, $LNF[i] \le LNF[i+1]+1$.

Proof. Assume to the contrary that for some i, $LNF[i] > LNF[i+1]+1$. Recall that $lnf[i]$ occurs at position $next[i]$ in X. This concludes that the string $Y = X[i+1..i+LNF[i]-1]$ of length $LNF[i]-1 > LNF[i+1]$ occurs at position $next[i]+1 > i+1$. This contradicts the definition of $LNF[i+1]$.   □

In the pseudocode of algorithm *Reconstruction* we use Lemma 7 on top of the reconstruction algorithm from the previous section, based on function *ComputeSingleSymbol*. The alphabet of the reconstructed string is $\{1, 2, \ldots\}$.

Algorithm *Reconstruction*
Input: $LCF[1..n]$ array
Output: the corresponding string $X[1..n]$
$bord := 0$;
for $i := n$ **downto** 1 **do**
    { Invariant: If $i < n$ then $bord = LNF[i+1]$. }

    if $LCF[i] = 1$ **then**
        $X[i] := NewLetter()$;
        $bord := 0$; { $LNF[i] = 0$ }
    else

        { Efficient implementation of *ComputeSingleSymbol(i)* }
        while $X[i+1..i+bord] \neq X[i+LCF[i]-bord..i+LCF[i]-1]$ **do**
            { $bord > border(X[i+1..i+LCF[i]-1])$ }
            $bord := bord - 1$;
        $X[i] := X[i+LCF[i]-1-bord]$;
        $bord := bord + 1$; { $LNF[i] = bord$ }

    return X;

Clearly, the total number of steps of the while-loop in the algorithm *Reconstruction* is at most n. Hence, the time complexity of the algorithm depends on how fast we can check equality of two factors of a string, with the letters of the string being appended on-line from right to left.

Theorem 8. *A string of length n over alphabet Σ can be reconstructed from its longest closed factor array with an $\mathcal{O}(n)$-time randomized algorithm or an $\mathcal{O}(n \min(\log |\Sigma|, \log \log n + \frac{\log^2 \log |\Sigma|}{\log \log \log |\Sigma|}))$-time deterministic algorithm.*

Proof. For the randomized algorithm, we use Karp-Rabin fingerprinting (see e.g. [6]) to check equality of factors of the string given on-line in $\mathcal{O}(1)$ time.

For the deterministic algorithm, we use one of the incremental suffix tree constructions. The first one is the algorithm by Blumer et al. [4] which computes suffix trees for growing suffixes of the string in $\mathcal{O}(n \log |\Sigma|)$ total time. The other comes from a recent paper by Fischer and Gawrychowski [9], where the authors show how to update the suffix tree in $O(\log \log n + \frac{\log^2 \log |\Sigma|}{\log \log \log |\Sigma|})$ time after prepending a character (see Corollary 4 and Theorem 5 in [9]). Finally, given a suffix tree, determining equality of factors reduces to LCA-computation. This can be done, however, with $\mathcal{O}(1)$ overhead using LCA queries for a dynamic tree; see [5].   □

Theorem 8 provides an efficient reconstruction algorithm from the LCF array only if the corresponding string exists. Otherwise the reconstruction algorithm may fail or reconstruct a string which does not have the given LCF array. In the latter case it suffices to construct its LCF array and check if it matches the input LCF array. We deal with this case in the following section.

4   LCF Array Computation and Verification

A successor of an integer x in a set X is defined as $succ(x, X) = \min\{y \in X : y > x\}$. For an integer array $A[1..n]$, a range successor query consists in computing $succ(x, \{A[i], \ldots, A[j]\})$ for any $1 \leq i \leq j \leq n$. Babenko et al. ([1], Section 2.5) show the following result:

Lemma 9 ([1]). *A collection of q range successor queries in an array of length n can be answered offline in $\mathcal{O}((n + q)\sqrt{\log n})$ time.*

As shown in [2], computation of the LCF array reduces to $\mathcal{O}(n)$ range successor queries in the suffix array of the string. Hence, we can use Lemma 9 to improve the running time of both LCF array computation and verification.

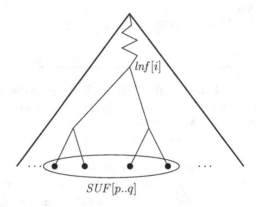

$$SUF[p..q]$$

Fig. 2. Illustration of the formula $next[i] = succ(i, SUF[p..q])$. To compute $next[i]$, we need to find the first occurrence of $lnf[i]$ after the position i.

Theorem 10.
(a) The LCF array of a string can be computed in $\mathcal{O}(n\sqrt{\log n})$ time.
(b) Verification of LCF array can be done in $\mathcal{O}(n\sqrt{\log n})$ randomized time or $\mathcal{O}(n(\sqrt{\log n} + \frac{\log^2 \log |\Sigma|}{\log \log \log |\Sigma|}))$ deterministic time.

Proof. For the start, recall that a suffix tree of a string (that is, a compact trie of its suffixes) over a linearly-sortable alphabet can be constructed in $\mathcal{O}(n)$ time [6].

We reformulate the algorithm of [2]. Recall that $LCF[i] = LNF[i] + next[i] - i$. The LNF array, together with the nodes of the suffix tree corresponding to $lnf[i]$, can be computed in $\mathcal{O}(n)$ time; see Lemma 6 in [2]. Let $SUF[p..q]$ be the sequence of leaves in the suffix tree corresponding to the subtree corresponding to $lnf[i]$. Then $next[i] = succ(i, \{SUF[p], \ldots, SUF[q]\})$; see Fig. 2.

Thus Lemma 9 implies an $\mathcal{O}(n\sqrt{\log n})$-time algorithm for computing LCF array. Part (b) follows from part (a) and Theorem 8. □

5   Final Remarks

For a string X of length n, the longest previous factor (LPF) array and the closest longest previous factor ($prev$) array are defined as follows. For $i = 1, \ldots, n$, $LPF[i]$ is the maximum ℓ such that $X[j..j + \ell - 1] = X[i..i + \ell - 1]$, for some $j < i$. For $i = 1, \ldots, n$, $prev[i]$ is the maximum $j < i$ such that $X[j..j + LPF[i] - 1] = X[i..i + LPF[i] - 1]$. Note that the LPF array is *not* the same as the LNF array of the reversed string.

The LPF array can be computed in $\mathcal{O}(n)$ time [7]. Using exactly the same approach as in Section 4 but with range predecessors instead of range successors, we obtain the following result:

Theorem 11. *The closest longest previous factor array prev of a string of length n can be computed in $\mathcal{O}(n\sqrt{\log n})$ time.*

Our Theorem 6 provides an example of a unique reconstruction of a string from its closed factors. Another example is Theorem 9 in [10], which states that every (finite or infinite) Sturmian word is uniquely determined, up to isomorphisms of the alphabet, by its sequence of open and closed prefixes. Both results are quite independent (none follows from the other).

Acknowledgments. The authors thank the participants of StringMasters workshops in 2014 in Prague and in 2015 in Warsaw, in particular Gabriele Fici, Ondrej Guth, and Jan Holub, for helpful discussions.

References

1. Babenko, M.A., Gawrychowski, P., Kociumaka, T., Starikovskaya, T.: Wavelet trees meet suffix trees. arXiv:1408.6182v4 (2015)
2. Badkobeh, G., Bannai, H., Goto, K., Tomohiro, I., Iliopoulos, C.S., Inenaga, S., Puglisi, S.J., Sugimoto, S.: Closed factorization. In: Holub, J., Žd'árek, J. (eds.) Prague Stringology Conference 2014, pp. 162–168 (2014)
3. Badkobeh, G., Fici, G., Lipták, Z.: On the number of closed factors in a word. In: Dediu, A.-H., Formenti, E., Martín-Vide, C., Truthe, B. (eds.) LATA 2015. LNCS, vol. 8977, pp. 381–390. Springer, Heidelberg (2015)

4. Blumer, A., Blumer, J., Haussler, D., Ehrenfeucht, A., Chen, M.T., Seiferas, J.I.: The smallest automaton recognizing the subwords of a text. Theor. Comput. Sci. **40**, 31–55 (1985)
5. Cole, R., Hariharan, R.: Dynamic LCA queries on trees. SIAM J. Comput. **34**(4), 894–923 (2005)
6. Crochemore, M., Hancart, C., Lecroq, T.: Algorithms on Strings. Cambridge University Press, Cambridge (2007)
7. Crochemore, M., Ilie, L., Iliopoulos, C.S., Kubica, M., Rytter, W., Waleń, T.: Computing the longest previous factor. Eur. J. of Comb. **34**(1), 15–26 (2013)
8. Fici, G.: A classification of trapezoidal words. In: Ambroz, P., Holub, S., Masáková, Z. (eds.) Combinatorics on Words - WORDS 2011. EPTCS, vol. 63, pp. 129–137 (2011)
9. Fischer, J., Gawrychowski, P.: Alphabet-dependent string searching with wexponential search trees. In: Cicalese, F., Porat, E., Vaccaro, U. (eds.) CPM 2015. LNCS, vol. 9133, pp. 160–171. Springer, Heidelberg (2015)
10. De Luca, A., Fici, G.: Open and closed prefixes of Sturmian words. In: Karhumäki, J., Lepistö, A., Zamboni, L. (eds.) WORDS 2013. LNCS, vol. 8079, pp. 132–142. Springer, Heidelberg (2013)

A Compact RDF Store Using Suffix Arrays

Nieves R. Brisaboa[1], Ana Cerdeira-Pena[1], Antonio Fariña[1(✉)],
and Gonzalo Navarro[2]

[1] Database Lab., University of A Coruña, Coruña, Spain
{brisabo,acerdeira,fari}@udc.es
[2] Department of Computer Science, University of Chile, Santiago, Chile
gnavarro@dcc.uchile.cl

Abstract. RDF has become a standard format to describe resources in
the Semantic Web and other scenarios. RDF data is composed of triples
(*subject, predicate, object*), referring respectively to a resource, a property of that resource, and the value of such property. Compact storage
schemes allow fitting larger datasets in main memory for faster processing. On the other hand, supporting efficient SPARQL queries on RDF
datasets requires index data structures to accompany the data, which
hampers compactness. As done for text collections, we introduce a *self-
index* for RDF data, which combines the data and its index in a single
representation that takes less space than the raw triples and efficiently
supports basic SPARQL queries. Our storage format, *RDFCSA*, builds
on compressed suffix arrays. Although there exist more compact representations of RDF data, *RDFCSA* uses about half of the space of the raw
data (and replaces it) and displays much more robust and predictable
query times around 1–2 microseconds per retrieved triple. *RDFCSA* is 3
orders of magnitude faster than representations like MonetDB or RDF-
3X, while using the same space as the former and 6 times less space than
the latter. It is also faster than the more compact representations on
most queries, in some cases by 2 orders of magnitude.

1 Introduction

The amount of data publicly available on the Web has been growing steadily over
the years. Many valuable resources are included in this gigantic repository, but
in many cases they are underutilized because of the lack of a common storage
format that allows those resources be automatically identified and accessed. The
Web of Data is an effort to structure the data published by resource providers in
a way that it can be discovered and used under a standard protocol in automatic
form. The Web of Data builds on the principles of the Semantic Web [9].

The *Resource Description Framework* (RDF) [19] provides a simple and powerful way to structure and link data. It uses triples (*subject, predicate, object*) to

Founded in part by Fondecyt 1-140796 (for Gonzalo Navarro); and, for the Spanish
group, by MINECO (PGE and FEDER) [TIN2013-46238-C4-3-R, TIN2013-47090-
C3-3-P]; CDTI, AGI, MINECO [CDTI-00064563/ITC-20133062]; ICT COST Action
IC1302; and by Xunta de Galicia (co-founded with FEDER) [GRC2013/053].

© Springer International Publishing Switzerland 2015
C. Iliopoulos et al. (Eds.): SPIRE 2015, LNCS 9309, pp. 103–115, 2015.
DOI: 10.1007/978-3-319-23826-5_11

model knowledge, in such a way that a value (*object*) for a property (*predicate*) of a given resource (*subject*) is represented. The adoption of RDF by the W3C as the recommended format to publish information [1] has boosted the growth of RDF repositories and RDF management systems that make up the basis of the current Web of Data. Those systems not only store the RDF data, but they also support queries on it via the SPARQL query language [23].

The increasing interest in the management of RDF repositories (also called RDF stores) is witnessed by the various storage schemes proposed in recent years, which go from those based on relational databases [25] to native solutions such as BITMAT [6], RDF-3X [22], HEXASTORE [26], MonetDB [2], or WaterFowl [12]. As the RDF repositories grow in size, scalability issues challenge the use of RDF storage schemes [18]. A recent work (*K2Triples*) [4] succeeded at reducing both the space usage of previous techniques and their performance to answer basic SPARQL queries: the so-called *basic graph patterns* that make up the primitive SPARQL operations and the algorithms for *merge* and *join*.

In this paper we introduce another storage scheme we call *RDFCSA*. It is based on Sadakane's *Compressed Suffix Array (CSA)* [24], which can represent a text collection in compressed space while supporting pattern searches on it. We modify the *CSA* so as to index a set of triples in a way that all the basic graph patterns of SPARQL boil down to pattern searches on the modified *CSA*. The result is a representation that uses about twice the space of *K2Triples*, but it is faster in most queries, up to 2 orders of magnitude in some cases, which include the most common ones in real-life SPARQL queries [5]. Compared to other representations, *RDFCSA* uses about the same space as MonetDB and 6 times less than RDF-3X, and it is 3 orders of magnitude faster than both.

2   Basic Concepts

2.1   State of the Art: *K2Triples*

A RDF dataset can be seen as a set $\mathcal{R}$ of triples (s, p, o) where s, p, and o are respectively a *subject*, a *predicate*, and an *object*. It can also be seen as a connected graph where *subjects* and *objects* are nodes that are connected via arcs labeled by a given *predicate* [19]. Figure 1 shows an example with (not really) fictitious data about the SPIRE conference and some attendants. In the left part, we show the source triples and the underlying RDF graph.

K2Triples [4] tackles the scalability problem of RDF datasets by focusing in reducing their space usage. The authors used two main areas:

(a) Reducing the size of the representation of the strings in the triples through a compressed string dictionary [20,6,14]. Each original triple is then represented by a triple of integer ids provided by the dictionary. The right part of Figure 1 depicts the dictionary organization used, and the final set of id-based triples.

(b) Representing the id-based triples in a compact (and indexed) way. The fact that the number of predicates (n_p) in a RDF dataset is typically very small is exploited by *K2Triples*, which resorts to vertical partitioning [3]: for each *predicate*, it stores the *subjects* that are connected to each *object*. Each such binary relation is generally sparse, so it is represented with a compact k²-tree data structure [10], which performs well on those relations. The k²-tree of each predicate can efficiently list the *subjects* related to a given *object* or the *objects* related to a given *subject*.

Simple graph patterns are the most basic SPARQL queries. They are triples where each component can be fully specified as a string (*S*, *P*, or *O*) or left unspecified or "unbounded" (?*S*, ?*P*, or ?*O*). Such a pattern matches all triples where the specified strings match. For example, in Figure 1, pattern (?*S*, attends, SPIRE) returns the 3rd, 4th and 5th triples listed on "Original RDF Triples".

Due to the vertical partitioning of *K2Triples*, patterns with a fixed *predicate*, that is, (S, P, O), $(?S, P, O)$, $(S, P, ?O)$, and $(?S, P, ?O)$, can be efficiently solved within a unique k²-tree, whereas patterns with unbounded *predicate* $((S, ?P, O)$, $(?S, ?P, O)$, $(S, ?P, ?O)$, and $(?S, ?P, ?O))$ would involve accessing all the n_p k²-trees. The *K2Triples* structure partially overcomes this issue by adding two auxiliary indexes, SP and OP, that respectively keep which *subjects* (s) or *objects* (o) occur in a triple related to each *predicate* p. Indexes SP and OP yield large speedups, while typically costing 20%–30% further space.

K2Triples is shown to improve the space of the best state-of-the-art alternatives by a factor of 1.5–12, whereas it matches or outperforms them all in simple graph patterns [4].

2.2 Compressed Suffix Arrays

Given a string $S[1, n]$ over alphabet $\Sigma = [1, \sigma]$, the *suffix array* $A[1, n]$ is a permutation of $[n]$ so that $S[A[i], n]$ is the ith lexicographically smallest suffix in S. Thus the range of suffixes starting with a search pattern $\alpha[1, m]$ (i.e., the occurrences of α in S) can be binary searched in A in time $O(m \log n)$.

Sadakane's *CSA* [24] represents S and A using two structures (plus others that we ignore in this paper). The first is a bitmap $D[1, n]$, where the 1s mark the first suffixes starting with each distinct symbol in A (i.e., $D[i] = 1$ iff $i = 1$ or $S[A[i]] \neq S[A[i-1]]$). The second *CSA* structure is the array $\Psi[1, n]$, where

RDF Graph   **Original RDF Triples**   **Dictionary Encoding**   **Id-based Triplets**

Fig. 1. Example of RDF graph and dictionary encoding in *K2Triples*. **SO** entries in the Dictionary represent terms that act as both *subjects* and *objects* in some triples.

$\Psi[i] = A^{-1}[(A[i] \bmod n) + 1]$. That is, if $A[i] = j$ points to the suffix $S[j, n]$, then $A[\Psi[i]] = j + 1$ points to the next text suffix, $S[j + 1, n]$.

In this paper we assume that every symbol in Σ appears at least once in S. Then $S[A[i]] = rank_1(D, i)$, where $rank_1(D, i)$ is the number of 1s in $D[1, i]$. Moreover, $S[A[i] + 1] = S[A[\Psi[i]]] = rank_1(D, \Psi[i])$, and in general $S[A[i] + j] = rank_1(D, \Psi^j[i])$. Operation $rank$ can be solved in constant time after building an $o(n)$-bit structure on D [11]. Therefore, D and Ψ are sufficient to extract any string $S[A[i], A[i] + m - 1]$ in time $O(m)$. As a result, the binary search on A can be simulated on D and Ψ in the same $O(m \log n)$ time, and the first ℓ symbols of any matching suffix can be extracted in $O(\ell)$ time as well. Array Ψ can be stored in $nH_0(S) + O(n \log H_0(S)) \le n \log \sigma + O(n \log \log \sigma)$ bits while supporting constant-time access, where $H_0(S)$ is the zero-order empirical entropy of S [24]. Array Ψ is compressible because it is formed by σ increasing subsequences, which can be differentially encoded using δ-codes. By giving special codes to the runs of consecutive 1s in the differences, the space gets closer to higher-order entropies of S [21]. Sampled Ψ values at regular intervals yield fast random access to Ψ.

Our *RDFCSA* is based on the *integer-based CSA* $(iCSA)$[1] [13], which is a variant Sadakane's *CSA* that is optimized for large (integer-valued) alphabets. The *iCSA* reaches the best compression when using truncated Huffman coding of differences and run lengths.

3   *RDFCSA*: A Compressed Suffix Array for RDF

An RDF collection is a set $\mathcal{R}$ of triples (s, p, o) where s, p, and o are respectively a *subject*, a *predicate*, and an *object*. We use the same dictionary encoding as in previous work [4] so that from now on the triple components s, p, and o are regarded as integer ids in the ranges $s \in [1, n_s]$, $p \in [1, n_p]$, and $o \in [1, n_o]$.

3.1   Structure

The first step to build our *RDFCSA* is to create an ordered list with the n triples from $\mathcal{R}$, and regarding it as a sequence $S_{id}[1, 3n]$ with $3n$ elements. Since the order is not relevant in a set of triples, we sort them by *object*, then by *predicate* and finally by *subject*. We obtain a sequence of integers $S_{id}[1, 3n] = \langle s_1, p_1, o_1, s_2, p_2, o_2, \ldots, s_n, p_n, o_n \rangle$.

To have disjoint subalphabets Σ_s, Σ_p, and Σ_o for the n_s subjects, the n_p *predicates*, and the n_o objects, we set an array $gaps[0, 2] = [0, n_s, n_s + n_p]$ and convert sequence $S_{id}[1, 3n]$ to $S[1, 3n]$, where $S[i] = S_{id}[i] + gaps[(i - 1) \bmod 3]$. Sequence S ranges over alphabet $\Sigma = [1, n_s + n_p + n_o]$, where values $[1, n_s]$ are reserved to *subjects*, $[n_s + 1, n_s + n_p]$ to *predicates*, and the rest to *objects*. We can obviously recover the original triples from S. Then, we build an *iCSA* on S.

Due to our alphabet mapping, every *subject* is smaller than every *predicate*, and this in turn is smaller than every *object*. Then, the suffix array A of S will

---

[1] http://vios.dc.fi.udc.es/indexing/wsi/

have three ranges: $A_s = A[1, n]$, $A_p = A[n + 1, 2n]$ and $A_o = A[2n + 1, 3n]$ where each range points to suffixes starting with a *subject*, a *predicate*, or an *object*, respectively. Array Ψ also has three separate ranges. Entries in $\Psi[1, n]$ will contain values in the range $[n+1, 2n]$ (corresponding to the range of *predicates*). Entries in $\Psi[n + 1, 2n]$ will contain values in the range $[2n + 1, 3n]$ (of *objects*). Finally, entries in $\Psi[2n+1, 3n]$ will contain values in the range $[1, n]$ (of *subjects*).

In a regular *CSA*, if $A[i]$, for $i \in [2n + 1, 3n]$, points to the *object* (third component) of the kth triple of S (i.e., $A[i] = 3k$), then $j = \Psi[i]$ will indicate the position such that $A[j]$ points to the *subject* (first component) of the $(k + 1)$th triple in S (i.e., $A[j] = 3k+1$). This is the key feature that allows traversing the string S virtually using Ψ.

For our purposes, it is more useful that Ψ cycles around the components of the same triple, instead of advancing to the next one. The *RDFCSA* modifies array Ψ so that values in $\Psi[2n + 1, 3n]$ point not to the *subject* of the *next* triple in S, but to the *subject* of the *same* triple. Given the way we have sorted the triples in S, it turns out that $A[i] = 3(i - 1) + 1$, and therefore all we have to do to make Ψ cycle through the same triples is to set $\Psi[i] \leftarrow \Psi[i] - 1$ for all $i \in [2n + 1, 3n]$ (or $\Psi[i] \leftarrow n$ if $\Psi[i] = 1$).

With this modified Ψ we can start at the position $A[i]$ pointing to any place inside a triple (s, p, o) and recover the triple by successive applications of Ψ. For example, if $A[i]$ points to p, then $p = rank_1(D, i)$, $o = rank_1(D, \Psi[i])$, and $s = rank_1(D, \Psi[\Psi[i]])$. If we take Ψ once more we return to $i = \Psi[\Psi[\Psi[i]]]$. In particular, we can retrieve the kth triple of S by starting the process from $A[i]$, which we know points to the *subject* because $i \in [1, n]$. This property will also allow us reduce any simple graph pattern query to the search for a short pattern in S using the *CSA*, and then extract the contents of the resulting triples.

Figure 2 shows the final structure of a *RDFCSA* created over the ten triples included in Figure 1. In this case we have $n = 10$, $n_s = 5$, $n_p = 6$, and $n_o = 5$. The first of the sorted set of source triples is $S_{id}[1, 3] = (1, 2, 5)$, the second is $S_{id}[4, 6]$, and so on. By adding $gaps[0, 2]$ to the triples in S_{id} we obtain $S[1, 30]$. We show the suffix array A built on S and the structures D and Ψ that make up the *RDFCSA* (Ψ is already modified from the original array, Ψ_{orig}, to cycle through each triple). We mark the boundaries of the three ranges $[1, 10], [11, 20]$, and $[21, 30]$. We verify that entries in $A[1, 10]$ point to positions in $S[3k + 1]$,

	1	2	3	4	5	6	7	8	9	10	11	12	13	14	15	16	17	18	19	20	21	22	23	24	25	26	27	28	29	30
S_{id}	1	2	5	2	3	1	3	1	2	3	4	3	4	1	2	4	4	5	4	6	1	5	1	2	5	4	5	5	5	4

gaps: | 0 | 1 | 2 | → | 0 | 5 | 11 | $S[i] = S_{id}[i] + gaps[(i-1) \bmod 3]$

	1	2	3	4	5	6	7	8	9	10	11	12	13	14	15	16	17	18	19	20	21	22	23	24	25	26	27	28	29	30
S	1	7	16	2	8	12	3	6	13	3	9	14	4	6	13	4	9	16	4	11	12	5	6	13	5	9	16	5	10	15
			subjects								predicates										objects									
A	1	4	7	10	13	16	19	22	25	28	8	14	23	2	5	11	17	26	29	20	6	21	9	15	24	12	30	3	18	27
D	1	1	1	0	1	0	0	1	0	0	1	0	0	1	1	1	0	0	1	1	1	0	1	0	0	1	1	1	0	0
Ψ	14	15	11	16	12	17	20	13	18	19	23	24	25	28	21	26	29	30	27	22	2	7	3	5	8	4	10	1	8	9
Ψ_{orig}	14	15	11	16	12	17	20	13	18	19	23	24	25	28	21	26	29	30	27	22	3	8	4	6	9	5	1	2	7	10

Fig. 2. Structures involved in the creation of a *RDFCSA* for the graph in Figure 1.

those in $A[11, 20]$ to $S[3k + 2]$, and those in $A[21, 30]$ to $S[3k]$. For example, $(rank_1(D, 1), rank_1(D, \Psi[1]), rank_1(D, \Psi[\Psi[1]])) = (1, 7, 16)$ recovers the triple in $S[1, 3]$. Also, the third source triple $S_{id}[7, 9]$ can be recovered by doing $S_{id}[7, 9] = (S[7] - gaps[0], S[8] - gaps[1], S[9] - gaps[2])$.

In the $RDFCSA$, the modified array Ψ is represented as in the $iCSA$ [13]. Bitvector D uses a fast $rank$ structure that uses $0.375n$ bits, also as in the $iCSA$. We will also need operation $select_1(D, j)$, which finds the position of the jth 1 in D. It is implemented by a binary search on the $rank$ directories.

We note that enforcing the property $\Psi^3[i] = i$ on our $RDFCSA$ is analogous to the more general *permuterm index* [17]. They index a set of strings as if they were circular, so that patterns of the form $\alpha * \beta$ can be found by searching for the substring $\beta\$\alpha$, where $\$$ is the string terminator. However, the permuterm index is built on an *FM-index* [15], which on large alphabets like our $[1, n_s + n_p + n_o]$ is implemented on a wavelet tree [16]. This implementation poses a time overhead factor $O(\log(n_s + n_p + n_o))$ for the operation equivalent to computing Ψ, which renders the *FM-index* inferior to the *CSA* on large alphabets [13]. We checked this by using the best $iSSA$ variant from [13] to represent sequence S. We tuned $iSSA$ to use the same space as $RDFCSA$ (around 60% the size of S regarded as 32-bit integers). Query time to solve (S, P, O) patterns was around $2.5 - 4$ times slower than in $RDFCSA$. Newer alternatives to wavelet trees on large alphabets are only slightly better when implementing FM-indexes [8]. This is why we opt for implementing the technique on top of the $iCSA$ for the case of RDF triples.

3.2   Supporting Basic Graph Pattern Queries in $RDFCSA$

Searching for triple patterns is the base to support more complex SPARQL queries on an RDF store. We first show how the 8 primitive operations (S, P, O), $(?S, P, O)$, $(S, ?P, O)$, $(S, P, ?O)$, $(?S, ?P, O)$, $(S, ?P, ?O)$, $(?S, P, ?O)$, $(?S, ?P, ?O)$ can be solved on $RDFCSA$. Then we discuss some RDF-specific optimizations.

The pattern $(?S, ?P, ?O)$ is treated differently because it retrieves all the triples in the dataset (thus it is not really useful as a query). If needed, it can be solved by retrieving every ith triples as described above. The other 7 patterns will be solved by an initial search followed by a traversal to recover the contents of the matching triples.

Binary $iCSA$ Search for Triple Patterns: As explained, the $iCSA$ can run a binary search for the range $A[l, r]$ pointing to the suffixes that start with any pattern $\alpha[1, m]$, so that α appears in S at positions $A[i]$ for $i \in [l, r]$. Then, it can use Ψ to recover the symbols $S[A[i], *]$ for any such i.

In our case, we can solve query (S, P, O) by searching for $\alpha[1, 3] = SPO$, thus determining if it exists in the dataset ($l = r$) or not ($l > r$). Further, we can solve queries $(S, P, ?O)$ and $(?S, P, O)$ by searching for $\alpha[1, 2] = SP$ and $\alpha[1, 2] = PO$, respectively. Because Ψ cycles over the triples, we can retrieve the resulting triples in either case, starting from each $i \in [l, r]$. Further, we can also solve queries $(S, ?P, O)$ by searching for $\alpha[1, 2] = OS$, since Ψ regards the triples as circular strings. When only S, P, or O are specified, we must simply search for $\alpha[1,1] = S$, $\alpha[1,1] = P$, or $\alpha[1,1] = O$. We give an example of each case:

- (S, P, O): We set $\alpha[1,3] = [S + gaps[0], P + gaps[1], O + gaps[2]]$, and obtain the range $[l, r]$ with the $iCSA$ binary search. If $l = r$ then (S, P, O) is in the set, otherwise it is not.
- $(S, ?P, O)$: We set $\alpha[1,2] = [O + gaps[2], S + gaps[0]]$, and find the interval $[l, r]$ with the $iCSA$. The number of answers is $r - l + 1$. For each $i \in [l, r]$, we return the triple $(S, rank_1(D, \Psi[\Psi[i]]) - gaps[1], O)$.
- $(?S, P, ?O)$: We set $\alpha[1,1] = [P + gaps[1]]$, and find the interval $[l, r]$ with the $iCSA$ (note that this does not require binary search on Ψ: $l = select_1(D, \alpha[1])$ and $r = select_1(D, \alpha[1] + 1) - 1$). The number of answers is $r - l + 1$ and for each $i \in [l, r]$, the triple $(rank_1(D, \Psi[\Psi[i]]) - gaps[0], P, rank_1(D, \Psi[i]) - gaps[2])$ is recovered.

By using binary search on the $iCSA$, all the triple pattern queries cost $O(r - l + \log n)$, where $r - l + 1$ is the number of occurrences retrieved. In practice, the compression of Ψ introduces important space/time tradeoffs. If the number of triples retrieved is large, the cost of the binary search is negligible. However, it becomes relevant when only one or a few triples are recovered (e.g., no triple recovering is needed for pattern (S, P, O)).

Our first optimization on the original $iCSA$ aims at improving the accesses to Ψ needed to retrieve the triples. Once $[l, r]$ is determined, we always have to compute $\Psi[i]$ and $\Psi[\Psi[i]]$ for all $i \in [l, r]$ (except on the pattern (S, P, O)). We have sped up the access to a range $\Psi[l, r]$ by sequentially decompressing that range of Ψ. Therefore, we only need to access once the sample preceding $\Psi[l]$ and reach position l; all the subsequent values are immediately decoded. This is especially fast if we are inside a run of consecutive values of Ψ. The remaining accesses to Ψ are random and are not be improved.

The other optimizations aim at decreasing the cost of the binary search for $[l, r]$. Two alternatives strategies, *D-select+forward-check* and *D-select+backward-check*, are discussed below.

D-select+forward-check Strategy: During the binary search, the comparison between α and $S[A[i], n]$ might be decided with the first integer comparison. Obtaining $S[A[i]] = rank_1(D, i)$ does not require the application of Ψ. At some moment, however, we start having $S[A[i]] = \alpha[1]$ and must compute Ψ to compare $\alpha[2]$ with $rank_1(D, \Psi[i])$. This isolated access to Ψ can be expensive. Instead, we can proceed as follows. Consider the triple pattern (S, P, O). We first find the intervals $R_s = [l_{S+gaps[0]}, r_{S+gaps[0]}]$, $R_p = [l_{P+gaps[1]}, r_{P+gaps[1]}]$, and $R_o = [l_{O+gaps[2]}, r_{O+gaps[2]}]$. These are computed with *select* on D: $l_c = select_1(D, c)$ and $r_c = select_1(D, c + 1) - 1$. Since Ψ is increasing within those intervals, for each i in R_s we can check if $\Psi[i] \in R_p$. The values i that do not pass this check can be discarded. For those that do, we still have to check if $\Psi[\Psi[i]] \in R_o$, in which case we report an occurrence of the searched triple.

Figure 3 (left) illustrates this scenario, where $R_s = [10, 12]$, $R_p = [200, 300]$, and $R_s = [600, 601]$. Neither $\Psi[10]$ nor $\Psi[12]$ map into range $[200, 300]$, only

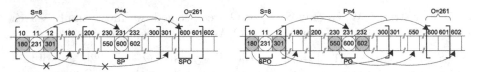

Fig. 3. D-select+forward-check (left) and D-select+backward-search (right) strategies for pattern $(S, P, O) = (8, 4, 261)$.

$\Psi[11]$ does. In addition $\Psi[\Psi[11]]$ maps into the range $[600, 601]$ corresponding to *object* 261. Hence, we report an occurrence of the triple $(8, 4, 261)$.

Computing all the values $\Psi[R_s]$ is much cheaper than computing $|R_s|$ isolated values of Ψ, because of the differential compression of Ψ. In general, there are more objects than subjects, and many more subjects than predicates. Thus, we expect that $|R_o| < |R_s| \ll |R_p|$. If the interval R_s is small enough, this technique may be faster than a standard binary search. Since our Ψ is cyclic, we can start the checking process in interval R_s, R_p, or R_o, so we start from the shortest one.

This procedure is not only applicable to pattern (S, P, O). If we have one unbounded term, we obtain the intervals R_x and R_y corresponding to the bounded terms x and y. Then, we use the same procedure to check whether after applying Ψ to the positions i in the starting interval R_x we fall into R_y or not. For pattern $(?S, P, O)$ we set $x = P, y = O$; for pattern $(S, ?P, O)$ we set $x = O, y = S$; and for pattern $(S, P, ?O)$ we set $x = S, y = P$. Finally, recall that patterns with only one bounded element are directly solved using *select* on D.

D-select+backward-check Strategy: Note those i in R_s that pass the check in the previous strategy form a subinterval of R_s, thus we can use binary search to find its limits instead of verifying every $i \in R_s$ one by one. The best way to proceed is known as the backward-search strategy [24]. We show how it can be carried out when searching for pattern (S, P, O). We start in interval $R_o = [l_o, r_o]$, and since Ψ is increasing within interval $R_p = [l_p, r_p]$, we binary search the limits of the subinterval $R_{po} = [l_{po}, r_{po}] \subseteq R_p$ such that $\Psi[i] \in R_o$ for all $i \in R_{po}$. If the subinterval is empty, no match exists. Otherwise, we repeat the same process to find the limits of the subinterval $R_{spo} = [l_{spo}, r_{spo}] \subseteq R_s$ that contain the entries $i \in R_s$ such that $\Psi[i] \in R_{po}$. The final answer is $[l, r] = R_{spo}$.

In Figure 3 (right) we can see that starting in range $R_o = [600, 601]$, when we binary search the interval $\Psi[200, 300]$ for the values that map into range $[600, 601]$, only the entry $\Psi[231]$ remains. Therefore, we obtain the subinterval $R_{po} = [231, 231]$. Now, we binary search the range $\Psi[10, 12]$ for the range that maps to 231 and find that $\Psi[11] = 231$. Then the final interval is $R_{spo} = [11, 11]$.

This strategy is also applicable to patterns $(S, P, ?O)$, $(S, ?P, O)$, $(?S, P, O)$. In the first case we find the subinterval $R_{sp} \subseteq R_s$ that maps via Ψ inside R_p. In the second, the subinterval $R_{os} \subseteq R_o$. In the third, the subinterval $R_{po} \subseteq R_p$.

4 Experimental Evaluation

Our experiments ran on an Amd Phenom-X4-955@3.2GHz CPU, with 8GB DDR2 RAM. The operating system was Ubuntu 12.04 (kernel 3.2.0-31-generic) and the compiler used was gcc 4.6.3 (option -O9). We measure elapsed times.

We evaluated the space/time performance of *RDFCSA* over *Dbpedia*[2], "the nucleus for a Web of Data" [7]. The size of this dataset is around 34GB, containing 232,542,405 triples (2,790,508,860 bytes when regarded as 32-bit integers). The number of different *subjects*, *predicates*, and *objects* is 18,425,128; 39,672; and 65,200,769; respectively. We compared *RDFCSA* with *K2Triples*, MonetDB, and RDF-3X. The recent WaterFowl [12] was not included. Yet, since it reports space 10 times smaller than RDF-3X and similar times [12], we expect it would obtain worse query times than *K2Triples* (see comparison with RDF-3X below) and similar space. Other systems do not run over a dataset of this size [4].

Figure 4 shows the space/time tradeoff of these RDF representations. For *K2Triples* we show two points, corresponding to *K2Triples* and *K2Triples+* [4] (the latter includes the indexes SP and OP that speed up searches with unbounded predicate, see Section 2.1) and we used the tuning recommended by the authors. In the case of *RDFCSA*, the lines connect four points that correspond to sampling Ψ every t_Ψ values: $t_\Psi \in \{16, 32, 64, 256\}$. MonetDB and RDF-3X store the index on disk; we measure the space they use to operate in memory, and run them in warm state, as in previous work [4].

Results clearly show that *K2Triples* (and even *K2Triples+*) use less space than *RDFCSA* (around a half in the case of *K2Triples*). Still, *RDFCSA* uses around half of the space of a raw representation of the triples (and can reproduce them, apart from supporting searches). This is about the same main memory space used by MonetDB, and 6 times less than RDF-3X.

On the other hand, *RDFCSA* obtains much more stable times than *K2Triples*, below 1–2 μsec per occurrence in all cases with a reasonable sampling. *RDFCSA* is in all cases at least 3 orders of magnitude faster than MonetDB and RDF-3X (the only exception is pattern $(?S, P, ?O)$, where RDF-3X is only twice as slow).

K2Triples still obtains the best time for (S, P, O) patterns, as it only needs to accesses the single cell (S, O) of the k^2-tree associated to the predicate P, and this is very fast on the k^2-tree. Instead, this is the worst case for *RDFCSA*, which must search for a pattern of length 3 and return at most one occurrence.

We can also see that, even though the performance of *K2Triples* is very poor when solving $(S, ?P, O)$ queries, the indexes SP and OP included in *K2Triples+* help solve $(S, ?P, O)$ queries very efficiently. This is because they discard many of the n_p k^2-trees that should be accessed otherwise. Only those predicates P that are related to subject S and also with object O must be considered.

On the remaining queries *RDFCSA* is typically faster than *K2Triples* and *K2Triples+*. In particular, for $(S, P, ?O)$, $(S, ?P, ?O)$, and $(?S, ?P, O)$, *RDFCSA* is up to 2 orders of magnitude faster. The first two of these are the most common queries in real-life SPARQL queries, according to an empirical study [5]. In the

---

[2] http://downloads.dbpedia.org/3.5.1/

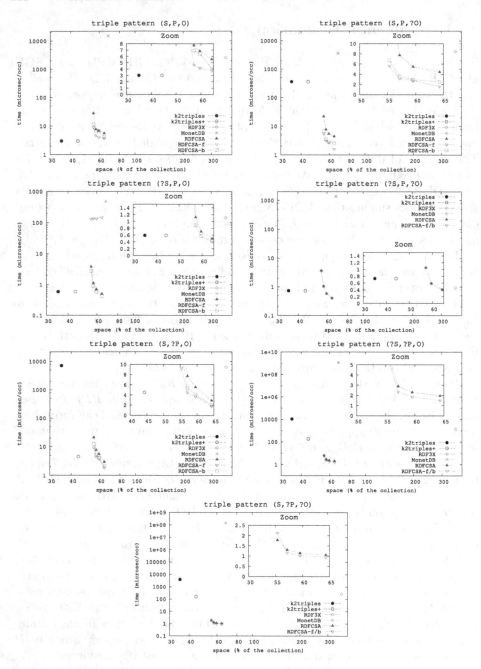

Fig. 4. Space/time tradeoff on the primitive SPARQL queries. Space is measured as the percentage of the size of the (in-memory) indexes with respect to the size of the source RDF triples represented as a sequence of 32-bit integers (the dictionary size is not considered here). Query time is the average (in μsec per occurrence) over 500 triple pattern queries of each type obtained from *K2Triples* authors website, http://dataweb.infor.uva.es/queries-k2triples.tgz. Note the logscale in the main plots; the zooms use linear scale.

case of $(?S, P, ?O)$ and $(?S, P, O)$, *RDFCSA* outperforms *K2Triples+* only when the denser samplings are used.

With respect to the optimizations discussed in Section 3, we can see that *D-select+forward-check* (RDFCSA-f in the plots) outperforms *D-select+backward-check* (RDFCSA-b in the plots) in all the triple patterns with one or no unbounded terms (with the exception of $(?S, P, O)$, where it performs very badly because it must sequentially traverse the generally long interval R_p). Except on the pattern (S, P, O), however, the improvement is not significant, and *D-select+backward-check* should be preferred for its more stable and guaranteed performance (of course, we can decide which strategy to use depending on the predicted performance, which can be easily estimated from the sizes of the intervals R_s, R_p, and R_o). In most cases (for example in (S, P, O)), those two strategies reduce the time of the regular *iCSA* binary search to a half or less. In other patterns, such as in $(S, ?P, O)$ those optimizations are the key to match (and even outperform) the performance of *K2Triples+*. Finally, in queries $(?S, ?P, O)$ and $(S, ?P, ?O)$ we can see the slight advantage obtained by performing two *select* operations instead of running the plain binary search with patterns of length 1 (for this experiment, the binary search strategy did not use that speedup).

5   Conclusions and Future Work

We have introduced *RDFCSA*, a competitive structure to self-index RDF data. It builds on an adaptation of the compressed suffix array of Sadakane [24], which is modified to index the RDF triples as cyclic strings, and then various domain-specific optimizations are studied on top of it.

RDFCSA uses about half the space required by the raw data and replaces it. It offers stable and predictable times to solve basic graph patterns (on which more sophisticated SPARQL queries are built), around 1–2 μsec per retrieved triple. Compared to literature standards, *RDFCSA* uses 6 times less space and runs most queries about 1000 times faster than RDF-3X. It uses about the same space as MonetDB, but this is even slower than RDF-3X. There are representations using around half the space of *RDFCSA* [4], their performance is less robust, being up to two orders of magnitude slower than *RDFCSA*, especially on the queries that appear most often in real applications.

Of course, *RDFCSA* handles only basic SPARQL queries, whereas RDF-3X and MonetDB are much more complete. Still, we believe this kernel can be extended to a wider functionality without sacrificing space and taking advantage of its speed when implementing more complex operations. We plan to start by extending *RDFCSA* functionality to handle the join and merge operations of SPARQL. As for the kernel functionality itself, we plan to further study the compressibility of the modified Ψ array (improvements considered in previous work [13] were already tried without success) and faster search algorithms for the particular case of triple patterns.

References

1. RDF 1.1 XML syntax, W3C recommendation (2004). http://www.w3.org/TR/rdf-syntax-grammar
2. MonetDB (2013). http://www.monetdb.org
3. Abadi, D.J., Marcus, A., Madden, S.R., Hollenbach, K.: Scalable semantic Web data management using vertical partitioning. In: Proc. VLDB, pp. 411–422 (2007)
4. Álvarez-García, S., Brisaboa, N., Fernández, J., Martínez-Prieto, M., Navarro, G.: Compressed vertical partitioning for efficient RDF management. Knowledge and Information Systems (2014) (to appear) preprint at www.dcc.uchile.cl/gnavarro/ps/kais14.pdf
5. Arias, M., Fernández, J.D., Martínez-Prieto, M.A., de la Fuente, P.: An empirical study of real-world SPARQL queries. CoRR abs/1103.5043 (2011). http://arxiv.org/abs/1103.5043
6. Atre, M., Chaoji, V., Zaki, M.J., Hendler, J.A.: Matrix "bit" loaded: A scalable lightweight join query processor for RDF data. In: Proc. WWW, pp. 41–50 (2010)
7. Auer, S., Bizer, C., Kobilarov, G., Lehmann, J., Cyganiak, R., Ives, Z.G.: DBpedia: A Nucleus for a Web of Open Data. In: Aberer, K., et al. (eds.) ASWC 2007 and ISWC 2007. LNCS, vol. 4825, pp. 722–735. Springer, Heidelberg (2007)
8. Barbay, J., Claude, F., Gagie, T., Navarro, G., Nekrich, Y.: Efficient fully-compressed sequence representations. Algorithmica **69**(1), 232–268 (2014)
9. Berners-Lee, T., Hendler, J., Lassila, O.: The semantic Web. Scientific American Magazine (2001)
10. Brisaboa, N., Ladra, S., Navarro, G.: Compact representation of Web graphs with extended functionality. Inf. Syst. **39**(1), 152–174 (2014)
11. Clark, D.: Compact PAT Trees. Ph.D. thesis, U. of Waterloo, Canada (1996)
12. Curé, O., Blin, G., Revuz, D., Faye, D.C.: WaterFowl: A Compact, Self-indexed and Inference-Enabled Immutable RDF Store. In: Presutti, V., d'Amato, C., Gandon, F., d'Aquin, M., Staab, S., Tordai, A. (eds.) ESWC 2014. LNCS, vol. 8465, pp. 302–316. Springer, Heidelberg (2014)
13. Fariña, A., Brisaboa, N., Navarro, G., Claude, F., Places, A., Rodríguez, E.: Word-based self-indexes for natural language text. ACM TOIS 30(1), article 1 (2012)
14. Fernández, J.D., Martínez-Prieto, M.A., Gutiérrez, C., Polleres, A., Arias, M.: Binary RDF representation for publication and exchange (HDT). Web Semantics **19**, 22–41 (2013)
15. Ferragina, P., Manzini, G.: Indexing compressed texts. J. ACM **52**(4), 552–581 (2005)
16. Ferragina, P., Manzini, G., Mäkinen, V., Navarro, G.: Compressed representations of sequences and full-text indexes. ACM Trans. Alg. 3(2), article 20 (2007)
17. Ferragina, P., Venturini, R.: The compressed permuterm index. ACM Trans. Alg. 7(1), article 10 (2010)
18. Jing, Y., Jeong, D., Baik, D.K.: SPARQL graph pattern rewriting for OWL-DL inference queries. Knowl. Inf. Syst. **20**(2), 243–262 (2009)
19. Manola, F., Miller, E., (eds): RDF primer, W3C recommendation. http://www.w3.org/TR/rdf-primer (2004)
20. Martínez-Prieto, M.A., Fernández, J.D., Cánovas, R.: Querying RDF dictionaries in compressed space. SIGAPP Appl. Comput. Rev. **12**(2), 64–77 (2012)
21. Navarro, G., Mäkinen, V.: Compressed full-text indexes. ACM Comp. Surv. 39(1), article 2 (2007)

22. Neumann, T., Weikum, G.: The RDF-3X engine for scalable management of RDF data. The VLDB J. **19**(1), 91–113 (2010)
23. Prud'hommeaux, E., Seaborne, A., (eds.): SPARQL query language for RDF, W3C recommendation. http://www.w3.org/TR/rdf-sparql-query (2008)
24. Sadakane, K.: New text indexing functionalities of the compressed suffix arrays. J. Algorithms **48**(2), 294–313 (2003)
25. Sakr, S., Al-Naymat, G.: Relational processing of RDF queries: A survey. SIGMOD Rec. **38**(4), 23–28 (2010)
26. Weiss, C., Karras, P., Bernstein, A.: Hexastore: Sextuple indexing for semantic web data management. Proc. VLDB **1**(1), 1008–1019 (2008)

Chaining Fragments in Sequences:
To Sweep or Not (Extended Abstract)

Julien Allali[1,2]([✉]), Cedric Chauve[2,3], and Laetitia Bourgeade[1]

[1] LaBRI, Université Bordeaux, Talence, France
{julien.allali,laetitia.bourgade}@labri.fr
[2] ENSEIRB-MATMECA, Bordeaux INP, Talence, France
[3] Department of Mathematics, Simon Fraser University, Burnaby, Canada
cedric.chauve@sfu.ca

Abstract. Computing an optimal chain of fragments is a classical problem in string algorithms, with important applications in computational biology. There exist two efficient dynamic programming algorithms solving this problem, based on different principles. In the present note, we show how it is possible to combine the principles of two of these algorithms in order to design a hybrid dynamic programming algorithm that combines the advantages of both algorithms.

1 Introduction

Sequence alignment is a fundamental task in bioinformatics, requiring very efficient algorithms, with subquadratic time complexity. One of the successful approaches is based on the technique of chaining fragments. Its principle is to first detect and score highly conserved factors, the *fragments* (also called *anchors* or *fragments*), then to compute a maximal score subset of fragments that are colinear and non-overlapping in both considered sequences, called an *optimal chain*. This optimal chain is then used as the backbone of a full alignment. Due to its applications, especially in computational biology, this problem has received a lot of attention from the algorithmic community [1,4,5,8–11]. We are interested in the problem of computing the score of an optimal chain of fragment from a given set of k fragments, for two sequences t and u of respective lengths n and m. This problem can be solved in $O(k + n \times m)$ time by using a simple dynamic programming (DP) algorithm (see [9] for example). However, in practical applications, the number k of fragments can be subquadratic, which motivated the design of algorithms whose complexity depends only of k and can run in $O(k \log k)$ worst-case time (see [5,8,10,12]). The later algorithms, known as Line Sweep (LS) algorithms, rely on geometric properties of the problem, where fragments can be seen as rectangles in the quarter plane, and geometric data structures that allow to retrieve and update efficiently (*i.e.* in logarithmic time) optimal subchains (see [12] for example). This raises the natural question of deciding which algorithm to use when comparing two sequences t and u. In particular, it can happen that the *density* of fragments differs depending on the location of the fragments in the considered

C. Iliopoulos et al. (Eds.): SPIRE 2015, LNCS 9309, pp. 116–123, 2015.
DOI: 10.1007/978-3-319-23826-5_12

sequences, due for example to the presence of repeats. In such cases, it might then be more efficient to rely on the DP algorithm in regions with high fragment density, while in regions of lower fragment density, the LS algorithm would be more efficient. This motivates the theoretical question we consider, that asks to design an efficient algorithm that relies on the classical DP principle when the density of fragments is high and switches to the LS principle when processing parts of the sequences with a low density of fragments. We show that this can be achieved, and we describe such a *hybrid* DP/LS algorithm for computing the score of an optimal chain of fragments between two sequences. Our algorithm achieves a theoretical complexity that is as good as both the DP and LS algorithm, *i.e.* that for any instance, our algorithm performs as at least as well, in terms of theoretical worst-case asymptotic time complexity, as both the DP and the LS algorithm. We refer the reader to [2] for an extended version of this abstract, including a more detailed analysis of our algorithm.

2   Preliminaries

Preliminary Definitions and Problem Statement. Let t and u be two sequences, of respective lengths n and m. We assume that positions index in sequences start at 0, so $t[0]$ is the first symbol in t and $t[n-1]$ its last symbol. By $t[i,j]$ we denote the substring of t composed of symbols in positions $i, i+1, \ldots, j$. A *fragment* is a factor which is common, possibly up to small variations, to t and u. Formally, a fragment s is defined by 5 elements $(s.\ell, s.r, s.t, s.b, s.s)$: the first four fields indicate that the corresponding substrings are $t[s.\ell, s.r]$ and $u[s.b, s.t]$, while the field $s.s$ is a *score* associated to the fragment. We call *borders* of s the coordinates $(s.\ell, s.b)$ and $(s.r, s.t)$. As usual in chaining problems, we see fragments as rectangles in the quarter plane, where the x-axis corresponds to t and the y-axis to u: $s.\ell$, $s.r$, $s.b$ and $s.t$ denote the *left* and *right* position of s over t and the *bottom* and *top* position of s over u ($s.\ell \leq s.r$ and $s.b \leq s.t$).

Let S denote a set of k fragments for t and u. A *chain* is a set of fragments $\{s_1, \ldots, s_\ell\}$ such that $s_i.r < s_{i+1}.\ell$ and $s_i.t < s_{i+1}.b$ for $i = 1, \ldots, \ell-1$; the score of a chain is the sum $\sum_{i=1}^{\ell} s_i.s$ of scores of the fragments it contains. A chain is optimal if there is no chain with a higher score. The problem we consider in the present work is to compute the score of an optimal chain, denoted by *MCS* (Maximum Chaining Score).

The Dynamic Programming (DP) and Line Sweep (LS) Algorithms. A first approach to compute the MCS is a dynamic programming (DP) algorithm where the MCS between prefixes $t[0,i]$ and $u[0,j]$ is computed as the maximum between (1) the MCS between $t[0, i-1]$ and $u[0,j]$, (2) the MCS between $t[0,i]$ and $u[0, j-1]$ and (3) the scores of the best chain that ends with a fragment s such that $s.r = i$ and $s.t = j$. Following classical DP methods, this algorithm can be implemented using a single column that is updated along t positions incrementally, and has a time complexity of $O(k + n \times m)$ as it fills a matrix of size $n \times m$ and each fragment requires a constant time treatment. We refer to [2] for the pseudocode of this algorithm.

A second algorithm relies on a geometric approach, and is known as the Line Sweep (LS) algorithm. We call a locally optimal chain (LOC) a chain C that ends with a fragment s such that, for all chains C' that ends with a fragment s' with $s'.r \leq s.r$, if $s'.t < s.t$, then the score of C' is at most the score of C. The LS algorithm considers fragments borders according to their fields $s.l$ and $s.r$ (positions along t) sorted increasingly (in case of equality, fragments are processed by increasing order of their right position); during this process, a data structure A (typically an AVL or RB-tree, see [12] for a discussion on such data structures.) stores the scores of all LOC built so far, so that, at the end of the algorithm, the MCS can be retrieved from A in constant time. During the LS algorithm, when the left position of a fragment s if reached, we compute the chain of highest score that ends on s by looking in A for the highest LOC below $s.b$. When the right position of s is reached, if its associated best chain previously computed is a LOC, its score is inserted into A and scores of LOC ending at highest position but smaller or equal score are removed from A. The time complexity of the LS algorithm is $O(k \log k)$ (see [2] and references there for a more detailed presentation of the LS algorithm). It can be reduced to $O(k \log \log k)$ if borders are sorted [12], but, for the sake of generality, we do not assume this here.

3 An Hybrid Algorithm

An instance of the chaining problem is said to be *compact*, if each position of t and each position of u contains at least one border. If an instance is not compact, then there exists a unique compact instance obtained by removing from t and from u all positions that do not contain a fragment border, leading to sequences t' and u', and updating the fragments borders according to the sequences t' and u', leading to a set $\mathcal{S}'$ of fragments. We denote by $(t', u', \mathcal{S}')$ the compact instance corresponding to $(t, u, \mathcal{S})$, and m' and n' the lengths of t' and u'. The compact instance $(t', u', \mathcal{S}')$ can be computed in time $O\left(k + \min(k \log(k), m) + \min(k \log(k), n)\right)$ and space $O(k + n + m)$, and from now, we assume that the compact instance has been computed and that it is the considered instance.

The *border density* $\mathcal{K}_p$ of a position p of t is the number of fragment borders (*i.e.* number of fragments extremities) located in $t[p]$. Our algorithm considers fragments in the same order than in the LS algorithmbut processes the fragments whose border in t' is in position i using either the DP approach if the density of fragments at $t'[i]$ is high, or the LS approach otherwise. Hence, the key requirement will be that (1) when using the DP approach, the previous column of the DP table is available, (2) when using the LS approach, a data structure with similar properties than the geometric data structure used in the LS algorithm is available.

We introduce now a data structure B that ensures that the above requirements are satisfied. The data structure B is essentially an array of m' entries augmented with a balanced binary search tree. Formally:

- We consider an array $\mathcal{B}$ of m' entries, such that $\mathcal{B}[i]$ contains chaining scores, and satisfies the following invariant: if s is the last processed fragment, for every $i = 1, \ldots, s.r$, $\mathcal{B}[i] \geq \mathcal{B}[i-1]$.
- We augment this array with a balanced binary search tree $\mathcal{C}$ whose leaves are the entries of $\mathcal{B}$ and whose internal nodes are labelled in order to satisfy the following invariant: a node x is labelled by the maximum of the labels of its right child and left child.

The data structure B will be used to answer the following queries: given $0 \leq p \leq m'$, find the optimal score of a partial chain whose last fragment s satisfies $s.t \leq p$. This principle is very similar to solutions recently proposed for handling dynamic minimum range query requests [3].

We describe now how we implement this data structure using an array. Let b be the smallest integer such that $m' \leq 2^b$. We encode B into an array of size 2^{b+1}, whose prefix of length $m' - 1$ contains the labels of the internal nodes of the binary tree $\mathcal{C}$ (so each cell contains a label and the indexes to two other cells, corresponding respectively to the left child and right child), ordered in breadth-first order, while the entries of $\mathcal{B}$ are stored in the suffix of length m' of the array (see figure 1). From now, we identify nodes of the binary tree and cells of the array, that we denote by B.

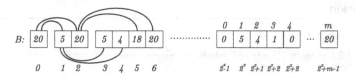

Fig. 1. Example of the implementation of the data structure B with an array.

Using this implementation, for a given node of the binary search tree, say encoded by the cell in position x in B (called node x from now), we can quickly obtain the position, in the array, of its left child, of its right child, but also of its parent (if $B[x]$ is not the root) and of its rightmost descendant, defined as the unique node reached by a maximal path of edges to right children, starting at x edges to a left (resp. right) child. Indeed, it is straightforward to verify that, the constraint of ordering the nodes of the binary tree in the array according to a breadth-first order implies that, for node x, if y is the largest integer such that $2^y \leq x + 1$ and $z = x - 2^y + 1$, then:

- if $x \geq 2^b - 1$, x is a leaf;
- $leftChild(x) = 2^{y+1} - 1 + 2 * x$ if x is not a leaf;
- $rightChild(x) = 2^{y+1} - 1 + 2 * z + 1$ if x is not a leaf;
- $parent(x) = -1$ if $x = 0$ (x is the root), and $2^{y-1} - 1 + \lfloor \frac{z}{2} \rfloor$ if $x \neq 0$;
- $rightmostChild(x) = 2^b - 1 + (z+1)2^{b-z} - 1$.

Implementing the DP and LS Algorithms with the Hybrid Data Structure. It is then easy to implement the DP algorithm using the data structure B, by using

$\mathcal{B}$ as the current column of the DP table (*i.e.* if the currently processed position of t' is i, $\mathcal{B}[j]$ is the score of the best partial chain included in the rectangle defined by $(0,0)$ and (i,j)), without updating the internal nodes of the binary search tree $\mathcal{C}$. To implement the LS algorithm, the key points are (1) to be able to update efficiently the data structure B, when a fragment s has been processed and (2) to be able to find the best score of a partial chain ending up at a position in u' strictly below p. Updating B can be done through the function *setScore* below, with parameters $p = s.t$ and $score = S[s]$, while the second task can be achieved by the function *getBestScore* described below.

Algorithm 1. Set a chaining score for a position p.

---

1 *setScore*$(B, p, score)$:
2      $index = 2^b - 1 + p$ // *start from leaf corresponding to p*
3   **while** $index! = -1$ && $B[index] < score$
4          $B[index] = score$
5          $index = parent(index)$

Algorithm 2. Retrieve the best chaining score for partial chains ending strictly below position p.

---

1 *getBestScore*(B, p) :
2   let b be the smallest integer s.t. $m' \leq 2^b$
3      $maxScore = 0$
4      $currentNode = 0$ // *the root node*
5      $indexOfP = 2^b - 1 + p$
6   **while** $rightmostChild(currentNode) > indexOfP$
7          $left = leftChild(currentNode)$
8   **if** $rightmostChild(left) >= indexOfP$ // *move left*
9              $ncurrentNode = left$
10   **else** // *move right*
11             $maxScore = max(maxScore, B[left])$
12             $currentNode = rightChild(currentNode)$
13   **return** $max(maxScore, B[currentNode])$

If all updates of B are done using the function *setScore*, then the two required invariants on B are satisfied. The time complexity of both *setScore* and *getBestScore* is in $O(\log(m'))$, due to the fact that the binary tree is balanced. We can then implement the LS algorithm on compact instances using the data structure B, with worst-case time complexity $O(k \log m')$.

LS/DP Update with the Hybrid Data Structure. In an hybrid algorithm that relies on the data structure B, when the algorithm switches approaches (from DP to LS, or LS to DP), the data structure B is assumed to be consistent for the current approach, and needs to be updated to become consistent for the next approach. So when switching from DP (say position $i - 1$, $i = 1, \ldots n'$) to LS (position i), we assume that $\mathcal{B}[j]$ ($j = 0, \ldots, m' - 1$) is the optimal score of a partial chain in the rectangle defined by $(0, 0)$ and $(i - 1, j)$, and we want to update B in such a way that the label of any internal node x of the binary tree is the maximum of both its children. As $\mathcal{B}$ are the leaves of the binary tree, this update can be done during a post-order traversal of the binary tree, so in time $O(m')$. When switching from LS to DP (say to use the DP approach on position i while the LS approach was used on position $i - 1$), we assume that for every leaf $\mathcal{B}[j]$ of the binary tree corresponding to a position at most $i - 1$, the value in $\mathcal{B}[j]$ is the optimal score of a partial chain whose last fragment ends in position $i - 1$; this follows immediately from the way labels of the leaves of the binary tree are inserted by the $setScore$ function. To update B, we want that in fact $\mathcal{B}[j]$ is the optimal score of a partial chain in the whose last fragment ends in position at most $i - 1$. So the update function needs only to give to $\mathcal{B}[j]$ the value $\max_{0 \leq j' \leq j} \mathcal{B}[j']$, which can again be done in time $O(m')$. It follows that updating the data structure B from DP to LS or LS to DP can be done in time $O(m')$. We denote by $update$ the function performing this update.

Deciding Between LS and DP Using the Fragment Density. Before we can finally introduce our algorithm, we need to address the key point of how to decide which paradigm (DP or LS) to use when processing the fragments having a border in the current position of t, say c. Let $\mathcal{K}_c$ be the number of fragments s such that $s.\ell = c$ or $s.r = c$. Using the DP approach, the cost of updating $\mathcal{B}$ (*i.e.* to compute the column c of the DP table) is $O(m' + \mathcal{K}_c)$. With the LS approach, the cost of updating B is in $O(\mathcal{K}_c \log m')$. So, if $\mathcal{K}_c > \frac{m'}{\log m' - 1}$, the asymptotic cost of the DP approach is better than the asymptotic cost of the LS approach, while it is the converse if $\mathcal{K}_c \leq \frac{m'}{\log m' - 1}$. So, prior to processing fragments, for each position i in t ($i = 0, \ldots, m' - 1$), we record in an array C if fragments borders in position i are processed using the DP approach ($C[i]$ contains DP) or the LS approach ($C[i]$ contains LS). This last observation leads to our main result, Algorithm 3 below.

Time and Space Complexity. In terms of space complexity, the algorithm, we avoid to use $O(k + n' \times m')$ space for storing the fragments borders in $n' \times m'$ lists (structure L of the DP algorithm) by using two lists: $L1[i]$ stores all fragments borders in position i of t', while $L2[j]$ stores all fragments borders in position i of t' and j of u', and is computed from $L[1]$. So the total space requirement is in $O(k + m' + n')$.

We now establish the time complexity of this algorithm. If the current position i of t is tagged as DP, the cost for updating the column is $O(m' + \mathcal{K}_i)$, including the cost of setting up $L2$ from $L1$, that is proportional to the number of fragments borders in the current position (line 14–24). If $C[i]$ is LS, the cost for computing chains scores on this position is $O(\mathcal{K}_i \log m')$ (line 25– 28). Thus, if we call P^1 the set of positions on t where we use the DP approach, P^2 the set

---

Algorithm 3. A hybrid algorithm for the fragment chaining problem.

---

 1 compute the compact instance $(t', u', \mathcal{S}')$
 2 $L1$: an array of $n' \times 2$ linked lists
 3 C: an binary array of size n'
 4 **foreach** s **in** $\mathcal{S}'$ **do**
 5   **if** $C[s.r]$ is DP **then** front insert $(s, end, s.t)$ into $L1[s.r][1]$
 6   **else** front insert (s, end) into $L1[s.r][0]$
 7   **if** $C[s.\ell]$ is DP **then** front insert $(s, begin, s.b)$ into $L1[s.\ell][1]$
 8   **else** front insert $(s, begin)$ into $L1[s.\ell][0]$
 9 $\mathcal{B}$: a binary tree **for** m' leafs (all nodes are set to zero)
10 B: refers to the m' leaves of $\mathcal{B}$
11 S: an array of integer of size k
12 **for** i from 0 to n' **do**
13   **if** $C[i] \neq C[i-1]$ **then** $update(B)$
14   **if** $C[i]$ is DP
15          $L2$: an array of m' linked lists
16   **for** each (s, t, j) in $L1[i][1]$ **do** front insert (s, t) into $L2[j]$
17          $left = 0, leftDown = 0$
18   **for** j from 0 to m' **do**
19              $maxC = 0$
20   **foreach** $(s, type)$ in $L2[j]$ **do**
21   **if** $type$ is $begin$ **then** $S[s] = s.s + leftDown$
22   **if** $type$ is end and $S[s] > maxC$ **then** $maxC = S[s]$
23              $leftDown = left, left = \mathcal{B}[j]$
24              $\mathcal{B}[j] = max(\mathcal{B}[j], \mathcal{B}[j-1], maxC)$
25   **else** // $C[i]$ is LS
26   **foreach** $(s, type)$ in $L1[i][0]$ **do**
27   **if** $type$ is $begin$ **then** $S[s] = s.s + getBestScore(B, s.b)$
28   **if** $type$ is end **then** $setScore(B, s.t, S[s])$
29 **if** $C[n'-1]$ is $direct$ **then return** $\mathcal{B}[m'-1]$
30 **else return** value of the root of B

---

of positions on t where we use the LS approach and $P = P^1 \cup P^2$, the time for the whole loop at line 12 is

$$O\left(\sum_{p \in P^1} (m' + \mathcal{K}_p) + \sum_{p \in P^2} \mathcal{K}_p \log m'\right)$$

We have $|P^1| + |P^2| = n'$, $\forall p \in P^1 : \mathcal{K}_p > \frac{m'}{\log m' - 1}$ and $\forall p \in P^2 : \mathcal{K}_p \leq \frac{m'}{\log m' - 1}$. Moreover, updating the data structure B from LS to DP or DP to LS (line 13) is done at most one more time then the size of P^1, so the total cost of this operation is $O\left(\sum_{p \in P^1} m'\right)$, and can thus be integrated, asymptotically, to the cost of processing the positions in P^1. This results in a worst-case time complexity

$$O\left(k + \min(k \log k, m) + \min(k \log k, n) + \sum_{p \in P^1} (m' + \mathcal{K}_p) + \log m' \sum_{p \in P^2} \mathcal{K}_p \right).$$

References

1. Abouelhoda, M.I., Ohlebusch, E.: Chaining algorithms for multiple genome comparison. J. Discrete Algorithms **3**, 321–341 (2005)
2. Allali, J., Bourgeade, L., Chauve, C.: Chaining fragments in sequences: to sweep or not. CoRR abs/1506.07458 (2015)
3. Arge, L., Fischer, J., Sanders, P., Sitchinava, N.: On (Dynamic) Range Minimum Queries in External Memory. In: Dehne, F., Solis-Oba, R., Sack, J.-R. (eds.) WADS 2013. LNCS, vol. 8037, pp. 37–48. Springer, Heidelberg (2013)
4. Eppstein, D., Galil, Z., Giancarlo, R., Italiano, G.F.: Sparse dynamic programming. I: linear cost functions; II: convex and concave cost functions. J. Assoc. Comput. Mach. **39**, 519–567 (1992)
5. Felsner, S., Müller, R., Wernisch, L.: Trapezoid graphs and generalizations, geometry and algorithms. Discrete Appl. Math. **74**, 13–32 (1997)
6. Gusfield, D.: Algorithms on Strings, Trees and Sequences. Cambridge University Press (1997)
7. Höhl, M., Kurtz, S., Ohlebusch, E.: Efficient multiple genome alignment. Bioinformatics **18**, S312–S320 (2002)
8. Joseph, D., Meidanis, J., Tiwari, P.: Determining DNA sequence similarity using maximum independent set algorithms for interval graphs. In: Nurmi, O., Ukkonen, E. (eds.) SWAT 1992. LNCS, vol. 621, pp. 326–337. Springer, Heidelberg (1992)
9. Morgenstern, B.: A simple and space-efficient fragment-chaining algorithm for alignment of DNA and protein sequences. Appl. Math. Lett. **15**, 11–16 (2002)
10. Myers, G., Miller, W.: Chaining multiple-alignmment fragments in sub-quadratic time. SODA **1995**, 38–47 (1995)
11. Myers, G., Huang, X.: An $O(N^2 \log N)$ restriction map comparison and search algorithm. Bull. Math. Biol. **54**, 599–618 (1992)
12. Ohlebusch, E., Abouelhoda, M.I.: Chaining Algorithms and Applications in Comparative Genomics. In: Aluru, S. (ed.) Handbook of Computational Molecular Biology. CRC Press (2005)

A Faster Algorithm for Computing Maximal α-gapped Repeats in a String

Yuka Tanimura[1], Yuta Fujishige[1], Tomohiro I[2], Shunsuke Inenaga[1]([⊠]),
Hideo Bannai[1], and Masayuki Takeda[1]

[1] Department of Informatics, Kyushu University, Fukuoka, Japan
{yuka.tanimura,yuta.fujishige,inenaga,bannai,takeda}@inf.kyushu-u.ac.jp
[2] Department of Computer Science, TU Dortmund, Dortmund, Germany
tomohiro.i@cs.tu-dortmund.de

Abstract. A string $x = uvu$ with both u, v being non-empty is called
a *gapped repeat with period* $p = |uv|$, and is denoted by pair (x, p). If
$p \leq \alpha(|x| - p)$ with $\alpha > 1$, then (x, p) is called an *α-gapped* repeat. An
occurrence $[i, i+|x|-1]$ of an α-gapped repeat (x, p) in a string w is called
a *maximal* α-gapped repeat of w, if it cannot be extended either to the
left or to the right in w with the same period p. Kolpakov et al. (CPM
2014) showed that, given a string of length n over a constant alphabet,
all the occurrences of maximal α-gapped repeats in the string can be
computed in $O(\alpha^2 n + occ)$ time, where occ is the number of occurrences.
In this paper, we propose a faster $O(\alpha n + occ)$-time algorithm to solve
this problem, improving the result of Kolpakov et al. by a factor of α.

1 Introduction

Finding repetitive substrings in a string has been a central task in stringology,
with various applications e.g., in bioinformatics [8]. A simplest form of a repetitive string is a *tandem repeat* (a.k.a. *square*): A string x is called a tandem
repeat if $x = uu$ with some non-empty string u. Gusfield and Stoye [9] proposed
an $O(n + t)$ time algorithm to compute all t occurrences of tandem repeats in a
given string of length n over a constant alphabet.

A natural generalization of tandem repeats is to allow some *gap* between the
two repeats. A string x is called a *gapped* repeat if $x = uvu$ with some non-empty
strings u and v. The leftmost and rightmost occurrences of u in x are called the
right and left copies of the gapped repeat, while v is called the gap. The *period*
of a gapped repeat uvu is the length $|uv|$ of uv. Notice that a string w can
correspond to gapped repeats with different periods, for instance, ababab is a
gapped repeat with period 4 taking $u = $ aba and $v = $ b, while it is also a gapped
repeat with period 6 taking $u = $ a and $v = $ babab. To distinguish them, a
gapped repeat x with period p will be denoted by a pair (x, p). An occurrence of
a gapped repeat (x, p) in a string w is called *maximal*, if it cannot be extended
either to the left or to the right preserving the period p. Gusfield [8] showed how
to compute all y occurrences of maximal gapped repeats with unbounded gap
length in a string of length n in $O(n + y)$ time. Since there is no restriction on

© Springer International Publishing Switzerland 2015
C. Iliopoulos et al. (Eds.): SPIRE 2015, LNCS 9309, pp. 124–136, 2015.
DOI: 10.1007/978-3-319-23826-5_13

the length of the gaps, his algorithm may output gapped repeats of which the left and the right copies are far apart in the string. Several attempts to bound the gap length have been introduced in literature. Kolpakov and Kucherov [11] described how to compute all f occurrences of maximal gapped repeats with fixed gap length d in $O(n \log d + f)$ time. Brodal et al. [4] proposed an $O(n \log n + r)$-time algorithm to compute all r occurrences of gapped repeats with gap length in a fixed range.

We study the following problem: Given a string w of length n over a constant alphabet and a real $\alpha > 1$, compute all maximal gapped repeats (x, p) in w which satisfy $p \leq \alpha(|x| - p)$. Namely, we would like to compute all occurrences of maximal gapped repeats of which the period is bounded by the length of the copies multiplied by α. These repeats are called *maximal α-gapped repeats* of w. Kolpakov et al. [10] presented an algorithm to compute all occ occurrences of maximal α-gapped repeats in w in $O(\alpha^2 n + occ)$ time and $O(n + occ)$ space. They also showed that $occ = O(\alpha^2 n)$, so the occ term in the above time complexity can be omitted (It is open whether the $O(\alpha^2 n)$ bound for occ is tight or not.).

In this paper, we present a faster $O(\alpha n + occ)$-time algorithm for finding all occ occurrences of maximal α-gapped repeats, improving the result of Kolpakov et al. by a factor of α. The space complexity of our algorithm is $O(n + occ)$. Our algorithm uses ideas similar to the algorithm of Badkobeh et al. [1], which computes gapped repeats of maximum exponents from a given overlap-free string. In more detail, we first compute a variant of the Lempel-Ziv 77 factorization (LZ77) [15] of the input string w, and construct the directed acyclic word graph (DAWG) [3] of a specific substring induced from each LZ77 factor. To achieve the $O(\alpha n + occ)$ bound, our algorithm utilizes combinatorial properties of DAWGs, together with non-trivial uses of several other data structures.

2   Preliminaries

Strings. Let Σ be an alphabet of size σ. An element of Σ^* is called a *string*. The length of a string w is denoted by $|w|$. The empty string ε is a string of length 0. For a string $w = xyz$, x, y and z are called a *prefix*, *substring*, and *suffix* of w, respectively. The set of substrings of w is denoted by $Substr(w)$. The i-th character of a string w is denoted by $w[i]$, for any $1 \leq i \leq |w|$. For a string w and two integers $1 \leq i \leq j \leq |w|$, let $w[i..j]$ denote the substring of w that begins at position i and ends at position j. For convenience, let $w[i..j] = \varepsilon$ when $i > j$. Let w^R denote the reversed string of w, namely, $w^R = w[|w|] \cdots w[1]$.

A string $x = uvu$ with non-empty strings u, v is called a *gapped repeat with period* $p = |uv|$, and is denoted by pair (x, p). The leftmost and rightmost occurrences of u in x are respectively called the left and right copies of u of the gapped repeat, and v is called the gap of the gapped repeat. If $p \leq \alpha(|x| - p)$ with a real $\alpha > 1$, then (x, p) is called an *α-gapped* repeat. An occurrence $[k..k + |x| - 1]$ of an α-gapped repeat (x, p) in string w is denoted by triplet $(k, k + |x| - 1, p)$. An α-gapped repeat $(k, k + |x| - 1, p)$ is called a *maximal α-gapped repeat* of w, if it can be extended neither to the left nor to the right in w with the same period

p. More formally, an α-gapped repeat $(k, k + |x| - 1, p)$ is a maximal α-gapped repeat if (1) $w[k-1] \neq w[k+p-1]$ or $k = 1$, and (2) $w[k+|x|-p] \neq w[k+|x|]$ or $k + |x| - 1 = |w|$. For any string w, let $MGR_\alpha(w)$ denote the set of all maximal α-gapped repeats in w.

In this paper, we deal with the following problem.

Problem 1. Given a string w of length n and a real $\alpha > 1$, compute the set $MGR_\alpha(w)$ of all maximal α-gapped repeats in w.

Tools. Here we list a number of data structures and algorithms which will be used as components of our efficient algorithm to solve Problem 1.

The *Lempel-Ziv 77 factorization without self-references* [15] of a string w, denoted $LZ(w)$, is a factorization $f_1, \ldots, f_m$ of w s.t. $w = f_1 \cdots f_m$, $f_1 = w[1]$, and for each $2 \leq i \leq m$, f_i is the longest prefix of $w[|f_1 \cdots f_{i-1}|+1..|w|]$ such that $f_i \in Substr(w[1..|f_1 \cdots f_{i-1}|]) \cup \Sigma$. Given a string w of length n, we can compute $LZ(w)$ in $O(n \log \sigma)$ time with $O(n)$ space, e.g., by the algorithm of [13].

The *directed acyclic word graph* (*DAWG*) [3] of a string z, denoted $DAWG(z)$, is the smallest partial DFA that accepts all suffixes of z. A more formal presentation of $DAWG(z)$ follows: For strings z, x, let $Endpos_z(x) = \{j \mid x = z[i..j], 1 \leq i \leq j \leq n\}$. For strings x, y, denote $x \equiv_z y$ iff $Endpos_z(x) = Endpos_z(y)$. We denote the equivalence class of string x w.r.t. $\equiv_z$ by $[x]_z$. When clear from the context, we abbreviate the above notations as $Endpos$, $\equiv$ and $[x]$, respectively. Note that for any two elements in $[x]_z$ with $x \in Substr(z)$, one is a suffix of the other. For $x \in Substr(z)$, we denote by $long([x]_z)$ the longest member of $[x]_z$. Then, the set V_z of nodes and the set E_z of edges of $DAWG(z)$ are defined by

$$V_z = \{[x] \mid x \in Substr(z)\} \text{ and}$$
$$E_z = \{([x], a, [xa]) \mid x, xa \in Substr(z), a \in \Sigma, x \not\equiv xa\}.$$

Define the set S_z of labeled reversed edges on V_z, called *suffix links*, by

$$S_z = \{([bx], b, [x]) \mid x, bx \in Substr(z), b \in \Sigma, x = long([x])\}.$$

For a suffix link $([bx], b, [x]) \in S_z$, we write $slink([bx]) = [x]$. Note that for any node of $DAWG(z)$ except for the source, it has exactly one out-going suffix link, while it can have at most σ in-coming suffix links.

Let \$ be a special character which appears only at the beginning of a string. For convenience, we consider the 0-based index for the positions of any string $\$z$ starting with \$. Namely, letting $z' = \$z$, $z'[0] = \$$ and $z'[i] = z[i]$ for $1 \leq i \leq |z|$. On this assumption, $Endpos_z(x) = Endpos_{\$z}(x)$ holds for any $x \in Substr(z)$. Now consider $DAWG(\$z) = (V_{\$z}, E_{\$z})$. Then, it is known that the suffix link tree $(V_{\$z}, S_{\$z})$ is isomorphic to the edge-reversed *suffix tree* [14] of the *reversed* string $z^R\$$. More precisely, in our definition of $S_{\$z}$, each edge label is truncated to the first character of the corresponding edge label of the edge-reversed suffix tree, as this is enough for our purposes. We denote this tree $(V_{\$z}, S_{\$z})$ by $STree(z^R\$)$. For $1 \leq i \leq |z|$, a leaf of $STree(z^R\$)$ stores integer i iff it represents $z[1..i]^R\$$.

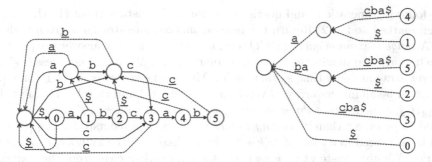

Fig. 1. (Left) $DAWG(\$z)$ with $z =$ **abcab** and its suffix links. The solid forward arcs represent the edges of $DAWG(\$z)$, while the broken reversed arcs represent the suffix links. A node numbered i corresponds to the prefix of z ending at position i. (Right) $STree(z^R\$)$ with $z^R\$ =$ **bacba$**. Note that $STree(z^R\$)$ is isomorphic to the suffix link tree induced from $DAWG(\$z)$. Each suffix link of $DAWG(\$z)$ is labeled with the first character of the label of the corresponding suffix tree edge, which is underlined.

The leaf representing $\$$ stores 0. This duality of these data structures is essential to the efficiency of our algorithm to be presented. See Fig. 1 for examples of $DAWG(\$z)$, the suffix links $S_{\$z}$, and $STree(z^R\$)$.

Blumer et al. [3] showed that for any string z, the number of nodes, edges, and suffix links of $DAWG(\$z)$ are $O(|z|)$, independent of the alphabet size σ. They also showed that $DAWG(\$z)$ together with $STree(z^R\$)$ can be constructed in $O(|z| \log \sigma)$ time and $O(|z|)$ space.

For any node $[x]$ of $DAWG(\$z)$, let $pos([x])$ denote the ending position of the rightmost occurrence of x in z, i.e., $pos([x]) = \max Endpos(x)$. We can compute $pos([x])$ for all nodes $[x]$ in $O(|z|)$ time, by a standard traversal on $STree(z^R\$)$.

Lemma 1 ([6,5]). *For a string z, assume we have $DAWG(\$z)$ together with the suffix links. Given a query string s, then for each $1 \le j \le |s|$, we can compute, in amortized $O(\log \sigma)$ time, the longest suffix of $s[1..j]$ which belongs to $Substr(z)$ and its rightmost occurrence in z.*

Let A be the sequence of leaves of $STree(z^R\$)$ obtained by a standard depth-first traversal of $STree(z^R\$)$. If the out-going edges of each node of $STree(z^R\$)$ are sorted in the lexicographical order of their labels, then A coincides with the *suffix array* [12] of $z^R\$$, denoted by $SA_{z^R\$}$. For any node y of $STree(z^R\$)$, let $\ell(y)$ and $r(y)$ be the position of the leftmost and rightmost leaves of the subtree rooted at y in the sequence A. In other words, $[\ell(y), r(y)]$ is the range on $SA_{z^R\$}$ such that for any $\ell(y) \le i \le r(y)$, $long(y)^R$ is a prefix of $z[1..SA_{z^R\$}[i]]^R\$$.

Consider a rooted tree T where each node is either marked or unmarked. A *nearest marked ancestor (NMA)* query on T is to answer the nearest marked ancestor of a given node. Gabow and Tarjan [7] showed that for a given static rooted tree T where all nodes are initially marked, there exists a data structure for NMA queries on T which requires space linear in the size of T, and supports the following operations and queries in amortized constant time: (1) Unmark a given marked node. (2) Return the nearest marked ancestor for a given node.

A *range maximum query* (*RMQ*) on an integer array is to answer the position which stores the maximum value in a query range. We can preprocess a given integer array in linear time so that each RMQ can be answered in $O(1)$ time [2].

Let A be an integer array. Given a range $[i, j]$ on A and threshold value τ, let $\mathrm{rtq}_A(i, j, \tau)$ be a *range threshold query* to return all values in $A[i..j]$ which are equal to or larger than τ, *in any order*. Using RMQs recursively, $\mathrm{rtq}_A(i, j, \tau)$ can be answered in a total of $O(k+1)$ time, where k is the number of values to output. We abbreviate $\mathrm{rtq}_A(i, j, \tau)$ by $\mathrm{rtq}(i, j, \tau)$ when clear from the context.

3 $O(\alpha^2 n + occ)$-time Algorithm by Kolpakov et al.

In this section, we briefly describe the framework of the algorithm of Kolpakov et al. [10] which solves Problem 1 of computing $MGR_\alpha(w)$ in $O(\alpha^2 n + occ)$ time.

For a string w, let $LZ(w) = f_1, \ldots, f_m$. For each $1 \le i \le m$, let $FMGR_\alpha(i)$ be the set of maximal α-gapped repeats in w which starts in $f_1 \cdots f_{i-1}$ and ends in f_i, i.e., those that cross the boundary between f_{i-1} and f_i. Also, let $SMGR_\alpha(i)$ be the set of maximal α-gapped repeats in w which are completely contained in f_i. Let $b_1 = 1$, and for each $2 \le i \le m$ let $b_i = |f_1 \cdots f_{i-1}| + 1$, i.e., b_i is the beginning position of f_i in w. More formally, we have

$$FMGR_\alpha(i) = \{(k, j, p) \in MGR_\alpha(w) \mid k < b_i \le j < b_{i+1}\},$$
$$SMGR_\alpha(i) = \{(k, j, p) \in MGR_\alpha(w) \mid b_i \le k < j < b_{i+1}\}.$$

It is clear that $FMGR_\alpha(i)$ and $SMGR_\alpha(i)$ are disjoint sets, and

$$MGR_\alpha(w) = \bigcup_{1 \le i \le m} (FMGR_\alpha(i) \cup SMGR_\alpha(i)).$$

$FMGR_\alpha(i)$ can further be divided into the following disjoint sets:

$$FMGR_\alpha^{\mathrm{rgt}}(i) = \{(k, j, p) \in FMGR_\alpha(i) \mid k + p - 1 < b_i\},$$
$$FMGR_\alpha^{\mathrm{mid}}(i) = \{(k, j, p) \in FMGR_\alpha(i) \mid j - p + 1 < b_i \le k + p - 1\},$$
$$FMGR_\alpha^{\mathrm{lft}}(i) = \{(k, j, p) \in FMGR_\alpha(i) \mid b_i \le j - p + 1\}.$$

Note that $j - p + 1$ and $k + p - 1$ are the beginning and ending positions of the gap of the maximal α-gapped repeat (k, j, p) in w, respectively. Hence, the right copy $w[k + p..j]$ of each element of $FMGR_\alpha^{\mathrm{rgt}}(i)$ covers or touches the boundary between f_{i-1} and f_i, the gap $w[j - p + 1..k + p - 1]$ of each element of $FMGR_\alpha^{\mathrm{mid}}(i)$ completely covers the boundary between f_{i-1} and f_i, and the left copy $w[1..j - p]$ of each element of $FMGR_\alpha^{\mathrm{lft}}(i)$ covers or touches the boundary between f_{i-1} and f_i. See also Fig. 2 for these notations. Let $FMGR_\alpha^{\mathrm{rgt}}(w) = \bigcup_{1 \le i \le m} FMGR_\alpha^{\mathrm{rgt}}(i)$, $FMGR_\alpha^{\mathrm{mid}}(w) = \bigcup_{1 \le i \le m} FMGR_\alpha^{\mathrm{mid}}(i)$, and $FMGR_\alpha^{\mathrm{lft}}(w) = \bigcup_{1 \le i \le m} FMGR_\alpha^{\mathrm{lft}}(i)$.

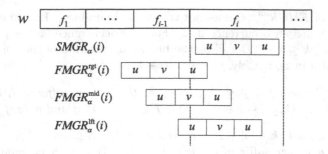

Fig. 2. Illustration for $SMGR_\alpha(i)$, $FMGR_\alpha^{\mathrm{rgt}}(i)$, $FMGR_\alpha^{\mathrm{mid}}(i)$, and $FMGR_\alpha^{\mathrm{lft}}(i)$.

Lemma 2 (Kolpakov et al. [10]). *Given a string w of length n over a constant alphabet and a real $\alpha > 1$, we can compute $FMGR_\alpha^{\mathrm{rgt}}(w)$ in $O(n)$ time with $O(n)$ space, $FMGR_\alpha^{\mathrm{mid}}(w)$ in $O(\alpha^2 n + occ_{\mathrm{mid}})$ time with $O(n + occ_{\mathrm{mid}})$ space, $FMGR_\alpha^{\mathrm{lft}}(w)$ in $O(\alpha n)$ time with $O(n)$ space, and $SMGR_\alpha(w)$ in $O(n + occ_{\mathrm{s}})$ time with $O(n + occ_{\mathrm{s}})$ space, where $occ_{\mathrm{mid}} = |FMGR_\alpha^{\mathrm{mid}}(w)|$ and $occ_{\mathrm{s}} = |SMGR_\alpha(w)|$.*

As can be seen in Lemma 2, the bottleneck of the algorithm of Kolpakov et al. [10] is their $O(\alpha^2 n + occ_{\mathrm{mid}})$-time solution for computing $FMGR_\alpha^{\mathrm{mid}}(w)$. In the next section, we propose a more efficient algorithm to compute $FMGR_\alpha^{\mathrm{mid}}(w)$, which runs in $O(\alpha n + occ_{\mathrm{mid}})$ time for a string over a constant alphabet. Our approach is significantly different from the one given by Kolpakov et al. [10].

4 Computing $FMGR_\alpha^{\mathrm{mid}}(w)$ in $O(\alpha n + occ_{\mathrm{mid}})$ Time

In this section, we present how to compute $FMGR_\alpha^{\mathrm{mid}}(w)$ in $O(\alpha n + occ_{\mathrm{mid}})$ time and $O(n + occ_{\mathrm{mid}})$ space when σ is constant, where $occ_{\mathrm{mid}} = |FMGR_\alpha^{\mathrm{mid}}(w)|$.

Assume we have computed $LZ(w) = f_1, \ldots, f_m$. Our algorithm consists of m steps, and we compute $FMGR_\alpha^{\mathrm{mid}}(i)$ in increasing order of i. In so doing, we use the following lemma, which is immediate by the definition of $FMGR_\alpha^{\mathrm{mid}}(i)$.

Lemma 3. *For any $(k, j, p) \in FMGR_\alpha^{\mathrm{mid}}(i)$, $b_i - \alpha(|f_i| - 1) < k < b_i$.*

For each position $b_i \le j \le b_{i+1} - 1$ in w, let j' be the relative position for j in f_i, namely, $j' = j - b_i + 1$. The basic strategy of our algorithm is to compute, for each position j' in f_i, the maximal α-gapped repeats ending at position j' in f_i. In so doing, we first compute the right copies of the maximal α-gapped repeats ending at position j' in f_i, and the left copies of the gapped repeats to the left of the boundary between f_{i-1} and f_i. We store the maximal α-gapped repeats of w in a table L of length n, such that $L[k]$ stores a list of the maximal α-gapped repeats starting at position k in w, sorted in increasing order of lengths.

By Lemma 3, at each step i of the algorithm it suffices to search substring $w[\max\{1, b_i - \alpha(|f_i| - 1)\}..b_i - 1]$ for the left copies. Let $z_i = w[\max\{1, b_i - \alpha(|f_i| - 1)\}..b_i - 1]$, and d_i the beginning position of z_i in w. We process $j' = 2, \ldots, |f_i|$ in increasing order (we do not need to consider $j' = 1$ since the right copy of any element of $FMGR_\alpha^{mid}(i)$ does not touch the boundary). Let u be a suffix of $f_i[2..j']$ which has an occurrence in z_i. For any occurrence $[k, k + |u| - 1]$ of u in w with $d_i \leq k \leq b_i - 1$, let k' be the beginning position of the corresponding occurrence of u in z_i, namely, $k' = k - d_i + 1$.

Lemma 4. *For any* $2 \leq j' \leq |f_i|$, *let* u_0 *be the longest suffix of* $f_i[1..j']$ *which also occurs in* z_i. $(k, j, j - k - |u| + 1) \in FMGR_\alpha^{mid}(i)$ *if and only if*

(1) (a) $u = u_0$, $w[k..k + |u| - 1] = u$, *and* $u_0 \neq f_i[1..j']$, *or*
    (b) u *is a proper suffix of* u_0, $w[k..k + |u| - 1] = u$, $u = long([u]_{z_i})$, *and*
      $w[k - 1] \neq w[j - |u|]$,
(2) $w[k + |u|] \neq w[j + 1]$, *and*
(3) $|z_i| - k' + j' - |u| + 1 \leq \alpha|u|$.

Proof. ($\Rightarrow$) First, we show Properties (1)-(a) and (1)-(b). Let $e = w[k - 1]$ if $k > 1$, and for convenience let $e = \#$ if $k = 1$, where $\#$ is a special character which does not appear in w. Also, let $a = w[j - |u|]$. Since $(k, j, j - k - |u| + 1) \in FMGR_\alpha^{mid}(i)$, $e \neq a$. If no occurrences of u in z_i are immediately preceded by a, then u is the longest suffix u_0 of $f_i[1..j']$ which also occurs in z_i. Also, by the definition of $FMGR_\alpha^{mid}(i)$, $u_0 \neq f_i[1..j']$. Hence Property (1)-(a) holds in this case. If some occurrence of u in z_i is immediately preceded by a, then u is a proper suffix of u_0. It holds that $u = long([u]_{z_i})$, since otherwise u cannot be the left/right copies of a maximal gapped repeat. Since $(k, j, j - k - |u| + 1)$ is a maximal gapped repeat, $w[k - 1] \neq w[j - |u|] = a$. Hence Property (1)-(b) holds in this case. Property (2) is due to the right-maximality of maximal gapped repeats, and Property (3) is due to the constraints to α-gapped repeats.

($\Leftarrow$) By definition, $k' + |u| - 1 \subset Endpos_{z_i}(u)$. Due to Property (1), $w[k - 1] = z_i[k' - 1] \neq f_i[j' - |u|] = w[j - |u|]$. In addition, due to Property (2) we have $w[k + |u|] \neq w[j + 1]$, and thus $(k, j, j - k - |u| + 1)$ is a maximal gapped repeat. By Property (3), $|z_i| - k' + j' - |u| + 1 = j - k - |u| + 1 \leq \alpha|u|$. Hence, $(k, j, j - k - |u| + 1) \in FMGR_\alpha^{mid}(i)$. $\square$

By Property (3) of Lemma 4, given a candidate u for the right copy of gapped repeats ending at position j' in f_i, we can limit the beginning position k' of the left copies in z_i by $k' \geq j' + |z_i| - (\alpha + 1)|u|$. In what follows, we denote this threshold $j' + |z_i| - (\alpha + 1)|u|$ by $\tau(j', u)$.

At each step i, we construct $DAWG(\$z_i)$ together with the suffix links $S_{\$z_i}$ (or equivalently $STree(z_i{}^R\$))$, where $\$$ is a special marker which does not appear in z_i. We also build $SA_{z_i{}^R\$}$ enhanced with the RMQ data structure.

Sections 4.1 and 4.2 show how to compute maximal α-gapped repeats which begin at position $k' > 1$ in z_i (or $k > d_i$ in w). Section 4.3, shows how to find maximal α-gapped repeats beginning at position $k' = 1$ in z_i (or $k = d_i$ in w).

4.1 When Copies are the Longest Suffix u_0 w.r.t. j'

By Property (1)-(a) of Lemma 4, we can start with the longest suffix u_0 of $f_i[1..j']$ which is also a substring of z_i. It follows from Lemma 1 that, for each

$j' = 2, \ldots, |f_i|$ in increasing order, we can compute u_0 in amortized $O(\log \sigma)$ time using $DAWG(\$z_i)$. The algorithm of Lemma 1 also gives us the locus of the node $[u_0]$ in $DAWG(\$z_i)$. If $u_0 = f_i[1..j']$, we proceed to proper suffixes of u_0 (see Section 4.2). In the rest of this subsection, we assume that $u_0 \neq f_i[1..j']$.

If $pos([u_0]) - |u_0| + 1 < \tau(j', u_0)$, then we immediately know that there are no maximal α-gapped repeats which end at position j' in f_i and have u_0 as the left and right copies. Otherwise (if $pos([u_0]) - |u_0| + 1 \geq \tau(j', u_0)$), let k ($> d_i$) be any beginning position of u_0 in z_i with $k \geq \tau(j', u_0)$. By Property (2) of Lemma 4, $(k, j, j - k - |u_0| + 1) \in FMGR_\alpha^{\mathrm{mid}}(i)$ only if node $[u_0]$ has an outgoing edge labeled with some character $c \neq b = w[j + 1]$. For each $c \in \Sigma \setminus \{b\}$, we go to the child $[u_0 c]$ of $[u_0]$ with out-going edge labeled by c, if it exists. We then perform range threshold query $\mathrm{rtq}(\ell([u_0 c]), r([u_0 c]), \tau(j', u_0) + |u_0|)$ on $SA_{z^R\$}$. Now, any answer h returned by the range threshold query is the ending position of the left copy u_0 of a maximal α-gapped repeat ending at position j' in f_i. Namely, letting $k = h - |u_0| + 1$, we have $(k, j, j - k - |u_0| + 1) \in FMGR_\alpha^{\mathrm{mid}}(i)$.

Overall, it takes $O(\sigma + s)$ total time to compute the maximal α-gapped repeats which have u_0 as their left and right copies and end at position j' in f_i, where s is the number of the such maximal α-gapped repeats.

4.2   When u is a Proper Suffix of the Longest Suffix u_0 w.r.t. j'

Basic Algorithm. Due to Property (1)-(b) of Lemma 4, we consider proper suffixes of u_0 as candidates of left/right copies of maximal α-gapped repeats. Let $b = w[j + 1]$ as in Section 4.1. For each $c \in \Sigma \setminus \{b\}$, let $[u_1 c], \ldots, [u_t c]$ be the sequence of nodes of $DAWG(\$z_i)$, where u_1 is the longest proper suffix of u_0 such that $u_1 c \not\equiv_{\$z_i} u_0 c$ and $u_1 c \in Substr(z_i)$, u_{g+1} is the longest proper suffix of u_g such that $slink([u_g c]) = [u_{g+1} c]$ for each $1 \leq g \leq t - 1$, and $[u_t c] = [c]$. Namely, $[u_1 c], \ldots, [u_t c]$ is the sequence of nodes in the suffix link path from $[u_1 c]$ to $[c]$, and $u_h c$ is the longest element of $[u_h c]$ for all $1 \leq h \leq t$.

After processing u_0 with the algorithm of Section 4.1, for each $c \in \Sigma \setminus \{b\}$ we move to node $[u_1 c]$ (we will later describe how to efficiently move there from $[u_0]$). Let q_1 be the non-empty string such that $q_1 u_1 = u_0$, and let $a_1 = q_1[|q_1|]$. Since $u_1 c$ is a proper suffix of $u_0 c$ and $u_1 c \not\equiv_{\$z_i} u_0 c$, $Endpos_{\$z_i}(u_0 c) \subset Endpos_{\$z_i}(u_1 c)$. Also, for any $r' \in Endpos_{\$z_i}(u_1 c) - Endpos_{\$z_i}(u_0 c)$, $z_i[r' - |u_1| - 1] \neq q_1[|q_1|]$. Hence, any $k' = r' - |u_1| + 1$ which satisfies Property (3) of Lemma 4 is the beginning position of the left copy of a maximal α-gapped repeat ending at position j'. To compute such occurrences, let e_1 be the label of any reversed suffix link from node $[u_1 c]$. We move from $[u_1 c]$ to $[e_1 u_1 c]$ by traversing the reversed suffix link. We then perform a range threshold query $\mathrm{rtq}(\ell([e_1 u_1 c]), r([e_1 u_1 c]), \tau(j', u_1) + |u_1|)$ on $SA_{z_i^R\$}$. Let h be any answer returned by the above queries for all reversed suffix links from $[u_1 c]$. Then, h is the ending position of the left copy u_1 of a maximal α-gapped repeat which ends at position j' in f_i, and is immediately followed by c. Thus, letting $k = h - |u_1| + 1$, we have $(k, j, j - k - |u_1| + 1) \in FMGR_\alpha^{\mathrm{mid}}(i)$. After processing node $[u_1 c]$ as above, we proceed to the next node $[u_2 c] = slink([u_1 c])$ in the suffix link path, and process it in an analogous way (exchanging u_2 with u_3, and u_1 with u_2). We continue this until we reach the last node $[u_t c]$ in the suffix link path w.r.t. character c. By conducting the above procedure for all characters $c \in \Sigma \setminus \{b\}$, we obtain the subset of $FMGR_\alpha^{\mathrm{mid}}(i)$ which end at position j' in f_i, and hence we can compute $FMGR_\alpha^{\mathrm{mid}}(i)$ by using the above algorithm for positions $j' = 2, \ldots, |f_i|$ in f_i.

The algorithm described above correctly computes $FMGR_\alpha^{\mathrm{mid}}(i)$, however, it is inefficient: For each position j' in f_i and for each character $c \in \Sigma \setminus \{b\}$, we traverse $O(n)$ reversed suffix links. Thus, it takes a total of $O(\sigma n^2)$ time.

Speeding up the Algorithm. To obtain an $O(\alpha n + occ_{\mathrm{mid}})$-time solution (for a constant alphabet), we need to consider the following points:

1. Given node $[u_0]$ for each j' in f_i, how can we find the first node $[u_1 c]$ in the suffix link path w.r.t. character $c \in \Sigma \setminus \{b\}$, where $b = w[j+1]$?
2. How can we bound the number of nodes to process in the suffix link path, by the number of maximal α-gapped repeats to output?

Firstly, we consider Point 1. If node $[u_0]$ has an out-going edge labeled by c, then we can easily obtain node $[u_1 c]$ in $O(\log \sigma)$ time, by moving from $[u_0]$ to $[u_0 c]$ by the edge $([u_0], c, [u_0 c])$ and then moving to node $[u_1 c]$ by the suffix link of $[u_0 c]$. To deal with the case where node $[u_0]$ does *not* have an out-going edge labeled by c, we pre-process $DAWG(\$z_i)$ as follows: For each node $[x]$ of $DAWG(\$z_i)$ and each character $c \in \Sigma$, we precompute the longest proper suffix sc of

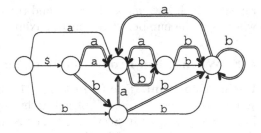

Fig. 3. $DAWG(\$abb)$ augmented with the shortcut links, which are depicted by the double arcs. The shortcut links for nodes $[\varepsilon]$, $[a]$, and $[b]$ are omitted as they are never used in the algorithm.

xc such that $sc \not\equiv xc$ and sc is a substring of z_i, and we create a shortcut link $([x], c, [sc])$. See also Fig. 3 for examples of shortcut links. We can conduct this preprocessing in a total of $O(\sigma |z_i|)$ time for all characters in Σ, based on a standard depth-first traversal on the suffix link tree of $DAWG(\$z_i)$, namely $STree(z_i{}^R\$)$, as follows: During the traversal, we maintain a table F of size σ. Assume that we just have arrived at a node $[y]$ in the tree during the traversal. Also, assume that for each $c \in \Sigma$, $F[c]$ stores all descendants $[r]$ of $[y]$ in the tree which do *not* have an out-going DAWG edge labeled by c. Now, if $[y]$ has an out-going DAWG edge labeled by c, then we create shortcut links $([r], c, [yc])$ from all nodes $[r]$ stored in $F[c]$, and we delete all the nodes in $F[c]$. We also create shortcut link $([y], c, [yc])$. Otherwise (if $[y]$ does not have an out-going DAWG edge labeled by c), node $[y]$ is added to $F[c]$. We then continue the traversal on the tree. For all characters c and all nodes $[y]$ in the tree, we can check whether node $[y]$ has an out-going DAWG edge labeled with c in a total of $O(\sigma |z_i|)$ time. For each character $c \in \Sigma$, each node is inserted to $F[c]$ at most once, and once it is deleted from $F[c]$, it will never be re-inserted to $F[c]$. Hence, the preprocessing takes a total of $O(\sigma|z_i|)$ time for all characters and nodes. The resulting data structure with the shortcut links occupy $O(\sigma|z_i|)$ space.

Secondly, we consider Point 2. Notice that some node in the suffix link path $[u_1c], \ldots, [u_tc]$ may *not* be associated to any occurrence of maximal α-gapped repeats ending at position j' in f_i. The basic algorithm performs at most $\sigma - 1$ range threshold queries at every node in the suffix link path. Thus, for each j', the basic algorithm takes a total of $O(\sigma t + \gamma_{j'})$ time, where $\gamma_{j'}$ denotes the number of maximal α-gapped repeats which have one of $u_1, \ldots, u_t$ as left/right copies and end at position j' in f_i. Note that the σt term in the above complexity comes from the range threshold queries with no answers. Since $t = O(j')$, if we simply run the basic algorithm for all $1 \le j' \le |f_i|$, then it will take at least $O(\sigma|f_i|^2) = O(\sigma n^2)$ time in the worst case.

To bound the number of nodes we process in the suffix link path traversal, we use the following monotonicity of α-gapped repeats.

Lemma 5. *For any $1 \le j' < l' \le |f_i|$, assume a string u is a suffix of both $f_i[2..j']$ and $f_i[2..l']$. Then $\tau(j', u) < \tau(l', u)$.*

Let $[u_1c], \ldots, [u_tc]$ be the sequence of nodes in the suffix link path, where each u_g is a candidate of the left and right copies of maximal α-gapped repeats which end at position j' in f_i and are immediately followed by c. By Lemma 5 and Property (3) of Lemma 4, if $(k, j, j - k - |u_g| + 1)$ is *not* an α-gapped repeat, then for any $l > j$, $(k, l, l - k - |u_g| + 1)$ cannot be an α-gapped repeat, either. A node $[x]$ is said to be *alive* at position j' if $pos([x]) \ge \tau(j', x) + |x| - 1$, and is said to be *dead* otherwise (if $pos([x]) < \tau(j', x) + |x| - 1$). At each node $[x]$ we maintain the set $C([x])$ of

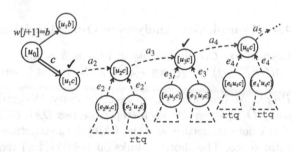

Fig. 4. Illustration for a traversal of the suffix link path $[u_1c], \ldots, [u_tc]$, where $e_g, e'_g \ne a_g$ for $1 < g \le t$, $C([u_2c]) = \{[e'_2u_2c]\}$, $C([u_3c]) = \emptyset$, $C([u_4c]) = \{[e_4u_4c], [e'_4u_4c]\}$. The checked nodes are the marked nodes in the suffix link path. Since $[u_1c]$ is marked, we move up to $[u_2c]$ and perform a range threshold query on the range corresponding to $[e'_2u_2c] \in C([u_2c])$. Then we perform an NMA query on $[u_1c]$ and obtain its NMA $[u_3c]$. We then move up to $[u_4c]$ and perform range threshold queries on the two ranges corresponding to $[e_4u_4c], [e'_4u_4c] \in C([u_4c])$.

suffix tree children of $[x]$ s.t. $[ax] \in C([x])$ with $a \in \Sigma$ iff node $[ax]$ is alive. The sets $C([x])$ for all nodes $[x]$ can be maintained in $O(|z_i| \log \sigma)$ time and $O(|z_i|)$ working space for all positions $j' = 2, \ldots, |f_i|$. Based on the sets C's above, we build the nearest marked ancestor (NMA) data structure by Gabow and Tarjan [7] on $STree(z_i{}^R)$. Initially, all nodes of $STree(z_i{}^R)$ are marked. A node $[x]$ will be unmarked, when all siblings of $[x]$ in $STree(z_i{}^R)$ become dead as j'

increases. Now we perform NMA queries on the suffix link path $[u_1 c], \ldots, [u_t c]$. At each node $[u_g c]$ returned by an NMA query in the suffix link path, we first move up to its suffix tree parent $[u_{g+1} c]$ in the path. We then access each suffix tree child $[e_{g+1} u_{g+1} c] \in C([u_{g+1} c])$ of $[u_{g+1} c]$, and perform a range threshold query on the range on $SA_{z_i R \$}$ which corresponds to $[e_{g+1} u_{g+1} c]$. Hence, the total time cost for traversing the suffix link paths is bounded by the number of outputs. See also Fig. 4 for illustration of the above procedure.

4.3   Finding Maximal α-gapped Repeats Starting at $k = d_i$

Consider a gapped repeat beginning at the first position of z_i, i.e., $k = d_i$ (or $k' = 1$). The left-maximality of such a gapped repeat cannot be checked locally in $z_i f_i$, and hence it has to be checked globally in w. Observe that such a gapped repeat can be found only if $u = f_i[2..|f_i|]$ or $z_i = w[1..|f_1 \cdots f_{i-1}|]$. In the latter case, the gapped repeat is always left-maximal. In the former case when the latter does not hold, then we simply compare the two characters $w[b_i]$ and $w[d_i - 1]$. If they are equal, then the gapped repeat $(d_i, b_{i+1} - 1, b_i - d_i) = (d_i, b_{i+1} - 1, |z_i| + 1)$ is not left-maximal, and if they are not equal, then it is left-maximal.

4.4   Complexity Analysis of Our Algorithm

Lemma 6. *Given a string w of length n and a real $\alpha > 1$, we can compute $FMGR_\alpha^{\mathrm{mid}}(w)$ in $O(\sigma \alpha n + occ_{\mathrm{mid}})$ time and $O(\sigma n + occ_{\mathrm{mid}})$ working space.*

Proof. We first show the space complexity. We can compute $LZ(w) = f_1, \ldots, f_m$ using $O(n)$ space. At each step i, we use $DAWG(\$z_i)$, $STree(z_i{}^R \$)$, $SA_{z_i R \$}$, the NMA data structure, and the RMQ data structure which require $O(|z_i|) = O(n)$ total space. The shortcut links on $DAWG(\$z_i)$ require $O(\sigma|z_i|) = O(\sigma n)$ space. As soon as step i has finished, we can discard all of them and will construct these data structures for the next step $i + 1$. During all the steps, we maintain a table L to store outputs, which requires $O(n + occ_{\mathrm{mid}})$ space. Thus the total space complexity for these data structures is $O(\sigma n + occ_{\mathrm{mid}})$ for all steps.

Let us analyze the preprocessing time. $LZ(w)$ can be computed in $O(n \log \sigma)$ time. At each step i of the algorithm, we spend $O(|z_i| \log \sigma)$ time to build $DAWG(\$z_i)$ and $STree(z_i{}^R \$)$, and $O(\sigma|z_i|)$ time to compute the shortcut links on $DAWG(\$z_i)$. The RMQ data structure can be constructed in $O(|z_i|)$ time. For all nodes $[x]$ of $DAWG(\$z_i)$, it takes $O(|z_i|)$ total time to precompute the smallest position j' at which node $[x]$ becomes dead. Hence, the total preprocessing time for all steps is $O(\sum_{i=1}^m \sigma|z_i|) = O(\sigma \alpha \sum_{i=1}^m |f_i|) = O(\sigma \alpha n)$.

Now we analyze the running time of the algorithm for each step i. On the NMA data structure, each node is unmarked at most once, and due to Lemma 5 any unmarked node will never be marked again. Since it takes $O(1)$ time to find the nearest marked ancestor of any given node, and since each occurrence of maximal α-gapped repeats can be obtained in $O(1)$ time, the total time cost for traversing the suffix link paths is $O(occ_{\mathrm{mid}}(i))$. What remains is how to store the outputs into the table L. For each $1 \leq k \leq n$, we need the elements stored in $L[k]$

to be sorted in increasing order of their lengths. This is important to compute maximal α-gapped repeats that belong to $SMGR_\alpha(w)$ in $O(n)$ total time (The elements of $SMGR_\alpha(i)$ that are completely contained in f_i are computed by copying maximal α-gapped repeats from a previous occurrence of f_i. See [10] for details). Since we compute maximal α-gapped repeats in increasing order of their ending positions j, for each $1 \le k \le n$ we can easily store the maximal α-gapped repeats starting at position k into $L[k]$ in decreasing order of their lengths, in time linear in the number of elements in $L[k]$. Thus this part of the algorithm takes $O(occ_{\text{mid}})$ total time.

Overall, the algorithm requires $O(\sigma \alpha n + occ_{\text{mid}})$ time and $O(\sigma n + occ_{\text{mid}})$ working space. This completes the proof.   □

The main result of this paper follows from Lemmas 2 and 6.

Theorem 1. *Given a string w of length n over a constant alphabet and a real $\alpha > 1$, we can compute the set $MGR_\alpha(w)$ of all maximal α-gapped repeats in w in $O(\alpha n + occ)$ time and $O(n + occ)$ space, where $occ = |MGR_\alpha(w)|$ is the number of occurrences of maximal α-gapped repeats in w.*

References

1. Badkobeh, G., Crochemore, M., Toopsuwan, C.: Computing the maximal-exponent repeats of an overlap-free string in linear time. In: Calderón-Benavides, L., González-Caro, C., Chávez, E., Ziviani, N. (eds.) SPIRE 2012. LNCS, vol. 7608, pp. 61–72. Springer, Heidelberg (2012)
2. Bender, M.A., Farach-Colton, M.: The LCA problem revisited. In: Gonnet, G.H., Viola, A. (eds.) LATIN 2000. LNCS, vol. 1776, pp. 88–94. Springer, Heidelberg (2000)
3. Blumer, A., Blumer, J., Haussler, D., Ehrenfeucht, A., Chen, M.T., Seiferas, J.: The smallest automaton recognizing the subwords of a text. TCS **40**, 31–55 (1985)
4. Brodal, G.S., Lyngsø, R.B., Pedersen, C.N.S., Stoye, J.: Finding maximal pairs with bounded gap. In: Crochemore, M., Paterson, M. (eds.) CPM 1999. LNCS, vol. 1645, pp. 134–149. Springer, Heidelberg (1999)
5. Crochemore, M., Rytter, W.: Text Algorithms. Oxford University Press, New York (1994)
6. Crochemore, M.: Transducers and repetitions. Theor. Comput. Sci. **45**(1), 63–86 (1986)
7. Gabow, H.N., Tarjan, R.E.: A linear-time algorithm for a special case of disjoint set union. Journal of Computer and System Sciences **30**, 209–221 (1985)
8. Gusfield, D.: Algorithms on Strings, Trees, and Sequences. Cambridge University Press (1997)
9. Gusfield, D., Stoye, J.: Linear time algorithms for finding and representing all the tandem repeats in a string. J. Comput. Syst. Sci. **69**(4), 525–546 (2004)
10. Kolpakov, R., Podolskiy, M., Posypkin, M., Khrapov, N.: Searching of gapped repeats and subrepetitions in a word. In: Kulikov, A.S., Kuznetsov, S.O., Pevzner, P. (eds.) CPM 2014. LNCS, vol. 8486, pp. 212–221. Springer, Heidelberg (2014)

11. Kolpakov, R.M., Kucherov, G.: Finding repeats with fixed gap. In: Proc. SPIRE 2000, pp. 162–168 (2000)
12. Manber, U., Myers, G.: Suffix arrays: A new method for on-line string searches. SIAM J. Computing **22**(5), 935–948 (1993)
13. Ukkonen, E.: On-line construction of suffix trees. Algorithmica **14**(3), 249–260 (1995)
14. Weiner, P.: Linear pattern-matching algorithms. In: Proc. of 14th IEEE Ann. Symp. on Switching and Automata Theory, pp. 1–11 (1973)
15. Ziv, J., Lempel, A.: A universal algorithm for sequential data compression. IEEE Transactions on Information Theory **IT–23**(3), 337–343 (1977)

Selective Labeling and Incomplete Label Mitigation for Low-Cost Evaluation

Kai Hui[(✉)] and Klaus Berberich

Max Planck Institute for Informatics, Saarbrücken, Germany
{khui,kberberi}@mpi-inf.mpg.de

Abstract. Information retrieval evaluation heavily relies on human effort to assess the relevance of result documents. Recent years have seen efforts and good progress to reduce the human effort and thus lower the cost of evaluation. Selective labeling strategies carefully choose a subset of result documents to label, for instance, based on their aggregate rank in results; strategies to mitigate incomplete labels seek to make up for missing labels, for instance, predicting them using machine learning methods. How different strategies interact, though, is unknown.

In this work, we study the interaction of several state-of-the-art strategies for selective labeling and incomplete label mitigation on four years of TREC Web Track data (2011–2014). Moreover, we propose and evaluate MaxRep as a novel selective labeling strategy, which has been designed so as to select effective training data for missing label prediction.

1 Introduction

Evaluation in information retrieval often relies on the Cranfield paradigm [10]. To establish the relative performance of several information retrieval systems, one agrees on a set of information needs (called *topics*), which are representative of the target workload. Each of these information needs is then formulated as a keyword query, and results are obtained from each of the information retrieval systems under comparison. Following that, human assessors label retrieved result documents with regard to their relevance. Finally, based on the collected labels, a retrieval effectiveness measure such as mean-average precision (MAP) or normalized discounted cumulative gain (nDCG) is computed to establish a relative order of the compared information retrieval systems according to their retrieval performance.

Manual labeling is laborious and costly, in particular when the number of topics and/or the number of compared systems is large. As a reaction, recent years have seen a fair amount of research that seeks to reduce the cost of information retrieval evaluation. *Selective labeling*, as a first direction, chooses a subset of returned result documents to label. Among the simplest strategies, depth-k pooling [16,17] only collects labels for documents returned in the top-k result of any of the compared systems. More sophisticated strategies leverage knowledge about the retrieval effectiveness measure used, for instance, Carterette and

© Springer International Publishing Switzerland 2015
C. Iliopoulos et al. (Eds.): SPIRE 2015, LNCS 9309, pp. 137–148, 2015.
DOI: 10.1007/978-3-319-23826-5_14

Allan [7] who label only documents with a potential effect on the relative order of any two systems. While cutting costs, selective labeling leads to result documents whose relevance label is not known. Such incomplete labels can also arise for other reasons, for example, when evaluating a novel information retrieval system that did not contribute to the original pool of result documents. *Mitigating incomplete labels*, as a second direction, seeks principled ways to make up for missing relevance assessments. The default of dealing with them is to assume that result documents are irrelevant if they have not been labeled. While this may appear pessimistic at first glance, it is not unreasonable given that most documents will be irrelevant to any specific information need. Alternative approaches have come up with novel effectiveness measures [3], removed documents without known label from consideration [15], and made use of machine learning methods to predict missing labels [4].

Contributions. What has received some prior attention but has not been fully explored, though, is how the different strategies for selective labeling and incomplete label mitigation interact with each other. As a *first contribution* of this paper we thus examine the interaction of state-of-the-art selective labeling and incomplete label mitigation strategies on four years of TREC Web Track data (2011–2014). The performance of different combinations is studied both in terms of approximating MAP scores (in terms of root mean square error) as well as system rankings (in terms of Kendall's τ). Also, strategies for selective labeling have typically been designed with no consideration of how incomplete labels are dealt with later on. Hence, as a *second contribution*, inspired by recent work in machine learning [19] and the cluster hypothesis [14], we propose MAXREP as a novel selective labeling strategy. MAXREP selects documents to label so as to maximize their representativeness of the pool of result documents, thus yielding effective training data for label prediction. MAXREP is formulated as an optimization problem, which permits efficient approximation.

Organization. The rest of this paper is organized as follows. Section 2 recaps existing strategies for selective labeling and incomplete label mitigation and puts our work in context. Section 3 puts forward our novel selective labeling strategy MAXREP. Our extensive experimental study is the subject of Section 4. Finally, in Section 5 we draw conclusions.

2 Technical Background and Related Work

In this section, we provide the technical background for our work by reviewing existing strategies for selective labeling and incomplete label mitigation. Moreover, we put our proposed MAXREP method in context with existing work.

2.1 Selective Labeling

Several efforts have looked into how, to reduce human effort and hence cost, only a subset of returned documents can be labeled, while still producing a reliable relative ranking of multiple information retrieval systems:

Pooling strategies merge the results returned by different systems to form a pool of result documents to be labeled by human assessors. The most common strategy, *depth-k pooling* as used by TREC, considers only documents that are returned within the top-k of any system. Cormack et al. [11], as an alternative, propose *move-to-front pooling* (MTF) as an iterative pooling procedure, requiring continued human effort, which systematically prioritizes documents returned by systems that have already returned relevant documents. Vu and Gallinari [17] make use of machine learning for pooling. Using documents from the top-5 pool as training data, they employ *learning-to-rank methods* to estimate the relevance of yet-unlabeled documents. Documents more likely to be relevant are then labeled with higher priority. Features, in their case, encode the rank at which the document was returned by different systems. Their approach thus requires two rounds of human interaction to label (i) documents in the top-5 pool as training data and (ii) a number of the remaining documents.

Aslam et al. [2] devise a biased sampling strategy that yields an unbiased estimator of MAP. A more practical sampling strategy with good empirical performance is described by Aslam and Pavlu [1]. The key idea here is to introduce a sampling distribution, so that documents ranked highly by many system, which are therefore expected to be relevant, are selected more often. The probability of selecting the document at rank r from a result list of length n is defined as

$$P[r] \approx \frac{1}{2n} \log \frac{n}{r} \,.$$

These per-system probabilities are aggregated, corresponding to choosing a system at uniform random, and documents are selected using stratified sampling.

Carterette et al. [7] propose the *minimal test collection* (MTC) method. For a specific retrieval effectiveness measure (e.g., MAP or nDCG), MTC iteratively selects discriminative documents to label until the relative order of systems has been determined. Requiring continued human interaction at every step, like MTF pooling described above, it is an active procedure.

Unlike all of the aforementioned strategies, which only take ranking information into account, our novel method MAXREP also considers document contents. Inspired by Yu et al. [19] and designed with label prediction in mind, MAXREP aims at selecting a representative set of documents from the pool of result documents to yield effective training data.

2.2   Incomplete Label Mitigation

Labels can be incomplete for different reasons, for instance, since they were collected only selectively or because the evaluated information retrieval system is novel and did not contribute to the initial result pool. Different strategies have been proposed as remedies:

As already mentioned above, a common way to deal with missing relevance labels, which is also used in TREC, is to *assume* that those documents are *irrelevant*. Given that most documents are irrelevant anyway for any specific information need, this can also be interpreted as label prediction with a simple majority classifier. More elaborate label prediction methods will be discussed below. Sakai [15], as an alternative, proposes to remove documents without known labels from consideration yielding *condensed result lists*. Both aforementioned incomplete label mitigation strategies are agnostic to the retrieval effectiveness measure used.

In contrast, Buckley and Voorhees [3] propose bpref as an *alternative retrieval effectiveness measure* mimicking mean-average precision (MAP). With R as the number of labeled relevant documents, it is defined as

$$\text{bpref} = \frac{1}{R} \sum_r \left(1 - \frac{|\text{ labeled irrelevant above rank } r \,|}{R}\right),$$

and the term in parenthesis can be interpreted as an estimator of precision at rank r. In their experiments, bpref proved robust and exhibited high rank correlation with MAP. However, in terms of numerical value, bpref may deviate from MAP if many labels are missing. Yilmaz and Aslman [18] describe two alternatives, based on sampling theory, that are closer to MAP. The first, induced average precision (indAP), removes documents with unknown label from consideration and can be seen as an application of the condensed list approach [15] to MAP. The second, inferred average precision (infAP), relies on the following improved estimator of precision at rank r

$$E[\text{precision at } r] = \frac{1}{r} + \frac{(r-1)}{r} \left(\frac{|\text{ labeled above rank } r \,|}{r-1} \cdot \frac{|\text{ labeled relevant }|}{|\text{ labeled }|}\right),$$

which also takes into account what fraction of documents has been labeled.

Another family of strategies uses machine learning methods to *predict missing relevance labels*. Carterette and Allan [6] use regularized logistic regression to predict the relevance of documents. Building on the cluster hypothesis [14], document features encode *tf.idf*-based cosine similarity with documents whose labels are known. Büttcher et al. [4], to the same end, explore two approaches, namely a simple classifier based on statistical language models and a support vector machine (SVM). For the latter, document features are *tf.idf*-weights for the 10^6 most common terms in the document collection. Given the good performance of the SVM-based label prediction in their experiments, we use this as one of the incomplete label mitigation strategies in our experiments.

3   Selecting Representative Documents to Label

We now describe MAXREP, our novel strategy for selective labeling. In contrast to existing strategies, MAXREP not only considers ranking information but also takes into account document contents. Intuitively, it aims at selecting a subset of documents that is representative, in particular of those documents expected to be relevant. MAXREP thus harvests effective training data for label prediction, since documents are representative of the overall pool of result documents, and it also makes up for the inherent bias against relevant documents.

Let $\mathcal{D}$ denote the pool of result documents for a specific topic. Our objective is to select a k-subset $\mathcal{L} \subseteq \mathcal{D}$ that best represents the pool of result documents. Intuitively, if two documents have similar contents, there is no need to label both of them, since their labels tend to be identical. We let $sim(d_i, d_j) \in [0, 1]$ denote a measure of *content similarity* between documents d_i and d_j. Further, we let $rel(d_i) \in [0, 1]$ denote a measure of *expected relevance* of document d_i

Our concrete implementation uses the cosine similarity between $tf.idf$-based document vectors as a measure of document content similarity. More precisely, with $tf(v, d)$ as the term frequency of term v in document d, $df(v)$ as its document frequency, and n as the total number of documents in the collection, the feature weight for term v in document vector $\boldsymbol{d}$ is

$$\boldsymbol{d}(v) = tf(v, d) \log \frac{n}{df(v)} \, ,$$

and we measure the similarity between documents d_i and d_j as

$$sim(d_i, d_j) = \frac{\boldsymbol{d}_i \cdot \boldsymbol{d}_j}{\|\boldsymbol{d}_i\| \, \|\boldsymbol{d}_j\|}$$

which ranges in $[0, 1]$ given that we only have non-negative feature weights. As in Büttcher et al. [4] our implementation only considers the 10^6 most frequent terms from the document collection. Moreover, in order to reduce noise, we ignore similarities below 0.8, setting them to zero, when choosing representative documents. As a measure of expected relevance our concrete implementation uses the probability according to the sampling distribution also used in Aslam and Pavlu [1] and described in Section 2.

We measure the representativeness of a document set $\mathcal{L}$ as

$$f(\mathcal{L}) = \sum_{d_i \in \mathcal{D}} rel(d_i) \max_{d_j \in \mathcal{L}} \left(sim(d_i, d_j) \right) . \tag{1}$$

This formulation rewards document sets that cover all documents from $\mathcal{D}$ that are expected to be relevant by including at least one similar document.

Building on this, we cast selecting the set of k most representative result documents into the following optimization problem

$$\underset{\mathcal{L}}{\text{argmax}} \; f(\mathcal{L}) \quad \text{s.t.} \quad |\mathcal{L}| = k$$

It turns out that the above optimization problem permits efficient approximation thanks to the submodularity of its objective function, which we state in the following lemma.

Lemma 1 (Submodularity). *Equation 1 defines a submodular function. Given two document sets $\mathcal{L}$ and $\mathcal{L}'$ with $\mathcal{L} \subseteq \mathcal{L}'$ and a document $d \in \mathcal{D}$, then*

$$f(\mathcal{L} \cup \{d\}) - f(\mathcal{L}) \geq f(\mathcal{L}' \cup \{d\}) - f(\mathcal{L}') .$$

Proof (of Lemma 1). We can rewrite for $\mathcal{X} \in \{\mathcal{L}, \mathcal{L}'\}$

$$f(\mathcal{X} \cup \{d\}) - f(\mathcal{X}) = \sum_{d_i \in \mathcal{D}} rel(d_i) \max\left(0, sim(d_i, d) - \max_{d_j \in \mathcal{X}} sim(d_i, d_j)\right) .$$

Now,

$$\mathcal{L} \subseteq \mathcal{L}' \quad \Rightarrow \quad \forall d_i \in \mathcal{D} : \max_{d_j \in \mathcal{L}} sim(d_i, d_j) \leq \max_{d_j \in \mathcal{L}'} sim(d_i, d_j)$$
$$\Rightarrow \quad f(\mathcal{L} \cup \{d\}) - f(\mathcal{L}) \geq f(\mathcal{L}' \cup \{d\}) - f(\mathcal{L}') .$$

$\square$

Having established the submodularity of our objective function, we can make use of the result by Nemhauser et al. [12] and greedily build up the set of representative documents $\mathcal{L}$. More precisely, starting from $\mathcal{L}_0 = \emptyset$, in the i-th iteration we include the document from $\mathcal{D} \backslash \mathcal{L}_{i-1}$ that maximizes $f(\mathcal{L}_i)$, and finally report $\mathcal{L}_k$ as a result. This greedy algorithm gives a $(1 - \frac{1}{e})$-approximation [12], guaranteeing the performance of the proposed greedy algorithm.

4   Experimental Evaluation

In this section, we describe our experimental evaluation. We report on the performance of different combinations of strategies for selective labeling, including MAXREP as the one proposed in this work, and incomplete label mitigation. This is done on four years' worth of participant data from the TREC Web Track (2011–2014), and we investigate how well combinations can approximate the system ranking, in terms of Kendall's τ, but also how well they can approximate MAP scores, in terms of root mean square error (RMSE).

4.1   Datasets

Our experiments are based on the CLUEWEB09[1] and CLUEWEB12[2] document collections. Queries and relevance labels are taken from the adhoc task of the TREC Web Track (2011–2014). This leaves us with a total of 200 queries (50 per year) and their corresponding relevance labels. We also obtained the runs submitted by participants of the TREC Web Track. There are 62 runs for 2011, 48 runs for 2012, 61 runs for 2013, and 42 runs for 2014. For each submitted run we consider the top-20 search results returned. In 2013 a subset of 21 queries was only labeled up to depth 10. For those queries we apply the condensed list approach, that is, for each system we consider the 20 highest-ranked labeled documents as its result.

---

[1] http://www.lemurproject.org/clueweb09.php/
[2] http://www.lemurproject.org/clueweb12.php/

4.2   Methods

We consider the following non-active strategies for *selective labeling*:

- **uniform random sampling**, as described by Buckley and Voorhees [3], we give the method an advantage by sampling retrospectively from relevant and irrelevant documents (we report averages based on 30 repetitions);
- **incremental pooling**, as described by Carterette [5,7], we select documents according to the best rank assigned by any system and break ties according to the average rank across all systems;
- **statAP**, as described by Aslam and Pavlu [1], with additional judgments obtained from pooling (we report averages based on 30 repetitions);
- **our method** MAXREP as described in Section 3.

To *mitigate incomplete labels*, we consider the following strategies:

- **trec-map** treats documents with unknown label as irrelevant;
- **bpref** [3] separates the labeled non-relevant documents from unlabeled documents;
- **indAP** [18] regards missing labels as non-existing
- **infAP** [18] relies on an improved estimator of precision at rank r
- **statAP** [1] computes AP with adjustments by inclusion probability from the document sampling phase
- **predict-map**, SVM-based label prediction approach [4], which we implemented using the scikit-learn [13] toolkit.

This gives us a total of 21 combinations to investigate. Given that statAP as a strategy for mitigating incomplete labels requiring inclusion probabilities as an input from selective labeling, we only compute statAP when labels have been selected with statAP itself.

4.3   Approximation of System Ranking and MAP Scores

Our first experiment studies how well different strategies can approximate the system ranking in terms of Kendall's τ and how well they can approximate the MAP scores of individual systems. To this end, we select a varying percentage, from 1% up to 95%, to label using the different strategies. Figure 1 shows the Kendall's τ value obtained for different selective labeling strategies on each of the four years (2011–2014) considered. Comparing the different incomplete label mitigation strategies, we observe that predict-map, the SVM-based label prediction approach, consistently achieves good performance, regardless of how documents to label are selected. In most plots, with as little as 20% of labeled documents, predict-map thus achieves a Kendall's τ value above 0.9, which indicates that the obtained system ranking is practically indistinguishable from the ground truth. Using trec-map and assuming that documents without known labels are irrelevant, totally mixing the labeled non-relevant and unlabeled documents, at the other extreme, performs worst in most plots. Not surprisingly, this is most

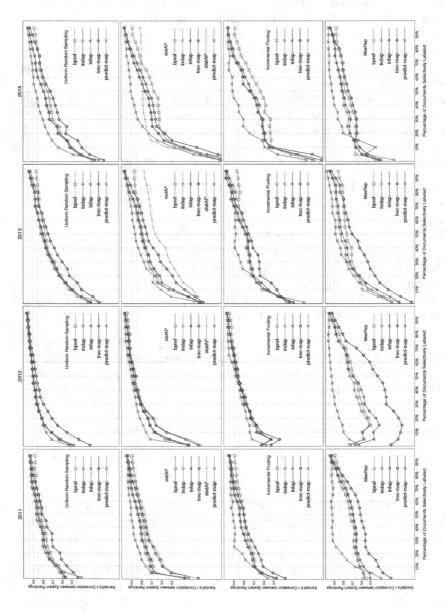

Fig. 1. Approximation of system rankings: Columns correspond to different years of the TREC Web Track. Rows correspond to different selective labeling strategies. X-axes indicate percentages of labeled documents. Y-axes indicate Kendall's τ correlation.

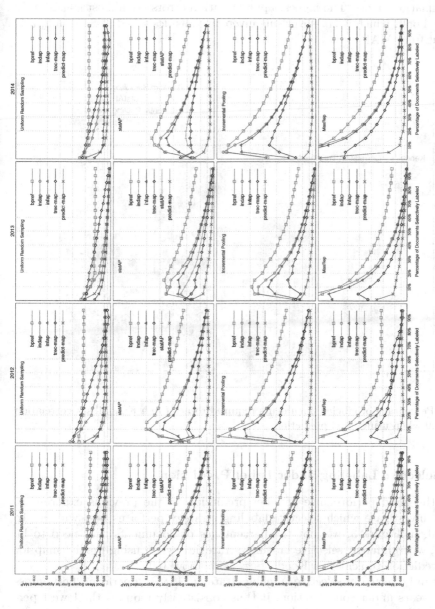

Fig. 2. Approximation of MAP scores: Columns correspond to different years of the TREC Web Track. Rows correspond to different selective labeling strategies. X-axes indicate percentages of labeled documents. Y-axes indicate root mean square error (RMSE).

pronounced when using our selective labeling strategy MAXREP. Figure 2 plots the corresponding root mean square error (RMSE), measuring how well the different combinations approximate MAP scores of individual systems. Predicting missing labels using predict-map again achieves the best result by yielding lowest approximation errors. The highest approximation errors are almost consistently seen for bpref, which is not surprising given that, as described in Section 2, it is different from MAP.

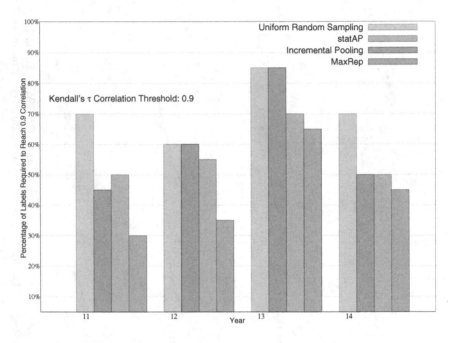

Fig. 3. Percentage of labeled documents required to achieve a Kendall's τ correlation above 0.9 when using label prediction.

4.4 Selective Labeling under Label Prediction

Given the good performance of label prediction in the previous experiment, we now investigate which selective labeling strategy performs best with it. To this end, in Figure 3, we plot the percentages of documents that need to be labeled, with different selective labeling strategies, when using predict-map for label prediction to achieve a Kendall's τ score above 0.9.

As can be seen, our selective labeling strategy MAXREP performs best across all four years under consideration. It thus consistently requires the lowest percentage of documents to be labeled to achieve a system ranking that is practically indistinguishable from the ground truth. Its relative advantage is clearest for the years 2011 and 2012 for which MAXREP requires as little as $30-35\%$ of labeled documents. Also in this experiment, uniform random sampling performs worst, typically requiring more than 60% of labeled documents to achieve a Kendall's τ value above the threshold. Additionally, we conduct paired two-tailed t-test between different baselines w.r.t. our method for these least percentage of labels required to get over 0.9 correlation, and our method outperform the uniform random sampling and incremental pooling at 95% significant level (p-value=.008 and .032), meanwhile outperform the statAP at 90% level (p-value=.063).

As for comparison on RMSE, from Figure 2, we can see that our method is comparable to other methods in terms of approximating MAP scores. However, no clear winner is observed among different selective labeling methods when combined with mitigation through label prediction.

5   Conclusion

Low-cost evaluation has been an active area of research within information retrieval for the past decade. In this work, we have investigated how different strategies for selective labeling and mitigating incomplete labels interact. To this end, we conducted a large-scale experimental evaluation on CLUEWEB09/12 with participant data from the adhoc task of TREC Web Track 2011–2014. We found that label prediction is a robust and viable strategy to mitigate incomplete labels, as long as at least 20% of documents have been labeled as training data. Moreover, with label prediction in mind, we proposed a novel strategy MAXREP for selective labeling. In contrast to existing strategies, it considers both ranking information and document contents and seeks to select a representative subset of documents to label. Our experiments confirmed that MAXREP is beneficial and outperforms other strategies when label prediction is used.

As part of our ongoing research, we investigate how strategies for selective labeling and incomplete label mitigation can be adapted for retrieval effectiveness measures such as α-nDCG [9] and ERR-IA [8] that capture novelty & diversity. Moreover, we study the reusability of this semi-automatically generated labeled collection, examining the reliability in evaluating systems without contributing to the initial document collection.

References

1. Aslam, J.A., Pavlu, V.: A practical sampling strategy for efficient retrieval evaluation. Report (May 2007)
2. Aslam, J.A., Pavlu, V., Yilmaz, E.: A statistical method for system evaluation using incomplete judgments. In: SIGIR, pp. 541–548 (2006)
3. Buckley, C., Voorhees, E.M.: Retrieval evaluation with incomplete information. In: SIGIR, pp. 25–32 (2004)
4. Büttcher, S., Clarke, C.L.A., Yeung, P.C.K., Soboroff, I.: Reliable information retrieval evaluation with incomplete and biased judgements. In: SIGIR, pp. 63–70 (2007)
5. Carterette, B.: Robust test collections for retrieval evaluation. In: SIGIR, pp. 55–62 (2007)

6. Carterette, B., Allan, J.: Semiautomatic evaluation of retrieval systems using document similarities. In: CIKM, pp. 873–876 (2007)
7. Carterette, B., Allan, J., Sitaraman, R.: Minimal test collections for retrieval evaluation. In: SIGIR, pp. 268–275 (2006)
8. Chapelle, O., Metlzer, D., Zhang, Y., Grinspan, P.: Expected reciprocal rank for graded relevance. In: CIKM, pp. 621–630 (2009)
9. Clarke, C.L., Kolla, M., Cormack, G.V., Vechtomova, O., Ashkan, A., Büttcher, S., MacKinnon, I.: Novelty and diversity in information retrieval evaluation. In: SIGIR, pp. 659–666 (2008)
10. Cleverdon, C.: The cranfield tests on index language devices. In: Aslib proceedings, vol. 19, pp. 173–194. MCB UP Ltd (1967)
11. Cormack, G.V., Palmer, C.R., Clarke, C.L.A.: Efficient construction of large test collections. In: SIGIR, pp. 282–289 (1998)
12. Nemhauser, G., Wolsey, L., Fisher, M.: An analysis of approximations for maximizing submodular set functions–i. Mathematical Programming 14, 265–294 (1978)
13. Pedregosa, F., Varoquaux, G., Gramfort, A., Michel, V., Thirion, B., Grisel, O., Blondel, M., Prettenhofer, P., Weiss, R., Dubourg, V., Vanderplas, J., Passos, A., Cournapeau, D., Brucher, M., Perrot, M., Duchesnay, E.: Scikit-learn: Machine learning in Python. Journal of Machine Learning Research 12, 2825–2830 (2011)
14. Rijsbergen, C.J.V.: Information Retrieval, 2nd edn. Butterworth-Heinemann, Newton (1979)
15. Sakai, T.: Alternatives to bpref. In: SIGIR, pp. 71–78 (2007)
16. Spärck Jones, K., Van Rijsbergen, K.: Information retrieval test collections. Journal of Documentation 32(1), 59–75 (1976)
17. Vu, H.-T., Gallinari, P.: A machine learning based approach to evaluating retrieval systems. In: HLT-NAACL, pp. 399–406 (2006)
18. Yilmaz, E., Aslam, J.A.: Estimating average precision with incomplete and imperfect judgments. In: CIKM, pp. 102–111 (2006)
19. Yu, K., Bi, J., Tresp, V.: Active learning via transductive experimental design. In: ICML, pp. 1081–1088 (2006)

Relative Select

Christina Boucher[1], Alexander Bowe[2], Travis Gagie[3],
Giovanni Manzini[4(✉)], and Jouni Sirén[5]

[1] Colorado State University, Fort Collins, CO, USA
[2] National Institute of Informatics, Tokyo, Japan
[3] Helsinki Institute for Information Technology,
University of Helsinki, Helsinki, Finland
[4] University of Eastern Piedmont, Alessandria, Italy
giovanni.manzini@unipmn.it
[5] Wellcome Trust Sanger Institute, Cambridge, UK

Abstract. Motivated by the problem of storing coloured de Bruijn graphs, we show how, if we can already support fast select queries on one string, then we can store a little extra information and support fairly fast select queries on a similar string.

1 Introduction

Many compressed data structures for strings rely on three fundamental queries: access, rank and select. The query $S.\mathsf{access}(i)$ on a string S returns its ith character; the query $S.\mathsf{rank}_a(i)$ returns the number of occurrences of character a in the prefix of S of length i; and the query $S.\mathsf{select}_a(j)$ returns the position of the jth leftmost occurrence of a in S. Suppose we have a data structure supporting these queries on a string S_1 and we want another data structure supporting them on a similar string S_2. It is not difficult to store S_2 in small space and support access to it via access to S_1. For example, we can find a longest common subsequence of S_1 and S_2, store two bitvectors with 1s marking their characters not in that subsequence, and store the characters marked in S_2. The total number of 1s in the two bitvectors is at most twice the standard edit distance d between S_1 and S_2 (i.e., the number of single-character insertions, deletions and substitutions needed to change one into the other) so we can store them in $\mathcal{O}(d)$ space and support rank and select on them using $\mathcal{O}(\log\log(|S_1| + |S_2|))$ time using a sparse-bitvector implementation [6]. To access $S_2[i]$, we check whether it appears in the common subsequence: if so, we use rank and select queries on the bitvectors to find the corresponding character in S_1, which access; if not, we find $S_2[i]$'s rank among characters marked in S_2 and look it up.

Last year, when describing their relative FM-index data structure, Belazzougui et al. [1] showed how to store $\mathcal{O}(d)$ extra words and support any rank

---

T. Gagie—Supported by grants 268324 and 250345 (CoECGR) from the Academy of Finland.

J. Siren—Supported by the Wellcome Trust grant 098051.

© Springer International Publishing Switzerland 2015
C. Iliopoulos et al. (Eds.): SPIRE 2015, LNCS 9309, pp. 149–155, 2015.
DOI: 10.1007/978-3-319-23826-5_15

query on S_2 using $\mathcal{O}(\log\log(|S_1| + |S_2|))$ time on top of a rank query on S_1. In this paper we show how to store $\mathcal{O}(d)$ extra words and support any select query on S_2 using $\mathcal{O}(\log\log(|S_1| + |S_2|))$ time on top of a select query on S_1. We call this *relative select* and we expect it to be useful when storing compressed data structures for navigating in coloured de Bruijn graphs [8].

Belazzougui et al. were interested in saving space when storing FM-indexes [5] for many genomes from the same species. An FM-index for a genome is essentially just a data structure supporting access and rank on the Burrows-Wheeler Transform [4] (BWT) of that genome. The BWT sorts the characters of a string into the lexicographic order of the suffixes that immediately follow them. The edit distance between two genomes from the same species tends to be small relative to their lengths and in practice the edit distance between their BWTs also tends to be small. Therefore, if we store the FM-index for one genome normally, we can use Belazzougui et al.'s result to save space when storing FM-indexes for other genomes from the same species (at the cost of higher query times).

It is possible to support nearly all the functionality of an FM-index without using select queries on the underlying BWT, so Belazzougui et al. did not consider relative select. When the FM-index is used in a compressed suffix tree, however, select queries are needed for computing suffix links and for certain other operations. Our interest in relative select comes from Bowe et al.'s [3] (see also [2]) compressed representation of de Bruijn graphs — which is based on something like an FM-index and uses select queries to find nodes' predecessors, and which we call the BOSS representation for the authors' initials — and the possibility of extending it to coloured de Bruijn graphs. Our plan for future work is to view a coloured de Bruijn graph as a union of normal de Bruijn graphs, and relatively compress the BOSS representations of those graphs. Due to space constraints, we provide a brief summary of the BOSS representation and coloured de Bruijn graphs as an appendix. In Section 2 we describe how we implement relative select, and in Section 3 we show experimentally that our implementation is practical. For simplicity and because we are interested mainly in working with DNA, we assume throughout that the size of the alphabet is constant, and we work in the word-RAM model with $\Omega(\log(|S_1| + |S_2|))$-bit words.

2   Design

Although our implementation of relative select is made up of steps that are individually very simple, the overall effect might be confusing. To mitigate this, we break our presentation into pieces: first, we consider the case when S_2 is a subsequence of S_1; then, we consider the case when S_2 is a supersequence of S_1; and finally, we combine our solutions for these special cases to obtain a general solution. We close this section with a small example.

Lemma 1. *Given a select data structure for a string S_1, and a subsequence S_2 of S_1, we can store $\mathcal{O}(|S_1| - |S_2|)$ extra words and support any select query on S_2 using $\mathcal{O}(\log\log|S_1|)$ time on top of a select query on S_1.*

Proof. We store a bitvector $B[1..|S_1|]$ with 1s marking the characters of S_1 that do not appear in S_2. For each distinct character x, we store a bitvector $B_x[1..\text{occ}(x, S_1)]$, where $\text{occ}(x, S_1)$ is the number of occurrences of x in S_1, with 1s marking the occurrences of x in S_1 that do not appear in S_2. We use the same sparse-bitvector implementation as in Section 1, so this takes a total of $\mathcal{O}(|S_1| - |S_2|)$ extra words and lets us compute

$$S_2.\text{select}_x(i) = B.\text{rank}_0(S_1.\text{select}_x(B_x.\text{select}_0(i)))$$

using $\mathcal{O}(\log \log |S_1|)$ time on top of a select query on S_1. To see why this equality holds, consider that $B_x.\text{select}_0(i)$ returns the rank in S_1 of the ith x that appears in S_2; $S_1.\text{select}_x(B_x.\text{select}_0(i))$ returns the position of that x in S_1; and $B.\text{rank}_0(S_1.\text{select}_x(B_x.\text{select}_0(i)))$ returns the position of that x in S_2. □

Lemma 2. *Given a select data structure for a string S_1, and a supersequence S_2 of S_1, we can store $\mathcal{O}(|S_2| - |S_1|)$ extra words and support any select query on S_2 using $\mathcal{O}(\log \log |S_2|)$ time on top of a select query on S_1.*

Proof. We store a bitvector $B[1..|S_2|]$ with 1s marking the characters of S_2 that do not appear in S_1, and a select data structure for the subsequence D of S_2 consisting of those marked characters. For each distinct character x, we store a bitvector $B_x[1..\text{occ}(x, S_2)]$ with 1s marking the occurrences of x in S_2 that do not appear in S_1. We use a sparse-bitvector implementation again, so this takes a total of $\mathcal{O}(|S_2| - |S_1|)$ extra words and lets us compute

$$S_2.\text{select}_x(i) = \begin{cases} B.\text{select}_0(S_1.\text{select}_x(B_x.\text{rank}_0(i))) & \text{if } B_x[i] = 0, \\ B.\text{select}_1(D.\text{select}_x(B_x.\text{rank}_1(i))) & \text{if } B_x[i] = 1. \end{cases}$$

using $\mathcal{O}(\log \log |S_2|)$ time on top of a select query on S_1. To see why this equality holds, suppose the ith x in S_2 also appears in S_1, so $B_x[i] = 0$. Consider that $B_x.\text{rank}_0(i)$ returns the rank of that x in S_1; $S_1.\text{select}_x(B_x.\text{rank}_0(i))$ returns the position of that x in S_1; and $B.\text{select}_0(S_1.\text{select}_x(B_x.\text{rank}_0(i)))$ returns the position of that x in S_2. Now suppose the ith x in S_2 does not appear in S_1, so $B_x[i] = 1$. Consider that $B_x.\text{rank}_1(i)$ returns the rank of that x in D; $D.\text{select}_x(B_x.\text{rank}_1(i))$ returns the position of that x in D; and $B.\text{select}_1(D.\text{select}_x(B_x.\text{rank}_1(i)))$ returns the position of that x in S_2. □

Theorem 1. *Given a select data structure for a string S_1, and another string S_2, we can store $\mathcal{O}(d)$ extra words, where d is the edit distance between S_1 and S_2, and support any select query on S_2 using $\mathcal{O}(\log \log(|S_1| + |S_2|))$ time on top of a select query on S_1.*

Proof. Consider a sequence of d single-character insertions, deletions and substitutions that turns S_1 into S_2. Let C be the common subsequence of S_1 and S_2 consisting of characters left unchanged by these d edits (or a longer common subsequence if we can find one). By Lemma 1, we can store $\mathcal{O}(d)$ extra words and support any select query on C using $\mathcal{O}(\log \log |S_1|)$ time on top of a select query on S_1. By Lemma 2, we can then store $\mathcal{O}(d)$ extra words and support any select query on S_2 using $\mathcal{O}(\log\log|S_2|)$ time on top of a select query on C. Therefore, we can store $\mathcal{O}(d)$ extra words on top of the select data structure for S_1 and support any select query on S_2 using $\mathcal{O}(\log\log(|S_1| + |S_2|))$ time on top of a select query on S_1.   □

For example, consider the strings S_1 = TCTGCGTAAAAGGTGC and S_2 = TGCTCGTAAAACGCG (the BWTs of GCACTTAGAGGTCAGT and GCACTA-GACGTCAGT, respectively, from the running example in Belazzougui et al.'s paper). Their edit distance is 5 and their longest common subsequence is C = TCTCGTAAAAGG. If we already have a select data structure for S_1 and we want one for S_2, we first add support for relative select on C by the bitvectors $B, B_A, \ldots, B_T$, shown below; then we add support for relative select on S_2 by storing bitvectors $B', B'_A, \ldots, B'_T$, also shown below, and a select data structure for D = GCC. We note that if we have a relative FM-index for S_2 with respect to S_1, then it already includes B, B' and D.

$$B[1..16] = 0001000000010101 \qquad B'[1..15] = 010000000001010$$

$$
\begin{aligned}
B_A[1..4] &= 0000 & B'_A[1..4] &= 0000 \\
B_C[1..3] &= 001 & B'_C[1..4] &= 0011 \\
B_G[1..5] &= 10100 & B'_G[1..4] &= 1000 \\
B_T[1..4] &= 0001 & B'_T[1..3] &= 000
\end{aligned}
$$

To compute $S_2.\text{select}_C(4)$, for instance, we check $B'_C[4]$ and see it is 1, meaning the fourth C in S_2 does not appear in C. Since $B'_C.\text{rank}_1(4) = 2$, it is the second C in D. Since $D.\text{select}_C(2) = 3$, it is the third character in D. Finally, since $B'_1.\text{select}_1(3) = 14$, it is the 14th character in S_2, meaning $S_2.\text{select}_C(4) = 14$.

To compute $S_2.\text{select}_G(3)$, we check $B'_G[3]$ and see it is 0, meaning the third G in S_2 also appears in C. Since $B'_G.\text{rank}_0(3) = 2$, it is the second G in C. Since

$$C.\text{select}_G(2) = B.\text{rank}_0(S_1.\text{select}_G(B_G.\text{select}_0(2))) = 11,$$

it is the 11th character in C. Finally, since $B'_1.\text{select}_0(11) = 13$, it is the 13th character in S_2, meaning $S_2.\text{select}_G(3) = 13$.

3   Experiments

We augmented our implementation of the Relative FM-index with the new select structure.[1] The implementation is written in C++ and based on the Succinct Data Structures Library 2.0 [6]. We used g++ version 4.8.1 to compile the code, and ran the experiments on a system with two 16-core AMD Opteron 6378 processors and Linux kernel 2.6.32. We used a single core for the query tests.

As our reference sequence, we used the 1000 Genomes Project's version of the GRCh37 human reference genome, both with (3.096 Gbp) and without (3.036 Gbp) chromosome Y. For a target sequence, we chose the maternal haplotypes of the 1000 Genomes Project's individual NA12878 (3.036 Gbp) [11]. We built

---

[1] https://github.com/jltsiren/relative-fm

Table 1. Average query times for 100 million random LF and Ψ queries on NA12878 stored relative to the human reference genome, with and without chromosome Y.

	FM-index			Relative FM-index			+ Relative Select	
ChrY	space	LF	Ψ	space	LF	Ψ	total space	Ψ
yes	1090 MB	0.55 s	1.22 s	218 MB	3.95 s	48.0 s	382 MB	6.11 s
no	1090 MB	0.55 s	1.11 s	181 MB	3.84 s	44.8 s	331 MB	6.12 s

a plain FM-index for the reference sequences and the target sequence, as well as relative FM-indexes for the target sequence relative to both references and with and without structures for relative select; the lengths of the common subsequences used were 2.992 Gbp and 2.991 Gbp, respectively. In all cases, we used plain bitvectors in the wavelet trees and entropy-compressed bitvectors [10] for marking the common subsequences.

To test the performance of relative select, we ran 100 million random $\Psi(i) = \mathsf{BWT.select}_c(i - \mathsf{C}[c])$ queries on the BWT of the target sequence, using a plain FM-index and Relative FM-indexes with and without relative select. (Character c is the ith character in the BWT in sorted order, while $\mathsf{C}[c]$ is the number of occurrences of characters smaller than c in the BWT.) The implementation of Ψ in the Relative FM-index without relative select was based on binary searching with rank queries. As a comparison, we also ran $\mathsf{LF}(i) = \mathsf{C}[\mathsf{BWT}[i]] + \mathsf{BWT.rank}_{\mathsf{BWT}[i]}(i)$ queries. Table 1 shows the results: the relative FM-indexes without relative select are each about a fifth the size of the normal FM-indexes but rank queries are about seven times slower and select queries are about forty times slower; the relative FM-indexes with relative select are about a third the size of the normal FM-indexes but select queries are only about five times slower (rank queries are unaffected).

References

1. Belazzougui, D., Gagie, T., Gog, S., Manzini, G., Sirén, J.: Relative FM-indexes. In: Moura, E., Crochemore, M. (eds.) SPIRE 2014. LNCS, vol. 8799, pp. 52–64. Springer, Heidelberg (2014)
2. Boucher, C., Bowe, A., Gagie, T., Puglisi, S., Sadakane, K.: Variable-order de Bruijn graphs. In: Proc. DCC (2015)
3. Bowe, A., Onodera, T., Sadakane, K., Shibuya, T.: Succinct de Bruijn graphs. In: Raphael, B., Tang, J. (eds.) WABI 2012. LNCS, vol. 7534, pp. 225–235. Springer, Heidelberg (2012)
4. Burrows, M., Wheeler, D.: A block sorting lossless data compression algorithm. Tech. Rep. 124, Digital Equipment Corporation (1994)
5. Ferragina, P., Manzini, G.: Indexing compressed text. J. ACM **52**, 552–581 (2005)
6. Gog, S., Beller, T., Moffat, A., Petri, M.: From theory to practice: plug and play with succinct data structures. In: Gudmundsson, J., Katajainen, J. (eds.) SEA 2014. LNCS, vol. 8504, pp. 326–337. Springer, Heidelberg (2014)
7. Idury, R., Waterman, M.: A new algorithm for DNA sequence assembly. J. Comput. Biol. **2**, 291–306 (1995)

8. Iqbal, Z., Caccamo, M., Turner, I., Flicek, P., McVean, G.: De novo assembly and genotyping of variants using colored de Bruijn graphs. Nature Genetics **44**, 226–232 (2012)
9. Pevzner, P., Tang, H., Waterman, M.: An Eulerian path approach to DNA fragment assembly. Proc. Nat. Acad. Sci. **98**, 9748–9753 (2001)
10. Raman, R., Raman, V., Rao Satti, S.: Succinct indexable dictionaries with applications to encoding k-ary trees, prefix sums and multisets. ACM Trans. Algorithms **3**, 43 (2007)
11. Rozowsky, J., et al.: AlleleSeq: analysis of allele-specific expression and binding in a network framework. Molecular Systems Biology **7**, 522 (2011)

A   de Bruijn Graphs

In biology, the (edge-centric) *kth-order de Bruijn graph* for a set of strings (e.g., DNA reads) is the graph whose nodes are those strings' k-mers (substrings of length k), with a directed edge (u, v) from u to v if at least one of the strings contains a substring of length $k + 1$ with u as a prefix and v as a suffix. We label (u, v) with the last character of v. Almost all state-of-the-art DNA assemblers build contigs via Eulerian assembly [7,9] on de Bruijn graphs, making their space- and time-efficient representation an important problem in bioinformatics.

Bowe et al. add certain dummy nodes and edges, sort the edges into the right-to-left lexicographic order of the nodes they leave, and take the last column of the matrix whose rows are the edges in sorted order (or, equivalently, take the last character in each edge). The result is like a BWT in which edges correspond to characters and nodes correspond to the substrings containing all their out-edges' characters. For example, for the string TACGTCGACGACT and $k = 3$, Bowe et al. derive the edge-BWT TCCGTGGATAA\$C. (This example is from [2].) With some auxiliary data structures, we can use rank and select queries on this edge-BWT to navigate forward and backward in the graph.

For the two strings TACGTCGACGACT and TACGACGCGACT and $k = 3$, the de Bruijn graph is 2 nodes larger than the graphs for strings separately. If we store whether each edge occurs in the first string, the second string, or both, then the result is a *coloured de Bruijn graph*. Coloured de Bruijn graphs were introduced by Iqbal et al. [8] for detecting variations between individuals' genomes, and are now also used in other areas of genomics. We can view the coloured de Bruijn graph as the union of each graph consisting of edges of the same colour. In a future paper we will show how to combine the BOSS representations of the individual de Bruijn graphs to obtain a representation of the coloured de Bruijn graph, and also how to relatively compress the auxiliary data structures for the BOSS representations of the individual graphs.

We can use Belazzougui et al.'s result to relatively compress the edge-BWTs of the individual graphs while still supporting rank over them. For example, the edge-BWTs for TACGTCGACGACT and TACGACGCGACT with $k = 3$ are TCCGTG-GATAA\$C and TCCGTGGACAA\$, respectively. They are so close — edit distance 2 — because most of the strings' 4-tuples are common to both and, thus, most of their de Bruijn graphs' edges are common to both. We note that, for reasonable values of k, most of the $(k+1)$-mers in genomes from the same species should also be common to most of the genomes. In this paper we showed how to support relative select on similar strings, which we will eventually need to navigate backward across edges in our representation of coloured de Bruijn graphs.

Temporal Query Classification
at Different Granularities

Dhruv Gupta[1,2](✉) and Klaus Berberich[1]

[1] Max Planck Institute for Informatics, Saarbrücken, Germany
[2] Saarbrücken Graduate School of Computer Science, Saarbrücken, Germany
{dhgupta,kberberi}@mpi-inf.mpg.de

Abstract. In this work, we consider the problem of classifying time-sensitive queries at different temporal granularities (day, month, and year). Our approach involves performing Bayesian analysis on time intervals of interest obtained from pseudo-relevant documents. Based on the Bayesian analysis we derive several effective features which are used to train a supervised machine learning algorithm for classification. We evaluate our method on a large temporal query workload to show that we can determine the temporal class of a query with high precision.

1 Introduction

Information needs conveyed in a time-sensitive query can only be served properly if the temporal class associated with it can be determined. Determining the temporal class of a query is an important stepping stone to larger components in a time-sensitive information retrieval system. For instance, selection of an appropriate retrieval model or deciding whether to diversify documents along time. Existing work in this direction has only relied on publication dates while ignoring temporal expressions in document contents. Temporal expressions allow us to analyze events in web collections which may not have reliable publication dates associated with them. This alleviates the problem of being restricted to the time period covered by the publication dates of the document collection. Analyzing the temporal class based on temporal expressions is challenging as (i) they are highly uncertain (e.g. `early 1990's`, `during last century`) and (ii) are present at multiple granularities (e.g., day, month, and year).

Determining the temporal class of a query has been studied before in approaches given in [2,5,6]. The approaches proposed in [2,6] however have three major problems. First, all approaches only use publication dates for a given a timestamped document collection. This may serve the purpose well for time-sensitive queries concerning only current events covered in the news. But it may be inadequate for queries covering historic events. Second, prior approaches ignore the fact that events described in a query may be periodic (e.g., `summer olympics` or `nobel prize physics`) or they may be aperiodic (e.g., `economic depression`). Third, temporal ambiguity is considered only at a single level of granularity. However temporal ambiguity may vary according to

© Springer International Publishing Switzerland 2015
C. Iliopoulos et al. (Eds.): SPIRE 2015, LNCS 9309, pp. 156–164, 2015.
DOI: 10.1007/978-3-319-23826-5_16

granularity. Consider, as a concrete example, the query `summer olympics tokyo athletics`. Relying only on publication dates this query would be incorrectly classified as temporally unambiguous; whereas it is temporally ambiguous at day granularity. Such an information need would be best served if these shortcomings can be overcome.

Hypothesis. By addressing the aforementioned problems we hypothesize we can improve upon the classification of time-sensitive queries containing: (i) historical events & entities; (ii) periodic events; and (iii) temporal ambiguity at a particular granularity.

We build on our earlier work [4] which suggests interesting time intervals using temporal expressions. For classifying queries we identify multiple features from Bayesian analysis of the time intervals of interest. We show the effectiveness of our proposed approach over prior work on a large testbed of time-sensitive queries.

Contributions made in this work are: (i) temporal class taxonomy taking into account multiple granularities and (a)periodicity of events (Section 4); (ii) determining time intervals as intents for temporally ambiguous queries (Section 5); (iii) effective features that outperform prior approaches (Section 6); and (iv) a large test bed of time-sensitive queries collected from previously available resources such as TREC time-sensitive queries [2], NTCIR Geo-Time queries [3] and other resources available on the Web (Section 7); which is made publicly available for future research.

2   Related Work

In this section, we describe the prior work in our context. Our work largely tries to overcome the shortcomings of work presented in [6]. The work by Jones and Diaz [6] describes a taxonomy of temporal classes for time-sensitive queries. They discuss various features derived from the distribution of document publication dates. Examples of these features are temporal clarity, kurtosis, and auto-correlation. We extend their taxonomy in our work to accomodate temporal ambiguity at different granularities, as well as (a)periodicity of events.

More recent efforts in the direction of temporal query classification have been described in works by Joho et al. [5] and Kanhabua et al. [7]. The *Temporalia* project described by Joho et al. [5] considers temporal query classification with a novel temporal taxonomy. The temporal classes they target are qualitatively labeled as *past, recency* and *future*. This has two major caveats. First, the qualitative classes leave room for ambiguity in temporal intents. For example, for `nba playoffs last week` the temporal class can either be *past* or *recent*. Second, quantitatively no information can be discerned about the *exact* time intervals the temporal class refers to. Both these problems are addressed in our work.

Detecting seasonality and periodicity associated with web-queries has also been explored by Kanhabua et al. [7]. They propose to use features acquired from web-query logs. Additionally, akin to existing approaches, they rely on features derived from signal processing on time series of publication dates from an external document collection. These may not be adequate to detect the temporal class at different granularities, as shown in our experiments.

3   Preliminaries

We now introduce the notation used throughout the paper and the approach for identifying time intervals of interest.

Notation. Consider a document collection D. Each document $d \in D$ consists of a bag of keywords d_{text} and a bag of temporal expressions d_{time}. We let $|d_{\text{text}}|$ and $|d_{\text{time}}|$ denote the cardinalities of these bags. A temporal expression is a four-tuple, $T = \langle b_l, b_u, e_l, e_u \rangle$. Each component of T is drawn from a time domain $\mathcal{T}$ (usually $\mathbb{N}$). A temporal expression T may refer to any time interval $[b, e] \in \mathcal{T} \times \mathcal{T}$ with $b_l \le b \le b_u$, $e_l \le e \le e_u$, and $b \le e$. We treat temporal expressions as a set of time intervals and let $|T|$ denote the number of time intervals that T may refer to.

   Time Intervals of Interest to the given keyword query q are identified using the approach proposed in [4]. In a nutshell, with R as the set of pseudo-relevant documents, the approach assigns the probability:

$$P([b, e] \mid q) = \sum_{d \in R} P([b, e] \mid d) P(d \mid q),$$

to time interval $[b, e]$. The first probability is estimated as

$$P([b, e] \mid d_{\text{time}}) = \frac{1}{|d_{\text{time}}|} \sum_{T \in d_{\text{time}}} \frac{\mathbb{1}([b, e] \in T)}{|T|},$$

following [1]. The second probability is estimated from the query likelihoods $P(q|d)$ under a unigram language model with Dirichlet smoothing, that is:

$$P(d \mid q) = \frac{P(q \mid d)}{\sum_{d' \in R} P(q \mid d')}.$$

4   Temporal Class Taxonomy

We propose a new taxonomy taking into account additional classes for periodicity, aperiodicity, and multiple granularities (day, month, and year). It builds on the existing taxonomy proposed by Jones and Diaz [6]. The taxonomy, depicted in Figure 1, is arrived at by noting the observations explained in this section.

   Atemporal queries as per [6] are time-invariant in nature. Thus, an atemporal query at year granularity also implies that it is atemporal at a finer level of granularity (day and month) and vice-versa.

   Temporally unambiguous queries are those with a unique time interval of interest associated with them. If a given query is identified to be unambiguous at day granularity then it will also be unambiguous at any coarser granularity.

For instance, an unambiguous query at day level `concorde crash` is also unambiguous at year level. However, this does *not* imply that an unambiguous query at year level may necessarily be unambiguous at month or day level.

Temporally ambiguous queries are those which may have multiple time intervals of interest associated with them. Ambiguity associated with a query may lie at different granularities. A temporally ambiguous query at a finer granularity may be unambiguous at coarser granularity. However, we make the distinction that a query ambiguous at *any* granularity be deemed temporally ambiguous at that level of granularity. For example the query `summer olympics 2000 rowing` is temporally ambiguous at day level granularity. Another aspect that we investigate is the (a)periodicity of keyword queries. For example the query `summer olympics` should be classified as a periodic temporally ambiguous query. Recurring events, such as `tropical storms`, which may not have fixed periodicity are classified as aperiodic. In this work, we limit ourselves to (a)periodicity at year level. However, approach described next is equally applicable to (a)periodicity at month and day granularity.

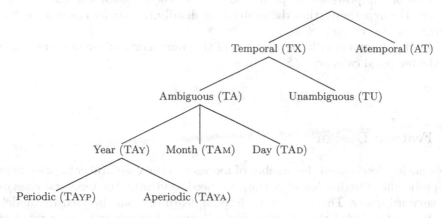

Fig. 1. Temporal class taxonomy with (a)periodicity and multiple granularity

5   Bayesian Analysis

To determine the temporal class of the keyword query q we first obtain the probability distribution of time intervals of interest at all three temporal granularities $P([b, e]|q)$. We consider time intervals of size equal to the granularity under consideration (e.g., for year granularity $[b, e]$ spans one year). We smooth $P([b, e]|q)$ with time intervals from the entire document collection D, in order to avoid the zero-probability problem:

$$\hat{P}([b, e]|q) = \lambda \cdot P([b, e]|q) + (1 - \lambda) \cdot P([b, e]|D),$$

where,

$$P([b, e]|D) = \frac{1}{|D_{\text{time}}|} \sum_{T \in D_{\text{time}}} \frac{\mathbb{1}([b, e] \in T)}{|T|}.$$

Detecting Multiple Modes. The distribution $\hat{P}([b, e]|q)$ is next analyzed for multi-modality. For this we utilize a *Bayesian Mixture Model* fitted using *reversible jump Markov chain Monte Carlo* (MCMC) procedure outlined by Xu et al. [11]. The approach fits an unknown probability distribution by approximating it as mixture of Gaussian distributions. Utilizing this approach has the advantage of performing both model selection and model fitting at the same time. That is, we do not need to know the number of components in the mixture apriori. The mixture model is described as follows:

$$\hat{P}([b, e]|q) = \sum_{i=1}^{k} w_i \cdot \mathcal{N}(\mu_i, \sigma_i),$$

such that $\sum_i^k w_i = 1$; μ_i and σ_i characterize the mean and standard deviation of the normal distribution $\mathcal{N}(\mu_i, \sigma_i)$. To assess confidence of our hypothesis whether $\hat{P}([b, e]|q)$ is multi-modal, we take Bayes factor as an objective. Bayes factor is the ratio of the posterior to prior odds. If the Bayes factor exceeds 100, we consider the hypothesis, that the probability distribution under observation has multiple modes, correct.

The time intervals with the means μ_i of the components of the mixture model are the temporal categories $(S^i_{[b,e]})$ of q:

$$S = \langle S^1_{[b,e]}, S^2_{[b,e]}, \dots, S^k_{[b,e]} \rangle.$$

6   Feature Design

After having determined the number of modes and the temporal categories from the probability distribution $\hat{P}([b, e]|q)$, we need to identify the temporal class of the keyword query. This is done by deriving features from the mixture model. The features encoded are: (i) modality, (ii) fuzzy feature, and (iii) p-value of randomness test. Next, we discuss the motivation behind the features.

Modality feature describes the number of modes identified by the Bayesian Mixture Model. The intuition is if $\hat{P}([b, e]|q)$ is unimodal ($|S| = 1$), then the temporal class should be temporally unambiguous. If the probability distribution $\hat{P}([b, e]|q)$ is multi-modal ($|S| > 1$), then it should be temporally ambiguous.

Fuzzy Feature. To analyze the temporally ambiguous query for periodicity we use the concept of fuzzy numbers. Fuzzy logic is used here to account for outlier cases in periodic events e.g. for `summer olympics` anomalous years would be $[1936, 1936]$ and $[1948, 1948]$. Specifically, we capture the membership value of the time lags between the time intervals associated with different modes against a fuzzy number around the mean of the time lags $(\hat{\Phi})$.

We first identify the time lags between ordered set temporal categories Φ:

$$\Phi^i_{[b,e]} = \langle t | t \in S^{i+1}_{[b,e]} - S^i_{[b,e]} \rangle \quad \text{with,} \quad \hat{\Phi} = \frac{\sum_{i=1}^{n} \Phi^i_{[b,e]}}{n}.$$

Difference between intervals is calculated element-wise. Next we construct a triangular fuzzy number with $\hat{\Phi}$ whose membership function is given by:

$$\mu(x) = \begin{cases} \frac{1}{1+x^2} & \text{if } x \neq \hat{\Phi} \\ 1 & \text{if } x = \hat{\Phi} \end{cases}$$

The motivation is: if $\mu(x) \neq 0 \; \forall x \in \Phi$, then issued query is a periodic query with period approximately equal to that of $\hat{\Phi}$. Otherwise if $\exists x \in \Phi$ for which $\mu(x) = 0$ then query could potentially be aperiodic.

Randomness Test. For atemporal queries we check $\hat{P}([b,e]|q)$ for randomness. For this we perform a a two-tailed runs up and down test for randomness [10] on time lags. We next note the p-value of this test as a feature. This feature thus captures if the time lags are randomly generated or not.

For a given query, we construct the feature vector at day, month, and year granularity. The feature data is then subsequently used for classification via a decision tree.

7 Experimental Evaluation

7.1 Datasets

Document Collection used was The New York Times Annotated [1] corpus. Temporal annotations for it were obtained from the authors of [1]; they used TARSQI [9]. TARSQI is able to identify both explicit and implicit temporal expressions in text.

Queries. The challenging aspect of evaluating our approach was compiling a list of queries for temporally ambiguous class at different granularities. To this end we use various previously published resources [4], TREC time-sensitive queries [2], NTCIR Geo-Time queries [3], and also manually compiled some of them from the Web. Table 1 summarizes the query workload. This dataset is publicly available with an accompanying description of how it was compiled at:

http://resources.mpi-inf.mpg.de/dhgupta/data/spire2015

Table 1. Query set sizes for our evaluation setup

Set Id				Description	Size
		TA$_Y$	TA$_{YP}$	Periodic and ambiguous at year	113
	TA		TA$_{YA}$	Aperiodic and ambiguous at year	118
TX		TA$_M$		Ambiguous at month	64
		TA$_D$		Ambiguous at day	74
		TU		Unambiguous	142
AT				Atemporal	154

---

[1] http://www.catalog.ldc.upenn.edu/LDC2008T19

Baseline. We use the approach proposed by Jones and Diaz [6] as a baseline. We selected the best-performing temporal features from [6] to build the baseline classifier. Temporal features considered were: first order autocorrelation, kurtosis, and features derived from a burst model. We consider these features at year level granularity for time intervals of interest. Since, we are considering time intervals of interest generated by the approach in [4]; we take into account temporal expressions and publication dates at year granularity for the baseline also.

7.2 Setup

We discus various aspects related to the experimental setup next.

Parameters. For identifying time intervals of interest we considered top-50 ($|R| = 50$) pseudo-relevant documents. The mixing parameter for smoothing the distribution was set to $\lambda = 0.70$. For modality assessment we performed reversible jump MCMC procedure with 2,200 iterations with initial 200 burn-in iterations.

Implementation. All methods for feature extraction were implemented in R, a statistical programming language. Procedure for reversible MCMC sampling was obtained from [11], also in R. The decision tree classifier based on the CART algorithm was utilized from the R package, *rpart* [8]. The generative model for time intervals of interest was programmed in Java.

Measures. For a classification task we report the standard measures for comparing performances – *Precision, Recall* and F_1. Statistical significance of our results is reported with the p-value calculated using McNemar's test. We also show an unweighted κ statistic for the classifiers. The κ statistic measures the agreement between the observed accuracy to the expected accuracy by chance. Higher value of κ indicates better discrimination between different classes.

7.3 Experimental Results

Below we report the results for each temporal class. In order to accurately gauge the performance we also report the confusion matrix for our classifier. Training and test set were constructed by sampling without replacement. Train and test set split was 80% to 20% of the combined query workload (665 queries). Baseline (B) and proposed approach (A) were trained on different random samples.

Discussion. For the *temporally ambiguous* class we can classify very accurately at all levels of granularity. For the *atemporal* case we can also discern the class with high precision. However, it is relatively difficult to identify *temporally unambiguous* queries. Another class that is hard to detect is *aperiodic*. Compared to the baseline our approach performs better in all classes.

Failure Analysis. There were two classes for which our approach didn't perform well: (i) temporally unambiguous and (ii) aperiodic.

   Temporally unambiguous may not have been classified precisely due to pseudo-relevance feedback. In pseudo-relevant documents it is inevitable to not

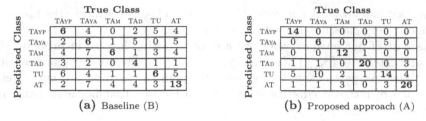

Fig. 2. Confusion matrix for decision tree

Table 2. Statistics by class for decision trees: baseline (B) and proposed approach (A)

	Statistics by Class					
	Precision		Recall		F_1	
Class	B	A	B	A	B	A
TX	0.81	**0.92**	0.79	**0.02**	0.80	**0.92**
TA	0.70	**0.87**	0.64	**0.71**	0.67	**0.78**
TAy	0.45	**0.80**	0.34	**0.51**	0.39	**0.62**
TAyp	0.29	**1.00**	0.26	**0.67**	0.27	**0.80**
TAya	0.32	**0.55**	0.20	**0.33**	0.24	**0.41**
TAm	0.24	**0.92**	0.50	**0.71**	0.32	**0.80**
TAd	0.36	**0.80**	0.22	**0.91**	0.28	**0.85**
TU	0.26	**0.39**	0.33	**0.64**	0.29	**0.48**
AT	0.38	**0.76**	0.41	**0.79**	0.39	**0.78**
Macroaverage	0.31	**0.74**	0.32	**0.67**	0.30	**0.69**
p-value	4.5e-2	2.2e-16				
κ-value	0.16	**0.62**				

consider other related events, which act as noise, for the keyword query in the distribution of time intervals. Some misclassified example queries are : `chernobyl soviet union` and `president nixon associated press orlando`.

Aperiodic queries were mostly misclassified as *unambiguous*. Most of the queries in the aperiodic query set comprise of famous personalities. Thus, the errors can be due to a very specific events in the corpus linked to the entity. Misclassified examples from this category are `george bush jnr`, `madrid bombing`, `muhammad ali`, and `ronald reagan`.

8   Conclusion and Future Work

We have proposed how to solve the problem of temporal query classification at multiple levels of granularity. Additionally, we can predict the periodicity of events with very high accuracy. We inspect both content temporal expressions as well as publication dates of pseudo-relevant documents given only a keyword query. Our approach considers features based on Bayesian analysis of the time intervals of interest. Experiments indicate that heuristics identified by us are able to predict the temporal class for ambiguous queries really well. In contrast, for *unambiguous* and *aperiodic* queries it is difficult to classify the class by looking at the pseudo-relevant documents. All in all, our classifier achieves the target of temporal query classification with good accuracy.

As part of our ongoing work; we are investigating how to incorporate the temporal categories ($S^i_{[b,e]}$) of given keyword query for diversifying search results along time. As part of our future work; we plan to carry out an end to end evaluation of retrieval effectiveness when considering disambiguated temporal categories.

References

1. Berberich, K., Bedathur, S., Alonso, O., Weikum, G.: A language modeling approach for temporal information needs. In: Gurrin, C., He, Y., Kazai, G., Kruschwitz, U., Little, S., Roelleke, T., Rüger, S., van Rijsbergen, K. (eds.) ECIR 2010. LNCS, vol. 5993, pp. 13–25. Springer, Heidelberg (2010)
2. Dakka, W., Gravano, L., Ipeirotis, P.G.: Answering general time-sensitive queries. IEEE Trans. Knowl. Data Eng. **24**(2), 220–235 (2012)
3. Gey, F., Larson, R., Kando, N., Machado, J., Sakai, T.: Ntcir-geotime overview: Evaluating geographic and temporal search. In: Proc. NTCIR-8 Workshop Meeting, pp. 147–153 (2010)
4. Gupta, D., Berberich, K.: Identifying time intervals of interest to queries. In: CIKM 2014 (2014)
5. Joho, H., Jatowt, A., Blanco, R.: NTCIR temporalia: a test collection for temporal information access research. WWW **2014**, 845–850 (2014)
6. Jones, R., Diaz, F.: Temporal profiles of queries. ACM Trans. Inf. Syst. **25**(3) (2007)
7. Kanhabua, N., Nguyen, T.N., Nejdl, W.: Learning to detect event-related queries for web search. TempWeb 2015 at WWW 2015 (2015)
8. Therneau, T., Atkinson, B., Ripley, B.: rpart: Recursive Partitioning and Regression Trees (2014). R package version 4.1-8
9. Verhagen, M., Mani, I., Sauri, R., Littman, J., Knippen, R., Jang, S.B., Rumshisky, A., Phillips, J., Pustejovsky, J.: Automating temporal annotation with TARSQI. In: ACL 2005 (2005)
10. Wald, A., Wolfowitz, J.: On a test whether two samples are from the same population. The Annals of Mathematical Statistics **11**(2), 147–162 (1940)
11. Xu, L., Bedrick, E.J., Hanson, T., Restrepo, C.: A comparison of statistical tools for identifying modality in body mass distributions. Journal of Data Science **12**(1), 175–196 (2014)

Prefix and Suffix Reversals on Strings

Guillaume Fertin$^{(\boxtimes)}$, Loïc Jankowiak, and Géraldine Jean

LINA UMR CNRS 6241, Université de Nantes, Nantes, France
{guillaume.fertin,geraldine.jean}@univ-nantes.fr,
loic.jankowiak@etu.univ-nantes.fr

Abstract. The SORTING BY PREFIX REVERSALS problem consists in sorting the elements of a given permutation π with a minimum number of prefix reversals, i.e. reversals that always imply the leftmost element of π. A natural extension of this problem is to consider strings (in which any letter may appear several times) rather than permutations. In strings, three different types of problems arise: *grouping* (starting from a string S, transform it so that all identical letters are consecutive), *sorting* (a constrained version of grouping, in which the target string must be lexicographically ordered) and *rearranging* (given two strings S and T, transform S into T). In this paper, we study these three problems, under an algorithmic viewpoint, in the setting where two operations (rather than one) are allowed: namely, *prefix and suffix* reversals - where a suffix reversal must always imply the rightmost element of the string. We first give elements of comparison between the "prefix reversals only" case and our case. The algorithmic results we obtain on these three problems depend on the size k of the alphabet on which the strings are built. In particular, we show that the grouping problem is in P for $k \in [2; 4]$ and when $n - k = O(1)$, where n is the length of the string. We also show that the grouping problem admits a PTAS for any constant k, and is 2-approximable for any k. Concerning sorting, it is in P for $k \in [2; 3]$, admits a PTAS for constant k, and is NP-hard for $k = n$. Finally, concerning the rearranging problem, we show that it is NP-hard, both for $k = O(1)$ and $k = n$. We also show that the three problems are FPT when the parameter is the maximum number of blocks over the source and target strings.

1 Introduction and Notations

A usual task in comparative genomics consists in comparing pairs of genomes, in order to define how (dis)similar they are. A genome is usually modeled as a permutation π or a string S, and the pairwise comparison of genomes is usually done by computing a distance between them. Such a distance is generally defined based on so-called "rearrangements", which are large scale operations involving contiguous segments of π (resp. S). The most famous example of such rearrangements is *reversals* (see e.g. [3,8]), which take a contiguous segment of π (resp. S), reverse its order, and reincorporate this new segment at the same location.

Supported by GRIOTE project, funded by Région Pays de la Loire

© Springer International Publishing Switzerland 2015
C. Iliopoulos et al. (Eds.): SPIRE 2015, LNCS 9309, pp. 165–176, 2015.
DOI: 10.1007/978-3-319-23826-5_17

More generally, given a set $\mathcal{S}$ of possible rearrangements, the distance between two permutations π and π' (resp. two strings S and T) is defined as the minimum number of operations from $\mathcal{S}$ that are required to obtain π' starting from π (resp. to obtain T starting from S). Many studies are concerned with the special case where $|\mathcal{S}| = 1$, i.e. only one type of rearrangement is authorized. There is a very abundant literature on the subject, and we refer the reader to [6] for a rather recent survey on the topic. In this paper, our object of study is strings, a generalization of permutations, in which any letter may occur several times. The set of rearrangements we authorize here is composed of two operations: *prefix reversals*, which necessarily involve the first letter of the string, and *suffix reversals*, in which the last letter of the string must be involved. When dealing with strings, three types of problems are usually considered: *grouping* (starting from a string S, transform it so that all identical letters are consecutive), *sorting* (a constrained version of grouping, in which the target string must be lexicographically ordered) and *rearranging* (given two strings S and T, transform S into T). In this paper, we study these three problems, under an algorithmic viewpoint.

Terminology and Basic Properties. Throughout the paper, and otherwise stated, any string S is a sequence of letters built on an alphabet Σ of cardinality k, and is of length n. Any string S is said to be *fully k-ary* if each letter of $\Sigma = \{0, 1 \ldots k - 1\}$ appears at least once in S. A letter l in S is said to be *lonely* whenever it appears only once in S, and *abundant* otherwise. When $k = n$, i.e. in the special case of permutations, the identity string is denoted Id_n. Two strings S and T are said to be *compatible* if they contain the same multiset of letters. If S is the concatenation of two strings S_1 and S_2, it is denoted $S = S_1.S_2$, or $S_1 S_2$ if clear from the context. If S is the concatenation of p occurrences of the same string s, this will be denoted $S = s^p$. For any $1 \leq i \leq j \leq n$, $S[i..j]$ denotes the substring of S starting at position i and ending at position j. A *block* in S is a maximal substring in which all letters are equal. Let $b(S)$ be the number of blocks in S. Given two compatible strings S and T, we let $b_{max} = \max\{b(S), b(T)\}$. Given two integers $1 \leq i \leq j \leq n$, a *reversal* $\rho(i, j)$ of a string S is the operation that reverses $S[i..j]$ and reincorporates it at the same location, i.e. $S = s_1 s_2 \ldots s_{i-1} \underline{s_i s_{i+1} \ldots s_{j-1} s_j} s_{j+1} \ldots s_n$ is transformed into $S' = s_1 s_2 \ldots s_{i-1} \underline{s_j s_{j-1} \ldots s_{i+1} s_i} s_{j+1} \ldots s_n$. For readability, the substring $S[i..j]$ to be reversed will often be underlined, as shown in the above example. If a string S is totally reversed, i.e. the reversal is $\rho(1, n)$, then we will denote the result as $\overline{S}$. A *prefix reversal* (resp. *suffix reversal*) is a reversal $\rho(1, j)$ (resp. $\rho(i, n)$), i.e. the first (resp. last) letter of S is involved in the reversal. We may use the terms *p-reversal* (resp. *s-reversal*) to denote a reversal that is prefix (resp. suffix).

We say that a fully k-ary string S is *grouped* if it contains exactly k blocks, i.e. if, for any letter l in Σ, all occurrences of l are consecutive. We say that S is *sorted* if S is grouped, and if S is lexicographically ordered. Given two compatible strings S and T, and given a set $\mathcal{S}$ of possible operations on strings, the *distance* between S and T is defined as the minimum number of operations from $\mathcal{S}$ that are required to obtain T, starting from S. In this paper, the two operations we allow are prefix reversals and suffix reversals. We are now ready to formally describe the three problems we are interested in, in their decision version.

GROUPING STRINGS BY PREFIX AND SUFFIX REVERSALS (GPSR)
Instance: A string S, an integer r
Question: Is there a sequence of at most r p- and s-reversals that transforms S into a grouped string T?

SORTING STRINGS BY PREFIX AND SUFFIX REVERSALS (SPSR)
Instance: A string S, an integer r
Question: Is there a sequence of at most r p- and s-reversals that transforms S into a sorted string T?

REARRANGING STRINGS BY PREFIX AND SUFFIX REVERSALS (RPSR)
Instance: Two compatible strings S and T, an integer r
Question: Is there a sequence of at most r p- and s-reversals that transforms S into T?

Given a string S, the least integer r for which the answer to GPSR is positive is called the *grouping distance*, and is denoted $d^g_{PS}(S)$. Similarly, the sorting and rearranging distances are denoted $d^s_{PS}(S)$ and $d^r_{PS}(S,T)$. Note that when $k = n$ (i.e. strings are in fact permutations), the grouping problem is trivial (S is already grouped), while the sorting and rearranging problems are equivalent (up to a relabeling of the input permutations). Note also that when dealing with the grouping (resp. sorting) problem, it is easy to show, by contradiction, that there always exist an optimal grouping (resp. sorting) algorithm for S that never cuts a block (otherwise, a more efficient algorithm would exist). Thus, the length of the blocks is irrelevant, and we can always assume that each block of S is of length 1 - we call such a string a *normalized* string. Consequently, any (prefix or suffix) reversal that makes two identical letters consecutive reduces the length of a normalized string by 1, and such a reversal is called a *1-flip*. A reversal that is not a 1-flip is called a *0-flip*.

Known results. Comparing pairs of strings by means of their rearrangement distance has mostly been studied in the case where only one type of operation is allowed, such as reversals [2,5,12], transpositions [5,12] or translocations [4]. When prefix reversals only are allowed, the main results are the following: first, the grouping problem is polynomial when $k = 2$ and $k = 3$, and admits a PTAS for constant $k \geq 4$ [9]. Concerning sorting, the problem is polynomial for $k = 2$ [4], and admits a PTAS for constant $k \geq 4$ [9]. Finally, concerning the rearrangement problem, it is NP-hard even for $k = 2$ [9] (and even for more constrained inputs [2]), it admits a PTAS for dense instances (which are instances for which the distance is cn with $c = O(1)$ [12]), and it is FPT in b_{max} [2]. When both prefix *and* suffix operations are permitted, recent works have either focused on the case of permutations [10,11], or on the case of transreversals on small-size alphabets ($k = 2$ and $k = 3$) [13].

Table 1. Our algorithmic results concerning problems GPSR, SPSR and RPSR.

$\|\Sigma\|$	Grouping	Sorting	Rearranging
2	in P		NP-hard
3	in P		NP-hard
4	in P	PTAS	NP-hard
$O(1)$	PTAS	PTAS	NP-hard
$n - c$, with $c = O(1)$	in P		
n	in P	NP-hard	NP-hard
any $2 \leq k \leq n$	2-approx. / FPT	FPT	FPT

Our Results. In this paper, we study the three problems GPSR, SPSR and RPSR, under an algorithmic viewpoint. The main algorithmic results we have obtained are summarized in Table 1. Due to space constraints, most proofs are omitted and deferred to the full version of this paper.

2   A Detour by Prefix Reversals

In this section, our main goal is to compare the prefix distance to the prefix/suffix distance for grouping, sorting and rearranging. Let $d_P^g(S)$ (resp. $d_P^s(S)$, $d_P^r(S,T)$) denote the grouping (resp. sorting, rearranging) distance when only prefix reversals are allowed. First, note that for any string S, any grouping/sorting/rearranging algorithm that uses only p-reversals is also a valid grouping algorithm using p- and s-reversals, thus, trivially, $d_{PS}^*(S) \leq d_P^*(S)$, for any $* \in \{g, s, r\}$. Notice also that any s-reversal can be realized by at most 3 p-reversals, which directly yields $d_P^*(S) \leq 3d_{PS}^*(S)$ for any $* \in \{g, s, r\}$. However, we can improve this upper bound, as shown below.

Proposition 1. *For any strings S and T, (i) $d_{PS}^g(S) \leq d_P^g(S) \leq 2d_{PS}^g(S)$, (ii) $d_{PS}^s(S) \leq d_P^s(S) \leq 2d_{PS}^s(S)+1$, (iii) $d_{PS}^r(S,T) \leq d_P^r(S,T) \leq 2d_{PS}^r(S,T)+1$.*

Another natural question that emerges is whether there exist examples of strings for which the prefix and prefix/suffix distances significantly differ. Below, we answer this question positively for sorting in permutations (Theorem 1), and for rearranging in strings built on constant-size alphabets (Theorem 3).

Theorem 1. *For infinitely many values of n, there exist a permutation π of length n such that $d_P^s(\pi) - d_{PS}^s(\pi) = \Omega(n)$.*

Proof. Take $n = 6q$, and consider the following permutation $\pi = s_0\, s_1 \ldots s_{q-1}$, where $s_i = 6i + 2\ 6i + 1\ 6i + 4\ 6i + 3\ 6i + 6\ 6i + 5$, $0 \leq i \leq q - 1$. Notice that here, $\Sigma = \{1, 2 \ldots 6q\}$. We will show that $d_P^s(\pi) \geq \frac{n}{2} + \frac{n}{12}$, while $d_{PS}^s(\pi) \leq \frac{n}{2}+1$, which proves that $d_P^s(\pi) - d_{PS}^s(\pi) = \Omega(n)$.
In order to show that $d_{PS}^s(\pi) \leq \frac{n}{2}+1$, let us distinguish two cases, depending on the parity of q. First, when q is even, we claim that π can be sorted by repeating

$\frac{n}{4}$ times the following two types of reversals: a p-reversal of length $n-2$ followed by an s-reversal of length $n-2$. Indeed, it can be shown by induction that for any $1 \le i \le \frac{n}{4}$, after the i-th s-reversal has been applied, the first $4i$ elements of the current permutation are sorted in increasing order, from $n - 4i + 1$ to n. Thus, when $i = \frac{n}{4}$, the permutation is in fact Id_n, and exactly $\frac{n}{2}$ reversals have been used. When q is odd, the reversals sequence is slightly different: first, repeat $\frac{n-2}{4}$ times the following two reversals: a p-reversal of length $n-2$ followed by an s-reversal of length $n-2$. Then, apply two final reversals: a p-reversal of length $n-2$, and a p-reversal of length n. As for the previous case, it can be shown by induction that, for any $1 \le i \le \frac{n+2}{4}$, after the i-th p-reversal of length $n-2$ has been applied, the last $4i - 2$ elements of the current permutation are sorted in decreasing order, from n to $n - 4i + 3$. Thus, when $i = \frac{n}{4}$, the permutation is in fact $\overline{Id_n}$, which is then totally reversed to obtain Id_n. Moreover, exactly $\frac{(n-2)+(n+2)}{4} + 1 = \frac{n}{2} + 1$ reversals have been used.

Now let us prove $d_P^s(\pi) \ge \frac{n}{2} + \frac{n}{12}$. We say that a *breakpoint* occurs in π whenever $|\pi_{i+1} - \pi_i| \ne 1$, $1 \le i \le n$, where we artificially assume that $\pi_{n+1} = n + 1$. Consider an optimal sequence of p-reversals that sorts π, and let us distinguish between *efficient* p-reversals, which strictly reduce the number of breakpoints, and *wastes*, which do not. Hence, we have $d_P^s(\pi) = e + w$, where e (resp. w) denotes the number of efficient p-reversals (resp. wastes) in an optimal solution. Since π contains $\frac{n}{2}$ breakpoints, we necessarily have $e \ge \frac{n}{2}$. The rest of the proof is then dedicated to proving $w \ge \frac{n}{12}$. For this, for each substring s_i, $0 \le i \le q-1$, let the *breaking p-reversal* for s_i (denoted $br(i)$) be the last p-reversal before which s_i or $\overline{s_i}$ existed, and does not exist anymore afterwards. Let us sort the $br(i)$s in their order of appearance. We thus obtain a string of strictly positive integers $(a_0, a_1 \dots a_{t-1})$, where a_j is the rank of the j-th breaking p-reversal. We first note that this string contains exactly $t = q$ elements, since there are q strings to break, and at most one string can be broken by a p-reversal. We will now prove that $w \ge \frac{n}{12}$, by showing that it takes at least one waste to break two strings of the form s_i or $\overline{s_i}$. Consider any integer $j \le q-1$, and let s be the string for which the a_j-th p-reversal is breaking. Let us first suppose $s = s_i$ (the case $s = \overline{s_i}$ is similar and leads to the same conclusion). Let π_0 be the permutation before the a_j-th p-reversal takes place. Then $\pi_0 = X.s_i.Y$. If X is empty, then $\pi_0 = s_i.Y$, and any breaking p-reversal for s_i is a waste. Otherwise, the only breaking p-reversal for s_i that is not a waste can occur if $X = (6i + 7).X'$, and we obtain $\pi_0' = 6i + 3 \ 6i + 4 \ 6i + 1 \ 6i + 2.\overline{X'}.6i + 7 \ 6i + 6 \ 6i + 5.Y$. But then any p-reversal that follows is necessarily a waste - however, this reversal may be breaking for another string s_l. We thus have $w \ge \frac{q}{2} = \frac{n}{12}$, which allows us to conclude that $d_P^s(\pi) \ge \frac{n}{2} + \frac{n}{12}$. $\qquad \square$

Since rearranging and sorting are equivalent when we deal with permutations, we have the following corollary.

Corollary 1. *For infinitely many values of n, there exist permutations π and π', of length n, such that $d_P^r(\pi, \pi') - d_{PS}^r(\pi, \pi') = \Omega(n)$.*

The following result, which is helpful in the remainder of the paper, is an extension of [9] (Theorem 10) to p- and s-reversals and to constant-size alphabets.

Theorem 2. *For any pair of compatible strings S and T built on Σ, and for any constant $k \geq 2$, there exist a pair of compatible strings S' and T' built on Σ', such that $d_P^r(S', T') = d_P^r(S, T)$, $d_{PS}^r(S', T') = d_{PS}^r(S, T)$ and $|\Sigma'| = k$.*

Putting together Theorem 2 and Corollary 1 above, we have the following result, which shows that rearranging by p- and s-reversals may be much more efficient than by p-reversals only, when dealing with constant-size alphabets.

Theorem 3. *For infinitely many values of n, and for any $k = O(1)$, there exist pairs (S, T) of compatible fully k-ary strings of length n such that $d_P^r(S, T) - d_{PS}^r(S, T) = \Omega(n^{\frac{1}{3}})$.*

3   Grouping Strings by Prefix and Suffix Reversals

This section describes our results concerning the GPSR problem. We begin by general bounds, before proving that GPSR is polynomial for any $k \in [2; 4]$. Here, n denotes the length of the normalized version of the considered string.

Proposition 2. *For any $k \geq 3$, and for any fully k-ary string S of length n, $n - k \leq d_{PS}^g(S) \leq \min\{n - 3, 2(n - k)\}$.*

Proof. Let S be a normalized fully k-ary string S of length n. First, it is easy to see that for any $k \geq 2$, $d_{PS}^g(S) \geq n-k$. Indeed, after S is grouped, its normalized string will be of length k. Since any reversal decreases the length of the string by at most 1, we conclude that it takes at least $n - k$ reversals to group S. Now, let us show that $d_{PS}^g(S) \leq min\{n-3, 2(n-k)\}$ for any $k \geq 3$. Let us first show that $d_{PS}^g(S) \leq n - 3$. For this, we apply the following algorithm, that we call $A_{PS}^g(S)$: while the last letter l of S is abundant, perform an s-reversal (which is in fact a 1-flip) to group l with another occurrence of it. Suppose we have made $n - k'$ such reversals; we then end up with a string of length $k \leq k' \leq n$, but whose last letter is lonely. This last letter will then be ignored, and we will group the first $k' - 1$ letters, using only p-reversals. By Lemma 2 of [9], we know that $(k' - 1) - 2$ reversals suffice to group this string by p-reversals. Thus, altogether, our grouping algorithm uses no more than $(n - k') + (k' - 3) = n - 3$ reversals. Now let us prove that $d_{PS}^g(S) \leq 2n - 2k$. Let $A_{PS}^{\prime g}(S)$ be the following algorithm. Let a (resp. b) be the first (resp. last) letter in S. If a or b is abundant in S, then there exist a 1-flip which groups two occurrences of a (or b), so we apply that reversal. Otherwise, if there exist an abundant letter c in S, we apply a 0-flip that brings c either at the first or last position in the string. We iterate this process while there exist abundant letters in the string. When the algorithm stops, the string is grouped, and in its normalized version, is of length k. It is not hard to see that our algorithm uses exactly $n - k$ 1-flips. Moreover, since each 0-flip is necessarily followed by a 1-flip, and since the last reversal is necessarily a 1-flip, we conclude that the number of 0-flips cannot exceed the number of 1-flips, that is $n - k$. Altogether, our $A_{PS}^{\prime g}(S)$ uses no more than $2(n - k)$ reversals.   □

Proposition 3. *For any fully binary string S of length n, $d^g_{PS}(S) = n - 2$, and the GPSR problem is in P.*

Theorem 4. *For any fully ternary string S of length n, $d^g_{PS}(S) = n - 3$ and the GPSR problem is in P.*

We know that $n - 4 \leq d^g_{PS}(S) \leq n - 3$ for any fully quaternary string S of length n (just invoke Proposition 2 with $k = 4$). Our goal here is to find a polynomial-time algorithm that determines $d^g_{PS}(S)$ for any such string, and that provides an optimal grouping scenario. To do so, we divide the set of all fully quaternary strings into two classes: *bad* strings which need a 0-flip to be grouped (leading to $d^g_{PS}(S) = n - 3$) and *good* strings that can be grouped using only 1-flips (leading to $d^g_{PS}(S) = n - 4$).

In the following, let $\Sigma = \{x, y, z, t\}$. A *H-string* is defined to be any fully quaternary string of length n having a lonely letter at one of its extremities. Thus, a H-string is of the form $S'.x$ (resp. $x.S'$) where S' is a fully ternary string of length $n - 1$ defined on $\Sigma' = \Sigma - \{x\}$. Since an s-reversal on a H-string of type $S'.x$ is necessary a 0-flip, we can use the results from [9] concerning the grouping distance of fully ternary strings if only p-reversals are allowed. Actually, H-string stands for "Hurkens-string", in reference to the first author in [9]. For the case $x.S'$, we can symmetrically apply the same results by inverting the roles of p- and s-reversals.

Definition 1. *Let S be a fully quaternary string. S is said to be a* bad *string if it belongs to one of the following categories:*

1. *alternating strings in which a letter x appears every two positions, i.e. strings of the form $x(\{y, z, t\}x)^a$, $(\{y, z, t\}x)^a$, $(x\{y, z, t\})^a$, $\{y, z, t\}(x\{y, z, t\})^b$ with $a \geq 3$ and $b \geq 2$.*
2. *semi-alternating strings which are not alternating strings and for which any 1-flip leads to an alternating string. They have the following form: $y(xz)^a xty(xz)^b xt$ with $a, b \geq 0$ and $a + b \geq 1$.*
3. *H-strings of the form $T.x$ and $x.\overline{T}$, where T is a bad string according Definition 1 in [9].*
4. *$Y_1 = tzxyzxt$, $Y_2 = yxztyzxt$, $Y_3 = tzyzxyzxzt$.*

Any other fully quaternary string is said to be good.

Lemma 1. *For any fully quaternary bad string S of length n, $d^g_{PS}(S) = n - 3$.*

Lemma 2. *For any fully quaternary good string S of length n, $d^g_{PS}(S) = n - 4$.*

Putting together Lemmas 1 and 2, we get the following result.

Theorem 5. *The GPSR problem is in P for $k = 4$, and there is an $O(n^3)$ algorithm that provides an optimal grouping scenario for S.*

We end this section with three algorithmic results concerning more general values of k: when $k = O(1)$ (Theorem 6), when $n - k = O(1)$ (Proposition 4), and for any value of k (Theorem 7).

Theorem 6. *The* GPSR *problem admits a PTAS for any fully k-ary string with $k \geq 5$ and $k = O(1)$.*

Proposition 4. *The* GPSR *problem is in P for any string such that $n - k = O(1)$.*

Proof. Suppose $k = n - c$ with $c = O(1)$. By Proposition 2, we know that $d_{PS}^g(S) \leq \min\{n-3, 2(n-k)\}$, that is $d_{PS}^g(S) \leq 2c$. Thus, consider the following algorithm: starting from S, branch on all the possible p- and s-reversals, up to a distance of $2c$, while stopping to explore a branch as soon as the string is grouped. Then, return the smallest number of reversals that have been found, for reaching a grouped string, starting from S. Note that, starting from any string of length n, there are at most $2n-3$ different possible reversals: $n-2$ of each kind (prefix or suffix), and reversing the whole string. Thus, exploring all the possible reversals up to a distance of $2c$ cannot create more than $(2n-3)^{2c}$ strings. By Proposition 2, we know that the optimal solution lies within this research space. Besides, since $c = O(1)$, the number of generated strings is polynomial, and consequently so is our algorithm.   □

Theorem 7. *For any $2 \leq k \leq n$,* GPSR *is 2-approximable.*

Proof. Algorithm $A_{PS}^{\prime g}$, described in proof of Proposition 2, runs in $O(n^2)$, since it uses at most $2(n - k)$ reversals and each reversal can be done in linear time. Let us write $OPT_{PS}^g(S)$ the minimum number of reversals needed to group S, and $a'(S)$ the number of reversals used by $A_{PS}^{\prime g}$. We know from Proposition 2 that $OPT_{PS}^g(S) \geq n - k \geq \frac{a'(S)}{2}$. Hence, $A_{PS}^{\prime g}$ is an approximation algorithm for GPSR, whose ratio is 2.   □

4   Sorting Strings by Prefix and Suffix Reversals

In this section, we describe our results concerning the SPSR problem. Here, n is the length of the considered string which has been normalized. First note that for any fully k-ary string S of length n, $d_{PS}^s(S) \geq d_{PS}^g(S)$. Since $d_{PS}^g(S) \geq n-k$ by Proposition 2, we obtain the following result.

Proposition 5. *For any $k \geq 2$, and for any fully k-ary string S of length n, $d_{PS}^s(S) \geq n - k$.*

Proposition 6. *For any fully binary string S of length n, $d_{PS}^s(S) = n - 2$.*

Proposition 7. *For any fully ternary string S of length $n \geq 4$, $d_{PS}^s(S) \leq n-2$.*

Propositions 5 and 7 show that when $k = 3$, $n - 3 \leq d_{PS}^s(S) \leq n - 2$ for any fully ternary string S. In the following, we will show that there exist a polynomial-time algorithm that determines $d_{PS}^s(S)$ for any fully ternary string S, and provides an optimal sorting scenario. Because the lower and upper bounds differ by one, determining $d_{PS}^s(S)$ consists in deciding whether S can be sorted using 1-flips only (leading to $d_{PS}^s(S) = n - 3$), or whether one 0-flip is necessary (leading to $d_{PS}^s(S) = n - 2$).

Table 2. Bad strings of length $n \geq 9$ for the SPSR problem with $k = 3$.

$n = 2p + 1$ is odd		$n = 2p$ is even			
Type	Form	Type	Form	Type	Form
1	$\{1,2\}(0\{1,2\})^p$	1	$(0\{1,2\})^p$	2	$(2\{0,1\})^p$
2	$\{0,1\}(2\{0,1\})^p$	3	$(\{0,1\}2)^p$	4	$(\{1,2\}0)^p$
3	$(0\{1,2\})^p0$	5	$0(12)^{p-1}0$	6	$1(02)^{p-1}1$
4	$2(\{0,1\}2)^p$	7	$1(20)^{p-1}1$	8	$2(01)^{p-1}2$
5	$2(10)^p$	9	$2(10)^{p-1}1$	10	$(12)^{p-1}10$
6	$(21)^p0$				

Note that for $n \leq 10$, by exhaustive search, it is possible to determine which strings of length n need $n - 3$ reversals to be sorted, and which ones need $n - 2$ reversals. Moreover, for each of them, an optimal sorting scenario can be provided. Now suppose $n \geq 9$, and let Table 2 describe a set of strings that we call *bad* strings.

Our goal is to show that, for any $n \geq 9$, the bad strings we have listed are the only ones having sorting distance $n - 2$. First, for $n = 9$ and $n = 10$, it is easy to verify that this is the case. We show the following lemma, which, combined with the base cases ($n = 9$ and $n = 10$) discussed above, shows that all the bad strings have sorting distance $n - 2$.

Lemma 3. *Let $n \geq 11$, and let S be a bad string of length n. Any 1-flip on S leads to a bad string S'.*

Because of Lemma 3 above and the base cases $n = 9$ and $n = 10$ previously discussed, an easy induction follows, which proves that any bad string has sorting distance $n - 2$. It now remains to show that no other string is bad. Let us call *good* any string which is not bad. We have the following lemma.

Lemma 4. *Let $n \geq 11$, and let S be a good string. If a 1-flip transforms S into a bad string S', then there exist another 1-flip that transforms S into a good string S''.*

Again, by induction, and based on Lemma 4, we conclude that any good string can be sorted by 1-flips only, and thus its sorting distance is $n - 3$. The only bad strings being the ones listed above, we have the following theorem.

Theorem 8. *Let S be a fully ternary string of length n. If S is bad, then $d_{PS}^s(S) = n - 2$, while if S is good, then $d_{PS}^s(S) = n - 3$. Hence, the SPSR problem is in P for $k = 3$. Moreover, there is an $O(n^3)$ algorithm that provides an optimal sorting scenario for S.*

Proposition 8. *For any fully quaternary string S of length $n \geq 4$, $d_{PS}^s(S) \leq n-1$.*

Theorem 9. *The SPSR problem admits a PTAS for any fully k-ary string with $k \geq 4$ and $k = O(1)$.*

5   Rearranging Strings by Prefix and Suffix Reversals

In this section, we describe our results concerning the RPSR problem. We start with the case where $n = k$, i.e. the case of permutations.

Theorem 10. *The* RPSR *problem is NP-hard for permutations.*

Proof. This result is obtained by a reduction from SORTING BY PREFIX REVERSALS in permutations, which has been proved to be NP-hard in [1]. More precisely, it is proved that it is NP-hard to decide whether a permutation π can be sorted using exactly $d_b(\pi)$ prefix reversals, where $d_b(\pi)$ is the number of breakpoints in π. Note that $d_b(\pi)$ is a lower bound on the number of p-reversals to sort π, since the number of breakpoints decreases by at most one each time a p-reversal is applied, and since $d_b(Id_n) = 0$. For any input permutation π of SORTING BY PREFIX REVERSALS, supposing π is built over the integers $1, 2 \ldots n$, we create the permutation $\pi' = \pi.(n+1).(n+2)$. We now claim that π can be sorted using $d_b(\pi)$ p-reversals iff π' can be sorted using $d_b(\pi)$ p- and s-reversals.

($\Rightarrow$) Clearly, if π can be sorted using $d_b(\pi)$ p-reversals, this is also a valid scenario for p- and s-reversals, that also sorts π'.

($\Leftarrow$) Suppose π' can be sorted using $d_b(\pi)$ p- and s-reversals. Because $d_b(\pi') = d_b(\pi)$, it can easily be seen that (i) no p-reversal ρ_p involves $(n+1)$ nor $(n+2)$ (otherwise, ρ_p would not decrease the number of breakpoints) and (ii) no s-reversal ρ_s is applied (otherwise, and because of (i), ρ_s would not decrease the number of breakpoints either). Hence, the sorting scenario for π' is also a valid p-reversals scenario for π, and it uses no more than $d_b(\pi)$ p-reversals. $\square$

A straightforward consequence of the above theorem is that the SPSR is also NP-hard, since sorting and rearranging are equivalent when we deal with permutations.

Corollary 2. *The* SPSR *problem is NP-hard for permutations.*

Combining Theorem 10 with Theorem 2, we get the following corollary.

Corollary 3. *The* RPSR *problem is NP-hard, even for strings built on constant-size alphabets.*

In the binary case, however, we are able to prove a stronger result: RPSR problem is NP-hard, even when all 0-blocks are of length 1 in S.

Theorem 11. *The* RPSR *problem is NP-hard, even for binary strings in which all 0-blocks are of length 1 in S.*

Proof. The reduction is from the NP-hard PARTITION problem [7]. PARTITION is a decision problem whose input is a multiset $A = \{a_1, a_2 \ldots a_q\}$ of positive integers, with $\sum_{a_i \in A} a_i = 2N$. The question is whether it is possible to partition A into A_1 and A_2 such that $\sum_{a_i \in A_1} a_i = \sum_{a_j \in A_2} a_j = N$. For any instance $A = \{a_1, a_2 \ldots a_q\}$ of PARTITION, we construct two binary strings S and T as follows: $S = 01^{a_1}01^{a_2}01\ldots 01^{a_q}0$, and $T = 1^N0^{q+1}1^N$. Now we claim that there is a positive solution for PARTITION iff $d^r_{PS}(S,T) = 2q - 2$.

($\Rightarrow$) First, suppose there exist a partition A_1, A_2 of A such that $\sum_{a_i \in A_1} a_i = \sum_{a_j \in A_2} = N$. Suppose $|A_1| = x_1$ and $|A_2| = x_2$, thus $x_1 + x_2 = q$. Then, starting from S, using two p-reversals, we can merge two blocks whose lengths are from A_1. We can iterate this process to merge a third block, etc. until the x_1 blocks of lengths corresponding to A_1 are merged. It thus takes $2(x_1 - 1)$ p-reversals to obtain one block of 1s of length N, x_2 blocks of 1s corresponding to the lengths in A_2, the string beginning and ending with 0. After another $2(x_2 - 1)$ p-reversals, we end up with $S' = 0^l1^N0^m1^N0^n$, with $l + m + n = q + 1$. It then takes 2 reversals (a suffix and a prefix) to obtain $T = 1^N0^{q+1}1^N$. Thus, a total of $2(x_1 - 1) + 2(x_2 - 1) + 2 = 2q - 2$ reversals suffice to rearrange S into T, since $x_1 + x_2 = q$.

($\Leftarrow$) Suppose that the answer to PARTITION is negative. This implies that, when rearranging from S to T, there will be at least one reversal which increases the number of blocks of 1s - let us call such a reversal a *breaking reversal*. However, S contains $b(S) = 2q + 1$ blocks, while T contains $b(T) = 3$ blocks. Since in one reversal, at most two blocks can be merged, necessarily at least $b(S) - b(T) = 2q - 2$ *merging reversals* are necessary. Thus, counting merging and breaking reversals together, we conclude that $d^r_{PS}(S,T) \geq 2q - 1$.   □

Theorem 12. *The RPSR problem is FPT, when parameterized by b_{max}.*

Theorem 12 above is adapted from [2]. Note also that it is valid regardless of the size of the alphabet. Another consequence is the following corollary; indeed, grouping and sorting are special cases of rearranging, in which the target string has $k = |\Sigma| \leq b_{max}$ blocks with $b_{max} = b(S)$.

Corollary 4. *The GPSR and SPSR problems are FPT, when parameterized by $b(S)$.*

6   Conclusion and Open Questions

In this paper, we have studied the three problems GPSR, SPSR and RPSR under an algorithmic viewpoint, and have provided a number of results, especially focused either on small-size alphabets ($k = O(1)$) and big alphabets ($k = n$, i.e. permutations). Some questions remain open, and we would like to end this paper by listing a few of them: first, is there a significant difference between the prefix and the prefix/suffix distance for the grouping problem? Second, what is the complexity of GPSR? Concerning the size of the alphabet, what can be said about SPSR and RPSR when $n - k = O(1)$? Can we prove that GPSR and SPSR are in P for $k = O(1)$? Finally, we note that the variant where strings are signed could also be studied.

References

1. Bulteau, L., Fertin, G., Rusu, I.: Pancake flipping is hard. In: Rovan, B., Sassone, V., Widmayer, P. (eds.) MFCS 2012. LNCS, vol. 7464, pp. 247–258. Springer, Heidelberg (2012)
2. Bulteau, L., Fertin, G., Komusiewicz, C.: Reversal distances for strings with few blocks or small alphabets. In: Kulikov, A.S., Kuznetsov, S.O., Pevzner, P. (eds.) CPM 2014. LNCS, vol. 8486, pp. 50–59. Springer, Heidelberg (2014)
3. Caprara, A.: Sorting by reversals is difficult. In: Proceedings of the First Annual International Conference on Computational Molecular Biology. RECOMB 1997, pp. 75–83. ACM, New York (1997)
4. Christie, D.: Genome Rearrangement Problems. Ph.D. thesis, University of Glasgow (1998)
5. Christie, D.A., Irving, R.W.: Sorting strings by reversals and by transpositions. SIAM J. Discret. Math. 14(2), 193–206 (2001)
6. Fertin, G., Labarre, A., Rusu, I., Tannier, E., Vialette, S.: Combinatorics of Genome Rearrangements. Computational Molecular Biology. MIT Press (2009)
7. Garey, M.R., Johnson, D.S.: Computers and Intractability; A Guide to the Theory of NP-Completeness. W. H. Freeman & Co., New York (1990)
8. Hannenhalli, S., Pevzner, P.A.: Transforming cabbage into turnip: Polynomial algorithm for sorting signed permutations by reversals. J. ACM 46(1), 1–27 (1999)
9. Hurkens, C.A.J., van Iersel, L., Keijsper, J., Kelk, S., Stougie, L., Tromp, J.: Prefix reversals on binary and ternary strings. SIAM J. Discrete Math. 21(3), 592–611 (2007)
10. Lintzmayer, C.N., Dias, Z.: Sorting permutations by prefix and suffix versions of reversals and transpositions. In: Pardo, A., Viola, A. (eds.) LATIN 2014. LNCS, vol. 8392, pp. 671–682. Springer, Heidelberg (2014)
11. Lintzmayer, C.N., Dias, Z.: On the diameter of rearrangement problems. In: Dediu, A.-H., Martín-Vide, C., Truthe, B. (eds.) AlCoB 2014. LNCS, vol. 8542, pp. 158–170. Springer, Heidelberg (2014)
12. Radcliffe, A.J., Scott, A.D., Wilmer, E.L.: Reversals and transpositions over finite alphabets. SIAM J. Discret. Math. (2005)
13. Rahman, M.K., Rahman, M.S.: Prefix and suffix transreversals on binary and ternary strings. Journal of Discrete Algorithms 33, 160–170 (2015)

Filtration Algorithms for Approximate Order-Preserving Matching

Tamanna Chhabra, Emanuele Giaquinta$^{(\boxtimes)}$, and Jorma Tarhio

Department of Computer Science, Aalto University,
P.O. Box 15400, 00076 Aalto, Finland
{tamanna.chhabra,emanuele.giaquinta,jorma.tarhio}@aalto.fi

Abstract. The exact order-preserving matching problem is to find all the substrings of a text T which have the same length and relative order as a pattern P. Like string maching, order-preserving matching can be generalized by allowing the match to be approximate. In approximate order-preserving matching two strings match if they have the same relative order after removing up to k elements in the same positions in both strings. In this paper we present practical solutions for this problem. The methods are based on filtration, and one of them is the first sublinear solution on average. We show by practical experiments that the new solutions are fast and efficient.

1 Introduction

The exact string matching problem consists in finding all the occurrences of a pattern string P of length m in a text string T of length n. A recent variant of this problem is the so called order-preserving matching problem [16,14,1,4,3]. In order-preserving matching, the task is to locate all the substrings of T which have the same length and relative order as P. This problem has applications in time series studies such as in the analysis of development of share prices in a stock market. Formally, two strings u and v over an ordered alphabet are *order-isomorphic* if they have the same length and $u_i \leq u_j \Leftrightarrow v_i \leq v_j$, for any $1 \leq i,j \leq |u|$. The term relative order refers to the numerical order of the numbers in the string. In $P = (3, 13, 5, 8, 21)$, the number 3 is the smallest number in the pattern, 5 is the second smallest, 8 is the third smallest number and so on. Therefore, the relative order of P is $1, 4, 2, 3, 5$. For instance, if $T = (6, 10, 55, 36, 45, 66, 6, 21, 28, 15, 36)$, then it can be observed that P and the substring of T starting at location 2 are order-isomorphic.

There exist various solutions for the exact order-preserving matching problem. Kubica et al. [16], Belazzougui et al. [1] and Kim et al. [14] presented generalizations of the Knuth–Morris–Pratt algorithm [15] which solve the problem in $O(n + m \log m)$ time, where m is the length of P and n is the length of T. Belazzougui et al. also presented a sublinear algorithm which runs in $O(\frac{n \log m}{m \log \log m})$ optimal time in the average case. Cho et al. [4] introduced a different sublinear solution based on a generalization of the Boyer–Moore–Horspool

© Springer International Publishing Switzerland 2015
C. Iliopoulos et al. (Eds.): SPIRE 2015, LNCS 9309, pp. 177–187, 2015.
DOI: 10.1007/978-3-319-23826-5_18

algorithm [11]. Independently, Chhabra and Tarhio [3] presented another sublinear solution based on filtration, which was proved to be faster than the previous solutions in practice. Recently, Crochemore et al. [5] presented a generalization of the suffix tree to the order-preserving case.

A natural generalization of the string matching problem can be obtained by allowing the matching to be approximate, so as to search for the substrings of the text T which are *similar* to the pattern P. One classical instance of this kind is the string matching with k mismatches problem, where the task is to find all the substrings of T that are at Hamming distance at most k from P, i.e., that match P with at most k mismatches. With respect to applications of order-preserving matching, approximate search seems more meaningful than exact search. Recently, Gawrychowski and Uznanski proposed a generalization of the order-preserving matching problem to the approximate case [7]. In this model, two strings are k-isomorphic if they have the same relative order after removing up to k elements in the same positions in both strings. In the previous example, for $k = 1$, we get two matches, at location 2 and 7. The algorithm presented by Gawrychowski and Uznanski [7] runs in $O(n(\log \log m + k \log \log k))$ time and it is the only existing solution for this problem to the best of our knowledge. The idea in their method is to quickly filter out positions in T which are non-matching by comparing signatures of the pattern and of the text substrings. As also acknowledged by the authors, this algorithm is rather theoretical and has not been implemented to date.

In this paper, we introduce two practical solutions for the approximate order-preserving matching problem, also based on filtration. Their worst-case time complexities are $O(nm(\lceil m/w \rceil + \log m))$ and $O(n(\lceil m/w \rceil \log \log w + m \log m))$, respectively, where w is the word size in bits, and the former is the first sublinear solution on average. We also present experimental results which show that the filtering is effective and the algorithms are considerably faster than the naive one where all the first $n - m + 1$ text positions are match candidates to be verified.

The paper is organized as follows. Section 2 contains the preliminaries, Section 3 outlines the previous solution for approximate order-preserving matching, Section 4 introduces our solutions based on filtration, Section 5 contains an analysis of the first solution, Section 6 presents the results of practical experiments, and Section 7 concludes the article.

2   Preliminaries

Let Σ be a finite alphabet of symbols and let Σ^* be the set of strings over Σ. Given a string x, we denote by $|x|$ the length of x and by x_i or $x[i]$ the i-th symbol of x, for $1 \leq i \leq |x|$. The concatenation of two strings x and y is denoted by xy. Given two strings x and y, y is a substring of x if there are indices $1 \leq i, j \leq |x|$ such that $y = x_i \ldots x_j$. We denote by $x^r = x_{|x|} x_{|x|-1} \ldots x_1$ the reverse of the string x. Given a string x and a permutation π of $\{1, 2, \ldots, |x|\}$ we denote by $\pi(x)$ the string $x_{\pi(1)} x_{\pi(2)} \cdots x_{\pi(|x|)}$.

Given two strings x and y of length m, the Hamming distance between x and y is $d_h(x, y) = |\{1 \leq i \leq m \mid x_i \neq y_i\}|$, and the matching statistics $M(x, y)$ is an array of $|x|$ integers where $M(x, y)[i]$ denotes the length of the longest substring of x starting at position i that exactly matches a substring of y. A factorization of a string x is a sequence $F_1, F_2, \ldots, F_r$ of strings such that $x = F_1 F_2 \ldots F_r$.

The RAM model is assumed, with words of size w in bits. We use some bitwise operations following the standard notation as in C language: $\&, |, \wedge, \sim,$ $\ll, \gg$ for and, or, xor, not, left shift and right shift, respectively.

Problem definition. Two strings u and v over Σ are *order-isomorphic with k mismatches* [7] or *k-isomorphic*, written $u \approx_k v$, if they have the same length and there exists a subset K of $\{1, 2, \ldots, |u|\}$ of size k at most, such that

$$u_i \leq u_j \Leftrightarrow v_i \leq v_j \text{ for } i, j \in \{1, 2, \ldots, |u|\} \setminus K.$$

The *order-preserving pattern matching with k mismatches* problem is to locate all the substrings of a text T which are k-isomorphic with a pattern P.

3 Previous Solution

This section describes the previous solution formulated for the approximate order-preserving matching problem. The method was proposed by Gawrychowski and Uznanski [7] and is based on the signature of a sequence. The signature $S(a_1, \ldots, a_m)$ of sequence $(a_1, \ldots, a_m)$ is $(1 - pred(1), \ldots, m - pred(m))$ where $pred(i)$ is the position where the predecessor of a_i occurs in the sequence. Its computation takes $O(m \log \log m)$ time by sorting. The key result is that if $(a_1, \ldots, a_m) \approx_k (b_1, \ldots, b_m)$ then the Hamming distance between $S(a_1, \ldots, a_m)$ and $S(b_1, \ldots, b_m)$ is at most $3k$. The algorithm iterates over each substring $(T_i, \ldots, T_{i+m-1})$ in the text T, determining its signature $S(T_i, \ldots, T_{i+m-1})$ in $O(\log \log m)$ time per position. For each position i, it checks if the Hamming distance between $S(T_i, \ldots, T_{i+m-1})$ and $S(P_1, \ldots, P_m)$ is greater than $3k$. This step can be done in $O(k + \log \log m)$ time. If the test is true, the position is discarded. Otherwise, the algorithm checks if $(T_i, \ldots, T_{i+m-1}) \approx_k (P_1, \ldots, P_m)$ by reducing the problem to the one of computing a heaviest increasing subsequence spanning at most $3(k + 1)$ elements. This step can be assessed in $O(k \log \log k)$ time. Therefore, the total time complexity is $O(n(\log \log m + k \log \log k))$.

4 Our Solutions

Given a string u, we denote by $\phi(u)$ the binary string of length $|u| - 1$ such that $\phi(u)_i$ is equal to 1, if $u_i < u_{i+1}$, and to 0 otherwise. The function ϕ is a linear approximation of the order for fast filtration. Observe that any position $2 \leq i < |u|$ in u covers two positions in $\phi(u)$, $i - 1$ and i. Let u and v be two strings and consider the mismatches between the strings $\phi(u)$ and $\phi(v)$. Each mismatch position i identifies a different relative order, in u and v, between the adjacent symbols at positions i and $i + 1$.

As the following Lemma shows, if $u \approx_k v$, then the Hamming distance between $\phi(u)$ and $\phi(v)$ is at most $2k$:

Lemma 1. *For any two strings u and v such that $u \approx_k v$, $d_h(\phi(u), \phi(v)) \leq 2k$.*

Proof. Suppose by contradiction that $d_h(\phi(u), \phi(v)) > 2k$ and let K be a subset of $\{1, 2, \ldots, |u|\}$ satisfying the definition of order-isomorphism with k mismatches for u and v. Observe that for any position i such that $\phi(u)_i \neq \phi(v)_i$ we have $K \cap \{i, i+1\} \neq \emptyset$, as $u_i < u_{i+1}$ and $v_i \geq v_{i+1}$ or *vice versa*. Hence, $|K| > k$, contradicting the hypothesis.   □

For example, if $u = (4, 1, 2, 3)$ and $v = (4, 5, 2, 3)$ we have $u \approx_1 v$, $\phi(u) = (0, 1, 1)$, $\phi(v) = (1, 0, 1)$, and $d_h(\phi(u), \phi(v)) = 2$. The following Lemma defines a distance measure d_o, based on the Hamming distance, which satisfies $d_o(\phi(u), \phi(v)) \leq k$:

Lemma 2. *Given two strings x and y of the same length, let $z_0, z_1, \ldots, z_{|x|}$ be integers such that $z_0 = 0$ and*

$$
z_i = \begin{cases} 1 & \text{if } x_i \neq y_i \wedge z_{i-1} = 0 \\ 0 & \text{otherwise} \end{cases}
$$

for $i = 1, \ldots, |x|$, and let also $H(x, y) = \{i : z_i = 1\}$. Then, for any two strings u and v such that $u \approx_k v$, $d_o(\phi(u), \phi(v)) = |H(\phi(u), \phi(v))| \leq k$.

Proof. Suppose by contradiction that $|H(\phi(u), \phi(v))| > k$ and let K be a subset of $\{1, 2, \ldots, |u|\}$ satisfying the definition of order-isomorphism with k mismatches for u and v. Observe that for any position $i \in H(\phi(u), \phi(v))$ we have $K \cap \{i, i+1\} \neq \emptyset$ and $H(\phi(u), \phi(v)) \cap \{i-1, i+1\} = \emptyset$. Hence, $|K| > k$, contradicting the hypothesis.   □

Informally, the set $H(x, y)$ is the largest subset of the mismatch positions between x and y such that no two positions are consecutive. Therefore, for any two strings u and v, there is no overlap between the positions in u and v covered by any two mismatches in $H(\phi(u), \phi(v))$. Our solution for approximate order-preserving matching consists of two parts: filtration and verification. First the text is filtered with an algorithm so as to locate all the potential matching locations and then the match candidates are verified using a checking routine. Lemma 2 gives a necessary condition for two strings to be k-isomorphic. The idea is to use it in the first phase to quickly filter out non-matching positions in T.

Filtration. For filtration, the consecutive numbers in the pattern P are compared pairwise in the preprocessing phase and transformed into the binary string $\phi(P)$ where a 1 bit means the successive element is greater than the current one and a 0 bit means the opposite. Thereafter, in the search phase, an algorithm is applied to filter the text T and find all the positions i in T such that $d_o(\phi(T_{i,m}), \phi(P)) \leq k$, where $T_{i,m} = T_i T_{i+1} \ldots T_{i+m-1}$ is the substring of T of length m starting at position i. The substrings $T_{i,m}$ are encoded into the binary string $\phi(T_{i,m})$ online in the same way as the pattern. The algorithm determines approximate matches of the transformed pattern $\phi(P)$ in the similarly transformed text $\phi(T)$. As these approximate matches are just the match candidates, they need to be verified using a checking routine.

Verification. For verification, we use the reduction, by Gawrychowski and Uznanski, of the problem of k-isomorphism to the one of computing an heaviest increasing subsequence (Lemma 8, [6]). To compute the heaviest increasing subsequence, we use the algorithm of Jacobson and Vo [13], which runs in $O(m \log m)$ time for a sequence of length m. If we use a sorting algorithm with $O(m \log m)$ worst-case time complexity, the total time complexity of the verification is also $O(m \log m)$. In theory, the time complexity can be reduced to $O(m \log \log m)$ by using Han's sorting algorithm [9] and plugging a data structure which supports predecessor search in $O(\log \log m)$ time, such as van Emde Boas trees, in Jacobson and Vo's algorithm. Observe that in the simpler case where there are no repeated elements in u and v, deciding whether $u \approx_k v$ can be reduced to computing the longest increasing subsequence of $\pi(v)$, where π is a sorting permutation of u.

We propose two filtration algorithms, which build on ideas from two algorithms for string matching with k mismatches, namely approximate SBNDM [10] and the GGF algorithm [8], respectively.

The first filtration algorithm, named AOPF1, is based on the following Lemma, which is a generalization of the method used by approximate SBNDM and first proposed by Chang and Lawler [2]:

Lemma 3. *Given two strings x and y of the same length, let $F_1 g_1 F_2 g_2 \ldots F_r g_r$ be the factorization of x such that $|F_i| = M(x,y)[1 + \sum_{j=1}^{i-1} |F_j g_j|]$, i.e., $F_i \in \Sigma^*$ is the longest substring of x starting at position $1 + \sum_{j=1}^{i-1} |F_j g_j|$ that matches a substring of y, $|g_i| = 2$, for $1 \le i < r$, and $|g_r| \le 2$. Then $r - 1 \le d_o(x,y)$.*

Proof. Let $s_i = |F_1 g_1 \ldots F_{i-1} g_{i-1} F_i| + 1$ be the position of symbol $g_i[1]$ in x, for $1 \le i < r$. For a given s_i, let j be a position in the interval $[s_i - |F_i|, s_i]$ such that $x_j \ne y_j$. Observe that such a position always exists, because $F_i g_i[1]$ is not a substring of y. Then, we have that either $z_j = 1$ or $z_{j-1} = 1$. In the latter case, observe that $j - 1 > s_{i-1}$, for $i > 1$, since $|g_{i-1}| = 2$ and $s_i - |F_i| = s_{i-1} + 2$. Hence, for each s_i we can find a distinct integer j such that $z_j = 1$. $\square$

Informally, the idea is to factorize x into substrings of y which cannot be extended to the right and are separated by 2-grams (pairs of symbols). Let $\hat{m} = |\phi(P)|$. The AOPF1 algorithm slides a window of size $\hat{m}$ along T, starting at position 1. For a given position i in T, the algorithm scans the substring $\phi(T_{i,m})$ from right to left and computes the factors F_j of $\phi(P)^r$ until either it has found $k + 2$ factors or it has scanned the whole substring. In the former case, by Lemma 3, the position is skipped. Otherwise the algorithm performs an additional filtration step, namely it computes $H(\psi(\pi(T_{i,m})), \psi(\pi(P)))$, where $\psi(u)$ is the the string of length $|u| - 1$ such that

$$\psi(u)_i = \begin{cases} 1 & \text{if } u_i < u_{i+1} \\ 2 & \text{if } u_i = u_{i+1} \\ 0 & \text{otherwise} \end{cases}$$

and π is a sorting permutation of P, computed in the preprocessing phase. The position is then verified only if $|H(\psi(\pi(T_{i,m})), \psi(\pi(P)))| \le k$. Indeed, Lemma 2 can be easily proved to hold also when using $\psi(\pi(u))$ and $\psi(\pi(v))$ in place of $\phi(u)$ and $\phi(v)$ (observe that, if $u \approx_k v$, then $\pi(u) \approx_k \pi(v)$). We permute the strings with π so as to obtain a permutation of P where repeated elements are clustered, which allows us to perform a finer filtering using the ψ function. Note that, in principle, this additional filtration works with any permutation and ordering of repeated elements. For example, if $u = (4,1,2,4)$, $v = (4,5,2,3)$ and π is the sorting permutation of u $2,3,1,4$, we have $\pi(u) = (1,2,4,4)$, $\pi(v) = (5,2,4,3)$, $\psi(\pi(u)) = (1,1,2)$, $\psi(\pi(v)) = (0,1,0)$. Note that $d_o(\psi(\pi(u)),\psi(\pi(v))) = 2$, while $d_o(\phi(u),\phi(v)) = d_o(\psi(u),\psi(v)) = 1$, as $\phi(u) = \psi(u) = (0,1,1)$ and $\phi(v) = \psi(v) = (1,0,1)$.

The factors F_j are computed using the nondeterministic factor automaton of $\phi(P)^r$, which is simulated using a modified version of the bit-parallel SBNDM algorithm [17,19]. The SBNDM algorithm is a slightly faster version of BNDM (Backward Nondeterministic DAWG Matching) [18] without bookkeeping of prefixes. The next scanned position is then $i + (\hat{m} - l) + 1$, where l is the length of the longest suffix of $\phi(T_{i,m})$ with at most $k+1$ factors. The worst-case time complexity of this algorithm is $O(nm(\lceil m/w \rceil + \log m))$.

The second filtration algorithm, named AOPF2, is based on the following Lemma:

Lemma 4. *Given two strings x and y of the same length, let $B_p = \{j : j \bmod 2 = p \wedge x_j \neq y_j\}$ and*

$$H'(x,y) = B_0 \cup B_1 \setminus (\{j-1 : j \in B_0\} \cup \{j+1 : j \in B_0\})$$

Then $|H'(x,y)| \leq d_o(x,y)$.

Proof. Let $i \in H'(x,y)$. Observe that, by definition, either i is even or $x_{i-1} = y_{i-1}$. Indeed, if $i-1$ is even and $x_{i-1} \neq y_{i-1}$ then $i \in \{j+1 : j \in B_0\}$ and $i \notin H'(x,y)$. Since $x_i \neq y_i$, we have $z_i = 1$ or $z_{i-1} = 1$. If $z_{i-1} = 1$ then i must be even and therefore $i-1 \in \{j-1 : j \in B_0\}$ and $i-1 \notin H'(x,y)$. Hence, for each $i \in H'(x,y)$ we can find a distinct integer j such that $z_j = 1$.   □

Informally, the set $H'(x,y)$ is the subset of the mismatch positions between x and y such that for each even position we exclude the two adjacent (odd) positions. For example, if $u = (4,1,2,3)$ and $v = (4,5,3,2)$ we have $u \approx_2 v$, $\phi(u) = (0,1,1)$, $\phi(v) = (1,0,0)$, $H(\phi(u),\phi(v)) = \{1,3\}$, $H'(\phi(u),\phi(v)) = \{2\}$. In the preprocessing, the AOPF2 algorithm computes the bit-vector X of $\hat{m}$ bits such that the i-th bit is set to 1 if $P_i < P_{i+1}$ and to 0 otherwise. In other words X is the bit-vector encoding of $\phi(P)$. The algorithm then scans the text from left to right and maintains the bit-vector encoding Y of $\phi(T_{i,m})$, for $i = 1, \ldots, |T|$. For a given position i in T, the bit-vector encodings of B_0 and B_1 are computed as $(X \wedge Y)$ & $01 \ldots 01$ and $(X \wedge Y)$ & $10 \ldots 10$, respectively. Then, we have that the bit-vector encoding of $H'(\phi(P),\phi(T_{i,m}))$ is equal to

$$B_0 \mid B_1 \text{ \& } \sim((B_0 \ll 1) \mid (B_0 \gg 1)).$$

The size of $H'(\phi(P),\phi(T_{i,m}))$ is computed using the sideways addition operation SA on each word of the resulting bit-vector. Given a word X, the sideways

$\text{AOPF1}(P, T, k)$	$\text{AOPF2}(P, T, k)$
1. $\hat{m} \leftarrow |P| - 1$	1. $\hat{m} \leftarrow |P| - 1$
2. $B[0] \leftarrow B[1] \leftarrow 0^{\hat{m}}$	2. $X \leftarrow Y \leftarrow 0^{\hat{m}}$
3. $E \leftarrow 1^{\hat{m}}$	3. $C[0] \leftarrow C[1] \leftarrow 0^{\hat{m}}$
4. **for** $i \leftarrow 1$ **to** $\hat{m}$ **do**	4. **for** $i \leftarrow 1$ **to** $\hat{m}$ **do**
5. $\quad c \leftarrow 0$	5. $\quad j \leftarrow i \bmod 2$
6. $\quad$ **if** $P_i < P_{i+1}$ **then** $c \leftarrow 1$	6. $\quad C[j] \leftarrow C[j] \mid (1 \ll (i-1))$
7. $\quad B[c] \leftarrow B[c] \mid (1 \ll (i-1))$	7. $\quad$ **if** $P_i < P_{i+1}$ **then**
8. $i \leftarrow \hat{m} + 1$	8. $\quad\quad X \leftarrow X \mid (1 \ll (i-1))$
9. **while** $i \leq |T|$ **do**	9. $\quad$ **if** $T_i < T_{i+1}$ **then**
10. $\quad (e, j, D) \leftarrow (0, 0, E)$	10. $\quad\quad Y \leftarrow Y \mid (1 \ll (i-1))$
11. $\quad$ **while** $e \leq k$ **and** $j < \hat{m}$ **do**	11. **for** $i \leftarrow \hat{m}$ **to** $|T| - 1$ **do**
12. $\quad\quad j \leftarrow j + 1$	12. $\quad$ **if** $T_i < T_{i+1}$ **then**
13. $\quad\quad c \leftarrow 0$	13. $\quad\quad Y \leftarrow Y \mid (1 \ll (\hat{m}-1))$
14. $\quad\quad$ **if** $T_{i-j} < T_{i-j+1}$ **then** $c \leftarrow 1$	14. $\quad B_0 \leftarrow (X \wedge Y) \mathbin{\&} C[0]$
15. $\quad\quad D \leftarrow (D \gg 1) \mathbin{\&} B[c]$	15. $\quad B_1 \leftarrow (X \wedge Y) \mathbin{\&} C[1]$
16. $\quad\quad$ **if** $D = 0^{\hat{m}}$ **then**	16. $\quad W \leftarrow (B_0 \ll 1) \mid (B_0 \gg 1)$
17. $\quad\quad\quad (e, j, D) \leftarrow (e+1, j+1, E)$	17. $\quad e \leftarrow \text{SA}(B_0 \mid B_1 \mathbin{\&} \sim W)$
18. $\quad$ **if** $j \geq \hat{m}$ **and** $e \leq k$ **then** **psi-filter**(P, T, i)	18. $\quad$ **if** $e \leq k$ **then** **verify**(P, T, i)
19. $\quad i \leftarrow i + (\hat{m} - \min(j, \hat{m})) + 1$	19. $\quad Y \leftarrow Y \gg 1$

Fig. 1. The AOPF1 and AOPF2 algorithms for the approximate order-preserving matching problem.

addition of X returns the number of bits set in X. This operation can be computed in $O(\log \log w)$ time in the word-RAM model [20] and is also available as a POPCNT instruction in recent processors of the x86 family. The worst-case time complexity of this algorithm is $O(n(\lceil m/w \rceil \log \log w + m \log m))$. The space complexity of both algorithms is $O(\lceil m/w \rceil)$. The pseudocode of the two algorithms is shown in Fig. 1. The **psi-filter** procedure called in AOPF1 at line 18 performs the additional filtration step based on the ψ function and calls the verification procedure, if necessary.

5   Analysis

In this section we analyze the average-case running time of the AOPF1 algorithm, and show that it is sublinear on average if k is not too large. Suppose that T is a uniformly random string over an alphabet Σ of size σ. The string $\phi(T)$ is not uniformly random in general as $Pr[\phi(T)_i = 1] = (\sigma + 1)/(2\sigma)$ and $Pr[\phi(T)_i = 0] = (\sigma - 1)/(2\sigma)$. We make the simplifying assumption that either all the symbols of T are distinct, in which case the distribution becomes uniform, or that the alphabet is large enough so that the distribution is arbitrarily close to uniform. Assume that $k < m/(\log_\sigma m + O(1))$ and let X_j be the random variable corresponding to the length of factor F_j. By the "Main Lemma" of Chang and Lawler [2] we obtain that

1. the probability $Pr[X_1 + X_2 + \ldots + X_{k+1} \geq m]$ of a verification using Lemma 3 is less than $1/m^3$;
2. $E[X_j] < \log_\sigma m + 3$;

Table 1. Execution times of the algorithms (in 10 of milliseconds) for Dow Jones data and Helsinki temperature data.

	Dow Jones						Helsinki Temperatures				
	$k = 1$						$k = 1$				
m	AOPF1	AOPF1b	AOPF2	AOPF2b	naive	m	AOPF1	AOPF1b	AOPF2	AOPF2b	naive
5	21.5	26.2	[16.9]	22.8	24.4	5	12.8	13.4	[9.5]	11.3	13.1
10	[4.1]	10.1	6.2	11.8	90.8	10	[2.1]	4.9	3.5	5.8	47.0
15	[1.7]	5.1	3.0	3.9	172.2	15	[0.9]	2.5	1.7	1.8	84.2
20	[1.0]	2.6	2.8	2.9	270.3	20	[0.5]	1.3	1.5	1.4	129.8
25	[0.7]	1.7	2.8	2.9	374.0	25	[0.3]	0.8	1.5	1.4	182.9
30	[0.6]	1.0	2.8	2.6	473.9	30	[0.2]	0.5	1.4	1.2	233.1
50	[0.3]	0.5	2.8	2.5	1069.5	50	[0.1]	0.2	1.4	1.2	522.6
	$k = 2$						$k = 2$				
m	AOPF1	AOPF1b	AOPF2	AOPF2b	naive	m	AOPF1	AOPF1b	AOPF2	AOPF2b	naive
5	30.3	34.3	28.9	31.6	[27.4]	5	16.8	17.3	15.6	15.7	[13.1]
10	[28.9]	36.0	31.3	54.1	96.5	10	[14.9]	16.8	16.2	27.5	46.7
15	9.7	21.7	[7.3]	19.4	172.0	15	5.0	10.4	[3.8]	9.8	84.0
20	3.8	19.6	[3.3]	5.5	261.5	20	2.0	9.7	[1.7]	2.9	129.8
25	[2.0]	11.9	3.0	3.2	372.3	25	[1.0]	6.0	1.6	1.5	182.8
30	[1.4]	6.1	3.1	2.7	465.8	30	[0.6]	2.8	1.4	1.3	225.2
50	[0.6]	1.5	3.1	2.7	1048.9	50	[0.2]	0.7	1.5	1.2	503.0
	$k = 3$						$k = 3$				
m	AOPF1	AOPF1b	AOPF2	AOPF2b	naive	m	AOPF1	AOPF1b	AOPF2	AOPF2b	naive
5	32.3	35.4	30.8	32.0	[28.0]	5	17.2	17.5	15.7	16.1	[13.0]
10	88.4	96.3	[80.6]	95.7	98.9	10	46.1	47.9	41.2	48.2	[39.6]
15	[31.0]	34.2	31.7	76.3	174.4	15	[14.5]	15.5	16.1	38.6	70.5
20	18.2	30.7	[7.9]	26.8	266.5	20	9.2	14.9	[3.9]	13.7	108.4
25	8.1	32.6	[3.6]	7.5	380.6	25	4.2	16.5	[1.8]	3.8	152.2
30	3.8	27.7	[3.0]	3.1	454.2	30	1.7	12.7	[1.4]	[1.4]	226.5
50	[1.1]	4.5	3.0	2.7	963.2	50	[0.5]	2.1	1.4	1.2	507.5

since skipping two symbols instead of one between each factor F_j does not invalidate the assumption that the variables X_j are independent and identically distributed. By (1), the total verification time is thus $O((n/m^3)m \log m)$. Instead, by (2), it follows that the average number of symbols scanned in a single window and the average shift length are equal to $(k + 1)(\log_\sigma m + 3)$ and $m - (k + 1)(\log_\sigma m + 3)$, respectively. From this we obtain that the average filtering time is $O((n/m)k \log_\sigma m)$ for the aformentioned choice of k. Hence, the running time of both phases is sublinear on average.

6   Experiments

We tested AOPF1 and AOPF2 against the following algorithms:

- AOPF1b: the filtration method based on the Hamming distance using Approximate SBNDM;
- AOPF2b: the filtration method based on the Hamming distance using the GGF algorithm;
- naive: the naive method where all the text positions are checked.

Note that the AOPF1b and AOPF2b algorithms must use $2k$ as bound on the number of mismatches. In the AOPF1b algorithm we employ the same additional filtration step used in AOPF1.

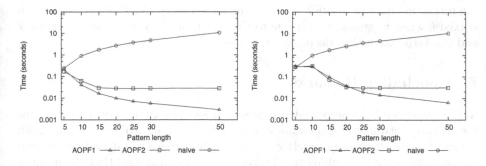

Fig. 2. Plot of the execution times of the algorithms for Dow Jones data with $k = 1$ (left) and $k = 2$ (right).

The tests were run in 64-bit mode on Intel 2.70 GHz i7 processor with 16 GB of memory running Ubuntu 12.10. All the algorithms were implemented in C and run in the testing framework of Hume and Sunday [12].

We performed the tests on two sequences of real data: the Dow Jones index and the Helsinki temperatures time series. The Dow Jones data contains $15,248$ integers corresponding to the daily values of the stock index in the years 1950–2011 and the Helsinki temperature data contains $6,818$ integers representing the daily mean temperatures in Fahrenheit (multiplied by ten) in Helsinki in the years 1995–2005. For each sequence we generated sets of 200 patterns, randomly extracted from the text, of fixed length $m \in \{5, 10, 15, 20, 25, 30, 50\}$. Table 1 shows the average execution times, over 99 repeated runs, of the algorithms for Dow Jones data and Helsinki temperature data in 10 of milliseconds for $k \in \{1, 2, 3\}$. In addition, a graph of the times for the Dow Jones data and $k = 1, 2$ (with logarithmic scale on the y axis) is shown in Fig 2. All the algorithms use the verification method described in Sect. 3.

From the results, we observe that i) AOPF1 and AOPF2 are significantly faster than the naive method, except for the cases $m = 5$ and also $m = 10$ for $k = 3$ where they are either faster or comparable; ii) AOPF1 is always faster than AOPF1b; iii) AOPF2 is either faster or comparable to AOPF2b. Consider the cases where the algorithms provide a significant speedup over the naive method, namely $m \geq 10$ for $k = 1, 2$ and $m \geq 15$ for $k = 3$. For $k = 1$, AOPF1 is the fastest algorithm, while for higher values of k either AOPF1 or AOPF2 is the fastest depending on the value of m. In particular, there is a region of m, $\{15, 20\}$ for $k = 2$ and $\{20, 25, 30\}$ for $k = 3$, where AOPF2 obtains the best running time. Note that the execution time of the naive algorithm is proportional to m, as it verifies all the positions. In the case of AOPF1 and AOPF2, the execution time drops notably after a threshold value for m which depends on k. In particular, for $k = 1$ the threshold value for m is 5 while for $k = 2, 3$ it is 10. The AOPF1b and AOPF2b also shows this behaviour, although the drop in the running time is not as significant. Note that although the filtration phase of AOPF2 is linear, the total time of AOPF2 decreases with a fixed k when m grows. This is due to the fact that the verification probability and the verification time decrease on average when m grows.

7   Concluding Remarks

In this paper we have presented two practical solutions, based on filtration, for the recently introduced approximate order-preserving matching problem. Both algorithms are effective in practice, as shown by experimental evaluation, and one of them is the first sublinear solution on average, provided that the number of errors is not too large.

References

1. Belazzougui, D., Pierrot, A., Raffinot, M., Vialette, S.: Single and multiple consecutive permutation motif search. In: Cai, L., Cheng, S.-W., Lam, T.-W. (eds.) Algorithms and Computation. LNCS, vol. 8283, pp. 66–77. Springer, Heidelberg (2013)
2. Chang, W.I., Lawler, E.L.: Sublinear approximate string matching and biological applications. Algorithmica 12(4/5), 327–344 (1994)
3. Chhabra, T., Tarhio, J.: Order-preserving matching with filtration. In: Gudmundsson, J., Katajainen, J. (eds.) SEA 2014. LNCS, vol. 8504, pp. 307–314. Springer, Heidelberg (2014)
4. Cho, S., Na, J.C., Park, K., Sim, J.S.: A fast algorithm for order-preserving pattern matching. Inf. Process. Lett. 115(2), 397–402 (2015)
5. Crochemore, M., Iliopoulos, C.S., Kociumaka, T., Kubica, M., Langiu, A., Pissis, S.P., Radoszewski, J., Rytter, W., Waleń, T.: Order-preserving incomplete suffix trees and order-preserving indexes. In: Kurland, O., Lewenstein, M., Porat, E. (eds.) SPIRE 2013. LNCS, vol. 8214, pp. 84–95. Springer, Heidelberg (2013)
6. Gawrychowski, P., Uznanski, P.: Order-preserving pattern matching with k mismatches. CoRR, abs/1309.6453 (2013)
7. Gawrychowski, P., Uznański, P.: Order-preserving pattern matching with k mismatches. In: Kulikov, A.S., Kuznetsov, S.O., Pevzner, P. (eds.) CPM 2014. LNCS, vol. 8486, pp. 130–139. Springer, Heidelberg (2014)
8. Giaquinta, E., Grabowski, S., Fredriksson, K.: Approximate pattern matching with k-mismatches in packed text. Inf. Process. Lett. 113(19–21), 693–697 (2013)
9. Han, Y.: Deterministic sorting in $O(n \log \log n)$ time and linear space. J. Algorithms 50(1), 96–105 (2004)
10. Hirvola, T., Tarhio, J.: Approximate online matching of circular strings. In: Gudmundsson, J., Katajainen, J. (eds.) SEA 2014. LNCS, vol. 8504, pp. 315–325. Springer, Heidelberg (2014)
11. Horspool, R.N.: Practical fast searching in strings. Softw. Pract. Exper. 10(6), 501–506 (1980)
12. Hume, A., Sunday, D.: Fast string searching. Softw. Pract. Exper. 21(11), 1221–1248 (1991)
13. Jacobson, G., Vo, K.: Heaviest increasing/common subsequence problems. In: Proceedings of the Combinatorial Pattern Matching, Third Annual Symposium, CPM 1992, Tucson, Arizona, USA, April 29–May 1, pp. 52–66 (1992)

14. Kim, J., Eades, P., Fleischer, R., Hong, S., Iliopoulos, C.S., Park, K., Puglisi, S.J., Tokuyama, T.: Order-preserving matching. Theor. Comput. Sci. **525**, 68–79 (2014)
15. Knuth Jr., D.E., Morris, J.H., Pratt, V.R.: Fast pattern matching in strings. SIAM J. Comput. **6**(2), 323–350 (1977)
16. Kubica, M., Kulczynski, T., Radoszewski, J., Rytter, W., Walen, T.: A linear time algorithm for consecutive permutation pattern matching. Inf. Process. Lett. **113**(12), 430–433 (2013)
17. Navarro, G.: Nr-grep: a fast and flexible pattern-matching tool. Softw. Pract. Exper. **31**(13), 1265–1312 (2001)
18. Navarro, G., Raffinot, M.: Fast and flexible string matching by combining bit-parallelism and suffix automata. ACM Journal of Experimental Algorithmics **5**, 4 (2000)
19. Peltola, H., Tarhio, J.: Alternative algorithms for bit-parallel string matching. In: Proceedings of the String Processing and Information Retrieval, 10th International Symposium, SPIRE 2003, Manaus, Brazil, October 8–10, pp. 80–94 (2003)
20. Vigna, S.: Broadword implementation of rank/select queries. In: McGeoch, C.C. (ed.) WEA 2008. LNCS, vol. 5038, pp. 154–168. Springer, Heidelberg (2008)

Fishing in Read Collections: Memory Efficient Indexing for Sequence Assembly

Vladimír Boža, Jakub Jursa, Broňa Brejová, and Tomáš Vinař[(✉)]

Faculty of Mathematics, Physics, and Informatics, Comenius University,
Mlynská dolina, 842 48 Bratislava, Slovakia
{boza,brejova,vinar}@fmph.uniba.sk

Abstract. In this paper, we present a memory efficient index for storing a large set of DNA sequencing reads. The index allows us to quickly retrieve the set of reads containing a certain query k-mer. Instead of the usual approach of treating each read as a separate string, we take an advantage of significant overlap between reads and compress the data by aligning the reads to an approximate superstring constructed specifically for this purpose in combination with several succinct data structures.

1 Introduction

The second generation sequencing technologies (such as Illumina) allow us to investigate DNA and RNA sequences at a previously unseen scale. A single sequencing run can produce vast amounts of sequencing reads of lengths 100–150bp that need to be processed by using efficient data structures and algorithms. In our work, we propose a new data structure, called *CR-index*, for read indexing without the reference genome. Our method improves on previously proposed compressed Gk-arrays [Välimäki and Rivals, 2013] by exploiting the internal structure of typical read collections.

Commonly applied first step in processing Illumina reads is to find (possibly approximate) occurrences of these reads in a reference genome. This task, also called *read mapping*, is often dependent on sophisticated indexes of the reference genome, such as uni-directional or bi-directional FM-index (see e.g. Langmead and Salzberg [2012]). After the reads are aligned to the reference genome, we can answer common queries, such as which or how many reads overlap a particular genomic position. These queries find many applications, including variant calling in population genetic analysis, locating transcription factor binding sites, assessing duplication structure of the genome, differential gene expression, and many more.

However, in many cases the reference genome is unknown or incomplete. In such cases, one would still want to preprocess large collections of reads so that similar queries can be processed efficiently. In particular, Philippe et al. [2011] introduced a problem of *read indexing*, where the task is to build an index which can be queried for all reads that contain a particular k-mer as a substring (maximum k is given beforehand). Their data structure, called Gk-array, is based on

© Springer International Publishing Switzerland 2015
C. Iliopoulos et al. (Eds.): SPIRE 2015, LNCS 9309, pp. 188–198, 2015.
DOI: 10.1007/978-3-319-23826-5_19

efficient indexing of a concatenation of all reads in the collection, and can answer these queries in $O(k \log n + |Q|)$ time, where n is the size of the read collection, and $|Q|$ is the number of reads in the answer. Recently, Välimäki and Rivals [2013] introduced compressed Gk-arrays, which decrease the memory use significantly by using compressed suffix arrays [Grossi et al., 2003]. These indexing structures find use in many practical applications involving read clustering, k-mer counting, and similarity search.

Our particular interest lies in *de novo* genome assembly based on maximum likelihood (GAML), which was introduced recently by Boža et al. [2014]. This method uses a likelihood score based on a simple probabilistic model of the sequencing process and attempts to find the highest likelihood assembly by simulated annealing. To this end, likelihoods of many candidate assemblies are evaluated with respect to several read collections. Here, reference-based index is impractical, since the reference genome changes in every iteration of the algorithm and realigning all reads to the new reference would be impractically slow. However, in each iteration, the assembly is modified only in a few places, and thus only a limited number of reads need to be realigned, as long as they can be located quickly without reference to a particular assembly. Thus read indexing with Gk-arrays can be used to focus realignment to those reads that overlap boundaries where the assembly has changed.

In our work, we build on the idea of compressed Gk-arrays, but instead of relying on the entropy-based compression of the whole read collection, we exploit the fact that the reads are likely random substrings of some unknown string, sampled at a high coverage. This assumption, quite reasonable for our intended application, as well as many other applications in bioinformatics, allows us to design a data structure that is much more memory efficient than compressed Gk-arrays as we demonstrate in the experimental evaluation.

2   CR-Index Overview

Our goal is to build a memory efficient data structure that can identify all reads from a given collection $R = (r_1, r_2, \ldots, r_n)$ containing a query k-mer x. Previously, this problem was treated as a standard pattern matching problem of finding x within string $T = r_1 \$_1 r_2 \$_2 \ldots \$_{n-1} r_n \$_n$ [Philippe et al., 2011].

These approaches were optimized for collections that consist of randomly generated strings. Yet, in many applications, the collection contains reads that have large overlaps, and there can even be many identical reads. Our approach is targeted at read collections that are randomly selected short substrings of a given template, sampled to high coverage, with only a few differences compared to the template (e.g., Illumina reads from a given genome at 50× coverage and 1% error rate).

The main idea of our new data structure, which we call *CR-index*, is to use a *guide superstring* G, which contains all reads r_i from the collection R as substrings. The guide superstring is supplemented by additional structures allowing identification of IDs of all reads that align to a particular position in G.

The guide superstring will be generally much shorter than the concatenation of all reads and since representation of this string accounts for most of the memory used in the previous indexing structures, it will be possible to reduce the memory footprint significantly on real data.

It may seem that the genome from which the reads originated may be the ideal guide superstring. At the same time, often we want to use CR-index to support the task of genome assembly from reads, so requiring a guide superstring as a prerequisite may seem somewhat circular. However, we do not require a guide string that would be a plausible interpretation of the read collection. Obtaining a plausible genome assembly means resolving sequence repeats correctly (including correct number of repeats and their organization) and joining as many contigs as possible into larger scaffolds, which is a very difficult task. In contrast, the guide string is only required to satisfy the above technical definition; it is allowed to be very fragmented, and the best guide strings will be over-collapsed, with each repeat included only once. Such guide strings can be easily obtained even through the simplest de Bruijn graph approaches or by simple approximation algorithms for building the shortest superstring.

However, there is a problem with errors contained in the reads. Any read that has been changed compared to the original template will likely not align to the original template and thus the guide string would have to be significantly enlarged to also include all the reads with errors. To avoid unnecessary enlargement of the guide string, we will relax our definition as follows:

Definition 1 (k-**guide Superstring**). *For a given read collection R and number k, a k-guide superstring is a string G such that for each read $r \in R$ there exists a substring of G or a reverse complement of a substring of G, denoted s_r, such that any two differences between s_r and r are located more than k bases apart.*

Note that in this work we allow only substitutions as differences between r and s_r. This relaxed definition however complicates the query algorithm, as illustrated in Fig. 1. For a given query k-mer x, we search the guide string for all strings at Hamming distance at most one from x and from the reverse complement of x. This bound on Hamming distance is sufficient, because differences between r and s_r are more than k bases apart, and thus the query will overlap at most one difference between the guide string and the target read. After recovering all potential matching reads, we verify that each of them actually contains the original query x as a substring.

Even though this search algorithm is somewhat complicated due to the relaxed definition of the k-guide superstring, we gain significant improvements in memory. For example on *E. coli* dataset, we were able to construct exact superstring of length 224 Mbp and k-guide string of length 108 Mbp.

3   Representing k-guide Superstring and Auxiliary Data Structures

As shown in Fig. 2, the CR-index consists of three main parts. The first part represents the k-guide superstring G and allows fast exact pattern matching.

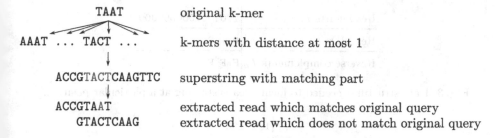

TAAT	original k-mer
AAAT ... TACT ...	k-mers with distance at most 1
ACCGTACTCAAGTTC	superstring with matching part
ACCGTAAT	extracted read which matches original query
GTACTCAAG	extracted read which does not match original query

Fig. 1. Overview of the algorithm for answering CR-index query.

Read 0	ACCGTAATCA
Read 1	GTACTCAAAG
Read 2	CTGAAAGTTC
Superstring G	ACCGTACTCAAAGTTC
Read positions	0: 0; 1: 3; 2: 6
Read substitutions	(0,6): A, (3,2): G

Fig. 2. Overview of the CR-index for three input reads shown at the top.

The second part represents the starting positions of individual reads in G and allows us to quickly locate reads covering a given region in G. Finally, the last part lists differences between the reads and the guide superstring, and allows us to quickly verify if a read matches our query k-mer x.

To implement the CR-index, we use succint data structures from the SDSL library [Gog et al., 2014]. In particular, the guide superstring is represented using FM-index [Ferragina and Manzini, 2000], which allows efficient exact pattern matching while maintaining a small memory footprint (linear in the length of the superstring). FM-index consists of a wavelet tree, which contains a BWT-transformed superstring G, and samples from suffix and inverse suffix arrays for G. The sampling rate influences memory usage and query time; a higher sampling rate results in faster queries but requires more memory.

The other two parts of the data structure are described in more details below, but both use sparse bit vectors represented as SDarrays [Okanohara and Sadakane, 2007]. In this representation, a bit vector of length n with m bits set to one occupies $m \lg \frac{n}{m} + 2m + o(m)$ bits of memory. The rank query (retrieving the number of bits set to 1 in a prefix of the vector of length i) works in $O(\log \frac{n}{m} + \log^4 m / \log n)$ time and the select query (retrieving the position of i-th bit set to 1) works in $O(\log^4 m / \log n)$ time.

Representation of reads starting at a given position. After locating a particular k-mer occurrence in the guide superstring G, we need to recover all reads that overlap this occurrence. First, assume that at most one read starts at each position of G. We will construct a bitvector P_b containing 1 at each position where

Read starts P_b	1001001000000000
Read IDs P_r	0 1 2
Reverse complements P_c	F F F

Fig. 3. Data structures needed to locate reads starting at a particular position.

P_b	1010100
P_r	0 3 1
L_b	11000
L_r	4 2

Fig. 4. Two reads ACG with IDs $0, 4$ start at position 0, one read GTT with ID 3 starts at position 2, and and two reads TAA with IDs $1, 2$ start at position 4. We omit arrays P_c and L_c for simplicity.

a read starts. This bitvector will be stored in an SDarray, and thus support fast rank and select queries. The read IDs will be stored in array P_r sorted by their position in the superstring. Finally, P_c is a bitvector indicating a strand of the read in the superstring. Fig. 3 demonstrates the use of these arrays. We can find the read located at position p by first checking whether $P_b[p] = 1$ and then using rank query to find the position of the read in P_r and P_c.

To accomodate multiple reads at the same position, we store reads mapping to the same position in a linked list. Our implementation of the linked lists is optimized for the case when most lists have length only one, which is usually the case unless the coverage is very high. The first item of each linked list is stored in arrays P_b, P_r, and P_c as before. To store the rest of the linked lists, we use a bitvector L_b which contains one at position r if read with ID r has successor in the linked list. Note that the length of L_b is the same as the total number of reads. We enhance this array to support rank queries. Read IDs of the remaining reads (not present in P_r) are in array L_r and their strand information is in bit vector L_c. Fig. 4 illustrates these structures.

Reads are ordered in L_r and L_c by the ID of their predecessor so that we can use the following algorithm to retrieve all reads starting at position p. We first use arrays P_b and P_r to find index i of the read which is the head of the linked list. The remaining reads are found as follows:

```
while L_b[i] == 1:
  rank = L_b.rank(i)
  output L_r[rank], L_c[rank]
  i = L_r[rank]
```

We store P_r and L_r as ordinary integer vectors with $\lceil \lg(n-1) \rceil + 1$ per integer. In total they take $n(\lceil \lg(n-1) \rceil + 1) + O(\lg n)$ bits. Arrays P_c and L_c

Bitvector D_b	0000001000 0000000000 0010000000
Original bases D_s	A G

Fig. 5. Representation of differences between the read and the guide.

are standard bitvectors and take $n + O(\lg n)$ bits in total. Vectors P_b and L_b are sparse, and thus represented as SDarrays.

Differences between the guide and the read. Since not all reads map to the guide string exactly, we need to be able to recover the differences between a particular read and the guide superstring. Let ℓ be the length of one read and n be the number of all reads. Bitvector D_b of length $n\ell$ will for each read and each position in the read store zero, if the read is identical to the guide at that particular position, or one otherwise. Vector D_s will store the differences corresponding to 1s in bitvector D_b. We can use rank and select queries to recover a particular difference. Fig. 5 illustrates the arrays. We can store the array D_s in $2d + O(\lg d)$ memory, where d is the total number of differences between the reads and the guide. D_b is a sparse bitvector represented by SDarrays.

Querying CR-index. The algorithm for querying the index follows the outline explained in Section 2. More specifically, given k-mer x we obtain the reads containing x using the following steps:

1. Construct reverse complement x^R of x and the set Q of all k-mers with Hamming distance of at most one from x or x^R.
2. For each $q \in Q$, find the set P of its positions in G using the FM index.
3. For each $p \in P$, find the set R of reads containing k-mer starting at p. This is achieved by retrieving the reads starting at positions $p - \ell + k, \ldots, p$, where ℓ is the length of a read.
4. For each $r \in R$: If $q = x$ or $q = x^R$ check if r does not contain any substitutions in the interval corresponding to q. Otherwise check if r contains exactly one substitution which is the same as the difference between q and x (x^R). Output the read, if it passes the test.

In the FM index, we search for $O(k)$ string from Q, each search taking time $O(k)$. If this search finds m matching positions in total, we spend $O(m)$ time to recover these positions. Let r be the number of reads overlapping these matches, s the length of G, t the total length of reads, and d the total number of differences. Extracting each read involves a constant number of rank and select queries in arrays P_b and L_b, which takes $O(r \lg \frac{s}{r} + \lg^4 r / \lg s)$ time. Extracting relevant differences takes one rank query in array D_b, which in total takes $O(r \lg \frac{d}{t} + \lg^4 d / \lg t)$.

In our algorithms, we search the data structure for all k-mers from Q. However, if we search for $q \in Q$ which differs from x or x^R by a particular substitution, we may find no matches, because this particular substitution does not occur in any read in the set. In real data, usually only few substitutions lead to a matching read. To reduce the number of useless queries, we implement a simple filter based on Bloom filters [Bloom, 1970]. In particular, we use a Bloom filter to store all k-mers from all reads that differ from their corresponding k-mers in G. For every substituion on a read compared to the superstring we thus store k strings. Before querying FM-index for occurrences of q, we first test whether the substitution in q compared to the query k-mer x or x^R is hashed in the Bloom filter.

4   Finding the k-guide Superstring

Our goal is to find as short k-guide superstring as possible for a given collection of reads R. This problem is a generalization of a well-known shortest superstring problem, which is NP-hard [Gallant et al., 1980]. In our work, we will use a heuristics based on commonly used sequence assembly tools in the following three steps:

1. **Read correction.** First, we identify the reads containing low-frequency substrings. Unique/low-frequency substrings would unnecessarily inflate the size of the guide string, and we attempt to remove these low-frequency substrings by introducing a small number of substitutions. Various formulations of such read correction problem have been studied in the context of sequence assemblers [Kelley et al., 2010]. We use the read correction algorithm from SGA [Simpson and Durbin, 2012], but we only accept the substitutions that are at least k bases apart (if SGA proposes substitutions that are too close, we greedily chose a subset satisfying our criterion).
2. **Finding read overlaps.** Again, this is a well studied problem in the context of sequence assembly. We use SGA overlapping algorithm with standard settings.
3. **Construction of the superstring.** The easiest approach would be to use a well-known greedy approximation shortest superstring algorithm, which repeatedly merges two strings with the largest overlap [Blum et al., 1994]. In practice, we have found that the guide construction can be speeded up by doing assembly first and only perform the greedy merge as a finishing step to incorporate reads that were not successfuly included in the assembly.
   Various modifications are possible. For example, using a simple string concatenation in the finishing step leads typically to about 25% inflation in the string length compared to the greedy merging.

In our experiments we found that our construction efficiently compressed around 90% of reads (those present in the assembly) and modestly compresses the remaining erroneous reads.

5   Experiments

We compare the performance of our data structure with compressed Gk-arrays [Välimäki and Rivals, 2013] on two datasets. The first data set is the set of 151bp

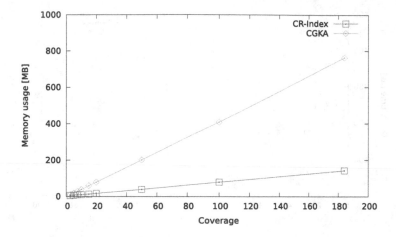

Fig. 6. Comparison of memory usage of CR-index and compressed Gk-arrays on the *E. coli* dataset with increasing coverage.

Illumina reads from *E. coli* strain MG1655 (genome length 4.7 Mbp, 184× coverage after removal of low-quality reads, 0.75% error rate) [Illumina, 2015]. The second data set is a set of 101bp Illumina reads from human chromosome 14 (sequence length 107 Mbp, 23× coverage, 1.5% error rate) from Genome Assembly Gold-Standard Evaluations (library 1; Salzberg et al. [2012]). In addition, we also use error-free reads of length 151bp simulated from the *E. coli* genome with coverage up to 50×. In all experiments, we have used query length $k = 15$.

First, we were interested in the guide string length and memory use when dealing with error-free reads from the *E. coli* genome. With increasing number of reads, the size of the guide string stayed almost the same (4.7 Mbp). The overall memory grew linearly with the number of reads, and at 50× coverage, the whole data structure required only 8.8 MB, or 5.9 bytes per read.

The situation is more complicated in the case of real reads containing errors (see Fig. 6). Memory usage of the CR-index still grows approximately linearly with the coverage, and the slope is much lower than for the compressed Gk-arrays. Whereas the length of the superstring is 6.9 Mbp at coverage 2.5 (1.7 times smaller than total length of reads), the size increases with the coverage up to 108 Mbp at coverage 184, which is 8× smaller than the length of all reads concatenated. The increase in length is due to the presence of reads with errors that are less than k positions apart, and consequently cannot be easily integrated into a smaller superstring. The largest dataset is represented in 142.4 MB of memory, of which 71% is taken by the FM index, 15% is taken by read positions, less than 2% is required to represent the substitutions in reads, and approximately 12% is taken by the Bloom filter.

Compressed Gk-array queries are faster at small coverage, but the relative differences become insignificant at higher coverage (Fig. 7).

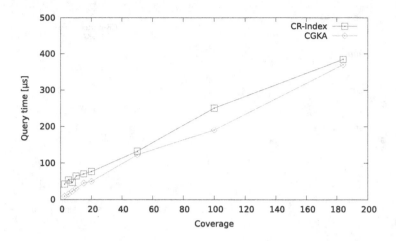

Fig. 7. Comparison of query time of CR-index and compressed Gk-arrays on the *E. coli* dataset with increasing coverage.

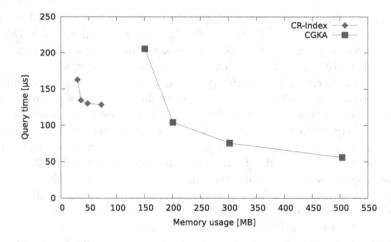

Fig. 8. Comparison of query time and memory usage of CR-index and compressed Gk-arrays on the *E. coli* dataset with coverage 50 for varying suffix array sampling rates.

On the human chromosome 14 (23× coverage of 107 Mbp sequence, 1.5% error rate), the CR-index requires only 571 MB of memory, while Gk-arrays require ≈ 1.7 GB. On the other hand, a typical query takes 214 ms in CR-index, or roughly 3× longer than in Gk-arrays (71 ms). We believe that this is due to higher error rate in the data which may cause the Bloom filter to filter out fewer queries.

Both CR-index and compressed Gk-arrays use sampling of the suffix array. This allows time vs. memory tradeoff, since by sampling more values, we get a better running time at the expense of higher memory usage. Figure 8 shows this tradeoff at coverage 50 in the *E. coli* dataset. CR-index is much less sensitive to sampling parameters than compressed Gk-arrays. This is due to compressed superstring representation which results in a smaller FM-index and a smaller number of hits in FM-index during the search.

6 Conclusion

We have presented a new compressed data structure, called CR-index, for indexing short sequencing reads. CR-index has a significantly smaller memory footprint than Gk-arrays [Philippe et al., 2011; Välimäki and Rivals, 2013] and still maintains a reasonable query time.

Our data structure directly supports queries for reads that contain a particular k-mer and for all occurrences of the k-mer in these reads (queries Q1 and Q3 in Välimäki and Rivals [2013]). In this work, we did not implement other queries supported by compressed Gk-arrays. In particular, while the number of reads and the number of occurrences in reads (queries Q2 and Q4) can also be easily recovered, an efficient implementation is more difficult. We could count the reads overlapping a given position in the guide string by rank queries, however these counts will also include reads that mismatch the guide string and these would need to be excluded, requiring additional operations proportional to the number of matching reads. In contrast, after locating the k-mer, compressed Gk-arrays can count the number of occurrences in constant time. Mismatches between the guide string and the reads also do not leave much space for non-trivial implementations of queries Q5-Q7 which require exactly one occurrence of the k-mer in a read. Perhaps additional auxiliary data structures would help to implement these queries in a more reasonable time.

In our work, we used standard algorithms for read correction and sequence assembly to construct the guide string. Yet, it may be possible to do better by developing more specialized algorithms, perhaps based on coding theory.

In our intended application, the same read is likely to be "fished-out" through many k-mer queries, and thus it may be argued that there should be an exact match of at least one k-mer without necessity to consider neighbourhoods to account for mismatches between the guide string and the reads.

Acknowledgments. This research was funded by VEGA grants 1/1085/12 (BB) and 1/0719/14 (TV), and a grant from the Slovak Research and Development Agency APVV-14-0253. We would like to thank authors of SDSL-lite for a functional and well-documented library.

References

Bloom, B.H.: Space/time trade-offs in hash coding with allowable errors. Communications of the ACM **13**(7), 422–426 (1970)

Blum, A., Jiang, T., Li, M., Tromp, J., Yannakakis, M.: Linear approximation of shortest superstrings. Journal of the ACM **41**(4), 630–647 (1994)

Boža, V., Brejová, B., Vinař, T.: GAML: Genome assembly by maximum likelihood. In: Brown, D., Morgenstern, B. (eds.) WABI 2014. LNCS, vol. 8701, pp. 122–134. Springer, Heidelberg (2014)

Ferragina, P., Manzini, G.: Opportunistic data structures with applications. In: Foundations of Computer Science (FOCS), pp. 390–398. IEEE (2000)

Gallant, J., Maier, D., Astorer, J.: On finding minimal length superstrings. Journal of Computer and System Sciences **20**(1), 50–58 (1980)

Gog, S., Beller, T., Moffat, A., Petri, M.: From theory to practice: plug and play with succinct data structures. In: Gudmundsson, J., Katajainen, J. (eds.) SEA 2014. LNCS, vol. 8504, pp. 326–337. Springer, Heidelberg (2014)

Grossi, R., Gupta, A., Vitter, J.S.: High-order entropy-compressed text indexes. In: Symposium on Discrete Algorithms (SODA), pp. 841–850. ACM/SIAM (2003)

Illumina (2015). E.coli MG1655 Illumina sequencing dataset. ftp://webdata: webdata@ussd-ftp.illumina.com/Data/SequencingRuns/MG1655/MiSeq_Ecoli_MG1655_110721_PF.bam (accessed: March 03, 2015)

Kelley, D.R., Schatz, M.C., Salzberg, S.L., et al.: Quake: Quality-aware detection and correction of sequencing errors. Genome Biology **11**(11), R116 (2010)

Langmead, B., Salzberg, S.L.: Fast gapped-read alignment with Bowtie 2. Nature Methods **9**(4), 357–359 (2012)

Okanohara, D., Sadakane, K.: Practical entropy-compressed rank/select dictionary. In: Workshop on Algorithms Engineering and Experiments (ALENEX), pp. 60–70. SIAM (2007)

Philippe, N., Salson, M., Lecroq, T., Leonard, M., Commes, T., Rivals, E.: Querying large read collections in main memory: a versatile data structure. BMC Bioinformatics **12**(1), 242 (2011)

Salzberg, S.L., Phillippy, A.M., et al.: GAGE: A critical evaluation of genome assemblies and assembly algorithms. Genome Research **22**(3), 557–567 (2012)

Simpson, J.T., Durbin, R.: Efficient de novo assembly of large genomes using compressed data structures. Genome Research **22**(3), 549–556 (2012)

Välimäki, N., Rivals, E.: Scalable and versatile k-mer indexing for high-throughput sequencing data. In: Cai, Z., Eulenstein, O., Janies, D., Schwartz, D. (eds.) ISBRA 2013. LNCS, vol. 7875, pp. 237–248. Springer, Heidelberg (2013)

How Big is that Genome?
Estimating Genome Size and Coverage
from k-mer Abundance Spectra

Michal Hozza, Tomáš Vinař, and Broňa Brejová[✉]

Faculty of Mathematics, Physics, and Informatics,
Comenius University, Mlynská dolina, 842 48 Bratislava, Slovakia
{hozza,vinar,brejova}@fmph.uniba.sk

Abstract. Many practical algorithms for sequence alignment, genome assembly and other tasks represent a sequence as a set of k-mers. Here, we address the problems of estimating genome size and sequencing coverage from sequencing reads, without the need for sequence assembly. Our estimates are based on a histogram of k-mer abundance in the input set of sequencing reads and on probabilistic modeling of distribution of k-mer abundance based on parameters related to the coverage, error rate and repeat structure of the genome. Our method provides reliable estimates even at coverage as low as 0.5 or at error rates as high as 10%.

1 Introduction

Two decades ago, sequencing even a short bacterial genome constituted a large project, potentially involving a collaboration of multiple research groups, its execution limited to several large sequencing centers. New sequencing technologies allow such a task to be accomplished even by a small research group at a reasonable cost. Often the biggest hurdle in this process is the lack of appropriate bioinformatics tools for various analyses of the obtained sequencing data.

In this paper, we provide a new method for estimating the size of the sequenced genome, which is one of the fundamental questions related to genome sequencing. Ideally, the result of genome sequencing would contain a complete sequence of each chromosome, and thus the size would be immediately apparent. However, sequencing technologies provide only short reads, and it is difficult, or often even impossible, to reconstruct the entire original sequence from such data due to sequence repeats and sequencing errors. The result of genome sequencing is thus typically a set of shorter sequences, contigs, which represent parts of chromosomes. A sequence repeat with multiple copies in the genome may be collapsed to a single copy, and some parts of the genome might be missing altogether. Therefore, even for a draft assembly, it is necessary to compensate for these issues when estimating total genome size.

For example, the panda genome was the first mammalian genome sequenced by the next-generation sequencing technologies [Li et al., 2010]. The authors

© Springer International Publishing Switzerland 2015
C. Iliopoulos et al. (Eds.): SPIRE 2015, LNCS 9309, pp. 199–209, 2015.
DOI: 10.1007/978-3-319-23826-5_20

have assembled around 2.3Gb of sequence to contigs and scaffolds, but using several methods, they estimate the genome size to be around 2.4Gb.

In our work, we concentrate on estimating the genome size directly from sequencing reads, without using the draft genome assembly. This approach has several advantages. Our method can produce size estimate faster, because we avoid a computationally intensive assembly process. By directly modeling the sequencing process, we avoid additional biases introduced by genome assemblers. Finally, as we show in our experiments, our approach can be also used on smaller set of reads that do not provide sufficient genome coverage for the assembly.

Genome size is closely related to the genome coverage, which is the average number of sequencing reads covering a single position of the genome. If one of these two quantities is known, the other can be easily computed because their product should be equal to the sum of the lengths of all reads (assuming that the fraction of reads corresponding to various artefacts unrelated to the genome is negligible). When planning a sequencing experiment, one needs an approximate genome size to determine the amount of data necessary to reach the desired sequence quality. Our method can be used already on preliminary data at a lower coverage to estimate the genome size and thus plan the overall experiment.

Similarly to the previous approaches to this problem [Li and Waterman, 2003; Williams et al., 2013], we start by counting the occurrences of all strings of a fixed length k (k-mers) in the set of all sequencing reads. There are several tools for counting k-mer frequencies efficiently [Kurtz et al., 2008; Marçais and Kingsford, 2011; Melsted and Pritchard, 2011; Zhang et al., 2014]. These tools are based on advanced text indexing, such as FM index [Ferragina and Manzini, 2000] or on hashing techniques, such as Bloom filters [Bloom, 1970]. We use only a summary statistics, called the k-mer abundance spectrum, which is a histogram containing for each j the number of distinct k-mers with exactly j occurrences in the data.

Approaches based on k-mer frequencies in a set of reads were previously used for several bioinformatics tasks, for example, to separate individual genomes when sequencing a metagenomic mixture of different microbes isolated from a certain environment [Wang et al., 2012; Wu and Ye, 2011] and to correct errors in sequencing reads prior to the assembly [Kelley et al., 2010; Pevzner et al., 2001].

Closely related to our work, methods of Li and Waterman [2003] and Williams et al. [2013] estimate the genome coverage from k-mer abundance spectra based on probabilistic models of the repeat content of a genome, but these authors do not explicitly consider sequencing errors in their models. However, even a small error rate may have a substantial impact on abundance spectra. Consider reads of length 100 with 1% error rate. The expected number of errors per read is then 1. If we consider $k = 21$, a single error may overlap 21 out of 80 k-mers. Thus, in this (rather typical) scenario, we may expect that up to 26% of k-mers will be erroneous, creating much more low-abundance k-mers than would be expected if there were no errors.

Williams et al. [2013] deal with this problem by removing low-abundance k-mers from the spectrum. We argue, that in doing so, they decrease the power of their models to estimate the genome size at low coverage and high error rates. In particular, our experiments show that their method generally does not work at coverage less than $10\times$ and for error rates greater than 3%. Yet, by including a simple error model in our probabilistic model, we were able to tolerate coverage as low as $0.5\times$ or error rates as high as 10%. Thus, we convincingly demonstrate that by explicit error modeling, we are able to significantly enlarge the application domain, particularly to the cases, where filtering or correcting erroneous reads will simply not work due to lack of data to build a consensus.

2   Models and Algorithms

Let us assume that we have a fixed value of k and an observed k-mer abundance spectrum $W = w_1, w_2, \ldots, w_m$, where w_j is the number of k-mers that occur in the input set of reads exactly j times, and m is the highest observed abundance.

In this section, we present several probabilistic models which represent a simplified version of the process that leads to the observed abundance spectrum. Table 1 shows the overview of the models and their parameters. Each model gives a formula for computing the likelihood $L(W|\theta)$ of observing spectrum W given model parameters θ, which always include the true genome coverage c. Then we try to estimate parameters θ that maximize this likelihood, and use the obtained coverage c as our estimate. Genome size is estimated by dividing the total amount of sequencing data by c.

Table 1. Overview of the models, their features and parameters. Parameters in parentheses are derived from other parameters.

Model	Features	Parameters θ
EF	error-free reads (read border effects)	real coverage c (k-mer coverage c_k)
SE	simple errors (abundance 1 only)	real coverage c error rate ϵ (erroneous k-mer prob. α) (effective k-mer coverage c')
E	full error model (higher abundances)	real coverage c error rate ϵ (expected abundance of k-mers w/ s errors λ_s) (probability k-mer has s errors α_s)
RE	repeats and errors	real coverage c error rate ϵ repeat distribution parameters q_1, q_2, q

Model EF: error free. In this simplest model, we assume that the genome does not contain any exact sequence repeats of length k or more and that there are no sequencing errors. Subsequent models will relax these assumptions, while using a similar overall modelling approach.

Recall that coverage c is the average number of reads covering a single base of the genome. However, when we look at a k-mer starting at a particular position, not all reads covering its start will cover its entire length. Therefore, if all reads have length r, the average coverage of a given k-mer will be only $c_k = c(r - k + 1)/r$. Our models will work with c_k, and subsequently use this relationship to obtain c.

Aside from this correction, our models do not take into account reads and rather consider sequencing data simply as a collection of k-mers. More precisely, we assume that the observed abundance of each k-mer is sampled uniformly independently from some distribution so that the probability of observing abundance j is p_j. Then the log-likelihood of observing spectrum W is $\log L(W|\theta) = \sum_{j=1}^{m} w_j \log p_j$.

In the EF model, we assume that the abundance of each k-mer in the genome is sampled from the Poisson distribution with mean c_k. It is important to note, however, that this random abundance is zero for some k-mers, and consequently we will never observe them in the spectrum. As a result, the probability p_j of observing abundance j among k-mers represented in the spectrum is

$$p_j = f(j; c_k) = \frac{c_k^j e^{-c_k}}{j!(1 - e^{-c_k})}$$

This distribution is called truncated Poisson. In effect, we renormalize probabilities of the Poisson distribution with mean c_k by $1/(1 - e^{-c_k})$ which is the probability of obtaining a value greater than zero in the Poisson distribution.

Model SE: simple errors. Our remaining models consider sequencing errors. We assume that during sequencing, each base is read correctly with probability $(1-\epsilon)$ and with probability $\epsilon/3$ it is substituted with one of the other three bases. We do not consider any sequence-specific error biases, nor do we explicitly model indels, although their influence on the k-mer abundance spectra would be to some extent similar to substitutions.

In the SE model, we assume that if a k-mer contains a sequencing error, it will always have abundance 1. The model is a mixture of two distributions. With probability α, a given k-mer contains an error and is thus assigned abundance one. With probability $(1 - \alpha)$, it is generated from the truncated Poisson distribution with mean c'. Thus, in this model, $p_1 = (1 - \alpha)f(1; c') + \alpha$ and $p_j = (1 - \alpha)f(j; c')$ for $j > 1$. Let $\gamma = (1 - \epsilon)^k$ be the probability that a sequenced k-mer contains no errors. Coverage c' is set to γc_k where c_k is the ideal k-mer coverage from the EF model. Parameter α is related to γ, but while γ is the probability that a single sequenced k-mer contains no errors, $1 - \alpha$ is proportion of error-free k-mers among all unique k-mers obtained by sequencing. A single k-mer in the genome gives rise to around c' error-free sequenced k-mers which together account for a single unique k-mer in our histogram. In contrast, under the assumptions of our model, each bad k-mer is unique. We will use the following simple formula to relate α to γ, and thus to ϵ: $\alpha = (1-\gamma)/(1-\gamma+\gamma/c')$.

For inference, we seek parameters c and ϵ such that when converted to c' and α, they maximize the likelihood of observed data W. As we describe below, we use a simple approximate parameter inference method in this model as a starting point for more complex models.

Model E: full error model. The previous model was based on the assumption that all errors produce k-mers with abundance one. As the coverage increases, the same error may happen multiple times, and as a result, errors will contribute also to k-mers with higher abundance, which we cover in the more complex model E. Let us consider k-mer x and another k-mer y, which is in Hamming distance s from x. The probability of obtaining k-mer y as a result of sequencing k-mer x is $\epsilon^s(1-\epsilon)^{k-s}3^{-s}$, assuming that all substitutions are equally likely. If the source k-mer x is read c_k times in expectation, the expected number of times we obtain as a result k-mer y is $\lambda_s = \epsilon^s(1-\epsilon)^{k-s}3^{-s}c_k$. The actual number of occurrences of y we then again model by the Poisson distribution with mean λ_s. Note that parameter λ_s will be close to zero for higher values of s, and consequently, we will never observe most of k-mers y as a result of sequencing x.

We will assume that each unique k-mer y from the histogram resulted only from one source k-mer x. In other words, two different source k-mers never result in the same erroneous k-mer, nor do we obtain a different source k-mer as a result of errors while reading k-mer x. As before, we will assume that each observed k-mer is generated from a mixture of different probability distributions, but this time the mixture will have $k+1$ components, one for each Hamming distance s. Component s will have weight α_s, and it will correspond to a truncated Poisson distribution with parameter λ_s, thus the overall probability of observing a k-mer with abundance j will be

$$p_j = \sum_{s=0}^{k} \alpha_s f(j; \lambda_s),$$

where f is the probability mass function of the truncated Poisson distribution, as in the previous models.

To relate weights α_s to the basic parameters of our model (c and ϵ), we first estimate n_s, which is the expected number of observed unique k-mers with Hamming distance s from their respective source k-mers. If the genome contains n different source k-mers, each can give rise to $\binom{k}{s}3^s$ different k-mers with s errors and the probability of observing such a k-mer with abundance at least one is $1 - e^{-\lambda_s}$. Therefore, we get

$$n_s = n\binom{k}{s}3^s(1 - e^{-\lambda_s}).$$

Mixture weights α_s are then computed by renormalizing n_s by their sum; the unknown parameter n conveniently cancels out.

Model RE: model with repeats and errors. In this model, we assume that a given k-mer can occur in the genome more than once. Let β_o be the fraction of k-mers with o occurrences in the genome. The overall probability distribution is again a mixture distribution $p_j = \sum_{o=1}^{\infty} \beta_o p_{o,j}$, where $p_{o,j}$ is the probability of abundance j for a k-mer with o occurrences. This probability is computed according to the E model, only coverage c_k is replaced with $o \cdot c_k$, because the k-mer is now o times more likely to be covered by a randomly placed read.

Related works [Li and Waterman, 2003; Williams et al., 2013] concentrate on accurate modelling of repeats, with mixture coefficients β_o being arbitrary values estimated from the data for small values of o. According to the authors of these studies, such an approach requires a relatively high coverage; ideally we would observe each multiple of c_k as a separate peak in the abundance spectrum, and the height of this peek would allow us to estimate β_o. However, at a small coverage, such peaks are difficult to distinguish, and therefore we instead model proportions β_o by the geometric distribution, which in our experience works well for higher values of o. We use three parameters q_1, q_2 and q such that $\beta_1 = q_1$, $\beta_2 = (1-q_1)q_2$, and $\beta_o = (1-q_1)(1-q_2)q(1-q)^{o-3}$ for $o \geq 3$. Our RE model thus uses previously introduced parameters c and ϵ, as well as three new parameters q_1, q_2 and q. The model of course also depends on the known values of r and k.

Optimization of model parameters. To estimate the coverage and the genome size, we start from a set of sequencing reads and compute the k-mer abundance spectrum using Jellyfish [Marçais and Kingsford, 2011]. Our algorithm then uses the spectrum to obtain the maximum likelihood estimates of model parameters.

We use $k = 21$ in all our experiments, as previously suggested by Williams et al. [2013]. Note that our models do not explicitly account for the reverse strand. However, k-mer counting tools generally add together counts for a k-mer and its reverse complement, and as a result, we may assume that each k-mer counted in the spectrum corresponds to a k-mer on the forward strand of the target DNA. Since we use an odd value of k, no k-mer is a reverse complement of itself.

Our parameter optimization routine proceeds in three steps. First, we use a heuristic approach based on the SE model to obtain the initial estimate of the parameter values. In the second step, we improve these estimates by the standard L-BFGS-B optimization algorithm [Zhu et al., 1997] as implemented in the Python's SciPy library. As we did not provide the optimization with the gradient function, the algorithm often fails to find a good local maximum. Thus, in the third step, we continue with a grid search to further improve these estimates. The second and the third steps use the E or RE models.

To obtain the initial parameter estimates, we use a heuristic based on the SE model which assumes that all k-mers with abundance at least two are correct. We start by computing the mean observed abundance c'' among these k-mers:

$$c'' = \frac{\sum_{j=2}^{m} j \cdot w_j}{\sum_{j=2}^{m} w_j}.$$

We would like to use c'' as an estimate of parameter c' for the Poisson distribution governing the abundance of error-free k-mers. However, c'' overestimates c', because we have not used correct k-mers with abundance of one or zero to compute the mean. We can relate c' and c'' by observing that c'' should converge to the mean of the Poisson distribution with parameter c' and values 0 and 1 truncated:

$$c'' = \frac{c' - c'e^{-c'}}{1 - e^{-c'} - c'e^{-c'}} = g(c').$$

We compute c' as inverse of $g(c')$ using the Newton method.

To compute c_k from c', we need to estimate α, the proportion of k-mers with errors among all observed unique k-mers. Let n_2 be the number of unique k-mers with abundance two or higher. We estimate the number of correct k-mers with abundance one and zero (denoted by n_1 and n_0) using the following formulas:

$$n_1 = \frac{n_2 c' e^{-c'}}{1 - e^{-c'} - c'e^{-c'}},$$

$$n_0 = \frac{n_2 e^{-c'}}{1 - e^{-c'} - c'e^{-c'}}.$$

Now we can estimate the number of erroneous k-mers as $n_e = w_1 - n_1$, and compute the parameters α and γ as follows:

$$\alpha = \frac{n_e}{n_0 + \sum_{i=1}^{m} w_i},$$

$$\gamma = \frac{c'(\alpha - 1)}{\alpha c' - \alpha - c'}.$$

Finally, we get initial estimates of c and ϵ using the formulas from the SE model:

$$c = \frac{c_k r}{r - k + 1} = \frac{c'r}{\gamma(r - k + 1)},$$

$$\epsilon = 1 - \gamma^{\frac{1}{k}}.$$

If there is not enough data for this estimate, we simply set $c = 1$ and $\epsilon = 0.5$ as the initial value. For optimization of the RE model, we also need initial values for parameters q_1, q_2 and q, which we simply set all to $1/2$.

To deal with cases when the L-BFGS-B algorithm fails to find a local maximum, we use a simple iterative grid search to further improve the model parameters. This algorithm maintains a current parameter estimate $\theta = (x_1, \ldots, x_\ell)$. In each iteration, it evaluates log likelihood at the grid points of a small grid around current θ and selects the estimate with the highest likelihood for the next iteration.

We use a fixed grid of size $G = 3$ with variable step S. To construct the grid, we first select a set of values for each parameter x_i separately as $R_i = \{S^d x_i \mid d \in \{-G, \ldots, -1\} \cup \{1, \ldots, G\}\}$. The grid is then simply the Cartesian product

$R = R_1 \times R_2 \times \cdots \times R_\ell$. In the E model, $\theta = \{c, \epsilon\}$, and thus the full grid size is 36. In the RE model, $\theta = \{c, \epsilon, q_1, q_2, q\}$, and the grid size is 7776.

If the log likelihood improves by less than one in a single iteration, we decrease step S. We finish the optimization when the log likelihood does not improve and the step is at most 1.001. In spite of its simplicity, the algorithm usually finds good parameters, as demonstrated in the next section.

To avoid numerical overflows and to speed up the computation, we omit some small terms from the likelihood function. First, we trim the input spectrum at $j = 100$. In our datasets, more than 99.999% of the spectrum was in the first 100 positions. Next, we consider only up to 8 errors per k-mer, as the probability of higher error counts was negligible for error rates in our dataset. In the RE model, we also consider only values of o for which $\beta_o \geq 10^{-8}$.

3   Experimental Evaluation

In our first set of experiments, we have evaluated our software CovEst on randomly generated genomes of length 1 Mbp with uniform base frequencies. Reads were generated at uniformly selected locations in the genome and errors in reads were generated with the same error probability at each position. In Fig. 1, we see the predicted coverage from the E model for inputs with various levels of genome coverage and error rates. As the coverage increases, CovEst estimates become very close to the true values. In the error-free data, we obtain reasonable estimates even at coverage 0.1, and for high coverage (≥ 10), we tolerate even 10% error rate.

We compare these results with the estimates from the KmerSpectrumAnalyzer (KSA) by Williams et al. [2013]. They concentrate on modeling repeats; errors are handled simply by ignoring k-mers with abundance lower than some threshold ℓ. Authors do not recommend KSA for coverage below 10. For coverage 30 and 50, we have used their default value $\ell = 10$; for lower values of coverage c, we use $\ell = c/2$. All other values were left at default settings. As we can see in Fig. 1, their estimates are in general farther from the true value. For lower coverage, their software crashed. Note also that we have computed the KSA coverage by dividing the total amount of sequencing data by the KSA estimate of the genome size. KSA also provides an estimate of sequence coverage, but these values are generally very low; perhaps they represent parameter c' of our SE model which measures the expected coverage of an entire k-mer by error-free portions of reads.

Overall this experiment demonstrates that the CovEst E model gives reliable coverage estimates at a broad range of parameters and works reliably even for very low coverage values, where methods based on discarding k-mers with low abundance do not work. Precise estimates of genome coverage automatically translate to precise estimates of genome size, as the genome size is simply the total amount of data divided by the coverage. For example at 1% error and 1× coverage, our five replicates yield genome size estimates between 0.95Mb and 1.15Mb with mean 1.04Mb, which is very close to the true value of 1 Mbp. Error

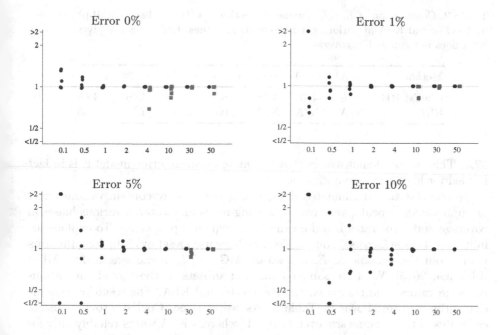

Fig. 1. Comparison of genome coverage estimates obtained by the CovEst E model (black circles) and KSA (red squares) on random genomes with various coverage levels and error rates. Columns in each plot correspond to eight different coverage levels; y-axis shows the ratio of predicted and real coverage on the logarithmic scale. Estimates more than a factor of two away from the real value are shown at the top or bottom of the plot. For each combination of parameters, we show five replicates. KSA did not produce answers for some inputs.

rate estimates are also generally very accurate; for inputs with coverage at least 2 and non-zero error rate, we have always obtained error rate estimates within 10% of the true value.

We have also run the RE model on some of these random inputs, to see whether it can detect the absence of repeats. Again, the coverage estimates were quite accurate; for coverage at least 2 and error at most 5%, we have obtained values within 5% of the true answer. Estimated proportion of k-mers with one occurrence in the genome (parameter q_1 of the RE model) was in all these cases estimated to be more than 99%.

The simpler SE model provides reasonable coverage estimates at medium coverage (for coverage between 2 and 10 and error at most 5% we obtain estimates within 15% of the correct value). However, as expected, it does not work well for high coverage where the assumption that all error k-mers have abundance one is no longer reasonable. For example, at 30× coverage and error rate 5%, the coverage estimates range around 18 and the estimated error rate is around

Table 2. Comparison of *E. coli* genome size estimates (true value 4.64 Mbp) provided by CovEst and KSA at various genome coverage values. Estimates are given in Mbp. KSA does not run at low coverage.

Method	$c = 0.5$	$c = 1$	$c = 2$	$c = 4$	$c = 10$	$c = 30$	$c = 50$
CovEst RE	4.16	4.70	4.58	4.63	4.71	4.69	4.68
KSA	N/A	N/A	N/A	6.03	4.61	4.59	4.58

62%. This clearly demonstrates that the more accurate error model E is in fact needed for high coverage data sets.

In contrast to our simulated data, real genomes may contain various exact or approximate repeats, and real sequencing reads may contain various biases in coverage and error rates based on the local sequence properties. To explore the influence of these factors on our models, we have used a set of 151bp long Illumina reads from the genome of *E. coli* strain MG1655 (genome length 4.64 Mbp) [Illumina, 2015]. We have sampled random subsets of these reads at various coverage values, and compared the RE model and KSA. The resulting genome size estimates are shown in Table 2. As we can see, CovEst provides useful estimates of the genome size even at $c = 1$, whereas KSA works reliably only for $c \geq 10$.

4   Conclusion

In this paper, we provide a new method for genome size estimation from raw sequencing reads. Our method is based on a compact representation of these reads in the form of k-mer abundance spectrum. Compared to previous methods, we combine a model of sequence repeats with a model of sequencing errors, which significantly shape the k-mer spectrum. According to our knowledge, this is the first method that can provide reasonable estimates of genome sizes from the data at 1× genome coverage or less; previous works [e.g. Williams et al., 2013] required genome coverage in excess of 10× in order to provide reliable estimates. Low coverage sequencing data arise in the study of very large plant genomes [e.g. Sveinsson et al., 2013] or due to pooling and barcoding multiple samples to save costs. Tools for reliable analysis of such data may allow cost-efficient sampling of a greater number of genomes than would be possible with standard high-coverage data sets.

Our models can be further improved by considering additional real-world phenomena, such as dependence of coverage and error rate on GC content and other sequence features. We also currently do not consider polymorphism; k-mers covering heterozygous loci would come in two forms, each at coverage $c/2$, which could be easily captured by our model, at the cost of adding another parameter characterizing the overall heterozygosity. We also plan to improve the optimization algorithms used in CovEst to improve the accuracy and decrease the running time.

Acknowledgments. This research was funded by VEGA grants 1/1085/12 (BB) and 1/0719/14 (TV), Comenius University grant UK/235/2015 (MH), and a grant from the Slovak Research and Development Agency APVV-14-0253.

References

Bloom, B.H.: Space/time trade-offs in hash coding with allowable errors. Communications of the ACM **13**(7), 422–426 (1970)

Ferragina, P., Manzini, G.: Opportunistic data structures with applications. In: Foundations of Computer Science (FOCS), pp. 390–398 (2000)

Illumina (2015). E.coli MG1655 Illumina sequencing dataset. ftp://webdata: webdata@ussd-ftp.illumina.com/Data/SequencingRuns/MG1655/MiSeq_Ecoli_ MG1655_110721_PF.bam (accessed: March 03, 2015)

Kelley, D.R., Schatz, M.C., Salzberg, S.L., et al.: Quake: Quality-aware detection and correction of sequencing errors. Genome Biology **11**(11), R116 (2010)

Kurtz, S., Narechania, A., Stein, J.C., Ware, D.: A new method to compute k-mer frequencies and its application to annotate large repetitive plant genomes. BMC Genomics **9**(1), 517 (2008)

Li, R., Fan, W., Tian, G., et al.: The sequence and de novo assembly of the giant panda genome. Nature **463**(7279), 311–317 (2010)

Li, X., Waterman, M.S.: Estimating the repeat structure and length of DNA sequences using ℓ-tuples. Genome Research **13**(8), 1916–1922 (2003)

Marçais, G., Kingsford, C.: A fast, lock-free approach for efficient parallel counting of occurrences of k-mers. Bioinformatics **27**(6), 764–770 (2011)

Melsted, P., Pritchard, J.K.: Efficient counting of k-mers in DNA sequences using a Bloom filter. BMC Bioinformatics **12**(1), 333 (2011)

Pevzner, P.A., Tang, H., Waterman, M.S.: An Eulerian path approach to DNA fragment assembly. Proceedings of the National Academy of Sciences **98**(17), 9748–9753 (2001)

Sveinsson, S., Gill, N., Kane, N.C., Cronk, Q.: Transposon fingerprinting using low coverage whole genome shotgun sequencing in Cacao (Theobroma cacao L.) and related species. BMC Genomics **14**(1), 502 (2013)

Wang, Y., Leung, H.C., Yiu, S.-M., Chin, F.Y.: MetaCluster 5.0: A two-round binning approach for metagenomic data for low-abundance species in a noisy sample. Bioinformatics **28**(18), i356–i362 (2012)

Williams, D., Trimble, W.L., Shilts, M., Meyer, F., Ochman, H.: Rapid quantification of sequence repeats to resolve the size, structure and contents of bacterial genomes. BMC Genomics **14**(1), 537 (2013)

Wu, Y.-W., Ye, Y.: A novel abundance-based algorithm for binning metagenomic sequences using l-tuples. Journal of Computational Biology **18**(3), 523–534 (2011)

Zhang, Q., Pell, J., Canino-Koning, R., Howe, A.C., Brown, C.T.: These are not the k-mers you are looking for: Efficient online k-mer counting using a probabilistic data structure. PloS One **9**(7), e101271 (2014)

Zhu, C., Byrd, R.H., Lu, P., Nocedal, J.: Algorithm 778: L-BFGS-B: Fortran subroutines for large-scale bound-constrained optimization. ACM Transactions on Mathematical Software **23**(4), 550–560 (1997)

Assessing the Efficiency of Suffix Stripping Approaches for Portuguese Stemming

Wadson Gomes Ferreira[⊠], Willian Antônio dos Santos,
Breno Macena Pereira de Souza, Tiago Matta Machado Zaidan,
and Wladmir Cardoso Brandão

Department of Computer Science,
Pontifical Catholic University of Minas Gerais, Belo Horizonte, Brazil
{wadson.gomes,willian.santos.838460,breno.macena,
tiago.zaidan}@sga.pucminas.br, wladmir@pucminas.br

Abstract. Stemming is the process of reducing inflected words to their root form, the stem. Search engines use stemming algorithms to conflate words in the same stem, reducing index size and improving recall. Suffix stripping is a strategy used by stemming algorithms to reduce words to stems by processing suffix rules suitable to address the constraints of each language. For Portuguese stemming, the RSLP was the first suffix stripping algorithm proposed in literature, and it is still widely used in commercial and open source search engines. Typically, the RSLP algorithm uses a list-based approach to process rules for suffix stripping. In this article, we introduce two suffix stripping approaches for Portuguese stemming. Particularly, we propose the hash-based and the automata-based approach, and we assess their efficiency by contrasting them with the state-of-the-art list-based approach. Complexity analysis shows that the automata-based approach is more efficient in time. In addition, experiments on two datasets attest the efficiency of our approaches. In particular, the hash-based and the automata-based approaches outperform the list-based approach, with reduction of up to 65.28% and 86.48% in stemming time, respectively.

1 Introduction

The Web comprises over 30 trillion uniquely addressable documents. Recently, the leading commercial search engine reported the processing of more than 3.3 billion user queries each day [4]. The massive-scale nature of the Web demands efficient tools to manage, retrieve and filter information [2]. In this environment, time and retrieval performance are critical for search engines to provide an effective search experience to their users.

Typically, search engines use stemming to reduce inflected words to their stem, i.e., the portion of the original word that remains after affixes removal [2]. For instance, the word "connect" is the stem of the words "connected", "connecting", "connection" and "connections". Stemming document corpus reduces the vocabulary size and hence, the time to process queries. In addition, query

© Springer International Publishing Switzerland 2015
C. Iliopoulos et al. (Eds.): SPIRE 2015, LNCS 9309, pp. 210–221, 2015.
DOI: 10.1007/978-3-319-23826-5_21

reformulation by stemming increases the match between the query and relevant documents in the corpus, ultimately improving the recall. Due to the increasing size of the Web and the increasing query traffic on search engines, a particularly challenging information retrieval problem is the development of effective stemming approaches. For Portuguese stemming, the RSLP [7] is an effective algorithm widely used in commercial and open source search engines. For instance, Lucene[1] uses the RSLP list-based approach for suffix stripping.

In this article, we propose two suffix stripping approaches for Portuguese stemming, the hash-based and the automata-based approaches. Additionally, we assess their efficiency by contrasting them with the state-of-the-art list-based approach [7]. In particular, the list-based approach compares word suffixes to a well defined list of suffix rules, iteratively reducing the word. At the end, the reduction performed by the last rule in the list produces the stem. However, usually only a few suffix rules need to be processed to reduce a word. From this observation, we propose the hash-based approach, which breaks the single big list of suffix rules in small ones, deciding which list should be processed based on the word suffixes, thus reducing the number of suffix rules to check. In addition, we also propose the automata-based approach which uses a deterministic finite automata (DFA) to iteratively check word suffixes to decide when to perform a reduction or not, avoiding unnecessary suffix rules checking.

Complexity analysis shows that the time complexity is constant for the automata-based approach, sub-linear for the hash-based approach, and linear for the list-based approach, considering the number of suffix rules to process. Additionally, the space complexity is constant for the automata-based approach and linear for the hash-based and list-based approaches. However, in addition to store the list of suffix rules, the hash-based approach stores a hash table to distribute the suffix rules of the list by word suffixes, thus consuming more space than the list-based approach.

Experiments using two datasets attest the efficiency of our approaches. The hash-based and the automata-based approaches outperform the list-based approach baseline with reduction of up to 65.28% and 86.48% in stemming time, respectively. The key contributions of this article can be summarised as follows:

- We propose two suffix stripping approaches for Portuguese stemming. The proposed approaches are simple and can be adapted to handle other languages, such as English and Spanish.
- We provide complexity analysis for the list-based approach baseline and our approaches, highlighting the differences in space and time complexity between them;
- We thoroughly evaluate our approaches contrasting them with the state-of-the-art baseline from literature [7].

The remainder of this article is organised as follows: Section 2 reviews the related literature on stemming algorithms. Section 3 describes our suffix stripping approaches for Portuguese stemming, as well as the baseline, presenting

---

[1] http://lucene.apache.org.

their complexity analysis. Section 4 describes the setup and results of the experimental evaluation of our approaches. Finally, Section 5 provides a summary of the contributions and the conclusions made throughout the other sections, presenting directions for future research.

2   Related Work

Stemming algorithms use four strategies to produce the stem for a word: table lookup, successor variety, n-grams, and affix removal [2]. The table lookup strategy performs a simple lookup of the word in a table to retrieve its stem. Despite simple, this strategy is strongly dependent on the language vocabulary. The sucessor variety is a complex strategy, which identifies morpheme boundaries within the language organization to produce the stem. The n-grams strategy is more a word clustering procedure than a stemming strategy, since it simply identifies bigrams and trigrams in the text. The affix removal strategy removes or replaces affixes of the word, following well defined affix rules for the language. In particular, the affix removal strategy is simple, intuitive, effective, and less dependent on the language vocabulary. Mostly, it includes suffix stripping.

For Portuguese, there are three suffix stripping algorithms reported in literature, an adapted version of the Porter stemmer for the English language [8], and two others specifically designed for the Portuguese language: the RSLP [7], and the STEMBR [1]. The Portuguese Porter stemmer is a translation of the English version for Portuguese, resulting in the absence of key reductions, such as feminine and degree. The STEMBR algorithm processes a set of rules for suffix removal in a specific order, defined from statistical analysis on a corpus of documents, with a balanced word distribution for a language. The performance of the algorithm depends on the quality of the corpus.

The RSLP was the first suffix stripping algorithm for Portuguese stemming. It processes a set of well defined suffix rules using an exception list to address language irregularities, ultimately providing an effective word reduction [6]. Different from STEMBR, the RSLP algorithm does not need to define an order to process the suffix rules from a corpus of documents. In addition, its set of suffix rules includes feminine and degree reduction. Thus, given the effectiveness and simplicity of the RSLP algorithm, the suffix stripping approaches we investigate in this article are based on it. The complexity analysis of the RSLP algorithm [3] shows that the time complexity, in the worst case, can be expressed by

$$n * W * \sum_{i=1}^{R}(S_i * E_i) \tag{1}$$

where n is the number of processed words, W is the size, in characters, of the longest word, R is the number of suffix rules, S_i is the size, in characters, of the suffix rule i, and E_i is the size of the exceptions list of i. Considering that lookups in the exceptions list are performed in constant time, the time complexity to reduce a word j can be expressed by:

$$W_j * \sum_{i=1}^{R} S_i \qquad (2)$$

From the Equation 2, we observe that the number of suffix rules that should be compared to the word suffix to perform reductions, in the worst case, is R, i.e., all the suffix rules should be compared to conclude that no reduction should be performed. In this case, the complexity is $O(W_j R)$, or simply $O(W_j)$, if we consider R constant. The point is that the value of R significantly impacts in time, since the greater the R, the greater the number of comparisons between characters in the suffix rules and word suffix must be performed. From this observation, in the next section we present three different approaches to perform suffix stripping, one corresponding to the list-based approach [3], and the other two corresponding to our proposed approaches, which improve the efficiency by reducing the amount of processed rules, i.e., the value of R.

3   Suffix Stripping Approaches

The RSLP algorithm is composed by eight steps, of which six are word reductions based on suffix stripping, i.e., removing or replacing a piece of the word, the suffix. Figure 1 presents the steps of the RSLP stemming algorithm.

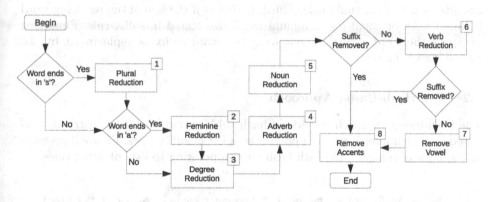

Fig. 1. The steps of the RSLP stemming algorithm for Portuguese

From Figure 1, we observe that suffix stripping are performed from step 1 to 6. The challenge in suffix stripping is not only to decide performing word reductions, but mostly avoid unnecessary decisions. In particular, reducing decisions are made by comparing word suffixes to suffix rules, where each rule is represented as a tuple composed by: i) the suffix to be replaced or removed from the word; ii) the minimum size of the word to perform the reduction; iii) the new suffix in case of replacement, or empty in case of removal, and; iv) the exceptions list, i.e., the list of words that should supposed be reduced by the rule, but should not be reduced.

Sometimes, there are more than one suffix rule for the same reduction. This occurs when one suffix is contained in another, e.g., for plural reduction in the word "mares", the suffix "s" is contained in the suffix "es". In this case, the suffix stripping approach should perform only the reduction related to the longest suffix [7]. Considering the aforementioned example, only the "es" suffix should be removed from the word, reducing it from "mares" to "mar".

3.1   The List-Based Approach

The list-based approach provides a single list of suffix rules to iteratively compare suffixes, one by one, to the word suffix and decides reducing the word or not. Figure 2 presents an example of a list of suffix rules for Portuguese. The complete list of suffix rules is presented by Orengo and Huyck [7].

Fig. 2. Example of a list of suffix rules for Portuguese

For each suffix rule in the list, the list-based approach first checks if the suffix of the rule matches the word suffix. Second, it checks if the word is not in the exceptions list of the suffix rule. Third, it checks if the size of the resulting word is larger than or equal the minimum word size stated in suffix rule. Finally, it performs the word reduction, removing the word suffix or replacing it by the new suffix in the rule.

3.2   The Hash-Based Approach

The hash-based approach provides a hash table, where each hash entry is composed by a character key, and a pointer to a list of suffix rules. Figure 3 presents an example of a hash table with hash entries pointing to lists of suffix rules.

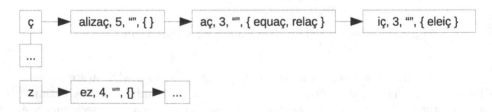

Fig. 3. Example of the data structure for the hash-based approach

From the word to be reduced, the hash-based approach takes the last character to lookup the hash table and, in case of a negative lookup no reduction is performed. In case of a positive lookup, the list of suffix rules pointed by the hash entry is recovered, and the approach iteratively compares the suffix rules, one by one, to the suffix of the remaining part of the word, i.e., the original word without the last character. The suffix rules comparisons performs likewise the list-based approach.

For example, considering to reduce the word "linguiç" using the hash table from Figure 3, there is a positive lookup for the last character "ç" of the word in the hash table. In this case, the list of suffix rules pointed to the hash entry "ç" is checked. For the first suffix rule, the suffix "alizaç" does not match the suffix of "linguiç", as well as for the second suffix rule. However, for the third suffix rule, "iç" matches the suffix of "linguiç" and, as "linguiç" is not in the exceptions list of the rule and its size is larger than 3, the reduction is performed, reducing "linguiç" to "lingu".

Different from the list-based approach which performs comparisons using all suffix rules, the hash-based approach only performs comparisons using a short list of suffix rules. However, for plural, feminine and degree reductions in Portuguese, the hash-based approach performs likewise the list-based approach, since the hash key entries are composed by only one character, i.e., "s" for plural, "a" for feminine, and "o" for degree. Thus the hash table has only one hash entry pointing to the same list of suffix rules of the list-based approach.

3.3   The Automata-Based Approach

Different from list-based and hash-based approaches that use lists of suffix rules and perform a significantly number of comparisons between rules and word suffixes, the automata-based approach uses a deterministic finite automata (DFA) to reduce the number of comparisons to the minimum. Figure 4 presents an example of a DFA used in noun reductions for words ending in "ç".

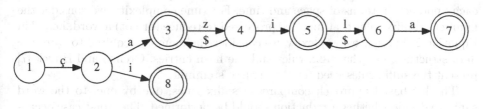

Fig. 4. Example of a DFA used by the automata-based approach

Note that, both word and suffix processing are performed backwards, so we must read the DFA starting by one of the accepting states (states 3, 5, 7 or 8 in Figure 4) and walk until the first state. For instance, if we start in the fifth state and walk until the first state we have the suffix "izaç". In the DFA, it is possible to check if there is a suffix inside another one just checking if an accepting state has a transition to another state. Different from the list-based and hash-based approaches, where suffix rules with larger suffixes must be checked first to avoid wrong reductions, in the automata-based approach the rules with the shorter suffix are checked first. Particularly, we use backtracking to insure the reduction by the longest possible suffix in a way that, if it is impossible to perform the transition for another state just reading a valid character, and there is a state which accepts part of the processed word, a transition to such accepting state is performed. For instance, consider the suffix "aç" (third state), where a suffix satisfying one rule was already found, but it is possible to perform a transition to another state with the character "z", i.e., an attempt to reduce with a longer suffix rule. If the transition to the third state is performed and it is impossible to reach the fourth state, by the backtracking condition, the transition to the third state is performed with the immediate accepting order, so there is no need to perform any character processing. Such accepting order is represented in the DFA with the symbol $ on transitions.

For performance, we consider that there is no need of the transition for the accepting state when there is no backtracking condition for it, e.g., states 7 and 8 in the Figure 4, making possible to perform the reduction at the state that checks the possible transition. So, the state 2 can perform the reduction right after it checks if a transition to state 8 is possible. Thus, we reduce the number of states in the DFA. In addition, we remove the $ transitions, performing rules checking in the states which performs the backtracking. For instance, rules checking can be performed at the states 4 and 6 in Figure 4, reducing the number of DFA transitions. Note that, the automata-based approach performs better when the difference between the size of the suffix rule and the word suffix is larger, resulting in a DFA with few possible state transitions.

3.4 Complexity Analysis

In this section, we present the complexity analysis for the list-based, hash-based and automata-based approaches. In particular, we present the worst case for each approach in terms of space and time. For time complexity, we consider the number of suffix rules that must be processed to reduce (or not) a word, i.e., R in Equation 2. For space complexity, we consider the memory required to store the data structure, i.e., the suffix rules and the hash entries. Orengo and Huyck [7] present the suffix rules used for Portuguese stemming.

The list-based approach compares n suffix rules, one by one, to the word suffix to decide whether a reduction should be performed. The worst case occurs when no reduction is performed and all the suffix rules are compared to the word suffix. Thus, the time complexity is linear. Moreover, the largest value of n occurs in the verb reduction, where $n = 89$.

The hash-based approach compares $m \ll n$ suffix rules, one by one, to the word suffix to decide whether a reduction should be performed. Particularly, the last character of the word is used to lookup the suitable hash entry pointing to the list of suffix rules to be compared. The worst case occurs when the hash entry points to the longest list of rules, comparing the term with all rules of this list. Thus, the time complexity is sub-linear. The largest value of m occurs in the verb reduction, where the hash entry for words ending in "o" has $m = 25$ rules.

The automata-based approach uses DFA to process the word backwards, and after each character comparison, the suitable reduction is performed by processing a very small number of suffix rules. The worst case occurs by checking the biggest path of the automata to compare $m = 4$ suffix rules for verb reduction. Thus, the time complexity is nearly constant.

The three approaches load all the n suffix rules into memory to use them when necessary, so the space complexity is linear for them. However, while the list-based approach requires space to store n suffix rules in memory, the hash-based approach requires an extra space to store the hash structure, i.e., the entry key and a pointer to the list of rules, and the automata-based approach stores the suffix rules as part of the source code.

4 Experiments

To validate our suffix stripping approaches, we contrast them with a state-of-the-art baseline on two different datasets. Particularly, in this section we answer the following research questions:

1. How efficient are the proposed approaches for suffix stripping?
2. How the proposed approaches perform in each step of suffix stripping?

4.1 Experimental Setup

To assess the efficiency of our suffix stripping approaches, we use the RSLP and WBR99 datasets. The RSLP dataset has been used to validate the effectiveness of Portuguese stemming algorithms [7]. It provides 198 Portuguese words into five categories: noun, verb, plural, feminine, degree (augmentative and diminutive). Table 1 presents the words distribution and the average word size (number of characters) by category in the RSLP dataset. The last column of the Table 1 shows the total number of words and the average word size for the entire dataset.

The WBR99[2] is a standard dataset used in information retrieval [9,10]. Indeed, it comprises different collections with almost 6 millions Web pages, crawled from the Brazilian Web (the .br domain). In our experiments, we use six different collections from the WBR99, with a vocabulary of more than 206 millions words, excluding numbers and large strings (more than 50 characters). Table 2 presents the number of words and the average word size by collection. The last row in Table 2 shows the total of words and the mean average word size.

Table 1. Words distribution and average sizes by category in the RSLP dataset

	Noun	Verb	Plural	Feminine	Degree	ALL
# of words	64	90	11	15	18	198
Avg. size	4.22	3.55	2.64	3.00	4.94	3.67

---

[2] Available at http://www.linguateca.pt/Repositorio/WBR-99.

Table 2. Words distribution and average sizes by WBR99 collection

Collection	# of words	Avg. word size
AmostRA-NILC	103,874	7.71
CETEMPúblico	201,658,990	9.23
Museu da Pessoa	1,518,881	8.06
ReLi	160,481	7.67
Tycho Brahe	471,504	7.53
Vercial	2,894,625	8.00
ALL	206,808,355	8.03

We contrast the efficiency of our hash-based approach (RSLP-HBA), described in Section 3.2, and automata-based approach (RSLP-ABA), described in Section 3.3, with the state-of-the-art list-based approach baseline (RSLP-LBA), described in Section 3.1. We report efficiency in terms of stemming time, i.e., the average elapsed time in microseconds (μs) for stemming a word, ignoring the time to load the suffix rules and hash entries in memory. The experimental results are averages on 10 trials and significance is verified with a two-tailed paired t-test [5], with the symbol ▲ (▼) denoting a significant increase (decrease) at the $p < 0.01$ level, and the symbol ● denoting no significant difference.

The experiments were carried out on a computer running a Linux kernel version 3.16, with a 64-bit Intel Core i3 2.13 GHz processor, 3 GB of main memory and one SATA II disk of 320 GB. The approaches were implemented using the C++ language.

4.2   Experimental Results

In this section, we present the results of the experiments we have carried out to evaluate our suffix stripping approaches. In particular, we address the two research questions stated previously, contrasting the efficiency of our approaches with the baseline.

Overall Performance. In this section, we address our first research question, by assessing the overall efficiency of our approaches on WBR99 collections. To this end, Table 3 shows the efficiency, in terms of stemming time, of each proposed approach and the baseline. Particularly, the percentage improvement compared to the baseline is shown. In addition, a first significance symbol denotes whether the improvements are statistically significant. For the RSLP-ABA, a second such symbol denotes significance with respect to RSLP-HBA. The best value in each column is highlighted in bold.

From Table 3, we observe that the RSLP-ABA significantly improves upon the RSLP-HBA and the RSLP-LBA baseline in all collections. In particular, the gains are up to 86.48%. Additionally, we also observe that RSLP-HBA outperforms the baseline in all collections with gains of up to 65.28%. Recalling our first research question, these observations attest the efficiency of our approaches for suffix stripping.

Table 3. Stemming time (μs) for suffix stripping on WBR99 collections

Collection	RSLP-LBA	RSLP-HBA	RSLP-ABA
AmostRA-NILC	1.37346	0.57674 (58.00%) ▼	**0.20779 (84.87%) ▼▼**
CETEMPúblico	0.03657	0.01338 (63.41%) ▼	**0.00504 (86.21%) ▼▼**
Museu da Pessoa	0.23419	0.09632 (58.87%) ▼	**0.03729 (84.07%) ▼▼**
ReLi	0.75644	0.31680 (58.11%) ▼	**0.12272 (83.77%) ▼▼**
Tycho Brahe	0.53799	0.19651 (63.47%) ▼	**0.07759 (85.57%) ▼▼**
Vercial	0.18901	0.06561 (65.28%) ▼	**0.02555 (86.48%) ▼▼**
All	0.52128	0.21090 (59.54%) ▼	**0.07933 (84.78%) ▼▼**

Performance by Category. In this section, we address our second research question, by assessing the efficiency of our approaches in different categories: noun, verb, plural, feminine, and degree. The aforementioned categories are the same categories in the RSLP dataset presented in Section 4.1, and also corresponds to the suffix stripping steps described in Section 3, except for the step 4, which is not considered, since the adverb reduction has only one suffix rule.

To this end, Table 4 shows the efficiency by category, in terms of stemming time, of each proposed approach and the baseline. Particularly, the percentage improvement compared to the baseline is shown. In addition, a first significance symbol denotes whether the improvements are statistically significant. For the RSLP-ABA, a second such symbol denotes significance with respect to RSLP-HBA. The best value in each column is highlighted in bold.

Table 4. Stemming time (μs) for suffix stripping by category on the RSLP dataset

Category (Step)	RSLP-LBA	RSLP-HBA	RSLP-ABA
Noun	2.20733	1.07529 (51.28%) ▼	**0.46008 (79.15%) ▼▼**
Verb	3.63152	1.54178 (57.54%) ▼	**0.68953 (81.01%) ▼▼**
Plural	**0.16546**	**0.16546 (00.00%) ●**	0.21505 (-29.97%) ▲▲
Feminine	0.28770	0.28770 (00.00%) ●	**0.20967 (27.12%) ▼▼**
Degree	0.22786	0.22786 (00.00%) ●	**0.18010 (19.64%) ▼▼**

From Table 4, we observe that for all categories, except plural, the RSLP-ABA significantly improves upon the RSLP-HBA and the RSLP-LBA baseline. In particular, the gains are up to 81.01%. Additionally, we also observe that RSLP-HBA outperforms the baseline for noun and verb with gains of up to 57.54%. Note that, for plural, feminine and degree, the RSLP-LBA and RSLP-HBA performs equally, since the suffix used to distribute suffix rules by hash entries, for each one of these categories, are composed by only one character, i.e., "s" for plural, "a" for feminine, and "o" for degree. Thus the hash table has only one hash entry pointing to the same list of suffix rules of the RSLP-LBA.

As mentioned is Section 3.3, the RSLP-ABA performs better when the difference between the size of the suffix rule and the word suffix is larger, resulting in a DFA with few possibilities of state transitions. From Table 4, we observe that RSLP-ABA presents greater gains for noun and verb than for feminine and degree. This performance gap occurs because for noun and verb, the difference between the size of the suffix rule and the word suffix is larger than for feminine and degree. Moreover, for plural, where the suffix rule and the word suffix have the same size, the RSLP-ABA uses a compact DFA with several possibilities of state transitions, resulting on the only case reported in Table 4 in which RSLP-LBA outperforms RSLP-ABA. Note that, plural reduction occurs in 10.60% of the total amount of reductions performed in our experiments. In summary, the RSLP-ABA outperforms the RSLP-LBA baseline for all categories, except plural. In addition, RSLP-HBA outperforms the RSLP-LBA baseline for noun and verb and presents the same performance for plural, feminine and degree. Recalling our second research question, these observations attest the efficiency of our approaches in each step of suffix stripping.

5   Conclusion

In this article we introduce two novel suffix stripping approaches for Portuguese stemming, the hash-based and the automata-based approaches. We also assess the efficiency of these approaches contrasting them with a state-of-the-art baseline from the literature, the list-based approach. Particularly, the list-based and hash-based approaches iteratively process lists of suffix rules to reduce a word to a stem. However, the list-based approach uses a single list of suffix rules while the hash-based approach uses multiple lists of rules grouped by word suffixes of the language. Different from the other two approaches, the automata-based approach uses a DFA to iteratively compare suffix characters to decide when to perform (or not) a reduction, avoiding unnecessary suffix rules processing, frequent in the other two approaches.

The proposed approaches are simple, effective and can be adapted to work with different languages, such as English and Spanish. In addition, the complexity analysis shows that the list-based approach presents linear complexity both in space and time on the number of suffix rules to process, while the hash-based approach presents linear complexity in space and sub-linear complexity in time, and the automata-based approach presents linear complexity in space and constant complexity in time. Moreover, we thoroughly evaluated the approaches using the RSLP and WBR99 datasets. The results of this evaluation attest the efficiency of the automata-based approach, with gains of up to 86.48% in stemming time over the baseline. By breaking down our analysis by categories, we demonstrated the robustness of the automata-based approach to perform reductions on noun, verb, feminine and degree words, with gains of up to 81.01% in stemming time.

For future work, we plan to deploy and assess the efficiency of a distributed version of the proposed approaches. Another plan is to adapt the proposed approaches to work with other languages.

Acknowledgments. We thank the partial support given by PUC Minas (grant FIP 2015/9396-S1), and CNPq Grant 444156/2014-3 (Wladmir C. Brandão).

References

1. Alvares, R.V., Garcia, A.C.B., Ferraz, I.N.: STEMBR: A stemming algorithm for the brazilian portuguese language. In: Bento, C., Cardoso, A., Dias, G. (eds.) EPIA 2005. LNCS (LNAI), vol. 3808, pp. 693–701. Springer, Heidelberg (2005)
2. Baeza-Yates, R., Ribeiro-Neto, B.: Modern information retrieval: the concepts and technology behind search, 2nd edn. Pearson Education, Harlow (2011)
3. Coelho, A.R.: Stemming for the Portuguese language: study, analysis and improvement of the RSLP algorithm. Universidade Federal do Rio Grande do Sul, Monography (2007)
4. Matt, C.: Spotlight keynote. In: Proceedings of Search Engines Strategies, San Francisco, CA, USA (2012)
5. Jain, R.: The art of computer systems performance analysis: Techniques for experimental design, measurement, simulation, and modeling. Wiley-Interscience, New York (1991)
6. Orengo, V.M., Buriol, L.S., Coelho, A.R.: A study on the use of stemming for monolingual Ad-Hoc portuguese information retrieval. In: Peters, C., Clough, P., Gey, F.C., Karlgren, J., Magnini, B., Oard, D.W., de Rijke, M., Stempfhuber, M. (eds.) CLEF 2006. LNCS, vol. 4730, pp. 91–98. Springer, Heidelberg (2007)
7. Orengo, V.M., Huyck, C.: A stemming algorithm for the portuguese language. In: Proceedings of the 8th International Symposium on String Processing and Information Retrieval (SPIRE), pp. 186–193 (2001)
8. Porter, M.F.: An algorithm for suffix stripping. Program: electronic library and information systems **40**, 211–218 (2006)
9. Bruno, P., Nivio Jr., Z., Meira, W., Ribeiro-Neto, B.A.: Set-based vector model: An efficient approach for correlation-based ranking. ACM Transactions on Information Systems **23**(4), 397–429 (2005)
10. Pôssas, B., Ziviani, N., Ribeiro-Neto, B.A., Meira Jr., W.: Processing conjunctive and phrase queries with the set-based model. In: Apostolico, A., Melucci, M. (eds.) SPIRE 2004. LNCS, vol. 3246, pp. 171–182. Springer, Heidelberg (2004)

Space-Efficient Detection of Unusual Words

Djamal Belazzougui[1,2] and Fabio Cunial[3(✉)]

[1] Department of Computer Science, University of Helsinki, Helsinki, Finland
Djamal.Belazzougui@cs.helsinki.fi
[2] Helsinki Institute for Information Technology, Helsinki, Finland
[3] Max Planck Institute of Molecular Cell Biology and Genetics, Dresden, Germany
cunial@mpi-cbg.de

Abstract. Detecting all the strings that occur in a text more frequently or less frequently than expected according to an IID or a Markov model is a basic problem in string mining, yet current algorithms are based on data structures that are either space-inefficient or incur large slowdowns, and current implementations cannot scale to genomes or metagenomes in practice. In this paper we engineer an algorithm based on the suffix tree of a string to use just a small data structure built on the Burrows-Wheeler transform, and a stack of $O(\sigma^2 \log^2 n)$ bits, where n is the length of the string and σ is the size of the alphabet. The size of the stack is $o(n)$ except for very large values of σ. We further improve the algorithm by removing its time dependency on σ, by reporting only a subset of the maximal repeats and of the minimal rare words of the string, and by detecting and scoring candidate under-represented strings that *do not occur* in the string. Our algorithms are practical and work directly on the BWT, thus they can be immediately applied to a number of existing datasets that are available in this form, returning this string mining problem to a manageable scale.

1 Introduction

Detecting all the patterns of a string whose number of occurrences matches some notion of statistical surprise is a fundamental requirement of the post-genome era, in which textual datasets grow faster than the ability to understand them, and in which over- and under-representation with respect to a statistical model is often an indicator of structure or function. The sheer volume of the available datasets makes even simple models of patterns and simple measures of statistical surprise useful in practice, if their detection scales to extremely long strings in reasonable time and space. In this paper we focus on the simplest possible model of a pattern – a string W, of any length, that occurs without mismatches $f_T(W)$ times in a text T of length n – and we consider measures of statistical surprise that score W according to the expected number $\mathbb{E}[f_T(W)]$ and to the variance $\mathbb{V}[f_T(W)]$ of the number of its occurrences in a random text of length $|T|$

This work was partially supported by Academy of Finland under grant 284598 (Center of Excellence in Cancer Genetics Research).

(see e.g. [1] and references therein). We assume that the random source is a given Markov chain, and for concreteness we set its order to zero, since this simple case already captures the computational structure of the problem [2]. Moreover, we focus on computing $\mathbb{V}[f_T(W)]$, since computing expectations with respect to a Markov chain of order zero is trivial (see e.g. [3]).

$\mathbb{V}[f_T(W)]$ enjoys the remarkable property that its computation can be carried out by iterating over all the proper *borders* of W, i.e. over all the nonempty substrings of W shorter than W that are at the same time prefix and suffix of W: see [3] for a detailed derivation. To make the paper self-contained, here we just recall that $\mathbb{V}[f_T(W)]$ can be computed in constant time from the functions $\phi(W)$ and $\gamma(W)$ defined below:

$$\phi(W) = \sum_{b \in \mathcal{B}(W)} (n - 2|W| + b + 1) \cdot \pi(W[b..|W| - 1])$$

$$\gamma(W) = \sum_{b \in \mathcal{B}(W)} \pi(W[b..|W| - 1])$$

where strings are indexed from zero, $\pi(W) = \prod_{i=0}^{|W|-1} \mathbb{P}[W_i]$, W_i is the ith character of string W, $\mathbb{P}[c]$ is the probability of character c according to the given zero-order Markov chain, and $\mathcal{B}(W)$ is the set of all border lengths of W. Removing the components of $\mathbb{V}[f_T(W)]$ that depend on borders can cause large relative errors in practice [2], so we focus on the exact computation of $\mathbb{V}[f_T(W)]$. It is well known that borders have a recursive structure, in the sense that the set of borders of W consists of the longest border V of W, and of all the borders of V. This observation enables one to map the computation of $\mathbb{V}[f_T(W)]$ for a given W onto the Morris-Pratt algorithm [17], thus achieving time $O(|W|)$ in the worst case [3]. Specifically:

$$\phi(W) = \delta(W) \cdot \left(\phi(B) - 2(|W| - |B|)\gamma(B) + n - 2|W| + |B| + 1\right)$$

$$\gamma(W) = \delta(W) \cdot \left(1 + \gamma(B)\right)$$

where B is the longest border of W, and $\delta(W) = \pi(W[|B|..|W| - 1])$. In practice, however, we are interested in extracting from a string T *all its substrings* W, *of any length*, such that a user-specified measure of surprise computed on $\mathbb{E}[f_T(W)]$ and $\mathbb{V}[f_T(W)]$ is, say, greater than a threshold. Even though computing $\mathbb{V}[f_T(W)]$ takes $O(|W|)$ time for a given W, enumerating and scoring in this way all substrings of a text T of length n takes $O(n^2)$ time.

Luckily, a number of statistical scores $z(W)$ enjoy the additional property that $z(XWY) \geq z(W)$ if $f_T(XWY) = f_T(W)$, where X and Y are strings [1]. Consider then a set $\mathcal{A}$ of substrings of T such that all substrings in the set have the same number of occurrences, and consider the partial order $\preceq$ on $\mathcal{A}$ such that $V \preceq W$ iff $W = XVY$ for (possibly empty) strings X and Y. If we display to the user just the maximal elements of $\mathcal{A}$ with respect to $\preceq$, we guarantee that every over-represented string $W \in \mathcal{A}$ that we do not output is a substring of a string $XWY \in \mathcal{A}$ in the output which has at least the same score. Symmetrically, if we display just the minimal elements of $\mathcal{A}$ with respect to $\preceq$, we guarantee that every under-represented string $XWY \in \mathcal{A}$ that we do not output is a superstring of a string $W \in \mathcal{A}$ in the output which has at most the same score. A possible choice for $\mathcal{A}$ is the set of all substrings that start at exactly the same positions in T: this class has a unique maximal element, which corresponds to a node of the suffix tree of T, and a unique minimal element, which corresponds to the right extension by a single character of a node of the suffix tree of T. Since the number of all such classes is $O(|T|)$, and since all such classes are connected to one another by a trie, known as the *suffix-link tree*, it is possible to devise an algorithm that computes $\mathbb{V}[f_T(W)]$ for all the maximal and minimal elements of all such classes, in a total amount of time that grows linearly in $|T|$ [2].

In this paper we engineer the algorithm described in [2] to use as its substrate the Burrows-Wheeler transform of T, rather than space-inefficient data structures like the (truncated) suffix tree of T, or space-efficient simulations of the suffix tree with $O(\log^\varepsilon n)$ slowdown, like compressed suffix trees (see e.g. [11] and references therein). We also observe that the time complexity of the algorithm described in [2] depends on the cardinality of the alphabet, and we remove this dependency. Moreover, we adapt the algorithm to work on the smallest possible set of equivalence classes, thus reducing time and output size in practice. Assuming an alphabet of size $\sigma \in o(\sqrt{n}/\log n)$, we can thus perform all the computations described in [2] in $O(n)$ time and in $o(n + \lambda\sqrt{n})$ bits of space, given the BWT of T and few additional data structures [7], where λ is the length of a longest repeat of T. For statistical reasons, the maximum length of a string to be reported is often $O(\log_\sigma n)$, thus our algorithm uses effectively $o(n)$ bits of space in addition to the input. Concatenating this setup to the BWT construction algorithm described in [5], we can discover all the over- and under-represented substrings of T, *directly from T itself*, in randomized $O(n)$ time and in $O(n \log \sigma)$ bits of space in addition to T itself. Finally, we extend the algorithm in [2] to consider potentially under-represented strings that do not occur in T, thus providing for the first time a way to score and rank the minimal absent words of T.

2   Preliminaries

2.1   Strings

Let $\Sigma = [1..\sigma]$ be an integer alphabet, let $\# = 0$ be a separator not in Σ, let $T = [1..\sigma]^{n-1}\#$ be a string, and let ε be the empty string. For reasons that will become clear in Section 2.2, we assume $\sigma \in o(\sqrt{n}/\log n)$ throughout the paper. We denote by $f_T(W)$ the number of (possibly overlapping) occurrences of a string W in the circular version of T. A *repeat* W is a string that satisfies $f_T(W) > 1$. We denote by $\Sigma_T^\ell(W)$ the set of characters $\{a \in [0..\sigma] : f_T(aW) > 0\}$ and by $\Sigma_T^r(W)$ the set of characters $\{b \in [0..\sigma] : f_T(Wb) > 0\}$. A repeat W is *right-maximal* (respectively, *left-maximal*) iff $|\Sigma_T^r(W)| > 1$ (respectively, iff $|\Sigma_T^\ell(W)| > 1$). It is well known that T can have at most $n - 1$ right-maximal substrings and at most $n - 1$ left-maximal substrings. A *maximal repeat* of T is a repeat that is both left- and right-maximal. Clearly a maximal repeat W of

T satisfies $f_T(aW) < f_T(W)$ and $f_T(Wb) < f_T(W)$ for any characters a and b in Σ. A repeat is *supermaximal* if it is not a proper substring of any other repeat. A *minimal rare word* of T is a string W that satisfies $f_T(W) < f_T(V)$ for every proper substring V of W. Clearly W must have the form aXb, where a and b are characters and X is a maximal repeat of T. If $f_T(W) > 0$, then aX is a right-maximal substring of T and Xb is a left-maximal substring of T. If $f_T(W) = 0$, then aX (respectively, Xb) must occur in T, but it is not necessarily right-maximal (respectively, left-maximal). A minimal rare word of T that does not occur in T is called *minimal absent word* (see e.g. [8,12] and references therein): the total number of such strings can be $\Theta(\sigma n)$ [9]. A minimal rare word of T that occurs exactly once in T is called *minimal unique substring* (see e.g. [14] and references therein). It is clear that the total number of minimal rare words of T that occur at least once in T is $O(n)$.

String $V \neq \varepsilon$ is a *proper border* of string W if $W = VX$ and $W = YV$ for nonempty strings X and Y. A string W can have zero, one, or multiple proper borders: we denote by $bord(W)$ the length of the *longest* border of W. Each border of W is followed by a character when it occurs as a prefix, and it is preceded by a character when it occurs as a suffix: we use $a|W$ to denote the length of the longest border of W that is preceded by character a when it occurs as a suffix, and we use $W|a$ to denote the length of the longest border of W that is followed by character a when it occurs as a prefix. Clearly both $a|W$ and $W|a$ can be zero. We denote by $\mathcal{B}(W)$ the set of lengths of all borders of W, by $\mathcal{B}^r(W)$ the set of pairs $\{(a, a|W) : a \in \sigma, a|W \neq 0\}$, and by $\mathcal{B}^\ell(W)$ the set of pairs $\{(a, W|a) : a \in \sigma, W|a \neq 0\}$. It is well known that $\mathcal{B}(W) = \{bord(W)\} \cup \mathcal{B}(V)$, where V is the longest border of W: see e.g. [10] and references therein. In this paper we will also use left_W to denote an array of size $|\Sigma_T^r(W)|$, indexed by the characters in $\Sigma_T^r(W)$ in lexicographic order, such that $\text{left}_W[c] = W|c$. Similarly, we will use right_W to denote an array of size $|\Sigma_T^\ell(W)|$, indexed by the characters in $\Sigma_T^\ell(W)$ in lexicographic order, such that $\text{right}_W[c] = c|W$. Set $\mathcal{B}(W)$ determines all possible ways in which W can overlap with itself: specifically, the maximum possible number of occurrences of W in a string of length n is $f^*(W, n) = \lceil (n - |W| + 1)/period(W) \rceil$, where $period(W) = |W| - bord(W)$. When $f_T(W)$ needs to be compared to $f_{T'}(W)$, where $|T'| \neq |T|$, it is customary to divide $f_T(W)$ by $f^*(W, |T|)$. It is easy to see that the longest border of a random string of length n generated by an IID source is expected to tend to a constant as n tends to infinity.

For reasons of space we assume the reader to be familiar with the notion of *suffix tree* ST_T of a string T, which we do not define here. We denote by $\ell(v)$ the string label of a node v in a suffix tree. It is well known that a substring W of T is right-maximal iff $W = \ell(v)$ for some internal node v of ST_T. We assume the reader to be familiar with the notion of *suffix link* connecting a node v with $\ell(v) = aW$ for some $a \in [0..\sigma]$ to a node w with $\ell(w) = W$: we say that $w = \text{suffixLink}(v)$ in this case. Here we just recall that suffix links and internal nodes of ST_T form a tree, called the *suffix-link tree* of T and denoted by SLT_T, and that inverting the direction of all suffix links yields the so-called *explicit Weiner links*. Given an internal node v and a symbol $a \in [0..\sigma]$, it might happen that string $a\ell(v)$ does occur in T, but that it is not right-maximal, i.e. it is not the label of any internal node of ST_T: all such left extensions of internal nodes that end in the middle of an edge are called *implicit Weiner links*. An internal node v of ST_T can have more than one outgoing Weiner link, and all such Weiner links have distinct labels: in this case, $\ell(v)$ is a maximal repeat. It is known that the number of suffix links (or, equivalently, of explicit Weiner links) is upper-bounded by $2n - 2$, and that the number of implicit Weiner links can be upper-bounded by $2n - 2$ as well.

If V is a nonempty proper border of W, then $\Sigma_T^\ell(W) \subseteq \Sigma_T^\ell(V)$ and $\Sigma_T^r(W) \subseteq \Sigma_T^r(V)$. Thus, if W is right-maximal (respectively, left-maximal) then V is right-maximal (respectively, left-maximal); if W is a maximal repeat, then V is a maximal repeat; and if $W = aX$ where $a \in \Sigma$ and X is a maximal repeat, then $V = aY$ where Y is a maximal repeat[1].

2.2   Enumerating Maximal Repeats and Minimal Rare Words

For reasons of space we assume the reader to be familiar with the notion and uses of the Burrows-Wheeler transform of T, including the C array, the rank function, and backward searching. In this paper we use BWT_T to denote the BWT of T, we use $\mathtt{range}(W) = [\mathtt{sp}(W)..\mathtt{ep}(W)]$ to denote the lexicographic interval of a string W in a BWT that is implicit from the context, and we use $\Sigma_{i,j}$ to denote the set of distinct characters that occur inside interval $[i..j]$ of a string that is implicit from the context. We also denote by $\mathtt{rangeDistinct}(i, j)$ the function that returns the set of tuples $\{(c, \mathtt{rank}(c, p_c), \mathtt{rank}(c, q_c)) : c \in \Sigma_{i,j}\}$, *in any order*, where p_c and q_c are the first and the last occurrence of character c inside interval $[i..j]$, respectively. Here we focus on a specific application of BWT_T: enumerating all the right-maximal substrings of T, or equivalently all the internal nodes of ST_T. In particular, we use the algorithm described in [5] (Section 4.1), which we sketch here for completeness.

Given a substring W of T, let $b_1 < b_2 < \cdots < b_k$ be the sorted sequence of all the distinct characters in $\Sigma_T^r(W)$, and let $a_1, a_2, \ldots, a_h$ be the sequence of all the characters in $\Sigma_T^\ell(W)$, not necessarily sorted. Assume that we represent a substring W of T as a pair $\mathtt{repr}(W) = (\mathtt{chars}[1..k], \mathtt{first}[1..k + 1])$, where $\mathtt{chars}[i] = b_i$, $\mathtt{range}(Wb_i) = [\mathtt{first}[i]..\mathtt{first}[i+1]-1]$ for $i \in [1..k]$, and function $\mathtt{range}()$ refers to BWT_T. Note that $\mathtt{range}(W) = [\mathtt{first}[1]..\mathtt{first}[k + 1] - 1]$, since it coincides with the concatenation of the intervals of the right extensions of W in lexicographic order. If W is not right-maximal, array $\mathtt{chars}$ in $\mathtt{repr}(W)$ has length one. Given a data structure that supports $\mathtt{rangeDistinct}$ queries on BWT_T, and given the C array of T, there is an algorithm that converts $\mathtt{repr}(W)$ into the sequence $a_1, \ldots, a_h$ and into the corresponding sequence

---

[1] Thus, maximal repeats connected by longest border relationships form a tree rooted at the empty string: the path from the root to a maximal repeat lists all its borders, and the internal nodes of this tree cannot be supermaximal repeats. Similarly, longest border relationships and strings aW (respectively, Wa) where W is a maximal repeat, form a tree rooted at the empty string.

$\text{repr}(a_1 W), \ldots, \text{repr}(a_h W)$, in $O(de)$ time and $O(\sigma^2 \log n)$ bits of space in addition to the input and the output [5], where d is the time taken by the rangeDistinct operation per element in its output, and e is the number of distinct strings $a_i W b_j$ that occur in the circular version of T, where $i \in [1..h]$ and $j \in [1..k]$. We encapsulate this algorithm into a function that we call extendLeft.

If $a_i W$ is right-maximal, i.e. if array chars in $\text{repr}(a_i W)$ has length greater than one, we push pair $(\text{repr}(a_i W), |W|+1)$ onto a stack S. In the next iteration we pop the representation of a string from the stack and we repeat the process, until the stack becomes empty. This process is equivalent to following all the explicit Weiner links from the node v of ST_T with $\ell(v) = W$, not necessarily in lexicographic order. Thus, running the algorithm from a stack initialized with $\text{repr}(\varepsilon)$ is equivalent to performing a depth-first preorder traversal of the suffix-link tree of T (but with an arbitrary exploration order on the children of each node), which guarantees to enumerate all the right-maximal substrings of T. Every operation performed by the algorithm can be charged to a distinct node or Weiner link of ST_T, thus the algorithm runs in $O(nd)$ time. We keep the depth of the stack to $O(\log n)$ rather than to $O(n)$ by using the folklore trick of pushing at every iteration the pair $(\text{repr}(a_i W), |a_i W|)$ with largest $\text{range}(a_i W)$ first (see e.g. [13]). Every suffix-link tree level in the stack contains at most σ pairs, and each pair takes at most $\sigma \log n$ bits of space, thus the total space used by the stack is $O(\sigma^2 \log^2 n)$ bits. The following theorem follows from our assumption that $\sigma \in o(\sqrt{n}/\log n)$:

Theorem 1 ([5]). *Let $T \in [1..\sigma]^{n-1}\#$ be a string. Given a data structure that supports rangeDistinct queries on BWT_T, we can enumerate all the right-maximal substrings W of T, and for each of them we can return $|W|$, $\text{repr}(W)$, the sequence $a_1, a_2, \ldots, a_h$ of all characters in $\Sigma_T^\ell(W)$ (not necessarily sorted), and the sequence $\text{repr}(a_1 W), \ldots, \text{repr}(a_h W)$, in $O(nd)$ time and in $O(\sigma^2 \log^2 n) = o(n)$ bits of space in addition to the input and the output, where d is the time taken by the rangeDistinct operation per element in its output.*

Theorem 1 does not specify the order in which the right-maximal substrings must be enumerated, nor the order in which the left extensions of a right-maximal substring must be returned. The algorithm we just described can be adapted to return all the maximal repeats of T, within the same bounds, by outputting a right-maximal string W iff $|\text{rangeDistinct}(\text{sp}(W), \text{ep}(W))| > 1$. Computing the minimal rare words that occur in T is also easy:

Lemma 1. *Let $T \in [1..\sigma]^{n-1}\#$ be a string. Given a data structure that supports rangeDistinct queries on BWT_T, we can enumerate all the minimal rare words W of T that occur at least once in T, and for each of them we can return $|W|$ and $\text{range}(W)$, in $O(nd)$ time and in $O(\sigma^2 \log^2 n) = o(n)$ bits of space in addition to the input and the output, where d is the time taken by the rangeDistinct operation per element in its output.*

Proof. We use a technique similar to the one described in [6]. Specifically, we adapt Theorem 1 to iterate over all maximal repeats of T, and we allocate a temporary array freq[$0..\sigma$], indexed by all characters in the alphabet. After having enumerated a maximal repeat W, we scan repr(W), we compute the number of occurrences of every right extension Wb of W using array first, and we write $f_T(Wb)$ in freq[b]. Then, for every $i \in [1..h]$, we check whether repr(a_iW) contains more than one character: if this is the case, then $f_T(a_iW) > f_T(a_iWb)$ for every $b \in [0..\sigma]$. Thus, we scan repr(a_iW) and for every b in its array chars we check whether $f_T(a_iWb) < f_T(Wb)$, by accessing freq[b]. If this is the case, then a_iWb is a minimal rare word, and its interval in BWT$_T$ can be derived in constant time from the array first of repr(a_iW). At the end of this process, we reset array freq to its initial state by scanning repr(W) again. □

Minimal rare words that do not occur in T can be enumerated using a slight variation of Lemma 1, as described in [6]:

Lemma 2 ([6]). *Let $T \in [1..\sigma]^{n-1}\#$ be a string. Given a data structure that supports* rangeDistinct *queries on* BWT$_T$, *we can enumerate all the minimal rare words aWb of T that do not occur in T, where a and b are characters and $W \in [1..\sigma]^*$, and for each of them we can return a, b, $|W|$, and* range(W), *in $O(nd + \text{occ})$ time and in $O(\sigma^2 \log^2 n) = o(n)$ bits of space in addition to the input and the output, where d is the time taken by the* rangeDistinct *operation per element in its output, and* occ *is the output size.*

For reasons of space, we assume throughout the paper that d is the time per element in the output of a rangeDistinct data structure that is implicit from the context.

3   Computing the Border of all Right-Maximal Substrings

As mentioned in the introduction, computing the exact variance of the number of occurrences of a string W in a random text of length n can be mapped to the computation of the longest border of all suffixes of W [3], and thus takes $O(|W|)$ time using the Morris-Pratt algorithm [17]. To compute the longest border of *all* right-maximal substrings of a text T, as well as of *all* substrings Wb of T such that $b \in [0..\sigma]$ and W is right-maximal, in overall linear time on $|T|$, we need the following algorithm described in [2], which we sketch here for completeness:

Theorem 2 ([2]). *Let $T \in [1..\sigma]^n$ be a string. There is an algorithm that computes* bord(W) *for all right-maximal substrings W of T, and for all substrings $W = Vb$ of T where $b \in [1..\sigma]$ and V is right-maximal, in $O(n)$ words of space. The running time of this algorithm is linear in n and depends on σ.*

Proof sketch. We build the suffix tree ST$_T$ of T, and we assume that every node v of ST$_T$ stores sets $\Sigma_T^\ell(\ell(v))$ and $\Sigma_T^r(\ell(v))$ as lexicographically sorted lists. We perform a depth-first traversal of the *suffix-link tree* of T, and we store in each node v with label $\ell(v) = W$ the arrays right$_W$ and left$_W$ described in Section 2.1: recall that right$_W$ (respectively, left$_W$) is indexed by all characters $a \in \Sigma_T^\ell(W)$ (respectively, $b \in \Sigma_T^r(W)$), and it stores value $a|W$ at position right$_W[a]$

(respectively, value $W|b$ at position $\text{left}_W[b]$). Clearly, if $\text{right}_W[a] > 0$, then $bord(aW) = \text{right}_W[a] + 1$. If $\text{right}_W[a] = 0$, then $bord(aW)$ is either one (if a matches the last character of W) or zero. Once we know $bord(aW)$, we compute array right_{aW} by exploiting the identity $\mathcal{B}(aW) = \{bord(aW)\} \cup \mathcal{B}(V)$, where V is the longest border of aW (see Section 2.1): specifically, for every character $c \in \Sigma_T^\ell(aW)$, we know that c belongs also to $\Sigma_T^\ell(V)$, thus we set $\text{right}_{aW}[c] = \text{right}_V[c]$ if $c \neq d$, and we set $\text{right}_{aW}[d] = bord(aW)$, where d is the character that precedes the suffix of aW of length $bord(aW)$. Since we know that every character $c \in \Sigma_T^r(aW)$ also belongs to $\Sigma_T^r(V)$, we can compute array left_{aW} from left_V in the same way.

Every cell of every array right_W can be charged to a (possibly implicit) Weiner link of ST_T, and every cell of every array left_W can be charged to an edge of ST_T, thus the algorithm uses $O(n)$ words of space overall in addition to ST_T. However, copying $\text{right}_{aW}[c]$ from $\text{right}_V[c]$, where V is the longest border of aW (respectively, $\text{left}_{aW}[c]$ from $\text{left}_V[c]$) requires retrieving c from the list of left extensions (respectively, right extensions) of V, or merging such list with the corresponding list of aW, which introduces a dependency on σ. $\square$

A dependency on σ can be problematic in data mining applications, where the alphabet could be the result of a dense discretization of a continuous range (see e.g. [15,16] and references therein). To make Theorem 2 independent of σ, we start from generalizing the algorithm described in [18] to a trie, counting *return arcs* that do not point to the root:

Lemma 3. *Let $T = (V, E, \sigma)$ be a trie on alphabet $[1..\sigma]$, let $\ell(v) = \ell(e_k) \cdot \ell(e_{k-1}) \cdots \ell(e_1)$ be the label of a node $v \in V$ that is reachable from the root with path $e_1, e_2, \ldots, e_k$, where $e_i \in E$ for all $i \in [1..k]$, and let $a|v = w \in V :$ $\ell(w) = a|\ell(v)$. The set of return arcs $E' = \{(v, a|v) : v \in V, a \in [1..\sigma], a|v \neq \emptyset\}$ satisfies $|E'| \leq 2|V|$.*

Proof. We say that a return arc $(v, a|v) \in E'$ is of *type 1* if there is an edge $e = (v, w) \in E$ with $\ell(e) = a$, and we say that it is of *type 2* otherwise. The total number of type-1 return arcs is at most $|V|$, thus we focus on type-2 return arcs. Let $A = \ell(v)$ and let $B = A[1..a|\ell(v)]$. We charge a type-2 return arc $(v, a|v)$ to the vertex w that satisfies $A = B \cdot \ell(w)$. Assume that two distinct type-2 return arcs (u_1, v_1), (u_2, v_2) are charged to the same vertex w. Clearly it must be $u_1 \neq u_2$. If u_1 and u_2 do not lie on the same path of T, then w must be the lowest common ancestor of u_1 and u_2, or one of its ancestors (excluding the root). Let W be the label of the path from w to the lowest common ancestor of u_1 and u_2: clearly $\ell(u_1) = X_1 \cdot a \cdot W \cdot \ell(w)$ and $\ell(u_2) = X_2 \cdot b \cdot W \cdot \ell(w)$, where a and b are distinct characters and X_1 and X_2 are strings of the same length, but at the same time it must be $X_1 \cdot a \cdot W = X_2 \cdot b \cdot W$, a contradiction. Assume thus that u_1 and u_2 lie on the same path of T: without loss of generality, let u_1 be an ancestor of u_2. Further assume that there is an edge $e = (u_1, v) \in E$ with $\ell(e) = a$. Then it must be that $\ell(u_1) = Y_1 \cdot \ell(w) = X_1 \cdot b \cdot Y_1$ and $\ell(u_2) = Y_2 \cdot a \cdot Y_1 \cdot \ell(w) = X_2 \cdot Y_2 \cdot a \cdot Y_1$ and $a \neq b$, but at the same time it must be that $a \cdot Y_1 = b \cdot Y_1$, a contradiction. $\square$

Lemma 4. *Let $T \in [1..\sigma]^n$ be a string. There is an algorithm that computes $bord(W)$ for all right-maximal substrings W of T in $O(n)$ time and words of space.*

Proof. We proceed as in Theorem 2, but at every node v of ST_T with $\ell(v) = W$, we store set $\mathcal{B}^r(W)$, sorted in lexicographic order, rather than right_W. Recall that $\mathcal{B}^r(W)$ is the set of pairs $\{(a, a|W) : a \in [1..\sigma], a|W \neq 0\}$, i.e. a representation of the return arcs of Lemma 3. At every node v we merge $\mathcal{B}^r(W)$ with the lexicographically sorted list of characters in $\Sigma_T^\ell(W)$, setting $bord(aW) = a|W$ (which might be zero) for every left extension aW. Once we know $bord(aW)$, we build $\mathcal{B}^r(aW)$ by copying the entire $\mathcal{B}^r(V)$, where V is the longest border of aW, and by updating or inserting pair $(d, d|aW)$ using a linear scan of $\mathcal{B}^r(V)$, where d is the character that precedes the suffix of aW of length $bord(aW)$. This process touches every return arc of Lemma 3 a constant number of times. □

Lemma 4 can be clearly applied to any trie $\mathcal{T}$ of size n on an alphabet of size σ, in which every node stores its children in lexicographic order: the space used by such algorithm in addition to the trie is $O(\min\{n, \lambda\sigma\})$, where λ is the length of a longest path in $\mathcal{T}$. However, Lemma 3 does not generalize to radix trees, thus we cannot use it to bound the construction time of `left` arrays in Theorem 2. To achieve this, it suffices to replace `left` arrays with suitable stacks:

Lemma 5. *Let $T \in [1..\sigma]^n$ be a string. There is an algorithm that computes $bord(W)$ for all right-maximal substrings W of T, and for all substrings $W = Vb$ of T where $b \in [0..\sigma]$ and V is right-maximal, in $O(n)$ time and words of space.*

Proof. We run the algorithm in Lemma 4, keeping σ stacks $S_1, S_2, \ldots, S_\sigma$. Assume that, when we visit the node v of ST_T with $\ell(v) = W$, the top of stack S_b for all $b \in \Sigma_T^r(V)$ stores value $W|b$ (which might be zero). Assume that, in the depth-first traversal of the suffix-link tree of T, we choose to visit the node w with $\ell(w) = aW$ next: then, we iterate over all characters $b \in \Sigma_T^r(aW)$, we compute $aW|b$ by accessing position $bord(aW)$ of stack S_b, and we push $aW|b$ on S_b. This works since, if aW can be extended to the right by character b, then every suffix of aW can be extended to the right with character b as well. This process takes overall $O(n)$ time, and the size of all stacks $S_1, \ldots, S_\sigma$ is $O(\min\{n, \lambda\sigma\})$, where λ is the length of a longest repeat of T, since every element in every stack can be charged to an edge of ST_T. □

The information stored by Lemma 5 is enough to compute in $O(n)$ time and space the longest border of all minimal rare words that occur at least once in T. Adapting Lemma 5 to compute the longest border of all minimal *absent* words of T is also easy:

Lemma 6. *Let $T \in [1..\sigma]^n$ be a string. There is an algorithm that computes $bord(W)$ for all minimal absent words W of T in $O(n + \mathsf{occ})$ time and words of space, where occ is the size of the output.*

Proof. We proceed as in Lemma 5, but when we visit the node v of ST_T with $\ell(v) = aW$ for some $a \in [1..\sigma]$, we assume that the top of stack S_b for all $b \in \Sigma_T^r(W)$ stores value $aW|b$ (which might be zero). Assume that, in the depth-first traversal of the suffix-link tree of T, we choose to visit the node w with $\ell(w) = caW$ next: then, we iterate over all characters $b \in \Sigma_T^r(aW)$, we compute $caW|b$ by accessing position $bord(caW)$ of stack S_b, and we push $caW|b$ on S_b. This works since, if aW can be extended to the right by character b, then the proper suffix of the longest border of caW can be extended to the right with character b as well. Every element in every stack can be charged either to an edge of ST_T or to a minimal absent word of T, thus the size of all stacks is $O(\min\{n + \mathsf{occ}, \lambda\sigma\})$, where λ is the length of a longest repeat of T. □

4   Detecting Unusual Words in Small Space

As mentioned in the introduction, in order to compute the exact variance of a substring W of a string T, we just need to compute functions $\phi(W)$ and $\gamma(W)$. Specifically, we just need to compute such functions on all right-maximal substrings of T, and on all substrings Wa of T such that W is right-maximal [2]. Since $\phi(W)$ and $\gamma(W)$ can be computed from $\phi(V)$ and $\gamma(V)$, where V is the longest border of W, it is easy to see that we can adapt Lemma 5 as follows. When we visit the node v of ST_T with $\ell(v) = W$, we store $\phi(W)$ and $\gamma(W)$, and the top of stack S_b for all $b \in \Sigma_T^r(W)$ stores the following values (which might be zero) in addition to $W|b$:

$$F_b(|W|) = D_b(|W|) \cdot \big(F_b(W|b) - 2(|W| - W|b) \cdot G_b(W|b) + |T| - 2|W| + W|b\big)$$
$$G_b(|W|) = D_b(|W|) \cdot \big(1 + G_b(W|b)\big)$$

where $D_b(|W|) = \pi\big(W[W|b + 1..|W| - 1]\big) \cdot \mathbb{P}[b]$. Note that such values are computed recursively. We derive $\pi\big(W[W|b + 1..|W| - 1]\big)$ by keeping an additional stack that stores $\pi(V)$ for every suffix V of W. The entire process can be implemented using just a data structure that supports `rangeDistinct` queries on BWT_T, by implementing Lemma 5 on top of the iterator described in Theorem 1:

Theorem 3. *Let $T \in [1..\sigma]^{n-1}\#$ be a string. Given a data structure that supports `rangeDistinct` queries on BWT_T, we can compute $\phi(W)$ and $\gamma(W)$ for all right-maximal substrings W of T, and for all substrings $W = Vb$ of T such that $b \in [1..\sigma]$ and V is right-maximal, in $O(nd)$ time and in $o(n + \lambda\sqrt{n})$ bits of space in addition to the input and the output, where d is the time taken by the `rangeDistinct` operation per element in its output and λ is the length of a longest repeat of T.*

Proof. Recall that the iterator of Theorem 1 returns, for every right-maximal substring W of T, the set of all its right extensions in lexicographic order, but the set of all its left extensions *in arbitrary order*. To implement Lemma 5 in this case, we just need a temporary array `buffer`$[1..\sigma]$ of $\sigma \log n$ bits that is initialized to all zeros at the beginning of the traversal. When the iterator visits substring

W, we scan $\mathcal{B}^r(W)$ and we set $\texttt{buffer}[a] = a|W$ for all $(a, a|W) \in \mathcal{B}^r(W)$. Then, for every left extension a_i of W provided by the iterator, we compute $bord(a_iW)$ by accessing $\texttt{buffer}[a_i]$, and we proceed as in Lemma 5. Once we have finished processing substring W, we reset $\texttt{buffer}$ to its previous state by setting $\texttt{buffer}[a] = 0$ for all $(a, a|W) \in \mathcal{B}^r(W)$. The claimed space complexity comes from Theorem 1 and from our assumption that $\sigma \in o(\sqrt{n}/\log n)$.   □

For statistical reasons only substrings of length $O(\log_\sigma n)$ are candidates for being over- or under-represented [1], so the space complexity of Theorem 3 is effectively $o(n)$, every surprising substring can be encoded in a constant number of machine words, and thus it can be printed in constant time[2]. The technique described in Theorem 3 can be used to apply Lemma 4 to tries whose nodes do not store the list of their children in lexicographic order. Moreover, it is easy to adapt Theorem 3 to *output* all candidate over- and under-represented substrings whose statistical score matches a user-specified criterion, by keeping an additional stack of characters of size $\lambda \log \sigma$ and by exploiting the fact that the iterator in Theorem 1 returns the length of every substring it visits.

Reducing the number of patterns displayed by a data mining algorithm is key for making it useful in practice. According to the monotonicity of the scores described in Section 1, we can limit the search for over-represented (respectively, under-represented) substrings to maximal repeats (respectively, to minimal rare words): this observation was already implicit in [1], and Theorem 3 can be easily adapted to consider only such candidates. Moreover, in a practical implementation we can compute and store $\phi(W)$ and $\gamma(W)$ just for maximal repeats, since the longest border of a maximal repeat is itself a maximal repeat. We can also avoid storing numbers $F_b(|W|)$, $G_b(|W|)$ and $W|b$ for all $b \in \Sigma_T^r(W)$ on the stacks of Lemma 5, whenever W is a right-maximal substring of T that cannot be written as aV for a character a and a maximal repeat V. Within the working space budget of Theorem 3, but in time $O(nd + \texttt{occ})$, we can also compute the border and the statistical scores of all $\texttt{occ}$ minimal absent words of T, by adapting Lemma 6 to work on Theorem 1: such strings are the only strings that do not occur in T which could be under-represented in T, however they were not reported in previous works [1,2]. The ability to assign a statistical score to minimal absent words could be useful also in other contexts, for example in choosing which minimal absent words should be displayed to the user, since their total number is $\Theta(n\sigma)$ in the worst case.

Acknowledgments. We thank Alberto Apostolico for illuminating the details of the algorithms in [2], and Stefano Lonardi for providing the source code of the implementation described in [4].

---

[2] We assume the word RAM model of computation with words of size $\Omega(\log n)$ bits, in which all standard operations including multiplication have unit cost.

References

1. Apostolico, A., Bock, M.E., Lonardi, S.: Monotony of surprise and large-scale quest for unusual words. Journal of Computational Biology **10**(3–4), 283–311 (2003)
2. Apostolico, A., Bock, M.E., Lonardi, S., Xu, X.: Efficient detection of unusual words. Journal of Computational Biology **7**(1–2), 71–94 (2000)
3. Apostolico, A., Bock, M.E., Xu, X.: Annotated statistical indices for sequence analysis. In: Proceedgins of Compression and Complexity of Sequences 1997, pp. 215–229. IEEE (1998)
4. Apostolico, A., Gong, F.-C., Lonardi, S.: Verbumculus and the discovery of unusual words. Journal of Computer Science and Technology **19**(1), 22–41 (2004)
5. Belazzougui, D.: Linear time construction of compressed text indices in compact space. In: Proceedings of the 46th Annual ACM Symposium on Theory of Computing, STOC 2014, pp. 148–193. ACM, New York (2014)
6. Belazzougui, D., Cunial, F.: A framework for space-efficient string kernels. In: Cicalese, F., Porat, E., Vaccaro, U. (eds.) CPM 2015. LNCS, vol. 9133, pp. 13–25. Springer, Heidelberg (2015)
7. Belazzougui, D., Navarro, G., Valenzuela, D.: Improved compressed indexes for full-text document retrieval. Journal of Discrete Algorithms **18**, 3–13 (2013)
8. Chairungsee, S., Crochemore, M.: Using minimal absent words to build phylogeny. Theoretical Computer Science **450**, 109–116 (2012)
9. Crochemore, M., Mignosi, F., Restivo, A.: Automata and forbidden words. Information Processing Letters **67**(3), 111–117 (1998)
10. Crochemore, M., Rytter, W.: Jewels of stringology. World Scientific (2002)
11. Gog, S.: Compressed suffix trees: design, construction, and applications. PhD thesis, University of Ulm, Germany (2011)
12. Herold, J., Kurtz, S., Giegerich, R.: Efficient computation of absent words in genomic sequences. BMC Bioinformatics **9**(1), 167 (2008)
13. Hoare, C.A.R.: Quicksort. The Computer Journal **5**(1), 10–16 (1962)
14. Ileri, A.M., Külekci, M.O., Xu, B.: A simple yet time-optimal and linear-space algorithm for shortest unique substring queries. Theoretical Computer Science **562**, 621–633 (2015)
15. Keogh, E., Lonardi, S., Chiu, B.Y.-C.: Finding surprising patterns in a time series database in linear time and space. In: Proceedings of the Eighth ACM SIGKDD International Conference on Knowledge Discovery and Data Mining, KDD 2002, pp. 550–556. ACM, New York (2002)
16. Lin, J., Keogh, E., Wei, L., Lonardi, S.: Experiencing SAX: a novel symbolic representation of time series. Data Mining and Knowledge Discovery **15**(2), 107–144 (2007)
17. Morris, J.H., Pratt, V.R.: A linear pattern-matching algorithm. Technical Report 40, University of California, Berkeley (1970)
18. Simon, I.: String matching algorithms and automata. In: First South American Workshop on String Processing, Belo Horizonte, Brazil, pp. 151–157 (1993)

Parallel Construction of Succinct Representations of Suffix Tree Topologies

Uwe Baier, Timo Beller, and Enno Ohlebusch$^{(\boxtimes)}$

Institute of Theoretical Computer Science, Ulm University, 89069 Ulm, Germany
{Uwe.Baier,Timo.Beller,Enno.Ohlebusch}@uni-ulm.de

Abstract. A compressed suffix tree usually consists of three components: a compressed suffix array, a compressed LCP-array, and a succinct representation of the suffix tree topology. There are parallel algorithms that construct the suffix array and the LCP-array, but none for the third component. In this paper, we present parallel algorithms on shared memory architectures that construct the enhanced balanced parentheses representation (BPR). The enhanced BPR is an implicit succinct representation of the suffix tree topology, which supports all navigational operations on the suffix tree. It can also be used to efficiently construct the BPS, an explicit succinct representation of the suffix tree topology.

1 Introduction

A suffix tree (ST) for a string S of length n is a compact trie storing all the suffixes of S, i.e., the concatenation of the edge labels on the path from the root to leaf i exactly spells out the i-th suffix of S; see Fig. 1 for an example. The ST is one of the most important data structures in string processing, with applications in other fields like bioinformatics or information retrieval. The drawback of STs is their huge space consumption: even carefully engineered implementations take up to 20 bytes per character. Suffix arrays reduce the space to about 4 bytes per character, but they do not support navigational operations such as suffix links. With the help of additional arrays, it is possible to support such operations, but this again results in a large memory requirement. In the last decade, research in that area therefore focused on compressed suffix trees (CSTs). A CST consists of three components: a compressed suffix array [10], a compressed LCP-array [15] and a succinct representation of the suffix tree topology. Sadakane [16] introduced the first CST representation that included a succinct representation of the tree topology and supported full suffix tree functionality. He used the traditional explicit representation, a balanced sequence of parentheses (BPS) that can be obtained by a depth-first traversal of the suffix tree [9]. It has been shown that the BPS can be constructed in $O(n)$ time using $O(n)$ bits of working space [5], but these algorithms are quite complex and involve large big-O-constants. Therefore, they are not used in existing implementations. Ohlebusch et al. [11–13] derive the tree topology *implicitly* from intervals in the LCP-array. They showed how to construct a so-called enhanced BPR from the Super-Cartesian tree of the LCP-array. This data structure has size $3n + o(n)$ bits and supports all navigational

C. Iliopoulos et al. (Eds.): SPIRE 2015, LNCS 9309, pp. 234–245, 2015.
DOI: 10.1007/978-3-319-23826-5_23

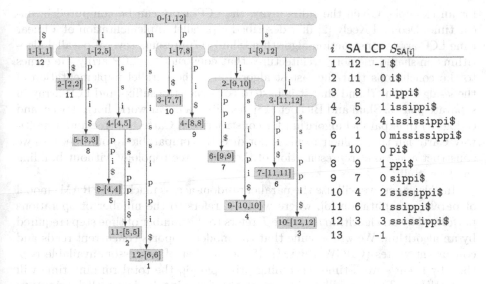

Fig. 1. Left: suffix tree for $S = $ mississippi$ (an annotation ℓ-$[i..j]$ within a node shows the corresponding LCP-interval, i.e., ℓ is the string-depth of the node and $[i..j]$ is the corresponding interval). Right: suffix- and LCP-array.

operations on the suffix tree; see [11, 12] for the algorithms that implement these navigational operations and their time complexities. We will recall in Section 3 that the enhanced BPR of the LCP-array can be constructed *without* knowing the Super-Cartesian tree itself. To the best of our knowledge, we here present the first parallel algorithms that construct a succinct representation of the suffix tree topology, namely the enhanced BPR. Gog [4, p.121] observed that the fastest way to construct the BPS goes as follows: first construct the enhanced BPR and then use it to traverse the (virtual) suffix tree in a depth-first manner. It is plausible that the same approach will work best for parallel implementations, but we did not investigate this yet.

2   Preliminaries and Related Work

In the following, S is a string of length n. Let S_i denote the i-th suffix of S. The suffix array SA specifies the lexicographic order of the suffixes of S. It is defined by $S_{\mathsf{SA}[1]} < S_{\mathsf{SA}[2]} < \cdots < S_{\mathsf{SA}[n]}$; see Fig. 1 for an example. A suffix array can be constructed in linear time; see e.g. the overview article [14]. Several parallel algorithms for suffix array construction exist (both on shared and distributed memory); see [2,17] and the references therein. The LCP-array stores the lengths of the *longest common prefixes* of lexicographically adjacent suffixes: $\mathsf{LCP}[1] = -1 = \mathsf{LCP}[n+1]$ and for $2 \leq i \leq n$, $\mathsf{LCP}[i] = \max\{k \geq 0 \mid S_{\mathsf{SA}[i]} \text{ and } S_{\mathsf{SA}[i-1]} \text{ share a prefix of length } k\}$; see Fig. 1

for an example. Given the suffix array, the LCP-array can be computed in linear time. Deo and Keely [2] first described a parallel implementation of a linear time LCP-array construction algorithm. Shun [17] presents several parallel algorithms on shared memory architectures that construct the LCP-array. He comes to the conclusions that the fastest algorithm is the parallel implementation of the Φ-algorithm [7] and that it is best to construct the suffix- and LCP-array in separate phases. Shun and Blelloch [18] describe "a linear work, linear space, and $O(\log^2 n)$ time parallel algorithm for constructing a Cartesian tree and a multiway Cartesian tree." Their result is not directly comparable to ours because we construct a *succinct* representation of the suffix tree topology without building the tree itself.

In this paper, we will use the parallel random-access machine (PRAM) model of parallel computation [6], where work W refers to the number of operations performed by an algorithm and time T refers to the number of time steps required by an algorithm. We will assume that the model supports concurrent reads and concurrent writes (CRCW PRAM). If the number of processors available is p, then by Brent's work-time scheduling principle [6], the total running time will be $O(W/p + T)$. We will make use of the following basic parallel primitive: Prefix sum takes a sequence A of length n, an associative binary operator $\oplus$, and an identity element id so that $id \oplus a = a$ for all a, and returns the sequence $(id, id \oplus A[1], id \oplus A[1] \oplus A[2], \ldots, id \oplus A[1] \oplus A[2] \oplus \cdots \oplus A[n])$. It requires $O(n)$ work and $O(\log n)$ time; see [6] for details.

3   Sequential Construction of the Enhanced BPR

As in [12], we introduce the enhanced BPR by means of the Super-Cartesian tree of the LCP-array. The latter is defined as follows.

Definition 1. *Let $A[l..r]$ be an array of elements of a totally ordered set $(M, \leq)$ and suppose that the minima of $A[l..r]$ appear at positions $p_1 < p_2 < \cdots < p_k$ for some $k \geq 1$. The* Super-Cartesian tree $C^{sup}(A[l..r])$ *of $A[l..r]$ is recursively constructed as follows:*

- *If $l > r$, then $C^{sup}(A[l..r])$ is the empty tree.*
- *Otherwise create k nodes $v_1, v_2, \ldots, v_k$, label each v_j with p_j, and for each j with $1 < j \leq k$ the node v_j is the right sibling of node v_{j-1} (in Fig. 2, node v_{j-1} is connected with v_j by a horizontal edge). Node v_1 is the root of $C^{sup}(A[l..r])$. Recursively construct $C_1 = C^{sup}(A[l..p_1-1])$, $C_2 = C^{sup}(A[p_1+1..p_2-1]), \ldots, C_{k+1} = C^{sup}(A[p_k+1..r])$. For each j with $1 \leq j < k$, the left child of v_j is the root of C_j. The left and right children of v_k are the roots of C_k and C_{k+1}, respectively.*

Fig. 2 depicts the Super-Cartesian tree of the LCP-array from Fig. 1. Ohlebusch and Gog [13] showed that the Super-Cartesian tree of the LCP-array can be represented by a sequence of balanced parentheses, called BPR (an acronym for balanced parentheses representation), and they gave a construction algorithm

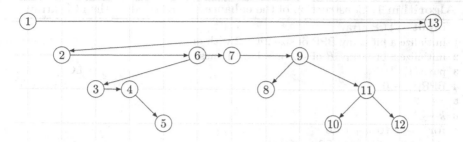

Fig. 2. The Super-Cartesian tree of the LCP-array from Fig. 1.

Fig. 3. Enhanced BPR of the Super-Cartesian tree of Fig. 2. The first row shows the bit string B and the second row depicts the BPR. The third row is only for illustrative purposes: The opening parentheses are numbered consecutively and a closing parenthesis has the number of its matching opening parenthesis. The last row shows the positions of the parentheses in the BPR.

that is solely based on the LCP-array. However, the sequence BPR lacks some information, as the cases "right child" and "right sibling" are treated in the same fashion, and the LCP-array itself is required to compensate for this. Ohlebusch et al. [12] compensate for the lack of information by enhancing the BPR with a bit string B. A closing parentheses (in the BPR) corresponding to a node that is a right sibling is marked with a 0, otherwise it is marked with a 1. The construction of the enhanced BPR of the Super-Cartesian tree of the LCP-array starts at the root of the tree and proceeds as follows (see Fig. 3 for an example):

1. Write the balanced parentheses sequence of the left child.
2. Write an opening parenthesis.
3. Write the balanced parentheses sequence of the right child/sibling.
4. Write a closing parenthesis. If the node under consideration is a right sibling, append 0 to B; otherwise append 1 to B.

The enhanced BPR of the LCP-array is a $3n$-bits representation because there are $2n + 2$ parentheses and the bit string B has length $n + 1$ (the LCP-array of the string S has size $n + 1$). Hence it uses $3n + o(n)$ bits to support navigational operations on the suffix tree. Fischer [3] showed that $2.54n + o(n)$ bits are enough for this task, but he did not provide an implementation of his data structure.

As already noted in [12], the enhanced BPR can be constructed *without* knowing the Super-Cartesian tree itself; see Algorithm 1. There, the array BPR is initialized with ones (an opening parenthesis is represented by '0', whereas a

238 U. Baier et al.

Algorithm 1. Construction of the enhanced BPR based on the LCP-array.

input : LCP-array
1 initialize a bit array BPR of size $2n + 2$ with ones
2 initialize a bit array B of size $n + 1$ with ones
3 push(1)   ▷ LCP$[1] = -1$
4 BPR$[1] \leftarrow 0$ ▷ write '('
5 $j \leftarrow 2$
6 $k \leftarrow 1$
7 **for** $i \leftarrow 2$ **to** $n + 1$ **do**
8   **while** LCP$[i] <$ LCP[top()] **do**
9         $last \leftarrow$ pop()
10         $j \leftarrow j + 1$ ▷ implicitly write ')'
11   **if** LCP$[last] =$ LCP[top()] **then**
12             $B[k] \leftarrow 0$
13         $k \leftarrow k + 1$
14   push(i)
15   BPR$[j] \leftarrow 0$ ▷ write '('
16     $j \leftarrow j + 1$
17 $B[n] \leftarrow 0$ ▷ write 01 for LCP$[1]$ and LCP$[n + 1]$
return : the arrays BPR and B

---

closing parenthesis is represented by '1'), and Algorithm 1 computes only the opening parentheses in the BPR. Similarly, the array B is initialized with ones and Algorithm 1 computes only the zero-entries in B.

For later purposes, we note that Algorithm 1 keeps the following invariants. Immediately after index i, $2 \leq i \leq n + 1$, was pushed onto the stack, the stack contents $i_1, \ldots, i_m$ satisfies the following properties:

1. $1 = i_1 < i_2 < \cdots < i_m = i$
2. LCP$[i_1] \leq$ LCP$[i_2] \leq \cdots \leq$ LCP$[i_m]$
3. for all q with $2 \leq q \leq m$ and for all p with $i_{q-1} < p < i_q$: LCP$[p] >$ LCP$[i_q]$

4 Parallel Construction of the Enhanced BPR

The parallelization of Algorithm 1 requires the knowledge at which position an opening parenthesis (and a bit in the array B) must be written. By construction $i - 1$ opening parentheses appear before the i-th opening parenthesis. So the i-th opening parenthesis appears at position $i - 1 + c(i) + 1$, where $c(i)$ is the number of closing parentheses before the i-th opening parenthesis. Obviously, $c(i)$ depends on the values in LCP$[1..i-1]$. Let $dist[i]$ be the number of elements on the stack immediately before i is pushed onto the stack. To put it differently, $dist[i]$ is the number of elements on the stack *below* i immediately after i was pushed onto the stack. Since $c(i)$ equals the number of elements that have been removed cumulatively from the stack before i is pushed onto it, we have

i	1	2	3	4	5	6	7	8	9	10	11	12	13
LCP$[i]$	-1	0	1	1	4	0	0	1	0	2	1	3	-1
NSV$[i]$	$\perp$	13	6	6	6	13	13	9	13	11	13	13	$\perp$
PSEV$[i]$	$\perp$	1	2	3	4	2	6	7	7	9	9	11	1
$dist[i]$	0	1	2	3	4	2	3	4	4	5	5	6	1

Fig. 4. The LCP-array from Fig. 1 and the corresponding arrays NSV, PSEV, and $dist$.

$c(i) = i - 1 - dist[i]$. So the task is to compute the array $dist$ in parallel, and we will show next how to do this.

The all nearest smaller values (ANSV) problem is defined as follows: for each element in a sequence of numbers, find the closest smaller element to the left of it and the closest smaller element to the right of it. Berkman et al. [1] showed that ANSVs can be computed in $O(n)$ work and $O(\log \log n)$ time on the CRCW PRAM. In other words, the arrays PSV (previous smaller value) and NSV (next smaller value), which are defined as follows for all i with $2 \leq i \leq n$

$$\mathsf{PSV}[i] = \max\{j \mid 1 \leq j < i \text{ and } \mathsf{LCP}[j] < \mathsf{LCP}[i]\}$$
$$\mathsf{NSV}[i] = \min\{j \mid i < j \leq n+1 \text{ and } \mathsf{LCP}[j] < \mathsf{LCP}[i]\}$$

can be computed in $O(n)$ work and $O(\log \log n)$ time on the CRCW PRAM. The method from [1] can be modified so that it computes the arrays PSEV and NSV instead of the arrays PSV and NSV, where PSEV is defined for all i with $2 \leq i \leq n+1$ by

$$\mathsf{PSEV}[i] = \max\{j \mid 1 \leq j < i \text{ and } \mathsf{LCP}[j] \leq \mathsf{LCP}[i]\}$$

Fig. 4 shows an example (an entry $\perp$ means that the value is undefined).

In Algorithm 1, suppose that the stack contents is $1 = i_1, i_2, \ldots, i_{m-1}, i_m = i$ immediately after index i was pushed onto the stack. It is not difficult to see that the invariants maintained by that algorithm (see end of Section 3) imply that $\mathsf{PSEV}[i_q] = i_{q-1}$ for all q with $2 \leq q \leq m$. (We shall say that there is a PSEV-pointer from i_q to i_{q-1}.) In other words, $dist[i]$ equals the number of PSEV-pointers that must be followed until the index 1 is reached (since LCP$[1] = -1$, all paths of PSEV-pointers end at index 1). Formally, for all i with $1 \leq i \leq n+1$, we have

$$dist[i] = \begin{cases} 0 & \text{if } i = 1 \\ 1 + dist[\mathsf{PSEV}[i]] & \text{otherwise} \end{cases}$$

Using a copy of the array PSEV, Algorithm 2 calculates the array $dist$ by "pointer jumping". It uses $O(n \log n)$ work and $O(\log n)$ time, but there is a parallel algorithm that requires only $O(n)$ work and $O(\log n)$ time for the same task; see [6, Section 3.1.2] for details.

Algorithm 3 calculates the enhanced BPR in parallel. It is easy to see that it uses $O(n)$ work and $O(1)$ time, but its correctness is not so obvious. In the following, let $_i($ denote the i-th opening parenthesis and let $)_i$ denote the closing parenthesis that matches it; cf. Fig. 3. As mentioned above, $_i($ must be placed

---

Algorithm 2. Parallel computation of the distance array.

---

   input   : a copy of the PSEV-array
1 create an integer array $dist$ of size $n + 1$
2 **foreach** $index$ i $with$ $2 \leq i \leq n + 1$ **do in parallel**
3    $\lfloor$ $dist[i] \leftarrow 1$ $\triangleright$ initialize $dist$ with ones

4 $dist[1] \leftarrow 0$
5 **while** $there$ is an $index$ i $with$ PSEV$[i] \neq \bot$ **do**
6   **foreach** $index$ i $with$ PSEV$[i] \neq \bot$ **do in parallel**
7        $dist[i] \leftarrow dist[i] + dist[$PSEV$[i]]$
8   PSEV$[i] \leftarrow$ PSEV$[$PSEV$[i]]$

   return : the $dist$ array

Algorithm 3. Parallel computation of the arrays BPR and B.

---

   input   : LCP, PSEV, NSV and $dist$
1 initialize a bit array BPR of size $2n + 2$ with ones (in parallel)
2 initialize a bit array B of size $n + 1$ with ones (in parallel)
3 **foreach** $index$ i $with$ $1 \leq i \leq n + 1$ **do in parallel**
4    $\lfloor$ BPR$[2i - 1 - dist[i]] \leftarrow 0$ $\triangleright$ write '('

5 **foreach** $index$ i $with$ $2 \leq i \leq n$ **do in parallel**
6   **if** LCP$[i]$ = LCP$[$PSEV$[i]]$ **then**
7      $\lfloor$ $B[$NSV$[i] - 1 - dist[i]] \leftarrow 0$

8 $B[n] \leftarrow 0$ $\triangleright$ write 01 for LCP$[1]$ and LCP$[n + 1]$
   return : the arrays BPR and B

---

at position $i - 1 + c(i) + 1$ in the BPR. Since $c(i) = i - 1 - dist[i]$, $_i($ must be written at position $2i - 1 - dist[i]$ and Algorithm 3 does this on line 4. To show that the assignment on line 7 in Algorithm 3 is correct, we have to prove that $)_i$ is the ℓ-th closing parenthesis in the BPR, where $\ell =$ NSV$[i] - 1 - dist[i]$. In Algorithm 1, the parenthesis $)_i$ is (implicitly) written when the index $q =$ NSV$[i]$ is reached. We know that there are $c(i) = i - 1 - dist[i]$ closing parentheses before $_i($. Additionally, all $q - i - 1$ opening parentheses between $_i($ and $_q($ must be closed, before $)_i$ is written. Therefore, there are $c(i) + q - i - 1 = i - 1 - dist[i] +$ NSV$[i] - i - 1 =$ NSV$[i] - dist[i] - 2$ closing parentheses before $)_i$. Hence $)_i$ must be written at the next position $\ell =$ NSV$[i] - 1 - dist[i]$. Consequently, Algorithm 3 correctly calculates the enhanced BPR.

To summarize, the enhanced BPR can be calculated using $O(n)$ work and $O(\log n)$ time on a CRCW PRAM.

5 A Heuristic Parallel Algorithm

In this section, we describe an algorithm that works very well in practice. To obtain a good load balance, we divide the subarray LCP$[1..n]$ into p blocks of size

$\lceil n/p \rceil$ (in fact, the last block may be smaller). Ideally, each processor is assigned to one block and computes the corresponding enhanced BPR. This, however, is only possible if the contents of the stack at the beginning of each block is known in advance. To achieve this, we use a threshold t and modify the start and end positions of the blocks as follows. For each original block k, say $\mathsf{LCP}[s_k..e_k]$, define $\ell_k = \min\{\mathsf{LCP}[i] \mid s_k \leq i \leq e_k\}$ and

$$startpos[k] = \begin{cases} \min\{i \mid s_k \leq i \leq e_k \text{ and } \mathsf{LCP}[i] = \ell_k\} & \text{if } \ell_k < t \\ 0 & \text{otherwise} \end{cases}$$

Suppose for a moment that $\ell_k < t$ for all k. In this case, we obtain p modified blocks: the k-th block starts at position $startpos[k]$ and ends at position $startpos[k+1]-1$. Note that the first block starts at position 1 because $\mathsf{LCP}[1] = -1$ and the last block ends at position $n+1$ if we define $startpos[p+1] = n+2$. For the LCP-array of Fig. 4 and $p = 3$, the original blocks are $\mathsf{LCP}[1..4]$, $\mathsf{LCP}[5..8]$, and $\mathsf{LCP}[9..12]$. In our example, for $t = 2$ we have $startpos[1..4] = [1, 6, 9, 14]$ and hence the modified blocks are $\mathsf{LCP}[1..5]$, $\mathsf{LCP}[6..8]$, and $\mathsf{LCP}[9..13]$. Each processor can sequentially compute the enhanced BPR of its block by Algorithm 1, provided that the stack contents is known immediately before $startpos[k]$ is pushed onto it. As we shall see, this can be achieved by combining the contents of local stacks: the contents of $stack[k]$, the local stack for block k, can be obtained by an application of Algorithm 1 to $\mathsf{LCP}[startpos[k]..startpos[k+1]-1]$. In contrast to Algorithm 1, a stack now contains LCP-values instead of indices (i.e., $\mathsf{LCP}[i]$ instead of i). Moreover, it stores only LCP-values that are strictly smaller than the threshold t. Thus, it can be represented by an array of size $t+1$, where the array index corresponds to an LCP-value ℓ and the value at the index is the number of occurrences of ℓ on the stack. Let us illustrate this with our running example. Since $t = 2$, $stack[1]$ is the array $[1, 1, 2]$, where the ℓ-th entry is the number of occurrences of the LCP-value $\ell - 2$ on the stack. For instance, $stack[1][1] = 1$ is the number of occurrences of the LCP-value -1 on the stack, and $stack[1]$ represents the stack with the elements $-1, 0, 1, 1$ (from bottom to top). The LCP-value 4 does not appear on top of the stack because $2 = t < 4$. Similarly, we obtain $stack[2] = [0, 2, 1]$. Lines 7-15 of Algorithm 4 compute the array $startpos$ and the local stacks.

To combine local stacks, we define a binary operator $\oplus$ on stacks. For two stacks A and B, where $m_B = \min\{i \mid 1 \leq i \leq t+1 \text{ and } B[i] \neq 0\}$, let $C = A \oplus B$ for all i with $1 \leq i \leq t+1$ be defined by

$$C[i] = \begin{cases} A[i] & \text{if } i < m_B \\ A[i] + B[i] & \text{if } i = m_B \\ B[i] & \text{if } i > m_B \end{cases}$$

For example $stack[1] \oplus stack[2] = [1, 1, 2] \oplus [0, 2, 1] = [1, 3, 1]$. It can be shown that $\oplus$ is associative and the empty stack $id = [0, \ldots, 0]$ is the identity element with respect to $\oplus$ (where $m_{id} = t + 2$). Moreover, if we would apply the modified version of Algorithm 1 (in which the stack contains LCP-values that are strictly smaller than t) to the k first blocks of the LCP-array

---

Algorithm 4. Computation of *startpos* and *stack* with p processors.

---

input : LCP-array

1   $blocksize \leftarrow \lceil n/p \rceil$

2   $t \leftarrow 2\log(n)$

3   create an array *stack* of size p consisting of pointers

4   **foreach** k *with* $1 \le k \le p$ **do in parallel**

5     $\lfloor$ $stack[k] \leftarrow$ new array of size $t + 1$, initialized with zeros

6   create an array *startpos* of size $p + 1$, initialized with zeros

7   **foreach** k *with* $1 \le k \le p$ **do in parallel**

8     $s \leftarrow 1 + (k-1)blocksize$

9     $e \leftarrow \min\{s + blocksize - 1, n\}$

10    $\ell \leftarrow t - 1$

11   **for** $i \leftarrow e$ **down to** s **do**

12   **if** $\mathsf{LCP}[i] \le \ell$ **then**

13        $\ell \leftarrow \mathsf{LCP}[i]$

14        $stack[k][\ell + 2] \leftarrow stack[k][\ell + 2] + 1$

15        $startpos[k] \leftarrow i$

16   $j \leftarrow 2$

17   **for** $k \leftarrow 2$ **to** p **do**

18   **if** $startpos[k] \ne 0$ **then**

19      $startpos[j] \leftarrow startpos[k]$

20      $stack[j] \leftarrow stack[k]$

21   **for** $\ell \leftarrow 1$ **to** $\mathsf{LCP}[startpos[j]] + 2$ **do**

22        $\lfloor$ $stack[j][\ell] \leftarrow stack[j][\ell] + stack[j-1][\ell]$

23      $j \leftarrow j + 1$

24   $p \leftarrow j - 1$

25   $startpos[p + 1] \leftarrow n + 2$

    return : the arrays *startpos* and *stack*.

(i.e., to $\mathsf{LCP}[1..startpos[k + 1] - 1]$), then the stack contents would be $id \oplus stack[1] \oplus stack[2] \oplus \cdots \oplus stack[k]$ (use induction to prove this). The prefix sums $(id, id \oplus stack[1], id \oplus stack[1] \oplus stack[2], \ldots, id \oplus stack[1] \oplus stack[2] \oplus \cdots \oplus stack[p])$ can be computed in parallel, requiring $O(p)$ work and $O(\log p)$ time. In practice, however, a sequential computation is faster; for an implementation see lines 17-23 of Algorithm 4. When Algorithm 4 returns the arrays *startpos* and *stack*, $stack[k]$ represents the stack contents after the modified version of Algorithm 1 was applied to $\mathsf{LCP}[1..startpos[k + 1] - 1]$. Consequently, the enhanced BPR can be computed in parallel: each of the p processors is assigned to one block and applies the sequential Algorithm 1 to it. Processor 1 can apply that algorithm to the first block without modification. Processor k, $2 \le k \le p$, must use the stack $stack[k - 1]$ from which all elements that are strictly smaller than $\mathsf{LCP}[startpos[k]]$ must be removed, because this is the correct stack contents immediately before $\mathsf{LCP}[startpos[k]]$ is pushed onto it. Because stacks are represented by arrays, one can avoid the removal of elements from $stack[k - 1]$

by just taking those components of $stack[k-1][\ell]$ into account that satisfy $\ell \leq \mathsf{LCP}[startpos[k]]+2$ (recall that the first two components of $stack[k-1]$ are the number of occurrences of -1 and 0, respectively). Let us denote the resulting stack by $stack_k$. If processor k wants to apply the sequential Algorithm 1, it must not only start with $stack_k$ but it must also know at which position the first opening parenthesis in the BPR and the first bit in B (corresponding to block k) must be written. Recall from Section 4 that the i-th opening parenthesis $_i($ must be written at position $2i-1-dist[i]$, where $dist[i]$ is the number of elements on the stack immediately before $\mathsf{LCP}[i]$ is pushed onto it. Moreover, $i-1-dist[i]$ closing parentheses appear before $_i($. For block k, it follows that the first opening parenthesis in the BPR must be written at position $2\,startpos[k]-1-size(stack_k)$, and $startpos[k]-size(stack_k)$ is the position of the first bit in B. In our example, since $\mathsf{LCP}[startpos[3]] = \mathsf{LCP}[9] = 0$, we just consider the first two components of $stack[2] = [1,3,1]$ (at this point in time, $stack[2]$ contains the result of $stack[1] \oplus stack[2]$). Thus, $stack_3$ contains the elements $-1,0,0,0$ (from bottom to top). For block 3, the first opening parenthesis in the BPR appears at position $2 \cdot 9 - 1 - 4 = 13$ and the first bit in B appears at position $9 - 4 = 5$.

Algorithm 4 depends on the threshold t: the smaller t the faster the stacks can be combined. However, t should not be too small: $\ell_k < t$ should hold for each original block k because otherwise we cannot compute the correct stack contents for any position in the block. In this case, the block must be combined with its preceding block; see lines 17-23 of Algorithm 4. For $t = 2\log n$, this did never happen for real-life data (see next section), but there are cases in which this cannot be avoided. The worst case for Algorithm 4 are LCP-arrays in which the values are strictly increasing (or decreasing), for instance the LCP-array of the string $S = \mathsf{A}^n$ (the character A repeated n times). In the worst case $\ell_k < t$ holds only for block $k = 1$, so all original blocks must be combined into one big block and the algorithm therefore takes time $O(n)$.

6   Implementation and Experimental Results

Our implementation of the algorithm described in Section 4, called PA-theo, uses Julian Shun's [19] implementation of the $O(n\log n)$ work and $O(\log n)$ time ANSV algorithm of Berkman et al. [1] to compute the arrays PSEV and NSV in parallel. (We could not find an implementation of the $O(n)$ work and $O(\log\log n)$ time ANSV algorithm—a much more complicated algorithm.) A direct implementation of Algorithm 2 turned out to be very slow, so we used the following alternative algorithm: For p processors, it divides the $dist$-array into p blocks of size $\lceil n/p \rceil$. For each block k, it computes $M[k]$—the last index i in the block so that $\mathsf{PSEV}[i]$ points out of the block. For all j in between $M[k]$ and the end of the block, the path of PSEV-pointers starting from $\mathsf{PSEV}[j]$ must end at $M[k]$. Thus local distances (distances relative to $M[k]$) can be computed for all indices to the right of $M[k]$. Now it is possible to compute the correct distances for all $M[k]$ based on the following observation: If $\mathsf{PSEV}[M[k]]$ points into a block k',

Table 1. The first column describes the input file, while the second column shows the algorithm. The remaining columns show the run-times in seconds (average of ten runs) for an increasing number of processors (specified in the first row).

file	algorithm	1	2	4	8	16	20
A^n	Seq	4.34	-	-	-	-	-
$\sigma = 1$	PA-theo	11.30	10.68	6.65	4.90	3.56	3.57
$n = 400$ MB	PA-heur	10.02	9.97	9.77	9.84	9.77	9.79
para	Seq	6.48	-	-	-	-	-
$\sigma = 5$	PA-theo	19.51	14.45	8.75	5.89	4.27	4.44
$n = 409$ MB	PA-heur	12.16	8.39	4.89	2.59	1.62	1.65
einstein	Seq	7.10	-	-	-	-	-
$\sigma = 139$	PA-theo	21.18	15.88	9.06	6.33	4.62	4.54
$n = 445$ MB	PA-heur	13.25	7.05	3.98	2.69	1.91	1.66
sources	Seq	3.10	-	-	-	-	-
$\sigma = 230$	PA-theo	9.37	8.11	4.70	3.16	2.53	2.32
$n = 200$ MB	PA-heur	5.65	3.17	1.77	1.51	1.02	0.95
english	Seq	31.69	-	-	-	-	-
$\sigma = 238$	PA-theo	125.79	76.30	46.03	33.23	26.69	25.98
$n = 2107$ MB	PA-heur	53.36	27.44	14.86	12.46	7.37	6.11
human genome	Seq	42.76	-	-	-	-	-
$\sigma = 5$	PA-theo	216.21	107.81	68.81	50.14	40.39	39.63
$n = 2945$ MB	PA-heur	80.05	53.42	31.70	16.92	9.48	8.83
mouse genome	Seq	38.55	-	-	-	-	-
$\sigma = 5$	PA-theo	160.57	94.36	60.58	41.70	32.46	32.60
$n = 2599$ MB	PA-heur	71.40	48.47	29.08	15.33	7.63	7.55

i.e., $j' = \mathsf{PSEV}[M[k]]$ belongs to block k', then $M[k'] \leq j'$ must hold. Given the distances for all $M[k]$, the *dist*-array can then be calculated.

In what follows, our implementation of the sequential Algorithm 1 will be called Seq, while the algorithm described in Section 5 will be called PA-heur. The threshold parameter $t = 2 \log n$ in PA-heur was chosen because Léonard et al. [8] studied the distribution of the average LCP-value when the size of the text grows and empirically showed that it follows a logarithmic function, not only for random texts, but also for very different types of texts: several DNA sequences of various repetitiveness and natural language texts.

The experiments were conducted on a 64 bit Ubuntu 14.04.1 LTS (Kernel 3.13) system equipped with two ten-core Intel Xeon processors E5-2680v2 with 2.8 GHz and 128GB of RAM. All programs were compiled with the same options using g++ (version 4.8.2). We tested our programs on four files from the Pizza&Chili corpus (http://pizzachili.dcc.uchile.cl): sources and english from the text collection as well as para and einstein from the repetitive corpus. Furthermore we used the human genome (hg38) and the mouse genome (mm10) from https://genome.ucsc.edu and removed all characters $\notin \{A, C, G, T, N\}$.

The experimental results show that PA-theo—although it is the best in theory—does not perform well in practice. As expected, PA-heur performs badly on its worst-case input A^n because it has to do most of the work with only one processor. On real-world data, however, PA-heur performs and scales quite well: with three processors it is already faster than Seq and with 20 processors it is roughly about 5 times faster than Seq on the large files.

References

1. Berkman, O., Schieber, B., Vishkin, U.: Optimal doubly logarithmic parallel algorithms based on finding all nearest smaller values. Journal of Algorithms 14(3), 344–370 (1993)
2. Deo, M., Keely, S.: Parallel suffix array and least common prefix for the GPU. ACM SIGPLAN Notices 48(8), 197–206 (2013)
3. Fischer, J.: Combined data structure for previous- and next-smaller-values. Theoretical Computer Science 412(22), 2451–2456 (2011)
4. Gog, S.: Compressed Suffix Trees: Design, Construction, and Applications. PhD thesis, University of Ulm, Germany (2011)
5. Hon, W.-K., Sadakane, K.: Space-Economical algorithms for finding maximal unique matches. In: Apostolico, A., Takeda, M. (eds.) CPM 2002. LNCS, vol. 2373, pp. 144–152. Springer, Heidelberg (2002)
6. Jaja, J.: Introduction to Parallel Algorithms. Addison-Wesley Professional (1992)
7. Kärkkäinen, J., Manzini, G., Puglisi, S.J.: Permuted longest-common-prefix array. In: Kucherov, G., Ukkonen, E. (eds.) CPM 2009. LNCS, vol. 5577, pp. 181–192. Springer, Heidelberg (2009)
8. Léonard, M., Mouchard, L., Salson, M.: On the number of elements to reorder when updating a suffix array. Journal of Discrete Algorithms 11, 87–99 (2012)
9. Munro, J.I., Raman, V.: Succinct representation of balanced parentheses and static trees. SIAM Journal on Computing 31(3), 762–776 (2001)
10. Navarro, G., Mäkinen, V.: Compressed full-text indexes. ACM Computing Surveys 39(1), Article 2 (2007)
11. Ohlebusch, E.: Bioinformatics Algorithms: Sequence Analysis, Genome Rearrangements, and Phylogenetic Reconstruction. Oldenbusch Verlag (2013)
12. Ohlebusch, E., Fischer, J., Gog, S.: CST++. In: Chavez, E., Lonardi, S. (eds.) SPIRE 2010. LNCS, vol. 6393, pp. 322–333. Springer, Heidelberg (2010)
13. Ohlebusch, E., Gog, S.: A compressed enhanced suffix array supporting fast string matching. In: Karlgren, J., Tarhio, J., Hyyrö, H. (eds.) SPIRE 2009. LNCS, vol. 5721, pp. 51–62. Springer, Heidelberg (2009)
14. Puglisi, S.J., Smyth, W.F., Turpin, A.: A taxonomy of suffix array construction algorithms. ACM Computing Surveys 39(2), Article 4 (2007)
15. Sadakane, K.: Succinct representations of lcp information and improvements in the compressed suffix arrays. In: Proc. 13th Annual ACM-SIAM Symposium on Discrete Algorithms, pp. 225–232 (2002)
16. Sadakane, K.: Compressed suffix trees with full functionality. Theory of Computing Systems 41, 589–607 (2007)
17. Shun, J.: Fast parallel computation of longest common prefixes. In: Proc. International Conference for High Performance Computing, Networking, Storage and Analysis, pp. 387–398. IEEE Press (2014)
18. Shun, J., Blelloch, G.E.: A simple parallel Cartesian tree algorithm and its application to suffix tree construction. ACM Transactions on Parallel Computing 1(1), Article 8 (2014)
19. Shun, J., Zhao, F.: Practical parallel Lempel-Ziv factorization. In: Proc. 23th Data Compression Conference, pp. 123–132. IEEE Computer Society (2013)

Computing the Longest Unbordered Substring

Paweł Gawrychowski[1], Gregory Kucherov[2], Benjamin Sach[3],
and Tatiana Starikovskaya[3]([✉])

[1] University of Warsaw, Warsaw, Poland
[2] Laboratoire d'Informatique Gaspard Monge,
Université Paris-Est and CNRS, Champs-sur-Marne, France
[3] University of Bristol, Bristol, England
tat.starikovskaya@gmail.com

Abstract. A substring of a string is *unbordered* if its only border is the empty string. The study of unbordered substrings goes back to the paper of Ehrenfeucht and Silberger [Discr. Math 26 (1979)]. The main focus of their and subsequent papers was to elucidate the relationship between the longest unbordered substring and the minimal period of strings. In this paper, we consider the algorithmic problem of computing the longest unbordered substring of a string. The problem was introduced recently by G. Kucherov et al. [CPM (2015)], where the authors showed that the average-case running time of the simple, border-array based algorithm can be bounded by $\mathcal{O}(\max\{n, n^2/\sigma^4\})$ for σ being the size of the alphabet. (The worst-case running time remained $\mathcal{O}(n^2)$.) Here we propose two algorithms, both presenting substantial theoretical improvements to the result of [11]. The first algorithm has $\mathcal{O}(n \log n)$ average-case running time and $\mathcal{O}(n^2)$ worst-case running time, and the second algorithm has $\mathcal{O}(n^{1.5})$ worst-case running time.

1 Introduction

A proper prefix of a string that is simultaneously its suffix is called a *border*. If the only border of a substring is the empty string, then this substring is called *unbordered*. The study of unbordered substrings commenced in the 1979 paper of Ehrenfeucht and Silberger [6]. The main focus of [6] and of subsequent papers [1, 4, 8] was to clarify the relationship between the maximal length of an unbordered substring of a string and its periodicity. As a result of this line of research, it was shown that in order to guarantee the equality between the maximal length of an unbordered substring and the minimal period, either the former should be smaller than 3/7 of the string length, or the latter should be smaller than 1/2 of the string length, where both bounds are tight. In this work, we focus on the computational problem that can be considered complementary to the previous study: Given a string T of length n, compute its unbordered substring of maximal length.

P. Gawrychowski—Currently holding a post-doc position at Warsaw Center of Mathematics and Computer Science.

© Springer International Publishing Switzerland 2015
C. Iliopoulos et al. (Eds.): SPIRE 2015, LNCS 9309, pp. 246–257, 2015.
DOI: 10.1007/978-3-319-23826-5_24

It is well-known that the minimal period can be easily computed on $\mathcal{O}(n)$ time by a variant of the Knuth-Morris-Pratt algorithm [12]. Note that if a string is periodic, i.e. its minimal period is at most half of the string length, then a longest unbordered substring can be found as an unbordered conjugate of string's root (which is a substring of minimal period length). This computation can be done in $\mathcal{O}(n)$ time as well [5].

However, this approach is not applicable in the general case. It can be easily seen that an unbordered substring cannot be longer than the minimal period of the string (as any substring longer than the period has a border). For most strings, the maximal length of an unbordered substring, and consequently their minimal period, are large. (More formally, it was shown recently that the average maximal length of an unbordered substring of a string of length n is at least $(1 - 8/\sigma^4)\ n$ for alphabets of size σ [11].) In this case, it is no longer possible to exploit the relation between unbordered substrings and the minimal period. A straightforward way to compute the longest unbordered substring is to compute the border array of each suffix of the string [12]. This algorithm has quadratic worst-case running time, and no better worst-case bound has been obtained, to the best of our knowledge. In [12], it was shown that the average-case running time of this algorithm is $\mathcal{O}(\max\{n, n^2/\sigma^4\})$. The average-case time complexity captures the 'typical' running time of the algorithm, rather than the running time on most hard problem instances [13]. Other problems on strings studied under this model include pattern matching, edit distance, suffix trees, and more (see e.g. [9]).

We give two algorithms for computing an unbordered substring of the maximal length: algorithm $\mathcal{A}$ and algorithm $\mathcal{B}$. Both algorithms have $\mathcal{O}(n)$ space complexity. The worst-case running time of algorithm $\mathcal{A}$ is $\mathcal{O}(n^2)$ – the same as of the simple, border-array based algorithm. However, its average-case time complexity, i.e. the time complexity averaged over all input strings of length n, is $\mathcal{O}(n \log n)$ which provides a considerable improvement to the bound of [11]. For algorithm $\mathcal{B}$, we show an $\mathcal{O}(n^{1.5})$ *worst-case* time bound. To our knowledge, this is the first sub-quadratic worst-case bound for this problem. We assume the word-RAM model of computation with $\Omega(\log n)$-bit words and an integer alphabet of polynomial size in n.

Both algorithms distinguish between two types of substrings that have a non-empty border: those having a 'short' border (shorter than a threshold τ) and those having only 'long' borders (longer than τ). For each position j, there are only τ possible short borders, which allows to identify the substrings $T[i..j]$ that have short borders quickly. On the other hand, we will show that the number of substrings $T[i..j]$ that have only long borders is small, which will also make it possible to identify them quickly.

2   Preliminaries

Let Σ be a finite *alphabet*. The elements of Σ are *letters*. A finite ordered sequence of letters (possibly empty) is called a *string*. Letters in a string are numbered starting from 1, that is, a string T of *length* n consists of letters $T[1], T[2], \ldots, T[n]$. The length n of T is denoted by $|T|$. For $1 \leq i \leq j \leq n$, $T[i..j]$ is a *substring* of T with endpoints i and j. A substring $T[1..j]$ is called a *prefix* of T, and a substring $T[i..n]$ is called a *suffix* of T. A prefix (or a suffix) of T different from T is called *proper*.

2.1 Borders and Periods

If a proper prefix of a string is simultaneously its suffix, then it is called a *border*. A string is called *unbordered* if the only border it has is the empty string. We define the *border array* B of T to contain the lengths of the longest borders of all prefixes of T, i.e. $B[i]$ is the length of the longest border of $T[1..i]$, $i = 1..n$. The last entry in the border array, $B[n]$, contains the length of the longest border of T. It is well-known that the border array and therefore the longest border of T can be computed in $\mathcal{O}(n)$ time and space [12].

We remark that the border array construction algorithm immediately gives an $\mathcal{O}(n^2)$-time algorithm for computing the longest unbordered substring of T: It suffices to build the border arrays of all suffixes of T. Then the longest unbordered substring starting at position i will correspond to the rightmost entry in the border array of $T[i..n]$ containing zero.

A *period* of T is a positive integer π such that for all i, $1 \leq i \leq n - \pi$, $T[i] = T[i + \pi]$. The smallest of all periods of T is called the *minimal* period of T. The minimal period of T is equal to $n - B[n]$, and hence can be computed in $\mathcal{O}(n)$ time. Note that if T is unbordered, then its smallest period is equal to its length. Note also that a border of a border of a string is again a border of that string, and that the shortest border is unbordered.

We will also exploit the Periodicity lemma:

Lemma 1. *If a string T has periods ρ and γ such that $\rho + \gamma \leq |T|$, then T has period $\gcd(\rho, \gamma)$, the greatest common divisor of ρ and γ.*

Finally, we will make use of the *shortest border array* B' of T which is defined to contain the lengths of the shortest borders of all prefixes of T. That is, for each $i = 1..n$, $B'[i]$ is the length of the shortest non-empty border of $T[1..i]$ if $T[1..i]$ has a non-empty border, and zero otherwise. It is not difficult to see that the shortest border array of T can be computed in linear time. It suffices to run the standard border array construction algorithm, then if $B[i]$ is the longest border of $T[1..i]$, the shortest border of $T[1..i]$ equals $B'[B[i]]$ and can be computed in $\mathcal{O}(1)$ time.

2.2 Suffix Trees and Auxiliary Data Structures

The (generalized) suffix tree of a set S of strings is a compacted trie of suffixes of the strings in S, where the suffixes of the i-th string are appended with a special letter $\$_i$ that does not belong to the alphabet Σ [14]. In this paper, we will consider the suffix trees for the following sets of strings:

- a singleton set containing T,
- a singleton set containing the reverse of T,
- a two-element set consisting of T and some substring S of T.

We assume that we know string depths of all branching nodes. The string depth of a node is the length of the string formed by the labels from the root to that node. We also assume that we have access to the corresponding suffix array for each suffix tree. The j-th entry in the suffix array gives the position i of the start of the j-th largest suffix in lexicographical order. We assume that each entry in the suffix array holds a pointer to the corresponding leaf in the suffix tree and vice versa. Furthermore we assume each internal node in the suffix tree holds a pointer to the leftmost and rightmost leaves in each subtree. We remark that the suffix array is not strictly required to achieve the claimed bounds in either algorithm but it will simplify the explanation.

As is relatively standard, we augment each tree with the *lowest common ancestor* (LCA) and the *range minimum query* (RMQ) data structures [2]. The RMQ data structure is built on top of the suffix array for the corresponding tree. We omit the definitions as these data structures are used only indirectly via Lemmas 2 and 3 below. On alphabets of size $|T|^{\mathcal{O}(1)}$ all suffix trees, as well as the suffix arrays, the LCA and the RMQ data structures, can be constructed in $\mathcal{O}(|T|)$ time and occupy $\mathcal{O}(|T|)$ space [2,7]. Augmented in this way, the suffix trees become a very powerful tool:

Lemma 2. *Using the augmented suffix trees/arrays of T and the reverse of T, the following queries can be answered in $\mathcal{O}(1)$ time:*

1. *Given endpoints of two substrings S_1 and S_2 of T, decide whether $S_1 = S_2$,*
2. *Given an interval in the suffix array of T find the suffix in the interval with the smallest starting position,*
3. *Given endpoints of two substrings S_1, S_2 of T, compute the longest common suffix of S_1 and S_2,*
4. *Given endpoints of two substrings S_1, S_2 of T compute the largest integer α and the longest suffix S of S_1 such that SS_1^α is a suffix of S_2. Here S_1^α denotes the string formed by α repetitions of S_1.*

Lemma 3. *Using the suffix tree of T and the suffix tree of T and some substring S of T, the following queries can be answered in $\mathcal{O}(|T|)$ time:*

1. *Retrieve all suffixes of T that are not prefixes of other suffixes of T, sorted in lexicographic order,*
2. *For each suffix $T[i..n]$ of T, compute the length of its longest prefix P_i that occurs in S and the first position of such an occurrence.*

Lemmas 2 and 3 are proved using standard suffix tree algorithms, perhaps with the only exception of Query 4 of Lemma 2. We answer this query in the following way. First, we find the longest common suffix of S_1 and S_2 (Query 3 of Lemma 2). If its length is smaller than $|S_1|$, we set α to zero and S to the suffix. Otherwise, S_1 is a suffix of S_2. Let $S_2 = T[i..j]$. Then $SS_1^{\alpha-1}$ is equal to the longest common suffix of $T[i..j]$ and $T[1..j - |S_1| + 1]$ and can be found in $\mathcal{O}(1)$ time by one more Query 3 of Lemma 2.

3   Algorithm $\mathcal{A}$

In this section we describe algorithm $\mathcal{A}$, which has $\mathcal{O}(n^2)$ *worst-case* time complexity and $\mathcal{O}(n \log n)$ *average-case* time complexity. Recall from the introduction, that both our algorithms set a threshold to distinguish between short and long borders. For algorithm $\mathcal{A}$, we set the threshold, τ to $6 \log n$[1]. That is, a non-empty border is short if its length is smaller than $6 \log n$, and long otherwise. We start with the following lemma which says that strings containing substrings with long borders are very rare. We can therefore afford to process them less efficiently and still achieve a good average-case time complexity.

Lemma 4. *Consider a random string T of length n with i.i.d. distribution of letters over a non-unary alphabet. The probability that T contains a substring with a long border is smaller than $\frac{1}{n}$.*

Proof. If T contains a substring with a long border, then there is a substring $T[i..j]$ with a border of length $6 \log n$ and consequently T contains a pair of equal substrings of length $6 \log n$. Furthermore, either these two equal substrings do not overlap, or they can be shortened to produce two non-overlapping substrings of length $3 \log n$, or the length of $T[i..j]$ is at most $9 \log n$. In the last case, the minimal period of $T[i..j]$ is at most $9 \log n - 6 \log n = 3 \log n$, and the prefix of length $3 \log n$ of $T[i..j]$ and a substring of the same length starting with the next full repetition of the period do not overlap. So in all the cases, there exist two non-overlapping equal substrings of length $3 \log n$.

Now we will show that the probability of a random string to have two non-overlapping equal substrings of length $3 \log n$ is small. Consider any two such substrings. Since they do not overlap and their letters are chosen uniformly and independently, the probability of the substrings being equal is at most $1/n^3$ (recall that the alphabet cardinality is at least 2). Since there are at most n^2 pairs of substrings of length $3 \log n$, by the union bound the probability of at least one such pair being equal is at most $1/n$.   □

The string, T contains a substring with a long border if and only if the suffix tree of T contains a branching node with string depth at least $6 \log n$. We can check whether this is true in $\mathcal{O}(n)$ time. If it is, we run the simple $\mathcal{O}(n^2)$-time algorithm to compute the longest unbordered substring. We can now proceed under the assumption that T contains no substring with a long border.

Algorithm $\mathcal{A}$ considers each position in the string T in turn and determines the largest unbordered substring that ends at that position. Consider an arbitrary position j and a substring $T[i..j]$. If $T[i..j]$ has a border, the border must be short and hence equal to one of $T[j - 6 \log n + 1..j], T[j - 6 \log n + 2..j], \ldots, T[j]$. Remember that our objective is to compute the smallest position i such that $T[i..j]$ is unbordered. It follows that we need to compute the smallest position i with no occurrence of $T[j - 6 \log n + 1..j], T[j - 6 \log n + 2..j], \ldots, T[j]$.

---

[1] We assume the logarithm base to be 2 throughout the paper.

Occurrences of any substring of T form an interval in the suffix array. This property is immediate from the lexicographical ordering. The positions where none of these substrings occur correspond exactly to the complement of these $\mathcal{O}(\log n)$ intervals. It therefore follows that these 'complementary' positions also form $\mathcal{O}(\log n)$ disjoint intervals in the suffix array. We can find these 'complementary' intervals by sorting the original intervals. The *smallest* position i where none of these substrings occur is the minimum position in any of the complementary intervals. We can compute the minimum in $\mathcal{O}(\log n)$ time by performing Query 2 on each of the intervals in $\mathcal{O}(1)$ time and reporting the minimum.

To find the intervals efficiently, we use the following lemma to retrieve the *locus* (the node labelled by the substring) of each substring $T[j - 6\log n + 1..j], T[j - 6\log n + 2..j], \ldots, T[j]$ in the suffix tree of T in constant time per substring. If the substring occurs only implicitly in the suffix tree, the locus is the node at the deeper end of the edge where the substring ends. We can then determine the suffix array interval corresponding to a locus in $\mathcal{O}(1)$ time by following the pointers to the leftmost and rightmost leaves in the subtree rooted at the locus and then following the pointers to the corresponding suffix array locations.

Lemma 5. *The suffix tree of T can be preprocessed in $\mathcal{O}(n \log n)$ time and $\mathcal{O}(n)$ space, so that the locus of any $T[j-\ell..j]$ such that $0 \leq \ell < 6\log n$ can be retrieved in constant time.*

Proof. For every leaf corresponding to a suffix $T[i..n]$ we store a bitvector of length $6\log n$, where the ℓ-th bit is set to $\mathbf{1}$ iff the leaf has an ancestor at string depth $0 \leq \ell < 6\log n$. The bitvectors can be constructed in $\mathcal{O}(n \log n)$ time in a straightforward manner and occupy $\mathcal{O}(n)$ (words of) space.

For each bitvector we build the rank/select data structure [10]. The data structures can be built in $\mathcal{O}(n \log n)$ time, occupy $\mathcal{O}(n)$ (words of) space, and allow to compute the number m of $\mathbf{1}$ in a prefix of a bitvector of given length ℓ and to find the m-th $\mathbf{1}$ in a bitvector $\mathcal{O}(1)$ time.

Next, we augment the suffix tree with the *level ancestor* data structure in $\mathcal{O}(n)$ time and space that given an integer d and a node allows to compute the node's ancestor of (node) depth d in $\mathcal{O}(1)$ time [3].

To retrieve the locus of $T[j - \ell..j]$ we use the rank/select data structure of the bitvector stored for $T[j - \ell..n]$ to compute the number m of ancestors of the corresponding leaf of string depths less than ℓ in constant time. The locus of $T[j - \ell..j]$ is the ancestor of the leaf of depth $d = m + 1$, and can be retrieved in $\mathcal{O}(1)$ time.   □

To sort the intervals efficiently, we build the suffix tree (and the suffix array) of the string $T_j = T[j-6\log n+1..j]$. This gives us the lexicographical ordering of the substrings $T[j-6\log n+1..j], T[j-6\log n+2..j], \ldots, T[j]$. We can use this to sort the corresponding intervals as follows. First we remove any substring $T[j - \ell_1+1..j]$ which has another substring $T[j-\ell_2+1..j]$ as a prefix. This cannot affect correctness as the interval in the suffix array for T corresponding to occurrences of $T[j - \ell_1 + 1..j]$ is completely contained within the interval corresponding to

$T[j - \ell_2 + 1..j]$. These substrings can be removed in $\mathcal{O}(\log n)$ time by applying Query 1 of Lemma 3. Finally, the key observation is that the remaining intervals do not intersect and the order of the remaining intervals within the suffix array of T corresponds to the lexicographical order of the corresponding substrings. This gives the desired $\mathcal{O}(\log n)$ time to compute the longest unbordered substring ending at a single position j.

Theorem 1. *The worst-case time complexity of algorithm $\mathcal{A}$ is $\mathcal{O}(n^2)$ and the average-case time complexity is $\mathcal{O}(n \log n)$. The space complexity of the algorithm is $\mathcal{O}(n)$.*

Proof. It is easy to see the bounds on the *worst-case* time complexity and the space complexity of the algorithm. We now show the average-case time complexity. If t is the running time in the case when there are no long borders (i.e. no branching nodes of string depth $\geq 6 \log n$ in the suffix tree of T), then by Lemma 4 the average-case time complexity is bounded by $\mathcal{O}(\frac{1}{n} \cdot n^2) + 1 \cdot t = \mathcal{O}(n) + t$.

It remains to show that $t = \mathcal{O}(n \log n)$. We start by building the suffix tree of T and preprocessing it according to Lemma 5 in $\mathcal{O}(n \log n)$ time. Then, for each $j = 1..n$ we build the suffix array and suffix tree of T_j and retrieve its suffixes that are not prefixes of other suffixes in $\mathcal{O}(\log n)$ time. For each of the retrieved suffixes we compute the interval of its occurrences. We then sort these intervals and take the complement of the intervals in $\mathcal{O}(\log n)$ time. The complement is a union of at most $6 \log n$ disjoint intervals and we compute the minimum in these intervals in $\mathcal{O}(\log n)$ time. The claim follows. $\square$

We remark that in the case of large alphabets (of size $n^{\Omega(1)}$) it is impossible to use the algorithm [7] to build the suffix trees of substrings $T[j - 6 \log n + 1..j]$ in linear time. To overcome this technicality, we apply the following alphabet reduction trick prior to constructing the trees. We partition T into $\mathcal{O}(n)$ blocks of length $12 \log n$ with overlaps of $6 \log n$ positions. We sort letters in each block and for each letter $T[i]$ store its rank in the block. Each $T[j - 6 \log n + 1..j]$ belongs to at least one of the blocks. To construct its suffix tree, we consider one of the blocks containing it and replace all letters with their ranks in the block. This reduces the size of the alphabet to $\mathcal{O}(\log n)$ and makes it possible to use the algorithm [7]. The alphabet reduction trick takes $\mathcal{O}(n \log n)$ extra time and does not affect the time complexity of the algorithm.

4   Algorithm $\mathcal{B}$

In this section we describe algorithm $\mathcal{B}$, which has $\mathcal{O}(n^{1.5})$ *worst-case* time complexity. As in Algorithm $\mathcal{A}$, we set a threshold to distinguish between short and long borders. For algorithm $\mathcal{B}$, we set the threshold, τ to $\sqrt{n}$. That is, a non-empty border is short if its length is smaller than $\sqrt{n}$, and long otherwise.

First note that we can compute the longest unbordered substring of length at most $4\sqrt{n}$ in $\mathcal{O}(n^{1.5})$ time by computing the border array of each substring of length $4\sqrt{n}$. From now on, we are only interested in unbordered substrings of length at least $4\sqrt{n}$. The algorithm will consist of $\sqrt{n}$ stages. At stage k it computes the longest unbordered substring that ends in an interval $J_k = [k\sqrt{n}+1, (k+1)\sqrt{n}]$. Let F_k^i, $i = 1..(k-3)\sqrt{n}$, be the set substrings of T that start at position i and end in J_k. The algorithm considers each $i = 1..(k-3)\sqrt{n}$ in order and either says that there is no unbordered substring in F_k^i or retrieves a substring $T[i..j] \in F_k^i$. We guarantee that $T[i..j]$ does not have short borders. Furthermore, if there are unbordered substrings in F_k^i, we guarantee that $T[i..j]$ is the longest of them. We refer to $T[i..j]$ as the candidate. After retrieving the candidate $T[i..j]$, the algorithm checks if it is unbordered. If it is, the algorithm updates the maximal length of unbordered substrings.

4.1 Candidates

Let P_i be the longest prefix of $T[i..n]$ that occurs in $T_k = T[(k-1)\sqrt{n}+1..(k+1)\sqrt{n}]$. If $T[i..j] \in F_k^i$ has a short border, then this border is a prefix of P_i. Moreover, if ℓ is the position of an occurrence of P_i in T_k and $\ell+|P_i|-1 < j$, then $T[\ell..j]$ has a non-empty border of length at most $|P_i|$. This simple observation will allow us to differentiate between substrings with short borders and without those. We explain the technical details below.

Preprocessing. We start by constructing the suffix tree for two strings T and T_k. With its help we compute, for each $i = 1..(k-3)\sqrt{n}$, the length and the position of an occurrence of the longest prefix P_i of $T[i..n]$ that occurs in T_k (Query 2 of Lemma 3).

Consider all conjugates $T[\ell..(k+1)\sqrt{n}]]$ \$ $T[(k-1)\sqrt{n}+1..\ell-1]$ of $T_k\$$, where \$ is a letter that does not belong to the main alphabet. We compute the shortest border array for each of the conjugates in $\mathcal{O}(n)$ time in total. Obviously, values in the arrays are bounded by $2\sqrt{n}$. We sort each array in $\mathcal{O}(\sqrt{n})$ time using bucket sort. Overall, it takes $\mathcal{O}(n)$ space and time. For $r \in [k\sqrt{n}+1, (k+1)\sqrt{n}]$ let

$$S_{\ell,r} = \begin{cases} T[\ell..r], \text{ if } r > \ell; \\ T[\ell..(k+1)\sqrt{n}] \ \$ \ T[(k-1)\sqrt{n}+1..r], \text{ if } r < \ell. \end{cases}$$

We define r_ℓ^p to be the largest position in $J_k \setminus [\ell, \ell+p-1]$ such that S_{ℓ,r_ℓ^p} is either unbordered or has the shortest border of length at least $p+1$. For a fixed ℓ, all values r_ℓ^p can be computed in $\mathcal{O}(\sqrt{n})$ time by scanning the (sorted) shortest border array for $T[\ell..(k+1)\sqrt{n}]]$ \$ $T[(k-1)\sqrt{n}+1..\ell-1]$.

Computing Candidates. Below we fix i and show how to compute the candidate in F_k^i. If P_i is the empty string, $T[i..(k+1)\sqrt{n}]$ is the longest, unbordered substring in F_k^i and we return it as the candidate. Otherwise, let ℓ be the position of an occurrence of P_i in T_k and let $p = |P_i|$.

Lemma 6. *If r_ℓ^p is not defined, then F_k^i contains no unbordered substrings.*

Proof. It suffices to show that $T[i..j]$ ends with a prefix of P_i for all $j \in J_k$. If $j \in [\ell, \ell + p - 1]$, the claim is obvious. Consider now $j \in J_k \setminus [\ell, \ell + p - 1]$. We know that $S_{\ell,j}$ has a border of length in $[1, p]$, which means that $S_{\ell,j}$ ends with a prefix of P_i. Since $S_{\ell,j}$ is a suffix of $T[i..j]$, we obtain that $T[i..j]$ also ends with a prefix of P_i.   □

If the condition of the lemma is satisfied, the algorithm says that F_k^i contains no unbordered substrings. Otherwise, let $j = r_\ell^p$. Note that $|S_{\ell,j}| \geq p + 1$ by definition.

Lemma 7. $T[i..j]$ *does not have a short border.*

Proof. The proof is by contradiction. Suppose that $T[i..j]$ has a short border B. As B is a prefix of $T[i..n]$ and occurs in T_k, it must be a prefix of P_i (not necessarily proper). Consequently, $S_{\ell,j}$ starts with B and ends with B. Hence, B is a border of $S_{\ell,j}$ of length $|B| \in [1, p]$, which contradicts the definition of j.   □

Lemma 8. *If* F_k^i *contains unbordered substrings, then* $T[i..j]$ *is the longest of them.*

Proof. Let us first show that if a substring $T[i..j'] \in F_k^i$ is unbordered, then $S_{\ell,j'}$ is either unbordered or has the shortest non-empty border of length at least $p + 1$. Suppose that the shortest non-empty border of $S_{\ell,j'}$ has length in $[1, p]$. This border is a prefix of P_i, i.e. $S_{\ell,j'}$ ends with a prefix of P_i. Consequently, $T[i..j']$ is not unbordered as it starts with P_i and ends with the prefix of P_i, a contradiction.

It follows that all unbordered substrings in F_k^i have length at most $|T[i..j]|$. It remains to show that if $T[i..j]$ has a long border, then all shorter substrings in F_k^i have a non-empty border. As $T[i..j]$ has a long border, for some $b \geq \sqrt{n}$ we have $T[i..i + b - 1] = T[j - b + 1..j]$. It follows that for all $j' \in [k\sqrt{n} + 1, j]$ we have $T[i..i + b + j - j' - 1] = T[j - b + 1..j']$, i.e. all substrings $T[i..j']$ shorter than $T[i..j]$ have a non-empty border.   □

The algorithm retrieves the candidate $T[i..j]$ in $\mathcal{O}(1)$ time. The last two lemmas guarantee that $T[i..j]$ does not have short borders and if there are unbordered substrings in F_k^i, then $T[i..j]$ is the longest of them. It remains to check if $T[i..j]$ is unbordered. As it has no short borders, it suffices to check if it has a long border.

4.2   Long Border Check

Let S_j be the shortest suffix of $T[1..j]$ such that its minimal period is larger than $\sqrt{n}/2$. We will show that every long border of $T[i..j]$ ends with an occurrence of S_j. During the long border check we will scan over a sorted list of occurrences of S_j to determine if one of them induces a long border of $T[i..j]$.

Preprocessing. Let S_j be the shortest suffix of $T[1..j]$ such that its minimal period is larger than $\sqrt{n}/2$. If there is no such suffix, S_j is undefined.

Lemma 9. *If $T[i..j]$ is unbordered or has only long borders, then S_j is defined and all long borders of $T[i..j]$ (if any) end with an occurrence of S_j.*

Proof. If $T[i..j]$ is unbordered, then its shortest period is equal to its length which is larger than $\sqrt{n}$. This shows that S_j is defined.

If $T[i..j]$ has only long borders, then consider the shortest of them. Since the shortest border is always unbordered, its shortest period is equal to its length which is larger than $\sqrt{n}$. This implies that S_j can only be shorter than this shortest long border. Thus, S_j is a suffix of the shortest long border, and therefore a suffix of any long border. □

Lemma 10. *For any j there are at most $2\sqrt{n}$ occurrences of S_j in T.*

Proof. Since the minimal period of S_j is larger than $\sqrt{n}/2$, any two occurrences of S_j are at least $\sqrt{n}/2$ positions apart. □

Lemma 11. *Any two occurrences of suffixes $S_{j_1} \neq S_{j_2}$ have distinct right endpoints.*

Proof. Assume the opposite. Let $|S_{j_1}| > |S_{j_2}|$. By the definition, S_{j_1} is the shortest suffix of $T[1..j_1]$ such that its minimal period at least $\sqrt{n}/2$. As S_{j_1} and S_{j_2} have occurrences with equal right endpoints, S_{j_2} is a suffix of S_{j_1}, and, as a corollary, of $T[1..j_1]$. But, S_{j_2} is shorter than S_{j_1} and its minimal period is larger than $\sqrt{n}/2$. A contradiction. □

The algorithm will make use of sorted lists of occurrences of distinct suffixes S_j. From above it follows that the length of one list is at most $2\sqrt{n}$, whereas the total length of the lists is at most n. We compute the lists in the following way. Suppose that each S_j is replaced with an integer $u_j \in [1, 2n]$ so that $S_{j_1} \neq S_{j_2}$ implies $u_{j_1} \neq u_{j_2}$. We create an array of $2n$ empty lists and scan positions of T from the left to the right. For each $j = 1..n$ we add position j to the list u_j. Thus, we can compute the lists in $\mathcal{O}(n)$ time. We now describe the replacement procedure.

Lemma 12. *Given j, the length of S_j can be computed in $\mathcal{O}(\sqrt{n})$ time.*

Proof. We start by computing the minimal periods of suffixes $T[j - \sqrt{n} + 1..j], T[j - \sqrt{n} + 2..j], \ldots, T[j]$. This can be done by constructing the border array B of the reverse of $T[j - \sqrt{n} + 1..j]$ in $\mathcal{O}(\sqrt{n})$ time: The minimal period of $T[j - \ell..j]$ will be equal to $\ell - B[\ell]$. If the minimal period π of $T[j - \sqrt{n} + 1..j]$ is at least $\sqrt{n}/2$, then S_j is one of the suffixes and we already know it. Suppose that $\pi \leq \sqrt{n}/2$. Let $T[k..j]$ be the longest suffix of $T[1..j]$ such that its minimal period equals π. We can compute $T[k..j]$ in constant time (Query 4 of Lemma 2). We claim that the minimal period γ of $T[k - 1..j]$ is larger than $\sqrt{n}/2$. Indeed, $T[k..j]$, as a suffix of $T[k - 1..j]$, is periodic with period γ. If $\gamma \leq \sqrt{n}/2$, we have $\pi + \gamma \leq \sqrt{n} \leq T[k..j]$. By the Periodicity lemma, $T[k..j]$ is periodic with period $\gcd(\pi, \gamma)$. Because of the minimality of π, γ must be a multiple of π. It follows that $T[k - 1..j]$ has period π, which contradicts the definition of $T[k..j]$. □

For each $j = 1..n$ we compute the length of S_j. We then build the suffix tree of the reverse of T. The suffix tree contains at most $2n$ nodes which we enumerate from 1 to $2n$. As we know, each position can be the right endpoint of an occurrence of at most one S_j. This suggests the following algorithm. We consider the leaves of the tree from left to right. Let the current leaf be labeled by $T[j]T[j-1]\ldots T[1]$. We follow the path from the leaf to the root to find the highest branching node of string depth $\geq |S_j|$. Leaves in the subtree of this node will correspond to the positions j' such that $S_{j'} = S_j$. We replace all $S_{j'}$ in the subtree with the order number of the node and proceed to the leftmost suffix outside the subtree. The replacement procedure takes $\mathcal{O}(n)$ time overall.

The Check. Throughout stage k we maintain, for all $j \in J_k$, a pointer to the last position in the list of u_j that has been explored by the algorithm. A long border (if any) of the candidate substring $T[i..j]$ must be induced by an occurrence of S_j in the interval $[i, j]$. The algorithm explores occurrences in the list of S_j in turn starting from the one it stopped at. For each occurrence $p \geq i$ the algorithm compares substrings $F_1 = T[i..p+|S_j|-1]$ and $F_2 = T[j-|F_1|+1..j]$ (Query 1 of Lemma 2). If they are equal, $T[i..j]$ has a long border. Otherwise, the algorithm proceeds to the next occurrence in the list. If no occurrence in the list induces a long border, $T[i..j]$ is unbordered.

4.3   Pseudocode and the Bounds

To summarize, we give pseudocode of stage k of algorithm $\mathcal{B}$. Preprocessing for long border checks (computation of suffixes S_j and their lists) is done before the first stage (not shown).

Algorithm 1. Stage k of Algorithm $\mathcal{B}$.

1: Build the suffix tree of T and $T_k = T[(k-1)\sqrt{n}+1..(k+1)\sqrt{n}]$
2: **for** $i = 1..n$ **do**
3:     Compute the longest prefix P_i of $T[i..n]$ that occurs in T_k
4: **for** $\ell = (k-1)\sqrt{n}+1..(k+1)\sqrt{n}$ **do**
5:     Compute the shortest border array of $T[\ell..(k+1)\sqrt{n}]$ \$ $T[(k-1)\sqrt{n}+1..\ell-1]$
6:     Sort the array
7:     Compute the values r_ℓ^p
8: **for** $i = 1..(k-3)\sqrt{n}$ **do**
9:     $\ell \leftarrow$ position of an occurrence of P_i in T_k
10:     $j \leftarrow r_\ell^{|P_i|}$
11:     **if** $T[i..j]$ does not have a long border **then**
12:         Update LongestUnbordered

Theorem 2. *The worst-case time complexity of algorithm $\mathcal{B}$ is $\mathcal{O}(n^{1.5})$. The space complexity of the algorithm is $\mathcal{O}(n)$.*

Proof. It suffices to show that one stage of the algorithm takes $\mathcal{O}(n)$ time. Pre-processing takes $\mathcal{O}(n)$ time. For each position $i = 1..n$ we spend constant time plus the time needed for the long border check. The total amount of time needed for the long border checks is linear in total length of the lists, which is at most n, as we never check an occurrence in a list twice for any position of T. $\square$

References

1. Assous, R., Pouzet, M.: Une caractérisation des mots périodiques. Journal of Discrete Mathematics **25**(1), 1–5 (1979)
2. Bender, M.A., Farach-Colton, M.: The LCA problem revisited. In: Gonnet, G.H., Viola, A. (eds.) LATIN 2000. LNCS, vol. 1776, pp. 88–94. Springer, Heidelberg (2000)
3. Bender, M.A., Farach-Colton, M.: The level ancestor problem simplified. In: Rajsbaum, S. (ed.) LATIN 2002. LNCS, vol. 2286, pp. 508–515. Springer, Heidelberg (2002)
4. Duval, J.-P.: Relationship between the period of a finite word and the length of its unbordered segments. Journal of Discrete Mathematics **40**(1), 31–44 (1982)
5. Duval, J.-P., Lecroq, T., Lefebvre, A.: Linear computation of unbordered conjugate on unordered alphabet. Journal of Theoretical Computer Science **522**, 77–84 (2014)
6. Ehrenfeucht, A., Silberger, D.M.: Periodicity and unbordered segments of words. Journal of Discrete Mathematics **26**(2), 101–109 (1979)
7. Farach-Colton, M.: Optimal suffix tree construction with large alphabets. In: Proceedings of the 38th Annual Symposium on Foundations of Computer Science, pp. 137–143. IEEE Computer Society (1997)
8. Holub, Š., Nowotka, D.: The Ehrenfeucht-Silberger problem. Journal of Combinatorial Theory, Series A **119**(3), 668–682 (2012)
9. Ilie, L., Navarro, G., Tinta, L.: The longest common extension problem revisited and applications to approximate string searching. Journal of Discrete Algorithms **8**(4), 418–428 (2010)
10. Jacobson, G.: Space-efficient static trees and graphs. In: Proceedings of the 30th Annual Symposium on Foundations of Computer Science, pp. 549–554 (October 1989)
11. Loptev, A., Kucherov, G., Starikovskaya, T.: On maximal unbordered factors. In: Cicalese, F., Porat, E., Vaccaro, U. (eds.) CPM 2015. LNCS, vol. 9133, pp. 343–354. Springer, Heidelberg (2015)
12. Morris, Jr., J.H., Pratt, V.R.: A linear pattern-matching algorithm, report 40. Technical report, University of California, Berkeley (1970)
13. Szpankowski, W.: Average Case Analysis of Algorithms on Sequences. John Wiley & Sons Inc. (2001)
14. Weiner, P.: Linear pattern matching algorithms. In: Proceedings of the 14th Annual IEEE Symposium on Foundations of Computer Science, pp. 1–11 (1973)

Online Self-Indexed Grammar Compression

Yoshimasa Takabatake[1], Yasuo Tabei[2], and Hiroshi Sakamoto[1](✉)

[1] Kyushu Institute of Technology, 680-4 Kawazu, Iizuka-shi,
Fukuoka 820-8502, Japan
{takabatake,hiroshi}@donald.ai.kyutech.ac.jp
[2] PRESTO, Japan Science and Technology Agency, 4-1-8 Honcho Kawaguchi,
Saitama 332-0012, Japan
tabei.y.aa@m.titech.ac.jp

Abstract. Although several grammar-based self-indexes have been proposed thus far, their applicability is limited to offline settings where whole input texts are prepared, thus requiring to rebuild index structures for given additional inputs, which is often the case in the big data era. In this paper, we present the first online self-indexed grammar compression named OESP-index that can gradually build the index structure by reading input characters one-by-one. Such a property is another advantage which enables saving a working space for construction, because we do not need to store input texts in memory. We experimentally test OESP-index on the ability to build index structures and search query texts, and we show OESP-index's efficiency, especially space-efficiency for building index structures.

1 Introduction

Text collections including many repetitions, so called highly repetitive texts, have become common. Version controlled software stores a large amount of documents with small differences. The current sequencing technology enables us to read individual genomes quickly and economically, which generates large databases of thousands of human genomes [3]. The genetic difference between individual human genomes is said to be approximately 0.1 percent, thus making the collection highly repetitive. There is therefore a strong need for developing powerful methods to store and process repetitive text collections on a large-scale.

Self-indexes aim at representing a collection of texts in a compressed format that supports the random access to any position and also provides query searches on the collection. Although grammar-based self-indexes are especially effective for processing highly repetitive texts and several grammar-based self-indexes have been proposed [1,2,5,6,15] (See Table 1), their applicability is limited to offline cases where all the text collections are given in advance, thus requiring to rebuild indexes when additional texts are given. Evenworse, they need to store whole input texts in memory for constructing indexes, which requires a large amount of working space. The problem is especially serious when we process

This work was supported by JSPS KAKENHI(24700140,26280088) and the JST PRESTO program

C. Iliopoulos et al. (Eds.): SPIRE 2015, LNCS 9309, pp. 258–269, 2015.
DOI: 10.1007/978-3-319-23826-5_25

Table 1. Comparison with offline methods. Construction time, search time and extraction time are presented in big O notation that is omitted for space limitations. N is the length of text, m is the length of query pattern, n is the number of variables in a grammar, σ is alphabet size, h is the height of the parse tree of the straight line program, z is the number of phrases in LZ77, d is the length of nesting in LZ77, occ is the number of occurrences of query pattern in a text, occ_q is the number of candidate appearances of query patterns, $\lg^*$ is the iterated logarithm, and $\alpha \in (0, 1]$ is a load factor for a hash table. $\lg$ stands for $\log_2$.

	Working space(bits)	Index size(bits)	Algorithm
LZ-index[14]	$O(N)$	$z \lg N + 5z \lg N$ $-z \lg z + o(N) + O(z)$	Offline
Gagie et al.[5]	$O(N)$	$2n \lg n + O(z \lg N$ $+ z \lg z \lg \lg z)$	Offline
SLP-index[1,2]	$O(N)$	$n \lg N + O(n \lg n)$	Offline
ESP-index[19]	$O(N)$	$n \lg N + n \lg n$ $+ 2n + o(n \lg n)$	Offline
OESP-index	$n \lg N + O((n + \sigma) \lg(n + \sigma))$	$n \lg N + O((n + \sigma) \lg(n + \sigma))$	Online

	Construction time	Search time	Extraction time
LZ-index [14]	$N \lg \sigma$	$m^2 d + (m + occ) \lg z$	md
Gagie et al. [5]	N	$m^2 + (m + occ) \lg \lg N$	$m + \lg \lg N$
SLP-index [1,2]	N	$m^2 + h(m + occ) \lg n$	$(m + h) \lg n$
ESP-index [19]	$\frac{1}{\alpha} N \lg^* N$ expected	$\lg \lg n (m + occ_q \lg m \lg N) \lg^* N$	$\lg \lg n (m + \lg N)$
OESP-index	$\frac{1}{\alpha} N \lg(n + \sigma) \lg^* N$ expected	$\lg(n + \sigma)(\frac{m}{\alpha} + occ_q (\lg N + \lg m \lg^* N))$ expected	$\lg(n + \sigma)(m + \lg N)$

massive collections of highly repetitive texts, which is ubiquitous in the big data era. An open challenge is to develop an online self-indexed grammar compression not only with a small working space for a large input but also with a functionality of updating data structures for building self-indexes from new additional texts.

Edit-sensitive parsing (ESP) [4] is an efficient parsing algorithm originally developed for approximately computing edit distances with moves between texts. ESP builds from a given text a parse tree that guarantees upper bounds of parsing discrepancies between different appearances of the same subtext. Maruyama et al. [11] presented a grammar-based self index called ESP-index on the notion of ESP and Takabatake et al. [19] improved ESP-index for fast query searches by using GMR's rank/select operations for general alphabet [7]. Unlike other grammar-based self-indexes, they perform top-down searches for finding candidate appearances of a query text on the data structure by leveraging the upper bounds of parsing discrepancies in ESP. However, their applicability is limited to offline cases.

In this paper, we present an online self-indexed grammar compression named OESP-index for building a self-index by reading input characters one-by-one. As far as we know, OESP-index is the first method for building grammar-based self-indexes in an online manner. OESP-index is built on the notion of ESP and its data structures are constructed by leveraging the idea behind fully-online

LCA (FOLCA) [12,13], an efficient online grammar compression that builds a context-free grammar (CFG) from an input text and encodes it into a succinct representation. We present a novel query search and random access algorithms for OESP-index and discuss their efficiency.

Experiments were performed on retrieving query texts from a benchmark collection of highly repetitive texts. The performance comparison with other algorithms demonstrates OESP-index's superiority.

2   Preliminaries

2.1   Basic Notations

Let Σ be a finite alphabet and $\sigma = |\Sigma|$. The length of string S is denoted by $|S|$. The set of all strings over et Σ is denoted by Σ^*. The set of all strings of length k is denoted by Σ^k. We assume a recursively enumerable set $\mathcal{X}$ of variables with $\Sigma \cap \mathcal{X} = \emptyset$. $S[i]$ and $S[i,j]$ denote the i-th symbol of string S and the substring from $S[i]$ to $S[j]$, respectively. lg stands for $\log_2$. Let $\lg^{(1)} u = \lg u$, $\lg^{(i+1)} u = \lg\lg^{(i)} u$, and $\lg^* u = \min\{i \mid \lg^{(i)} u \leq 1\}$. Practically $\lg^* u = O(1)$ since $\lg^* u \leq 5$ for $u \leq 2^{65536}$.

2.2   Straight-Line Program (SLP)

A context-free grammar (CFG) in Chomsky normal form is a quadruple $G = (\Sigma, V, D, X_s)$ where V is a finite subset of $\mathcal{X}$, D is a finite subset of $V \times (V \cup \Sigma)^2$ and $X_s \in V$ is the start symbol. An element in D is called production rule. A variable in V is called nonterminal symbol. $val(X_i)$ denotes the string derived from $X_i \in V$. For $X_1, X_2, ..., X_k \in V$, let $val(X_1, X_2, ..., X_k) = val(X_1)val(X_2)...val(X_k)$. A grammar compression of S is a CFG that derives S and only S. The size of a CFG is the number of variables, i.e., $|V|$ and let $n = |V|$.

The parse tree of G is a rooted ordered binary tree such that (i) internal nodes are labeled by variables in V and (ii) leaves are labeled by symbols in Σ, i.e., the label sequence in leaves is equal to input string S. In a parse tree, any internal node Z corresponds to a production rule $Z \to XY$ and has a left child with label X and a right child with label Y. A partial parse tree [18] is an ordered tree formed by traversing the parsing tree in a depth-first manner and pruning out all descendants under every node of variables appearing no less than twice.

Straight-line program (SLP) [10] is defined as a grammar compression over $\Sigma \cup V$ and its production rules are in the form of $X_k \to X_i X_j$ where $X_k, X_i, X_j \in \Sigma \cup V$ and $1 \leq i, j < k \leq n + \sigma$.

2.3   Phrase Dictionary and Reverse Dictionary

A phrase dictionary is a data structure for directly accessing a digram $X_i X_j$ from a given X_k if $X_k \to X_i X_j \in D$. It is typically implemented by an array

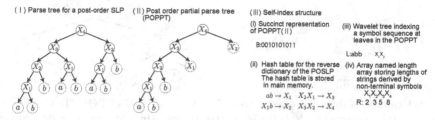

Fig. 1. Example of parse tree, post order partial parse tree and self-index structure. The self-index structure consists of four data structures which are directly built from the parse tree.

requiring $2n \log(n + \sigma)$ bits for storing n production rules. A reverse dictionary D^{-1} is a mapping from a digram to an associated variable. $D^{-1}(XY)$ returns the variable Z if $Z \rightarrow XY \in D$; otherwise, it creates a new variable $Z' \notin V$ and returns Z'.

2.4 Succinct Data Structures

We use the fully indexable dictionary (FID) for indexing bit strings. Our method represents CFGs using a rank/select dictionary, a succinct data structure for a bit string B [9] supporting the following queries: $\text{rank}_c(B, i)$ returns the number of occurrences of $c \in \{0, 1\}$ in $B[0, i]$; $\text{select}_c(B, i)$ returns the position of the i-th occurrence of $c \in \{0, 1\}$ in B; $\text{access}(B, i)$ returns i-th bit in B. Data structures with $|B| + o(|B|)$ bit storage to achieve $O(1)$ time rank and select queries [17] have been presented.

For online grammar compression, we adopt the dynamic range min/max tree (DRMMT) [16] for online construction of parse tree. We can obtain $parent(B, i)$, the parent of node i of DRMMT B in $O(\frac{\lg n}{\lg \lg n})$ time where n is the number of nodes of the tree. We consider the wavelet tree (WT) [8], an extension of FID for general alphabet. A WT is a data structure for a string over finite alphabets, and it can compute the rank and select queries on a string S over Σ^* in $O(\log \sigma)$ time and using $|S| \log \sigma(1 + o(1))$ bits.

3 Edit Sensitive Parsing (ESP) and Fully-Online LCA (FOLCA)

We review the ESP algorithm [4] and its online variant named FOLCA [13] in this section. The original ESP is an offline algorithm and builds a parse tree named ESP-tree from a given string. ESP-trees are complete, balanced binary trees each subtree of which is 2-tree in the form of $X \rightarrow AB$ or 2-2-tree in the form of $X \rightarrow AY$ and $Y \rightarrow BC$. The algorithm partitions a string S into non-overlapping substrings $S_1 S_2 \cdots S_\ell$ each of which belongs to one of three substring types. Type1 is a substring of a repeated symbol, i.e., a^k for $a \in \Sigma$ and $k > 1$;

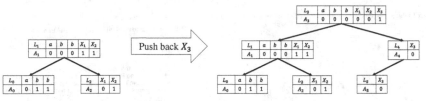

Fig. 2. Example of dynamic wavelet tree. L is the leaf label of POPPT; Code is the integer representation of L; B_i is the bit vector representing elements in code; Only A_i at each node is stored.

type2 is a substring longer than $2 \lg^* |S|$ not including type1 substrings; type3 is a substring that is neither type1 nor type2.

The parsing algorithm parses each substring S_i according to three substring types. For type1 and type3 substrings, the algorithm performs the left aligned parsing as follows. If $|S_i|$ is even, the algorithm builds a 2-tree from $S_i[2j-1, 2j]$ for each $j \in \{1, 2, ..., |S_i|/2\}$; Otherwise, the algorithm builds a 2-tree from $S_i[2j - 1, 2j]$ for each $j \in \{1, 2, ..., \lfloor (|S_i| - 3)/2 \rfloor\}$, and it builds a 2-2-tree from the last trigram $S_i[|S_i| - 2, |S_i|]$. For type2 substrings, the algorithm further partitions substring S_i into short substrings of length two or three by using an efficient string partitioning procedure named *alphabet reduction* [4], and it builds 2-trees for substrings of length two and 2-2-trees for substrings of length three. The parsing algorithm generates a new shorter string S' of length from $|S|/3$ to $|S|/2$, and it parses S'. The above process iterates on the new sequence until its length is one.

FOLCA is an online algorithm that builds ESP-tree as *a post-order partial parse tree (POPPT)* using the parsing rule in the ESP algorithm from a given string in an online manner. A POPPT and the *post-order SLP (POSLP)* corresponding to a POPPT are defined as follows.

Definition 1 (POPPT and POSLP [13]). *A POPPT is a partial parse tree whose internal nodes have post-order variables. A POSLP is an SLP whose partial parse tree is a POPPT.*

Figure 1-(I) and -(II) show an example of parse tree and POPPT.

Since FOLCA builds POPPT using the rules in the ESP algorithm, it can exploit advantages existing in both SLP and ESP-tree. Given a string S, FOLCA builds the POPPT of height $O(\lg |S|)$ in $O(|S| \lg^* |S|)$ time. FOLCA's worst-case approximation ratio to the smallest CFG is $O(\lg^* |S|)$. OESP-index directly encodes FOLCA'S POPPT into a succinct representation and build an index structure in an online manner for fast query searches and substring extractions, which is explained in the next section.

4   Index Structure of OESP-Index

OESP-index's succinct representation consists of four data structures: (i) B : succinct tree of POPPT, (ii) H : hash table (iii) L : non-negative integer array indexed by wavelet tree and (iv) R : non-negative integer array.

Succinct tree of POPPT B is a bit string made by traversing POPPT in post-order, and putting '0' if a node is a leaf and '1' otherwise. The last bit '1' in B represents a virtual node and B is indexed by the DRMMT [16]. The succinct tree supports the following three operations: $parent(B, i)$ return the parent node of a node i; $left\_child(B, i)$ returns the left child of a node i; $right\_child(B, i)$ returns the right child of a node i; They are computed in $O(\lg n / (\lg \lg n))$ time. The space for our succinct tree is at most $2n + o(n)$ bits.

A reverse dictionary $H : (V \cup \Sigma) \times (V \cup \Sigma) \to V$ is implemented by a chaining hash table. Let α be a constant called a load factor. The hash table has αn entries and each entry stores a list of integers i representing the left hand side of a rule $X_i \to X_j X_k$. The size of the data structure is $\alpha n \lg(n+\sigma)$ bits for the hash table and $n \lg(n + \sigma)$ bits for the lists. Thus, the total size is $n(1+\alpha) \lg(n + \sigma))$ bits. The access time is expected $O(1/\alpha)$ time.

Non-negative integer array L stores symbols at leaves from the leftmost leaf to the rightmost leaf in the POPPT. L is indexed by *dynamic wavelet tree (DWT)* that is presented in the next subsection. Each element of non-negative integer array R is the length of the string derived from a variable, i.e., $|val(X_i)|$ for $X_i \in V$. The size of R is $n \lg |S|$ bits. Figure 1-(III)-(iv) shows an example of data structures.

4.1 Dynamic Wavelet Tree (DWT)

Our DWT is a wavelet tree supporting a operation of adding an element to the tail of a sequence. Such a operation is called a pushback that is necessary for implementing DWT. A wavelet tree for sequence L over range of alphabet and variables $[1..(n + \sigma)]$ can be recursively described over sub-range $[a..b] \subseteq [1..(n+\sigma)]$. A wavelet tree over range $[a..b]$ is a binary balanced tree with $b - a + 1$ leaves. If $a = b$, the tree is just a leaf labeled a. Otherwise it has an internal root node that represents L. The root has a bitstring $A_{root}[1, |S|]$ defined as follows: if $L[i] \le (a + b)/2$ then $A_{root}[i] = 0$, else $A_{root}[i] = 1$. We define $L_0[1, \ell_0]$ as the subsequence of L formed by the symbols $c \le (a + b)/2$, and $L_1[1, \ell_1]$ as the subsequence of L formed by the symbols $c > (a + b)/2$. Then, the left child of the root is a wavelet tree for $L_0[1, \ell_0]$ over range $[a..\lfloor(a + b)/2\rfloor]$ and the right child of the root is a wavelet tree for $L_1[1, \ell_1]$ over range $[1 + \lfloor(a + b)/2\rfloor..b]$.

Implementing WTs without pointers uses a small space of $n \lg(n + \sigma) + o(n \lg(n + \sigma))$ bits, but supporting the pushback operation is difficult. Thus, we implement DWTs using pointers where the binary tree is explicitly represented. When a new symbol exceeding the representation ability of the current binary tree in DWT is added to DWT, DWT adds new nodes to the binary tree, resulting in increasing the height of the tree. The space of DWT uses $(3n + 2\sigma) \lg(n + \sigma) + o(n \lg(n + \sigma))$ bits. Figure 2 shows an example of DWT.

4.2 Complexity for Building OESP-Index

Theorem 1. *The size of OESP-index is $n \lg |S| + O((n + \sigma) \lg(n + \sigma))$ bits. The construction time is $O(\frac{1}{\alpha} |S| \lg(n + \sigma) \lg^* |S|)$ and the memory consumption*

Algorithm 1. NEXTCORE on implicit parse tree POPPT(B, L)

```
 1: v: the leftmost occurrence node of maximal core, p: empty stack
 2: function NEXTCORE(v, p)
 3:     if v ≠ root then
 4:         if v is the left child of parent(B, v) then              ▷
 5:             p.push(left)
 6:         else
 7:             p.push(right)
 8:         end if
 9:         NEXTCORE(parent(B, v), p)
10:         i ← 1
11:         p.pop()
12:         while (u = select_v(L, i)) ≠ NULL do    ▷ (u, p): the next occurrence on explicit tree
13:             if u is left child of parent(B, u) then
14:                 p.push(left)
15:             else
16:                 p.push(right)
17:             end if
18:             NEXTCORE(parent(B, u), p)
19:             p.pop()
20:             i ← i + 1
21:         end while
22:     end if
23: end function
```

is the same as the index size, where S is an input string, n is the number of variables, α is a load factor of the hash table, and we assume the size of alphabet is constant. The update time for the next input symbol is $O(\frac{1}{\alpha} \lg(n + \sigma) \lg^* |S|)$.

Proof. The size of the length array R is $n \lg |S|$ bits for n variables. The size of B is $2n + o(n)$ bits and the size of L and H are $O((n + \sigma) \lg(n + \sigma))$ bits each. We can access $Z = H(XY)$ in $O(1/\alpha)$ time for a load factor $\alpha \in (0, 1]$. The alphabet reduction is iterated at most $\lg^* |S|$ times for each symbol. The time to get the parent and left/right children of a node in the partial parse tree is $O(\lg(n + \sigma))$ using the rank/select over the DWT for L. Thus, the construction time of the parse tree is $O(\frac{1}{\alpha} |S| \lg(n + \sigma) \lg^* |S|)$. Analogously, the update time is clear.

4.3 Query Search and Substring Extraction

For a node v of a parse tree of the string $S \in \Sigma^*$, and $yield(v_1 \cdots v_k) = yield(v_1) \cdots yield(v_k)$. $Label(v)$ denotes the label of v and $Label(v_1 \cdots v_k) = Label(v_1) \cdots Label(v_k)$. If $Label(v) = X$, $yield(X)$ is identical to $yield(v)$. $lca(u, v)$ is the lowest common ancestor of u, v. For a pattern $P \in \Sigma^*$, nodes $\{v_1, \ldots, v_k\}$ such that $yield(v_1 \cdots v_k) = P$ are called *embedding nodes* of P. For embedding nodes $\{v_1, \ldots, v_k\}$, string $Q = Label(v_1 \cdots v_k)$ is called an evidence of pattern P. Since the trivial evidence Q identical to P always exists, the notion of evidence is well-defined. In addition, for embedding nodes $\{v_1, \ldots, v_k\}$, a node z such that $z = lca(v_1, v_k)$ is called an *occurrence node* of P

The next theorem tells that we can find shorter evidence depending on $|P|$.

Theorem 2. *([11]) There exists an evidence $Q = Q_1 \cdots Q_t$ of P such that each Q_i is a maximal repetition or a symbol and $t = O(\lg |P| \lg^* |S|)$.*

The time to find the evidence Q of pattern P is bounded by the construction time of the parsing tree of P. In our data structure of OESP, the time to find the evidence Q is estimated as follows.

Theorem 3. *The time to find Q is $O(\frac{1}{\alpha}|P| \lg(n+\sigma) \lg^* |S|)$.*

Proof. This bound is clear by Theorem 1.

Let us consider the simple case that $|Q_i| = 1$ for any i. In this case, Q contains no repetition such that $Q = q_1 \cdots q_t \in \Sigma^t$. A symbol q_k is called a maximal core if $|yield(q_k)| \geq |yield(q_i)|$ for any i. For an internal node v of the parse tree T of S with $Label(v) = q_k$, an ancestor z of v is the occurrence node of P iff all $q_1, \ldots, q_{k-1}$ and $q_{k+1}, \ldots, q_t$ can be embedded around v. Moreover, any occurrence node of P is restricted by the case $Label(v) = q_k$. For the general case $Q_i = a^\ell$ ($\ell \geq 2$), i.e., Q_i is a repetition, we can reduce the embedding of a^ℓ to the embedding of a string $AB \cdots C$ of length at most $O(\lg \ell)$ such that $yield(AB \cdots C) = a^\ell$. Thus, the embedding of type1 string is easier than others, and then, without loss of generality, we can assume $|Q_i| = 1$ for any i.

The remaining task of the search problem is the random access to all occurrences of the maximal core q_k over the POPPT, the pruned parse tree. By the definition of POPPT, the internal node with rank k is the leftmost occurrence of the symbol q_k itself. In the previous indexes [11,19], a next occurrence of q_k is obtained using a data structure based on the renaming variables in a lexicographic order. This data structure is not, however, dynamically constructable. Therefore, we develop the search algorithm NEXTCORE (Algorithm 1) for the OESP-index.

The NEXTCORE visits all occurrences of the maximal core on the parse tree T using its implicit POPPT T'. When NEXTCORE receives a candidate node v containing a maximal core q as its descendant, it computes the pair (u, p) where u is the next occurrence of v in T' and p is the path from u to q. Thus, (u, p) indicates the occurrence of q in the explicit parse tree. We show the correctness of this algorithm and its complexity.

Lemma 1. *The NEXTCORE find any occurrence of the maximal core exactly once. The amortized time to find a next occurrence is $O(\frac{\lg n \lg |S|}{\lg \lg n})$.*

Proof. Let T be the parse tree and T' be the POPPT (B, L). By the definition of T', any internal node x of B is the variable itself, i.e., $Label(x) = x$. For the maximal core q, let $v_1 > v_2 > \cdots > v_k$ be the post-order of its occurrences in T. We show that the algorithm finds any v_i as (u, p) by induction on i. Given q, the internal node q of B represents the leftmost occurrence of q itself. Then, for the base case $i = 1$, the occurrence is obtained v_1 as (q, p) with $|p| = 0$. Assume the induction hypothesis on some i. Since the node v_{i+1} was pruned in T', let u be the leaf of T' corresponding to the root of the pruned maximal subtree containing v_{i+1}. For $Label(u) = u'$, there is the leftmost occurrence of u' as an internal node of B. The subtree on the node u' contains an occurrence of q because the two subtrees on u and u' in T are identical each other. Let p be

Table 2. Index size in mega bytes(MB).

	OESP-index	ESP-index	SLP-index	LZ-index	FM-index
einstein	22.84	1.76	2.28	177.02	942.85
cere	364.92	27.40	45.74	438.05	806.52

the path from u' to v' for some $v' \in \{v_1, \ldots, v_i\}$. By the induction hypothesis, the algorithm finds v' as (u', p). Then, v_{i+1} can be also found as (u, p). On the other hand, any (u, p) is unique, then the algorithm finds any occurrence of q exactly once. For the time complexity, the number of executed select operations is bounded by the number of different (u, p) that is $O(occ_q \log |S|)$ where occ_q is the number of occurrences of q. Each select operation on L and parent operation on B take $O(\lg(n+\sigma))$ and $O(\frac{\lg n}{\lg \lg n})$ time, respectively. Therefore, the total time is $O(occ_q(\lg(n + \sigma) + \frac{\lg n \lg |S|}{\lg \lg n})) = O(occ_q(\frac{\lg n \lg |S|}{\lg \lg n}))$ and the amortized time to find a next occurrence of q is $O(\frac{\lg n \lg |S|}{\lg \lg n})$.

Theorem 4. *The counting/locating time of pattern and extraction time are* $O(\lg(n + \sigma)(\frac{|P|}{\alpha} + occ_q(\lg |S| + \lg |P| \lg^* |S|)))$ *and* $O(\lg(n + \sigma)(|P| + \lg |S|))$, *respectively, where P is a query pattern and occ_q is the number of occurrences of the maximal core of P in the parse tree.*

Proof. Since we can get the length of the substring encoded by any variable in $O(1)$ time, the locating time is same as the counting time. Given the pattern P, as previously shown, the evidence Q of P is found in $O(\frac{1}{\alpha}|P| \lg(n + \sigma) \lg^* |S|)$ time. For each occurrence of a maximal core, we can check if the sequence of symbols of length $O(\lg |P| \lg^* |S|)$ is embedded around the core in $O(\lg(n + \sigma)(\lg |S| + \lg |P| \lg^* |S|))$ time. Therefore, by Lemma 1, the total counting time of pattern is

$$O\left(\frac{|P|}{\alpha} \lg(n + \sigma) \lg^* |S| + occ_q \lg(n + \sigma)(\lg |S| + \lg |P| \lg^* |S|) + occ_q \frac{\lg n \lg |S|}{\lg \lg n}\right)$$

$$= O(\lg(n + \sigma)(\frac{|P|}{\alpha} + occ_q(\lg |S| + \lg |P| \lg^* |S|))).$$

On the other hand, for any $S[i, j]$ of length m, we can find $S[i]$ in $O(\lg |S|)$ time and visit all leaves in $S[i, j]$ in $O(|P|)$ time because the parsing tree is balanced. This follows the extraction time.

5   Experiments

We evaluated the actual performance of OESP-index for real data[1]. The environment is Intel(R) Core(TM)i7-2620M CPU(2.7GHz) machine with 16GB memory.

---

[1] http://pizzachili.dcc.uchile.cl/repcorpus/real/

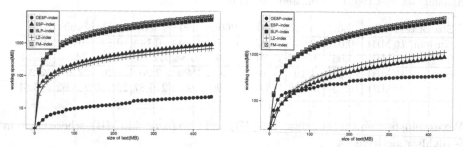

Fig. 3. Working memory of each method in megabytes for einstein(left) and cere(right).

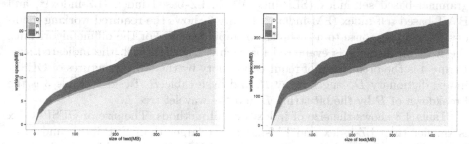

Fig. 4. Working space of dictionary D, length array R and hash table H for einstein (left) and cere (right).

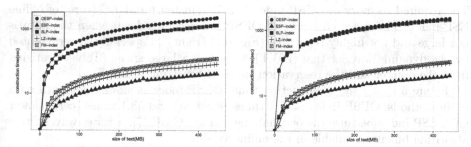

Fig. 5. Construction time of each method in seconds for einstein(left) and cere(right).

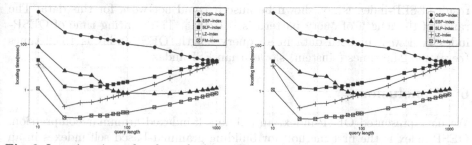

Fig. 6. Locating time of each method in milliseconds for einstein(left) and cere(right).

Table 3. Working memory of dictionary D consisting of the bit string B and the dynamic wavelet tree L for einstein and cere.

Size of text(MB)		50	100	150	200	250	300	350	400	all
einstein	B(MB)	0.04	0.07	0.07	0.07	0.14	0.14	0.14	0.14	0.14
	L(MB)	5.06	6.63	7.84	8.99	10.88	12.23	13.38	14.37	15.14
cere	B(MB)	1.10	1.10	1.10	2.20	2.20	2.20	2.20	2.20	2.20
	L(MB)	102.34	131.58	164.05	179.90	199.08	216.87	225.70	235.62	245.81

We use einstein.en.txt (einstein, 446 MB) and cere (cere, 440 MB), where einstein is highly repetitive.

Compared self-indexes are offline version of ESP-index (ESP-index)[19], other grammar-based self-index (SLP-index)[1,2], LZ-based index (LZ-index)[2], and BWT-based self-index (FM-index)[3]. Figure 3 shows the required working memory (MB) in response to an increase of input string. For the offline algorithms, the working memory is evaluated for each static data with the indicated size. Figure 4 is the breakdown of required memory by the data structures of OESP-index: dictionary D, length array R, and hash table H. Besides, Table 3 is the breakdown of D by the bit string B and the wavelet tree L.

Table 4.3 shows the size of indexes of all methods. The size of OESP-index is smaller than LZ-index and FM-index but larger than ESP-index and SLP-index. The increase of index size arise from DWT. Reducing this data size is an important future work.

The memory consumption of OESP-index is smallest for both type of data. The required memory of OESP-index is 2.5%(einstein) and 40%(cere) of offline ESP-index. The space efficiency of OESP-index comes down when the data is not large and not highly repetitive (Figure 4 (right)). Especially, L represented by the dynamic wavelet tree (DWT) consumes a large space (Table 3) arising from the pointer and the reservation space of bit string in DWT.

Figure 5 shows the construction time. OESP-index is slowest for both data in all methods. OESP-index is 57.1 times (einstein) and 58.1 times (cere) slower than ESP-index because the original one can use GMR [7], a faster wavelet tree algorithm but not available in the online version.

Figure 6 shows the search time. Here the search time means the locating time since the counting time is almost same to the locating time. We note that the result of SLP-index is not shown because it could not work for this data. The range of the length of query pattern is [10, 1000]. The locating time of OESP-index is slowest in both data in all query length. OESP-index is 163.2 times (cere) and 24.9 times (einstein) slower than ESP-index.

6   Conclusion

We have presented OESP-index, an online self-indexed grammar compression. OESP-index is the first method for building grammar-based self-indexes in an

---

[2] http://pizzachili.dcc.uchile.cl/indexes/LZ-index/LZ-index1
[3] https://code.google.com/p/fmindex-plus-plus/

online manner. Experimental results demonstrated OESP-index's potential for processing a large collection of highly repetitive texts. Future work is to make OESP-index scalable to massive collections of the same type, which is required in the big data era.

References

1. Claude, F., Navarro, G.: Self-indexed grammar-based compression. Fundam. Inform. **111**, 313–337 (2010)
2. Claude, F., Navarro, G.: Improved grammar-based compressed indexes. In: Calderón-Benavides, L., González-Caro, C., Chávez, E., Ziviani, N. (eds.) SPIRE 2012. LNCS, vol. 7608, pp. 180–192. Springer, Heidelberg (2012)
3. Genomes Project Consortium: A map of human genome variation from population-scale sequencing. Nature **467**, 1061–1073 (2010)
4. Cormode, G., Muthukrishnan, S.: The string edit distance matching problem with moves. TALG **3**, 2:1–2:19 (2007)
5. Gagie, T., Gawrychowski, P., Kärkkäinen, J., Nekrich, Y., Puglisi, S.J.: A faster grammar-based self-index. In: Dediu, A.-H., Martín-Vide, C. (eds.) LATA 2012. LNCS, vol. 7183, pp. 240–251. Springer, Heidelberg (2012)
6. Gagie, T., Gawrychowski, P., Kärkkäinen, J., Nekrich, Y., Puglisi, S.J.: LZ77-based self-indexing with faster pattern matching. In: Pardo, A., Viola, A. (eds.) LATIN 2014. LNCS, vol. 8392, pp. 731–742. Springer, Heidelberg (2014)
7. Golynski, A., Munro, J.I., Rao, S.S.: Rank/select operations on large alphabets: a tool for text indexing. In: SODA, pp. 368–373 (2006)
8. Grossi, R., Gupta, A., Vitter, J.S.: High-order entropy-compressed text indexes. In: SODA, pp. 636–645 (2003)
9. Jacobson, G.: Space-efficient static trees and graphs. In: FOCS, pp. 549–554 (1989)
10. Karpinski, M., Rytter, W., Shinohara, A.: An efficient pattern-matching algorithm for strings with short descriptions. Nord. J. of Comp. **4**, 172–186 (1997)
11. Maruyama, S., Nakahara, M., Kishiue, N., Sakamoto, H.: ESP-Index: A compressed index based on edit-sensitive parsing. JDA **18**, 100–112 (2013)
12. Maruyama, S., Tabei, Y.: Fully-online grammar compression in constant space. In: DCC, pp. 218–229 (2014)
13. Maruyama, S., Tabei, Y., Sakamoto, H., Sadakane, K.: Fully-online grammar compression. In: Kurland, O., Lewenstein, M., Porat, E. (eds.) SPIRE 2013. LNCS, vol. 8214, pp. 218–229. Springer, Heidelberg (2013)
14. Navarro, G.: Indexing text using the Ziv-Lempel tire. JDA **2**(1), 87–114 (2004)
15. Navarro, G.: Implementing the LZ-index: Theory versus practice. ACM Journal of Experimental Algorithmics **13** Article No. 1.2 (2008)
16. Navarro, G., Sadakane, K.: Fully-functional static and dynamic succinct trees. ACM Transactions on Algorithms (2012). Accepted. A preliminary version appeared in SODA 2010
17. Raman, R., Raman, V., Rao, S.S.: Succinct indexable dictionaries with applications to encoding k-ary trees, prefix sums and multisets. TALG **3** (2007)
18. Rytter, W.: Application of Lempel-Ziv factorization to the approximation of grammar-based compression. Theoretical Computer Science **302**(1–3), 211–222 (2003)
19. Takabatake, Y., Tabei, Y., Sakamoto, H.: Improved ESP-index: a practical self-index for highly repetitive texts. In: Gudmundsson, J., Katajainen, J. (eds.) SEA 2014. LNCS, vol. 8504, pp. 338–350. Springer, Heidelberg (2014)

Tight Bound for the Number of Distinct Palindromes in a Tree

Paweł Gawrychowski, Tomasz Kociumaka, Wojciech Rytter,
and Tomasz Waleń[✉]

Faculty of Mathematics, Informatics and Mechanics,
University of Warsaw, Warsaw, Poland
{gawry,kociumaka,rytter,walen}@mimuw.edu.pl

Abstract. For an undirected tree with n edges labelled by single letters, we consider its substrings, which are labels of the simple paths between pairs of nodes. We prove that there are $\mathcal{O}(n^{1.5})$ different palindromic substrings. This solves an open problem of Brlek, Lafrenière and Provençal (DLT 2015), who gave a matching lower-bound construction. Hence, we settle the tight bound of $\Theta(n^{1.5})$ for the maximum palindromic complexity of trees. For standard strings, i.e., for paths, the palindromic complexity is $n + 1$.

1 Introduction

Regularities in words are extensively studied in combinatorics and text algorithms. One of the basic type of such structures are palindromes: words which are the same when read in both directions. The *palindromic complexity* of a word is the number of distinct palindromic substrings in the word. An elegant argument shows that the palindromic complexity of a word of length n does not exceed $n + 1$ [5], which is already attained by a unary word $\mathbf{a}^n$. Therefore the problem of palindromic complexity for words is completely settled, and the natural next step is to generalize it to trees.

In this paper we consider the palindromic complexity of undirected trees with edges labelled by single letters. We define substrings of such a tree as the labels of simple paths between arbitrary two nodes. Each label is the concatenation of the labels of all edges on the path. Fig. 1 illustrates palindromic substrings in a sample tree. Note that palindromes in a word of length n naturally correspond to palindromic substrings in a path of n edges.

P. Gawrychowski—Work done while the author held a post-doctoral position at Warsaw Center of Mathematics and Computer Science.

T. Kociumaka—Supported by Polish budget funds for science in 2013-2017 as a research project under the 'Diamond Grant' program.

W. Rytter—This work was supported by the Polish National Science Center, grant no NCN2014/13/B/ST6/00770.

T. Waleń—Supported by the Polish Ministry of Science and Higher Education under the 'Iuventus Plus' program in 2015-2016 grant no 0392/IP3/2015/73.

C. Iliopoulos et al. (Eds.): SPIRE 2015, LNCS 9309, pp. 270–276, 2015.
DOI: 10.1007/978-3-319-23826-5_26

The study of the palindromic complexity of trees was recently initiated by Brlek, Lafrenière and Provençal [3], who constructed a family of trees with n edges containing $\Theta(n^{1.5})$ distinct palindromic substrings. They conjectured that there are no trees with asymptotically larger palindromic complexity and proved this claim for a restricted case of trees in which the label of every path consists of up to 4 blocks (runs) of equal letters.

Our Result. We show that the number of distinct palindromic substrings in a tree with n edges is $\mathcal{O}(n^{1.5})$. This bound is tight by the construction given in [3]; hence we completely settle the maximum palindromic complexity for trees.

Related Work. Palindromic complexity of words was studied in various aspects. This includes algorithms determining the complexity [7], bounds on the average complexity [1] or generalizations to circular words [9]. Finite and infinite palindrome-rich words received particularly high attention; see e.g. [2,5,6]. This class contains, for example, all episturmian and thus all Sturmian words [5].

In the setting of labelled trees other kinds of regularities were also studied. It has been shown that a tree with n edges contains $\mathcal{O}(n^{4/3})$ distinct squares [4] and $\mathcal{O}(n \log n)$ distinct cubes [8]. The former bound is known to be tight. Interestingly, the lower bound construction resembles that for palindromes [3].

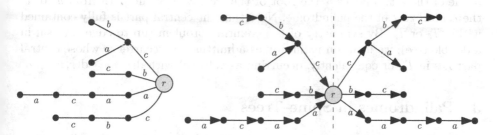

Fig. 1. To the left: an example undirected tree with 9 nontrivial palindromic substrings *bcb, bccb, aca, cbc, caac, cc, cbcbc, aa, acaaca*. To the right: deterministic double tree obtained after rooting the tree at r, merging both subtrees connected to r with edges labelled by c, and duplicating the resulting tree.

2   Preliminaries

A word w is a sequence of characters $w[1], w[2], \ldots, w[|w|] \in \Sigma$, often denoted $w[1..|w|]$. A substring of w is any word of the form $w[i..j]$, and if $i = 1$ ($j = |w|$) it is called a prefix (suffix). A period of w is any integer p, $1 \le p \le |w|$, such that $w[i] = w[i+p]$ for $i = 1, 2, \ldots, |w| - p$. The shortest period of w, denoted $\mathrm{per}(w)$, is the smallest such p. The following fact is a straightforward consequence of the periodicity lemma.

Fact 1. *Suppose a word v is a substring of a longer word u which has a period $p \leq \frac{1}{2}|v|$. Then $\text{per}(u) = \text{per}(v)$.*

A palindrome is a word w such that $w = w^R$, where w^R denotes the reverse of w. We have the following connection between periods and palindromes.

Fact 2. *Suppose a palindrome v is a suffix of a longer palindrome u. Then v is a prefix of u and thus $|u| - |v|$ is a period of u and of v.*

Define a *double tree* $\mathcal{D} = (T_\ell, T_r, r)$ as a labelled tree consisting of two trees T_ℓ and T_r sharing a common root r but otherwise disjoint. The edges of T_ℓ and T_r are directed to and from r, respectively. The size of $\mathcal{D}$ is defined as $|\mathcal{D}| = |T_\ell| + |T_r|$. For any $u, v \in \mathcal{D}$, we use $\text{val}(u, v)$ to denote the sequence of the labels of edges on the path from u to v. A *substring* of $\mathcal{D}$ is a word $\text{val}(u, v)$ such that $u \in T_\ell$ and $v \in T_r$. Also, let $\text{d}(u, v) = |\text{val}(u, v)|$ and $\text{per}(u, v) = \text{per}(\text{val}(u, v))$.

We consider only *deterministic* double trees, meaning that all the edges outgoing from a node have distinct labels, and similarly all the edges incoming into a node have distinct labels. An example of such a double tree is shown in Fig. 1. Symmetry of palindromic substrings $\text{val}(u, v)$, where $u \in T_\ell, v \in T_r$ gives a natural pairing of nodes on the path from u to v, where u is paired with v (and, if the path consists of an odd number of nodes, the central node is paired with itself). For any two paired nodes u', v' on such path, $\text{val}(u', v')$ is a palindrome; if one of these two nodes is the root of the tree, we call the path from u' to v' the *central part* of the palindrome. Note that the central part is fully contained within T_ℓ or T_r. By symmetry of the counting problem (up to edge reversal in a double tree), we focus on palindromes admitting an occurrence whose central part lies in T_ℓ, or equivalently, occurring as $\text{val}(u, v)$ with $\text{d}(u, r) \geq \text{d}(r, v)$.

3   Palindromes in Spine-Trees

A *spine-tree* is a deterministic double tree with a distinguished path, called *spine*, joining vertices $s_\ell \in T_\ell$ and $s_r \in T_r$. Additionally, we insist that this path cannot be extended preserving the period $p = \text{per}(s_\ell, s_r)$. A palindromic substring is *induced* by such a spine-tree if its central part is a fragment of the spine of length at least p; see Fig. 2 for an example.

For a node u of the spine-tree let $s(u)$ denote the nearest node of the spine (if u is already on the spine, then $u = s(u)$). Since the spine-tree is deterministic, it satisfies the following property.

Fact 3. *For any induced palindrome $\text{val}(u, v)$, the path $\text{val}(s(u), s(v))$ is an inclusion-maximal fragment of $\text{val}(u, v)$ admitting period p.*

Lemma 4. *There are up to $n\sqrt{n}$ distinct palindromic substrings induced by a spine-tree of size n.*

Proof. Define the *label* $L(u)$ for a node $u \in T_\ell$ as the prefix of $\text{val}(u, s_r)$ of length $\text{d}(u, s(u)) + p$. Similarly, the label $L(v)$ of a node $v \in T_r$ is the reversed suffix

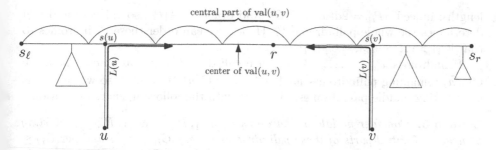

Fig. 2. A spine-tree, whose spine is the path from s_ℓ to s_r, with an induced palindrome $\mathrm{val}(u,v)$. Observe that $L(u) = L(v)$ is a prefix of the palindrome. Note that $d(s(u),r) \geq p$ but $d(r,s(v))$ might be smaller than p.

of $\mathrm{val}(s_\ell,v)$ of length $p + d(s(v),v)$. We leave the label undefined if $\mathrm{val}(u,s_r)$ or $\mathrm{val}(s_\ell,v)$ is not sufficiently long, i.e., if $d(s(u),s_r) < p$ or $d(s_\ell,s(v)) < p$.

Consider a palindrome $\mathrm{val}(u,v)$ induced by the spine-tree. Fact 3 implies that $\mathrm{val}(s(u),s(v))$ is a maximal fragment of $\mathrm{val}(u,v)$ with period p. Since the central part of the palindrome is of length at least p and lies within this fragment, the fragment must be symmetric, i.e., we must have $d(u,s(u)) = d(s(v),v)$, and the labels of both u and v are defined. Consequently, $|L(u)| = |L(v)|$ and actually the labels $L(u)$ and $L(v)$ are equal. Hence, to bound the number of distinct palindromes, we group together nodes with the same labels. Let V_L be the set of vertices of $T_\ell \cup T_r$ with label L. We have the following claim.

Claim. For any label L, there are at most $\min(|V_L|^2, n)$ distinct palindromes with endpoints in V_L.

Proof. Consider all distinct induced palindromes $\mathrm{val}(u,v)$ such that $L(u) = L(v) = L$. A substring is uniquely determined by the endpoints of its occurrence, so $|V_L|^2$ is an upper bound on the number of these palindromes. We claim that every such palindrome is also uniquely determined by its length, which immediately gives the upper bound of n. Indeed $d(u,s(u)) = d(s(v),v) = |L| - p$ and $\mathrm{val}(s(u),s(v))$ has period p, so if the length is known, $\mathrm{val}(s(u),s(v))$ can be recovered from its prefix of length p, i.e., the suffix of L of length p.   □

The sets V_L are disjoint, so by the above claim and using the inequality $\min(x,y) \leq \sqrt{xy}$ the number of distinct palindromes induced by the spine-tree is at most:

$$\sum_L \min(|V_L|^2, n) \leq \sum_L \sqrt{|V_L|^2 \cdot n} \leq \sqrt{n} \cdot \sum_L |V_L| \leq n^{1.5}.   □$$

4   Palindromes in General Deterministic Double Trees

Consider a node $u \in T_\ell$ and all distinct palindromes $P_1, \ldots, P_k$ with an occurrence starting at u. Observe that their central parts $C_1, \ldots, C_k$ have distinct lengths: indeed, $|P_i| = 2d(u,r) - |C_i|$ and $d(u,r) \geq \frac{1}{2}|P_i|$, so $\mathrm{val}(u,r)$ and $|C_i|$ determines the whole palindrome P_i. Hence, we can order these palindromes so that $|C_1| > \ldots > |C_k|$, (i.e., $|P_1| < \ldots < |P_k|$).

Palindromes $P_{4\sqrt{n}+1}, \ldots, P_{k-2\sqrt{n}}$ are called *middle palindromes*. There are $\mathcal{O}(\sqrt{n})$ remaining palindromes for fixed u and $\mathcal{O}(n^{1.5})$ in total, so we can focus on counting middle palindromes. We start with the following characterization.

Lemma 5. *Consider middle palindromes $P_{4\sqrt{n}+1}, P_{4\sqrt{n}+2}, \ldots, P_{k-2\sqrt{n}}$ starting at node u. Central parts of these palindromes satisfy $|C_i| \geq 2\sqrt{n}$ and $\mathrm{per}(C_i) \leq \frac{1}{2}\sqrt{n}$. Moreover, for each P_i extending the central part C_i by $2\sqrt{n}$ characters in each direction preserves the shortest period.*

Proof. Since we excluded the $2\sqrt{n}$ palindromes with the shortest central parts, the middle palindromes clearly have central parts of length at least $2\sqrt{n}$.

First, let us prove that $\mathrm{per}(C_{2\sqrt{n}}) \leq \frac{1}{2}\sqrt{n}$. By Fact 2, $|C_j| - |C_{j+1}|$ is a period of C_j for $1 \leq j \leq 2\sqrt{n}$. Since $\sum_{j=1}^{2\sqrt{n}}(|C_j| - |C_{j+1}|) < |C_1| \leq n$, for some j we have $\mathrm{per}(C_j) \leq |C_j| - |C_{j+1}| \leq \frac{1}{2}\sqrt{n}$. Moreover, $C_{2\sqrt{n}}$ is a suffix of C_j, so the claim follows.

For $i > 4\sqrt{n}$, in particular if P_i is a middle palindrome, C_i is a suffix of $C_{2\sqrt{n}}$. Hence, Fact 1 implies that $\mathrm{per}(C_i) = \mathrm{per}(C_{2\sqrt{n}})$. Moreover, $|C_i| \leq |C_{2\sqrt{n}}| + 2\sqrt{n} - i < |C_{2\sqrt{n}}| - 2\sqrt{n}$, so extending C_i by $2\sqrt{n}$ characters to the left preserves the period. By symmetry of P_i, extension to the right also preserves the period.   □

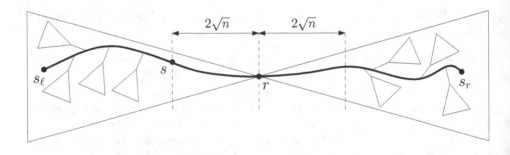

Fig. 3. A spine-tree constructed for a vertex s in a deterministic double tree. Note that we do not attach subtrees at distance less than $2\sqrt{n}$ from the root.

Let us choose any $s \in T_\ell$ such that $d(s,r) = 2\sqrt{n}$ and $\mathrm{per}(s,r) \leq \frac{1}{2}\sqrt{n}$. Then, extend the period of $\mathrm{val}(s,r)$ to the left and to the right as far as possible, arriving at nodes s_ℓ and s_r, respectively. We create a spine-tree with spine corresponding to the path from s_ℓ to s_r as shown on Fig. 3. We attach to the spine all subtrees hanging off the original path at distance at least $2\sqrt{n}$ from the root. In other words, a vertex $u \in T_\ell$ which does not belong the spine is added to the spine-tree if $d(s(u),r) \geq 2\sqrt{n}$ and a vertex $v \in T_r$ — if $d(r,s(v)) \geq 2\sqrt{n}$. If $d(r,s_r) < 2\sqrt{n}$ this leaves no subtrees hanging in T_r so we do not create any spine-tree for s.

Now consider a middle palindrome. By Lemma 5 its central part satisfies $|C| \geq 2\sqrt{n}$ and $\text{per}(C) \leq \frac{1}{2}\sqrt{n}$. Moreover, by Fact 1 we have $\text{per}(C) = \text{per}(s,r)$ for the unique node $s \in T_\ell$ at distance $2\sqrt{n}$ from the root within C. Consequently, C lies on the spine of the spine-tree created for s and u belongs to a subtree attached to the spine. Additionally, since C can be extended by $2\sqrt{n}$ characters in each direction preserving the period, the other endpoint v must also belong to such a subtree in T_r (that is, we have $\text{d}(r, s(v)) \geq 2\sqrt{n}$). Hence, any of the middle palindromic substrings is induced by some spine-tree.

The spine-trees are not disjoint, but nevertheless their total size is small.

Lemma 6. *The sizes* $n_1, \ldots, n_k$ *of the created spine-trees satisfy* $\sum_i n_i \leq 2n$.

Proof. We claim that at least $n_i - 2\sqrt{n}$ nodes of the i-th spine-tree are disjoint from all the other spine-trees. Let c_i be the node on the spine of the i-th spine-tree such that $\text{d}(c_i, r) = \sqrt{n}$ and similarly let s_i satisfy $\text{d}(s_i, r) = 2\sqrt{n}$. Recall that $\text{per}(s_i, r) \leq \frac{1}{2}\sqrt{n}$. Thus, by Fact 1 $\text{per}(s_i, r) = \text{per}(c_i, r)$. Since the tree is deterministic, c_i uniquely determines s_i and hence the whole spine-tree. Thus, the nodes c_i are all distinct and so are their predecessors on the spines and all attached subtrees. A similar argument shows that all nodes d_i on the spine of the i-th spine-tree such that $\text{d}(r, d_i) = \sqrt{n}$ are also all distinct. Therefore, we proved $\sum_i n_i - 2\sqrt{n} \leq n$. Each spine-tree has at least $4\sqrt{n}$ vertices on the spine, so this yields $n_i \geq 4\sqrt{n}$ and thus we obtain $\sum_i n_i \leq 2\sum_i(n_i - 2\sqrt{n}) \leq 2n$. □

By Lemma 4, the number of palindromes induced by the i-th spine-tree is at most $n_i^{1.5}$. Accounting the $\mathcal{O}(n^{1.5})$ palindromes which do not occur as middle palindromes, we have $\mathcal{O}(n^{1.5}) + \sum_i n_i^{1.5} \leq \mathcal{O}(n^{1.5}) + \sum_i n_i\sqrt{n} = \mathcal{O}(n^{1.5})$ palindromes in total.

Lemma 7. *Every deterministic double tree of size n has $\mathcal{O}(n^{1.5})$ distinct palindromic substrings.*

5   Main Result

To derive the final theorem, we follow the approach from [4]. We use the folklore fact that every tree T on n edges contains a *centroid* node r such that every component of $T \setminus \{r\}$ is of size at most $\frac{n}{2}$. We separately count palindromic substrings corresponding to the paths going through the centroid r and paths fully contained in a single component of $T \setminus \{r\}$. To bound the former, we root T at r directing all the edges so that they point towards the root, and then determinize the resulting tree by gluing together two children of the same node whenever their edges have the same label. Finally, we create a deterministic double tree by duplicating the tree and changing the directions of the edges in the second copy.

It is easy to see that for any simple path from u to v going through r in the original tree we can find $u' \in T_\ell$ and $v' \in T_r$ such that $\text{val}(u,v) = \text{val}(u',v')$. Hence, the number of distinct palindromic substrings corresponding to such paths, by Lemma 7, is $\mathcal{O}(n^{1.5})$. Finally, we obtain the following recurrence for $\mathsf{pal}(n)$, the maximum number of palindromes in a tree with n edges:

$$\mathsf{pal}(n) = \mathcal{O}(n^{1.5}) + \max\left\{\sum_i \mathsf{pal}(n_i) : \forall_i \, n_i \leq \frac{n}{2} \wedge \sum_i n_i < n\right\}$$

which solves to $\mathsf{pal}(n) = \mathcal{O}(n^{1.5})$.

Theorem 8. *A tree with n edges contains $\mathcal{O}(n^{1.5})$ distinct palindromic substrings.*

References

1. Anisiu, M., Anisiu, V., Kása, Z.: Total palindrome complexity of finite words. Discrete Mathematics **310**(1), 109–114 (2010)
2. Brlek, S., Hamel, S., Nivat, M., Reutenauer, C.: On the palindromic complexity of infinite words. Int. J. Found. Comput. Sci. **15**(2), 293–306 (2004)
3. Brlek, S., Lafrenière, N., Provençal, X.: Palindromic complexity of trees. In: Potapov, I. (ed.) DLT 2015. LNCS, vol. 9168, pp. 155–166. Springer, Heidelberg (2015). arXiv:1505.02695
4. Crochemore, M., Iliopoulos, C.S., Kociumaka, T., Kubica, M., Radoszewski, J., Rytter, W., Tyczyński, W., Waleń, T.: The maximum number of squares in a tree. In: Kärkkäinen, J., Stoye, J. (eds.) CPM 2012. LNCS, vol. 7354, pp. 27–40. Springer, Heidelberg (2012)
5. Droubay, X., Justin, J., Pirillo, G.: Episturmian words and some constructions of de Luca and Rauzy. Theor. Comput. Sci. **255**(1–2), 539–553 (2001)
6. Glen, A., Justin, J., Widmer, S., Zamboni, L.Q.: Palindromic richness. Eur. J. Comb. **30**(2), 510–531 (2009)
7. Groult, R., Prieur, É., Richomme, G.: Counting distinct palindromes in a word in linear time. Inf. Process. Lett. **110**(20), 908–912 (2010)
8. Kociumaka, T., Radoszewski, J., Rytter, W., Waleń, T.: String powers in trees. In: Cicalese, F., Porat, E., Vaccaro, U. (eds.) CPM 2015. LNCS, vol. 9133, pp. 284–294. Springer, Heidelberg (2015)
9. Simpson, J.: Palindromes in circular words. Theor. Comput. Sci. **550**, 66–78 (2014)

Beyond the Runs Theorem

Johannes Fischer[1], Štěpán Holub[2], Tomohiro I[1(✉)], and Moshe Lewenstein[3]

[1] Department of Computer Science, TU Dortmund, Dortmund, Germany
tomohiro.i@cs.tu-dortmund.de
[2] Department of Algebra, Charles University, Praha, Czech Republic
[3] Department of Computer Science, Bar-Ilan University, Ramat Gan, Israel

Abstract. In [3], a short and elegant proof was presented showing that a word of length n contains at most $n - 3$ runs. Here we show, using the same technique and a computer search, that the number of runs in a binary word of length n is at most $\frac{22}{23}n < 0.957n$.

Keywords: Runs · Lyndon words · Combinatorics on words

1 Introduction

Repetitions in words are one of the most basic and well studied characteristics of words, with various theoretical and practical applications (see [6,20,21] for surveys). The most notable notion of repetitions would be *runs* [12] (a.k.a. maximal periodicities [15] or maximal repetitions [13]) as runs can capture all periodic sub-intervals in a word compactly.

The research on the possible (maximal) number $\rho(n)$ of runs in a word of length n dates back at least to [13], where they showed that $\rho(n) = O(n)$. Since then, there were two types of efforts: finding words rich of runs [9,14,16,19], and proving an upper bound (exact coefficient) of $\rho(n)$ [3–5,7,10,11,17,18]. For upper bounds, an (at least psychologically) important barrier was the question whether the number of runs can be larger than the length of the word, and the negative answer was known as "the runs conjecture". The barrier was broken, turning the conjecture into a theorem, by a remarkably simple and computer-free proof in [3]. The proof is based on the characterization of runs by Lyndon words established in [4]. Note that before [4], the approach to get upper bounds was completely different, and the bound $\rho(n) < 1.029n$ in [7] required a heavy use of computer search.

Although the conjecture was solved, the quest continues towards the optimal value of $\lim_{n\to\infty}(\rho(n)/n)$, which is known to exist [10]. In this paper we show how to lower the upper bound by adding again some computer backing to the technique that led to the proof of the Runs Theorem in [3]. We show that the

Štěpán Holub is supported by the Czech Science Foundation grant number 13-01832S.

J. Fischer and M. Lewenstein are supported by a Grant from the GIF, the German-Israeli Foundation for Scientific Research and Development.

© Springer International Publishing Switzerland 2015
C. Iliopoulos et al. (Eds.): SPIRE 2015, LNCS 9309, pp. 277–286, 2015.
DOI: 10.1007/978-3-319-23826-5_27

number of runs in a binary word of length n is at most $\frac{22}{23}n < 0.957n$. The bound we provide is obviously not optimal. Our paper is therefore just another step in the extensive effort to understand the behaviour of runs. For the more detailed description of the history of the problem, see for example [3,7].

2   Runs and Lyndon Roots

For any word u, an integer p with $1 \le p \le |u|$ is said to be a *period* of u if $u[i] = u[i+p]$ for all $1 \le i \le |u| - p$. Especially, the smallest period of u is called *the period* of u. A prefix v of u that is also a suffix of u is said to be a *border* of u. The empty word and u are trivial borders of u. We call u *unbordered* if there is no border other than trivial ones.

Given a word w, we say that an interval $[i..j]$ with $1 \le i \le j \le |w|$ is *period-maximal in* w if $w[i..j]$ has no extension in w with the same period. That is, if $1 \le i' \le i \le j \le j' \le |w|$ is such that $w[i..j]$ and $w[i'..j']$ have the same period, then $i = i'$ and $j = j'$. A period-maximal interval is said to be *left-open* if $i = 1$, otherwise it is *right-closed*. Similarly, a period-maximal interval is *right-open* or *left-closed* depending on whether or not $j = |w|$. If $1 < i$ and $j < |w|$, the interval is said to be *closed*. A period-maximal interval is a *run* if its length is at least double of the period p of $w[i..j]$, that is $j - i + 1 \ge 2p$.

We shall work with the two-letter alphabet $\{0, 1\}$, which allows two lexicographic orders: $\prec_0$ is defined by $0 \prec_0 1$, and $\prec_1$ by $1 \prec_1 0$. We shall write $\bar{a} = 1 - a$.

Remark 1. Our paper is restricted to the binary alphabet. Although some claims can be easily extended to arbitrary alphabet size, we do not do so for sake of simplicity, since the computer search is performed for the binary case anyway. For a result on more letters, see for example [8]. Note that it is an open question of interest whether binary words attain $\rho(n)$ although it is believed so (see for example [2]).

A word v is said to be a *Lyndon word* with respect to some order $\prec$ if and only if $w \prec u$ for any nonempty proper suffix u of w. In particular, Lyndon words are unbordered. We say that a Lyndon word v is a *Lyndon root* of w if v is a factor of w and $|v|$ is the period of w.

A right-closed period-maximal interval $[i..j]$ of w is said to be *a-broken* in w, if $a = w[j + 1]$. We will also say, a bit imprecisely, that the period of $w[i..j]$ is broken by a.

Let $\rho(n, 2)$ denote the maximal number of runs in a binary word of length n.

The basic idea of [3] is to associate an a-broken run $r = [i..j]$ with the set $\Lambda(r)$ of intervals corresponding to the Lyndon root of r with respect to the order $a \prec \bar{a}$, excluding from $\Lambda(r)$, if necessary, the interval starting at the beginning of r. This definition has to be completed to cover also runs that are not broken, that is, right-open runs. For those runs, the set $\Lambda(r)$ can be defined as consisting of Lyndon roots with respect to both orders. In [3], the case of unbroken runs is solved by appending a special symbol $ to the end of w, which is equivalent to arbitrarily choosing one of the orders (the order $0 \prec 1$ in their case).

Let $\mathtt{Beg}(S)$ denote the set of starting positions of intervals in the set S, and let $B(r) = \mathtt{Beg}(\Lambda(r))$ for any run r. Note that $B(r)$ is nonempty if r is a run. The crucial fact, implying instantaneously that there are at most $|w| - 1$ runs, is that $B(r)$ and $B(r')$ are disjoint for $r \neq r'$. At no cost, it is possible to make this basic tool a bit stronger. For sake of clarity, let us first give a formal definition.

Definition 1. *Let w be a binary word. Let $s = [i..j]$ be a period-maximal interval in w with period p. Then $\Lambda(s)$ denotes the set of all intervals $[i'..j']$ of length p such that $i < i' \leq j' \leq j$ and $w[i'..j']$ is a Lyndon word with respect to an order $\prec$ satisfying the following condition: if $j < |w|$ and $[i..j]$ is a-broken in w, then $a \prec \bar{a}$ (the condition being empty if $[i..j]$ is not broken). Also, let $B(s) = \mathtt{Beg}(\Lambda(s))$.*

Example 1. Take a word $w = 1110101101$ of length 10. For a period-maximal interval $s_1 = [1..3]$ with period 1, $\Lambda(s_1) = \{[2..2], [3..3]\}$. For a period-maximal interval $s_2 = [3..7]$ with period 2, $\Lambda(s_2) = \{[5..6]\}$. For a period-maximal interval $s_3 = [5..10]$ with period 3, $\Lambda(s_3) = \{[6..8], [7..9]\}$. Note that $w[6..8] = 011$ and $w[7..9] = 110$ are Lyndon words w.r.t. $\prec_0$ and $\prec_1$, respectively. For a period-maximal interval $s_4 = [2..10]$ with period 5, $\Lambda(s_4) = \{[4..8]\}$.

The following lemma is now stronger than the binary case of the corresponding [3, Lemma 8] in two ways. First, it applies also to period-maximal intervals that are not runs, and second, as noted above, $\Lambda(r)$ is defined more generously for unbroken runs. The proof, however, is the same.

Lemma 1. *Let s and t be two distinct period-maximal intervals in w. Then $B(s)$ and $B(t)$ are disjoint.*

Proof. Let $s = [i_s..j_s]$ and $t = [i_t..j_t]$. Suppose that $k \in B(s) \cap B(t)$, and let $[k..m_s] \in \Lambda(s)$ and $[k..m_t] \in \Lambda(t)$. If $m_s = m_t$, then s and t have the same period and $s = t$. We can therefore, w.l.o.g., suppose that $m_s < m_t$. Then $w[k..j_s]$ has a smaller period than the unbordered $w[k..m_t]$, which implies that $j_s < m_t$. Therefore s is a-broken with $a = w[j_s + 1]$.

Since a breaks the period of $w[k..j_s]$, we have $w[m_s + 1..j_s + 1] \prec_a w[k..j_s + 1]$. Since both $w[m_s + 1..j_s + 1]$ and $w[k..j_s + 1]$ are factors of $w[k..j_t]$, we deduce that $w[k..m_t]$ is Lyndon w.r.t. $\prec_{\bar{a}}$. Note that $w[k..j_t]$ contains both letters. Therefore $w[k] = \bar{a}$, and the $\prec_a$-minimality of $w[k..m_s]$ implies that $w[i_s..j_s] \in \bar{a}^+$. The definition of $\Lambda(s)$ yields $i_s < k$ and $i_t < k$, which leads to a contradiction with $\prec_{\bar{a}}$-minimality of $w[k..m_t]$.

Example 2. It is worth noting that the appearance of $\bar{a}^+$ in the previous proof is significant, and it is the place where we use the prohibition of the very first position of a run. Without this condition, Lemma 1 would not hold. Consider the word 1101011 and position 2, which is the starting point of the Lyndon root 1 of the run 11 and the starting point of the Lyndon root 10 of 10101, the latter being excluded by the prohibition.

Lemma 1 implies that for each position k there is at most one period-maximal interval s such that $k \in B(s)$. Such an s can be found using the following rules.

Lemma 2. *Let $k > 1$ be a position of w such that $w[k] = a$ and $w[k-1..|w|] \neq \bar{a}a^+$. Then $k \in B(s)$ where $s = [i..j]$ is the period-maximal extension of*

- *$[k..k]$, if $w[k] = w[k-1]$;*
- *$[k..k']$, where $w[k..k']$ is the longest Lyndon word with respect to $\prec_a$ starting at the position k, otherwise.*

Proof. If $w[k] = w[k-1]$, then s is a run with period one containing the position k with $i < k$. Hence, $k \in B(s)$ immediately follows from Definition 1.

Let $w[k-1] = \bar{a}$, and let $w[k..k']$ be the longest Lyndon word with respect to $\prec_a$ starting at the position k. From $w[k..|w|] \neq a^+$, it is easy to see that $k' \neq k$ and $w[k'] = \bar{a}$, which implies $i < k$. If s is right-open, we are done: $k \in B(s)$ since the condition on $\prec$ is empty (see Definition 1). It remains to show that s is a-broken if it is broken. Assume to the contrary that s is $\bar{a}$-broken. We show that $w[k..j+1]$ is a Lyndon word with respect to $\prec_a$. Let p denote the length of the Lyndon word $w[k..k']$, that is, $p = k' - k + 1$. Let first $k < h \le k'$. Since $w[k..k']$ is a Lyndon word with respect to $\prec_a$, we have $w[k..k'] \prec_a w[h..k']$, and thus also $w[k..j+1] \prec_a w[h..j+1]$. Let now $k' < h \le j+1$. As above, $w[k..k'] \prec_a w[h-p+1..h] \prec_a w[h-p+1..j+1]$. (Which covers also the possibility $w[k..k'] = w[h-p+1..h]$.) Also $w[h-p+1..j+1] \prec_a w[h..j+1]$, since $w[h..j+1] = w[h..j]\bar{a}$ and $w[h..j]a$ is a prefix of $w[h-p+1..j+1]$. Therefore, $w[k..j+1]$ is a Lyndon word, which contradicts that $w[k..k']$ is the longest Lyndon word starting at the position k.

Note that for the position k with $w[k-1..|w|] = \bar{a}a^+$, there is no period-maximal interval s with $k \in B(s)$. An algorithm computing for all positions the longest Lyndon words starting there is discussed in [3, Section 4.1].

3 Idle Positions

In order to make explicit the relation between runs and positions, we associate with a run r the position $\max B(r)$ and say that such a position is *charged* (by r). We repeat that the Runs Theorem was proved in [3] by pointing out that charging is an injective mapping, which is a corollary of Lemma 1. This also yields an obvious strategy for further lowering the upper bound on the number of runs. One has to find positions that are not charged in an arbitrary word. We shall call such positions *idle*. Equivalently, we want to identify a position i satisfying either of the following two conditions.

1. i is not contained in $B(r)$ for any run, or
2. i is in $B(r) \setminus \{\max B(r)\}$ for some run r.

3.1   Idle Positions that are Resistant to Extensions

In order to be able to estimate the number of idle positions locally, we are interested in idle positions that remain idle in any extension of w. One obvious fact is that closed period-maximal intervals are not affected by extensions. For example, the third position in the word 1010011 remains idle for any extensions. That is because the period three of 1001 is broken by 1, and the period-maximal extension of 1001 is $s = [2..6]$ that is closed, but s is not a run, and Definition 1 and Lemma 1 yield that the position is idle.

Also, it is easy to see that runs r with $|B(r)| > 1$ that are right-closed preserve this property in any extension. However, we have to be careful with right-open runs since some positions in $B(r)$ may disappear when the run r gets broken by a right-extension. To clarify this case, let $\Lambda_a(r)$ denote the set of Lyndon roots in $\Lambda(r)$ that are Lyndon words with respect to $\prec_a$, and let $B_a(r) = \text{Beg}(\Lambda_a(r))$. Note that $B_a(r) = B_{\bar{a}}(r)$ if and only if r is a run with period one. Now we consider the set $D(w)$ of idle positions k in a word w falling into one of the following cases:

(a) $k \in B(s)$, where s is a closed period-maximal interval that is not a run.
(b) $k \in (B_a(r) \setminus \{\max B_a(r)\})$, where r is an a-broken run.
(c) $k \in (B_a(r) \setminus \{\max B_a(r)\})$, where r is a right-open run and a is chosen such that $\min B_a(r) \geq \min B_{\bar{a}}(r)$ ($a \in \{0, 1\}$ is arbitrary if its period is 1).

By $D(w)$ we intend to say that, for any $k \in D(w)$, the position $|u| + k$ in uwv is idle for any extensions u and v. The only exception is the case (c) in which the position $|u| + k$ may not be idle if r is $\bar{a}$-broken in the extension. But even in this case we have that at least one of the positions $|u| + k$ and $|u| + k - g$ of uwv is idle, where $g = \min B_a(r) - \min B_{\bar{a}}(r)$. Therefore the number of idle positions does not decrease for any extensions. This is formulated in the following claim.

Claim 1. *Let w, u and v be arbitrary binary words. Then*

$$\left| D(uwv) \cap [|u| + 2..|uw| - 1] \right| \geq |D(w)|.$$

Proof. We examine $k \in D(w)$ of each case:

- For Case (a). Since $s = [i..j]$ is closed, we have a closed period-maximal interval $s' = [|u| + i..|u| + j]$ in uwv. Since s' is not a run, $|u| + k$ is in $D(uwv)$.
- For Case (b). Since $r = [i..j]$ is an a-broken run, we have an a-broken run $r' = [i'..|u| + j]$ with $i' \leq |u| + i$ in uwv. Since any Lyndon root in $\Lambda(r)$ appears in $\Lambda(r')$ (with shift $|u|$), $|u| + k$ is in $D(uwv)$.
- For Case (c). Let $r = [i..|w|]$ be a right-open run in w. We have a run $r' = [i'..j']$ with $i' \leq |u| + i$ and $|uw| \leq j'$ in uwv. Note that $k - g \in B_{\bar{a}}(r)$, where $g = \min B_a(r) - \min B_{\bar{a}}(r)$.
  - If r' is still open or a-broken, any Lyndon root in $\Lambda_a(r)$ appears in $\Lambda_a(r')$ (with shift $|u|$), and hence, $|u| + k$ is in $D(uwv)$.

- If r' is $\bar{a}$-broken, any Lyndon root in $\Lambda_{\bar{a}}(r)$ appears in $\Lambda_{\bar{a}}(r')$ (with shift $|u|$), and hence, $|u| + k - g$ is in $D(uwv)$.

We have described an injective map from $D(w)$ to $D(uwv) \cap [|u| + 1..|uw|]$. The map always assigns, for some a and some r, a position k in $[1..|w|] \cap B_a(r)$ to the position $k + |u|$. Note that $1 \notin D(w)$ since 1 is not in $B(s)$ for any s. Also, $|w| \notin D(w)$ follows from the definition of $D(w)$. This completes the proof. □

This yields the following lemma:

Lemma 3. If $|D(w)| \geq d$ for any binary word w of length m, then

$$\lim_{n \to \infty} \left(\frac{\rho(n, 2)}{n} \right) \leq \frac{m - 2 - d}{m - 2}.$$

Proof. Let $y = a y_1 y_2 \cdots$ be an infinite binary word, where a is a letter, and $|y_i| = m - 2$ for each i. By Claim 1, each interval corresponding to a factor y_i in y contains at least d idle positions. The statement follows. □

3.2   Idle Positions that are Resistant to Left Extensions

We further identify positions that remain idle when we consider "only" left extensions, which only comes into play in Section 5 to estimate the number of idle positions in a suffix of a word. Formally, for any word w we define the set $D'(w)$ of idle positions k in w falling into one of the following cases:

(A) $k \in \max B(s)$, where s is a left-closed period-maximal interval that is not a run.
(B) $k \in (B(s) \setminus \{\max B(s)\})$, where s is a period-maximal interval (which is possibly a run).
(C) $w[k - 1..|w|] = \bar{a}a^+$ holds.

Note that $D(w) \subseteq D'(w)$. Since we do not consider right-extensions, we can show the following claim, which is a bit stronger than Claim 1 for D.

Claim 2. Let w and u be arbitrary binary words. For any $k \in D'(w)$, $|u| + k \in D'(uw)$.

Proof. We examine $k \in D'(w)$ of each case:

- For Case (A). Since $s = [i..j]$ is left-closed, we have a left-closed period-maximal interval $s' = [|u| + i..|u| + j]$ in uw. Since s' is not a run, $|u| + k$ is in $D'(uw)$.
- For Case (B). Let $s = [i..j]$. We have a period-maximal interval $s' = [i'..|u| + j]$ with $i' \leq |u| + i$ in uw. Since any Lyndon root in $\Lambda(s)$ appears in $\Lambda(s')$ (with shift $|u|$), $|u| + k$ is in $D'(uw)$.
- For Case (C). Since $\bar{a}a^+$ stays in a suffix of uw, $|u| + k$ is in $D'(uw)$.

Considering that $1 \notin D'(w)$, we get:

Claim 3. Let w and u be arbitrary binary words. Then

$$\left| D'(uw) \cap [|u| + 2..|uw|] \right| \geq |D'(w)|.$$

Algorithm 1. Computing m_d.

Input: A positive integer d.
Output: Return m_d.

```
1 m ← 0; // Let m be a global variable.
2 Extend(0);
3 return m;

  procedure Extend(w);
1 if |w| > m then m ← |w|;
2 compute D(w);
3 if |D(w)| ≥ d then return
4 foreach a ∈ {0,1} do
5 ⌊ Extend(wa);
```

---

4   Computer Search

Given a positive integer d, Algorithm 1 computes the minimum integer m_d such that $|D(w)| \geq d$ for any binary word w of length m_d. The algorithm traverses words by appending characters to the right. If $|D(w)| \geq d$, we stop the extension since $|D(wv)| \geq d$ for any word v.

If we already know the value $m_{d'}$ for some $d' < d$, then the following pruning of the search space can be employed: If $|D(w) \cap [1..m - m_{d'} + 1]| \geq d - d'$, then we stop the extension. That is because for any word z of length $m_{d'}$, $D(z)$ contains at least d' positions (and $1 \notin D(z)$), and hence, for any word v of length $m - |w|$, $D(wv)$ contains at least $d - d'$ positions in $[1..m - m_{d'} + 1]$ and at least d' positions in $[m - m_{d'} + 2..m]$. Namely, $D(wv)$ contains at least $d - d' + d' = d$ positions, and hence, any right extension of the current w cannot lead to an update of m.

Furthermore, we employ the following optimization trick: Since our pruning method works more effectively as m becomes larger (during the traversal), it would be beneficial that we explore some promising words and enlarge m to some extent before starting an exhaustive search. For example, it is likely that a longest word w_{d-1} obtained for $d - 1$ can be extended by one or more for d. Based on this idea, at the very beginning of the computation of m_d, we explore the words having a prefix of w_{d-1} of length $|w_{d-1}| - c$, where c is a parameter (non-negative integer) set to explore some "neighbors" of w_{d-1}.

By computing m_d and using Lemma 3, we obtained upper bounds for

$$\lim_{n \to \infty} (\rho(n, 2)/n)$$

given in Table 1. It also shows approximate time (in seconds) required to compute m_d using the naive approach, and using the described pruning method with parameter $c = 100$. [1] The experiments were conducted on a machine with a

---

[1] Since the times for $d = 19, 20$ were obtained by an older (slower) version of the program, they are not straightforwardly comparable with others.

Table 1. Upper bounds of $\lim_{n\to\infty}(\rho(n,2)/n)$.

d	m_d	$\lim_{n\to\infty}(\rho(n,2)/n)$	$m_d - m_{d-1}$	naive (s)	pruning (s)
1	63	0.98360655737...	63	0	0
2	96	0.97872340425...	33	0	0
3	126	0.97580645161...	30	8	0
4	150	0.97297297297...	24	135	1
5	172	0.97058823529...	22	1905	5
6	194	0.96875	22	23 199	15
7	216	0.96728971962...	22	258 552	45
8	237	0.96595744680...	21	2 652 099	122
9	258	0.96484375	21		286
10	274	0.96323529411...	16		1 451
11	295	0.96245733788...	21		2 750
12	314	0.96153846153...	19		6 773
13	332	0.96060606060...	18		18 717
14	351	0.95988538681...	19		38 869
15	369	0.95912806539...	18		94 733
16	388	0.95854922279...	19		177 540
17	407	0.95802469135...	19		321 412
18	425	0.95744680851...	18		689 423
19	444	0.95701357466...	19		1 654 611
20	462	0.95652173913...	18		6 583 840

3.40GHz Intel Core i5-4670 CPU with 6144KB L2 cache and 16GB Memory running Linux (Ubuntu 12.04, 64bit). The program was compiled using g++ 4.6.4 with -O3 option for optimization.

5   Upper Bound for Finite Words

We now prove that we can omit the limit in the bounds in Table 1. That is, we verify that, for any $d \le 20$, $\rho(n,2)/n \le (m_d - 2 - d)/(m_d - 2)$ does hold for any n.

Let y be a finite word and let $p_1, p_2, \ldots, p_\ell$ be the list of idle positions of y. Note that $p_1 = 1$. For a given d we define

$$s_k = s_k(y,d) := [p_{(k-1)d+1}..p_{kd+1} - 1] \qquad \text{for } k = 1, 2, \ldots, \lceil \ell/d \rceil - 1,$$
$$s_k = s_k(y,d) := [p_{(k-1)d+1}..|y|] \qquad \text{for } k = \lceil \ell/d \rceil.$$

In other words, we make a disjoint decomposition of the interval $[1..|y|]$ into subintervals s_k such that each s_k starts with an idle position of y, and each s_k, except maybe the last one, contains exactly d idle positions.

We first claim that all intervals s_k, $k < \lceil \ell/d \rceil$, have length at most $m_d - 2$. Suppose that the length of some $s_k = [i..j]$ is at least $m_d - 1$ and consider the word $y[i..j + 1]$ of length m_d. By the definition of m_d and by Claim 1, the cardinality of $D(y) \cap [i + 1..j]$ is at least d which means that $[i..j]$ contains at least $d + 1$ idle positions, a contradiction.

It remains to count idle positions in the tail of the word y, that is, in the interval $s_{\lceil \ell/d \rceil}$. By an argument similar to the one above, one can see that the length of the interval is at most $m_d - 1$. Let z denote the suffix in question, that is, $z = y[p_{(\lceil \ell/d \rceil -1)d+1}..|y|]$. Since we only have to consider left-extensions of z, we now use $D'(z)$ to estimate the number of idle positions. Since $1 \notin D'(z)$ and the first position of $s_{\lceil \ell/d \rceil}$ is idle in y, our goal is to show

$$\frac{|z| - |D'(z)| - 1}{|z|} < \frac{m_d - 2 - d}{m_d - 2}. \tag{*}$$

Let $d = 20$. Then the right hand of $(*)$ is $22/23$. We first note that $(x-1)/x < 22/23$ for each $x < 23$. Therefore, we can assume $|z| \geq 23$.

A simple computer search verified that $|D'(w)| \geq 3$ for any word w with $|w| \geq 13$, which means there are at least 3 idle positions in the last 12 positions of w that are resistant to left extensions. Now, let $z = z_1 z_2$ with $|z_2| = 12$. If $m_i - 1 \leq |z_1| < m_{i+1} - 1$ (where $m_0 := 0$), then z has at least i idle positions in $[2..|z_1|]$ by Claim 1, and hence, $|D'(z)| \geq i + 3$. Using the results in Table 1, a direct calculation yields that, for each $i = 0, 1, \ldots, 19$, if $m_i - 1 \leq |z_1| < m_{i+1} - 1$, then

$$\frac{|z| - |D'(z)| - 1}{|z|} \leq \frac{(m_{i+1} - 2 + 12) - i - 3 - 1}{m_{i+1} - 2 + 12} < \frac{22}{23}.$$

Therefore we get the following result.

$$\rho(n, 2)/n < \frac{22}{23} = 0.\overline{9565217391304347826086}.$$

6   Conclusion

Search for words with high number of runs in the literature yields words with approximately $0.944n$ runs, where $n = |w|$, see [1,14,16,19]. Therefore, the optimal multiplicative constant is somewhere between 0.944 and 0.957. The lower bound corresponds to words where on average about every 18th position is idle. This seems to fit very well with the eventual distances between m_{d-1} and m_d in Table 1. It is therefore reasonable to expect that the optimal density of runs is close to the lower bound, maybe around $1 - 1/18.5 \approx 0.946$.

References

1. http://www.shino.ecei.tohoku.ac.jp/runs/
2. Bannai, H., Giraud, M., Kusano, K., Matsubara, W., Shinohara, A., Simpson, J.: The number of runs in a ternary word. In: Proc. PSC, pp. 178–181 (2010)
3. Bannai, H., I, T., Inenaga, S., Nakashima, Y., Takeda, M., Tsuruta, K.: The "runs" theorem (2014). arXiv:1406.0263 [cs.DM]
4. Bannai, H., I, T., Inenaga, S., Nakashima, Y., Takeda, M., Tsuruta, K.: A new characterization of maximal repetitions by Lyndon trees. In: Proc. SODA, pp. 562–571 (2015)

5. Crochemore, M., Ilie, L.: Maximal repetitions in strings. Journal of Computer and System Sciences, 796–807 (2008)
6. Crochemore, M., Ilie, L., Rytter, W.: Repetitions in strings: Algorithms and combinatorics. Theor. Comput. Sci. **410**(50), 5227–5235 (2009)
7. Crochemore, M., Ilie, L., Tinta, L.: The "runs" conjecture. Theor. Comput. Sci. **412**(27), 2931–2941 (2011)
8. Deza, A., Franek, F.: Bannai et al. method proves the d-step conjecture for strings (2015)
9. Franek, F., Yang, Q.: An asymptotic lower bound for the maximal number of runs in a string. International Journal of Foundations of Computer Science **1**(195), 195–203 (2008)
10. Giraud, M.: Not so many runs in strings. In: Martín-Vide, C., Otto, F., Fernau, H. (eds.) LATA 2008. LNCS, vol. 5196, pp. 232–239. Springer, Heidelberg (2008)
11. Giraud, M.: Asymptotic behavior of the numbers of runs and microruns. Information and Computation **207**(11), 1221–1228 (2009)
12. Iliopoulos, C.S., Moore, D., Smyth, W.: A characterization of the squares in a fibonacci string. Theor. Comput. Sci. **172**(1–2), 281–291 (1997)
13. Kolpakov, R.M., Kucherov, G.: Finding maximal repetitions in a word in linear time. In: Proc. FOCS, pp. 596–604 (1999)
14. Kusano, K., Narisawa, K., Shinohara, A.: On morphisms generating run-rich strings. In: Proc. PSC, pp. 35–47 (2013)
15. Main, M.G.: Detecting leftmost maximal periodicities. Discrete Applied Mathematics **25**(1–2), 145–153 (1989)
16. Matsubara, W., Kusano, K., Ishino, A., Bannai, H., Shinohara, A.: New lower bounds for the maximum number of runs in a string. In: Proc. PSC, pp. 140–145 (2008)
17. Puglisi, S.J., Simpson, J., Smyth, W.F.: How many runs can a string contain? Theor. Comput. Sci. **401**, 165–171 (2006)
18. Rytter, W.: The number of runs in a string: improved analysis of the linear upper bound. In: Durand, B., Thomas, W. (eds.) STACS 2006. LNCS, vol. 3884, pp. 184–195. Springer, Heidelberg (2006)
19. Simpson, J.: Modified Padovan words and the maximum number of runs in a word. Australas. J. Comb. **46**, 129–145 (2010)
20. Smyth, W.F.: Repetitive perhaps, but certainly not boring. Theor. Comput. Sci. **249**(2), 343–355 (2000)
21. Smyth, W.F.: Computing regularities in strings: A survey. Eur. J. Comb. **34**(1), 3–14 (2013)

Sampling the Suffix Array with Minimizers

Szymon Grabowski[(✉)] and Marcin Raniszewski

Institute of Applied Computer Science, Lodz University of Technology,
Al. Politechniki 11, 90–924 Łódź, Poland
{sgrabow,mranisz}@kis.p.lodz.pl

Abstract. Sampling (evenly) the suffixes from the suffix array is an old idea trading the pattern search time for reduced index space. A few years ago Claude et al. showed an alphabet sampling scheme allowing for more efficient pattern searches compared to the sparse suffix array, for long enough patterns. A drawback of their approach is the requirement that sought patterns need to contain at least one character from the chosen subalphabet. In this work we propose an alternative suffix sampling approach with only a minimum pattern length as a requirement, which seems more convenient in practice. Experiments show that our algorithm achieves competitive time-space tradeoffs on most standard benchmark data. As a side result, we show that n' arbitrarily selected suffixes from a text of length n, where $n' < n$, over an integer alphabet, can be sorted in $O(n)$ time using $O(n')$ words of space.

1 Introduction

Full-text indexes built over a text of length n can roughly be divided into two categories: those requiring at least $n \log_2 n$ bits and the more compact ones. Classical representatives of the first group are the suffix tree and the suffix array. Succinct solutions, often employing the Burrows–Wheeler transform and other ingenious mechanisms (compressed rank/select data structures, wavelet trees, etc.), are object of vivid interest in theoretical computer science [19], but their practical performance does not quite deliver; in particular, the locate query is significantly slower than with the suffix array [8,10,20].

A very simple, yet rather practical alternative to both compressed indexes and the standard suffix array is the *sparse suffix array* (SpaSA) [16]. This data structure stores only the suffixes at regular positions, namely those being a multiple of q ($q > 1$ is a construction-time parameter). The main drawback of SpaSA is that instead of one (binary) search over the plain SA it has to perform q searches, in $q - 1$ cases of which followed by verification of the omitted prefix against the text. If, for example, the pattern $P[1 \ldots 6]$ is tomcat and $q = 4$, we need to search for tomcat, omcat, mcat and cat, and 3 of these 4 searches will be followed by verification. Obviously, the pattern length must be at least q and this approach generally works better for longer patterns.

The *sampled suffix array* (SamSA) by Claude et al. [3] is an ingenious alternative to SpaSA. They choose a subset of the alphabet and build a sorted array

© Springer International Publishing Switzerland 2015
C. Iliopoulos et al. (Eds.): SPIRE 2015, LNCS 9309, pp. 287–298, 2015.
DOI: 10.1007/978-3-319-23826-5_28

over only those suffixes which start with a symbol from the chosen subalphabet. The search starts with finding the first (leftmost) sampled symbol of the pattern, let us say at position j, and then the pattern suffix $P[j \ldots m]$ is sought in the sampled suffix array with standard means. After that, each occurrence of the pattern suffix must be verified in the text with the previous $j - 1$ symbols. A great advantage of SamSA over SpaSA is that it performs only one binary search. On the other hand, a problem is that the pattern must contain at least one symbol from the sampled subalphabet. It was shown however that a careful selection of the subalphabet allows for leaving out over 80% suffixes and still almost preserving the pattern search speed for the standard array, if the patterns are long (50–100).

An idea most similar to ours was presented more than a decade ago by Crescenzi et al. [4,5] and was called text sparsification via local maxima. Using local maxima, that is, symbols in text which are lexicographically not smaller than the symbol just before them and lexicographically greater than the next symbol, has been recognized even earlier as a useful technique in string matching and dynamic data structures, for problems like indexing dynamic texts [1], maintaining dynamic sequences under equality tests [18] or parallel construction of suffix trees [22]. Crescenzi et al., like us, build a suffix array on sampled suffixes, yet in their experiments (only on DNA) the index compression by factor about 3 requires patterns of length at least about 150 (otherwise at least a small number of matches are lost). Our solution does not suffer a similar limitation, that is, the minimum pattern lengths with practical parameter settings are much smaller.

We use the standard notation throughout the paper. The pattern $P[1 \ldots m]$ is sought over the text $T[1 \ldots n]$. Both strings are composed of symbols from a common integer alphabet $\Sigma = \{1, \ldots, \sigma\}$. The suffix array $SA[1 \ldots n]$ built for the text T is a permutation of the indexes $1, 2, \ldots, n$ such that $T[SA[i] \ldots n] \prec T[SA[i+1] \ldots n]$ for all $1 \leq i < n$, where the "$\prec$" relation is the lexicographical order. All logarithms are in base 2.

2   Our Algorithm

2.1   The Idea

Our goal is to combine the benefits of the sparse suffix array (searching any patterns of length at least the sampling parameter q) and the sampled suffix array (one binary search).

The actual problem may be stated as follows. For each substring $T[i \ldots i + q - 1]$, $1 \leq i \leq n - q + 1$, for some fixed $q > 1$ we apply a deterministic function f selecting a substring $T[i' \ldots i' + p - 1]$, $i \leq i' \leq i + q - p$, where $1 \leq p \leq q$ is also chosen beforehand and $f(T[i \ldots i + q - 1])$ depends only on the string $T[i \ldots i + q - 1]$, not on the value of i. By f being deterministic we understand that for any two substrings of T, s_1 and s_2, of length q, if $s_1 = s_2$, then $f(s_1) = f(s_2)$. We want to have the sequence of starting positions i' of the selected substrings minimized. Note it cannot be less than $(n - q + 1)/(q - p + 1)$.

While we don't know the optimal solution for this problem, the so-called minimizers seem to be a feasible heuristic. The idea of minimizers was proposed in 2004 by Roberts et al. [21] and seemingly first overlooked in the bioinformatics (or string matching) community, only to be revived in the last years [2,6,11,17, 23]. The minimizer for a sequence s of length r is the lexicographically smallest of its all $(r-p+1)$ p-grams (or p-mers, in the term commonly used in computational biology); usually it is assumed that $p \ll r$. For a simple example, note that two DNA sequencing reads with a large overlap are likely to share the same minimizer, so they can be clustered together. That is, the smallest p-mer may be the identifier of the bucket into which the read is then dispatched.

Coming back to our algorithm: in the construction phase, we pass a sliding window of length q over T and calculate the lexicographically smallest substring of length p in each window (i.e., its minimizer). Ties are resolved in favor of the leftmost of the smallest substrings. The positions of minimizers are start positions of the sampled suffixes, which are then lexicographically sorted, like for a standard suffix array. The values of q and p, $p \le q$, are construction-time parameters.

In the actual construction, we build a standard suffix array and in extra pass over the sorted suffix indexes copy the sampled ones into a new array. This requires an extra bit array of size n for storing the sampled suffixes and in total may take $O(n)$ time and $O(n)$ words of space.

The search is simple: in the prefix $P[1 \ldots q]$ of the pattern its minimizer is first found, at some position $1 \le j \le q - p + 1$, and then we binary search the pattern suffix $P[j \ldots m]$, verifying each tentative match with its truncated $(j - 1)$-symbol prefix in the text.

Note that any other pattern window $P[i \ldots i + q - 1]$, $2 \le i \le m - q + 1$, could be chosen to find its minimizer and continue the search over the sampled suffix array, but using no such window can result in a narrower range of suffixes to verify than the one obtained from the pattern prefix. This is because for any non-empty string s with occ_s occurrences in text T, we have $occ_s \ge occ_{xs}$, where xs is the concatenation of a non-empty string x and string s.

We dub our algorithm the *sampled suffix array with minimizers* (SamSAMi).

2.2   Parameter Selection

There are two free parameters in SamSAMi, the window length q and the minimizer length p, $p \le q$. Naturally, the case of $p = q$ is trivial (all suffixes sampled, i.e. the standard suffix array obtained). For a settled p choosing a larger q has a major benefit: the expected number of selected suffixes diminishes, which reduces the space for the structure. On the other hand, it has two disadvantages: q is also the minimum pattern length, which excludes searches for shorter patterns, and for a given pattern length $m \ge q$ the average length of its sought suffix $P[j \ldots m]$ decreases, which implies more occurrence verifications. Note also, that in the worst case the number of selected suffixes is $n - q + 1$, as this happens for text $T = aa \ldots a$.

For a settled q the optimal choice of p is not easy; too small value (e.g., 1) may result in frequent changes of the minimizer, especially for a small alphabet, but on the other hand its too large value has the same effect, since a minimizer can be unchanged over at most $q - p + 1$ successive windows. Yet, the pattern suffix to be sought has in the worst case exactly p symbols, which may suggest that p should not be very small. Table 1 shows the fraction of suffixes sampled for a given (q, p) pair and four 50 MB Pizza & Chili datasets.

Table 1. The percentage of suffixes that are sampled using the idea of minimizers with the parameters q and p

q	p	dna50	english50	proteins50	xml50
4	1	46.1	39.7	40.5	45.8
4	2	55.2	51.0	51.0	54.1
5	1	40.9	32.3	34.0	39.3
5	2	44.9	39.9	40.8	45.9
6	1	37.6	27.7	29.4	32.5
6	2	38.0	32.3	34.1	39.3
8	1	33.7	22.1	23.2	22.0
8	2	29.5	23.8	25.5	26.6
10	1	31.8	19.3	19.4	17.1
10	2	24.5	18.5	20.5	18.5
10	3	25.8	20.8	22.7	21.9
12	1	30.7	17.9	16.8	13.7
12	2	21.2	15.4	17.1	15.1
12	3	21.4	16.8	18.6	17.0
16	1	29.7	16.4	13.7	11.0
16	2	17.1	12.0	12.9	11.3
16	3	16.1	12.6	13.7	11.9
24	2	13.3	8.4	8.7	7.1
24	3	11.1	8.7	9.0	7.4
32	2	11.7	6.5	6.6	5.1
32	3	8.7	6.7	6.7	5.4
40	2	10.8	5.3	5.3	4.2
40	3	7.3	5.4	5.3	4.3
64	2	9.8	2.9	3.4	3.1
64	3	5.4	3.0	3.3	2.6
64	4	4.4	3.1	3.4	2.7
80	2	9.6	1.9	2.7	2.9
80	3	4.8	1.8	2.7	2.2
80	4	3.7	1.9	2.7	2.2

2.3   Faster Verification with Previous Minimizer Position

For some texts and large value of q the number of verifications on the pattern prefix symbols tends to be large. Worse, each such verification requires a lookup to the text with a likely cache miss. We propose a simple idea reducing the references to the text.

To this end, we add an extra 4 bits to each SamSAMi offset. They store the distance to the previous sampled minimizer in the text. In other words, the list of distances corresponds to the differences between successive SamSAMi offsets in text order. For the first sampled minimizer in the text and any case where the difference exceeds 15 (i.e., could not fit 4 bits), we use the value 0. To give an example, if the sampled text positions are: 3, 10, 12, 15, 20, then the list of differences is: 0, 7, 2, 3, 5. In our application the extra 4 bits are kept in the 4 most significant bits of the offset, which restricts the 32-bit offset usage to texts up to 256 MB.

In the search phase, we start with finding the minimizer for $P[1 \ldots q]$, at some position $1 \leq \ell \leq q - p + 1$, and for each corresponding suffix from the index we read the distance to the previous minimizer in the text. If its position is aligned in front of the pattern, or the read 4 bits hold the value 0, we cannot draw any conclusion and follow with a standard verification. If however the previous minimizer falls into the area of the (aligned) pattern, in some cases we can conclude that the previous $\ell - 1$ symbols from the text do not match the $\ell - 1$ long prefix of the pattern. Let us present an example. Let $P = \text{ctgccact}$, $q = 5$, $p = 2$. The minimizer in the q long prefix of P is cc, starting at position $P[4]$. Assume that P is aligned with a match in T. If we shift the text window left by 1 symbol and consider its minimizer, it may be either ct (corresponding to $P[1 \ldots 2]$), or ?c, where ? is an unknown symbol aligned just before the pattern and c aligned with $P[1]$, if ?c happens to be lexicographically smaller than ct. If, however, the distance written on the 4 bits associated with the suffix ccact... of T is 1 or 2, we know that we have a mismatch on the pattern prefix and the verification is over (without looking up the text), since neither gc or tg cannot be the previous minimizer. Finally, if the read value is either 0 or at least 4, we cannot make use of this information and invoke a standard verification.

2.4   Faster Verification with Text Symbol Sketches

In this subsection we present another heuristic idea to reduce the number of verifications. The solution is to keep with each sampled suffix some rudimentary information about its preceding symbols in the text. More precisely, if we denote the SamSAMi structure with SA' and $SA'[beg], \ldots, SA'[end]$ is the found range of offsets to be verified, for each $SA'[i]$, $beg \leq i \leq end$, we keep a *sketch* (hash value) of k bits for each of the u symbols: $T[SA'[i] - u], \ldots, T[SA'[i] - 1]$ (with a proviso for the rare case when $SA'[i] - u < 1$). We set $k = 2$ and $u = 8$ and as the hash function we used the mapping of the symbol's byte value to a value from $\{0, 1, 2, 3\}$ dictated by its "most discriminating" pair of successive bits. For example, as the four DNA symbols, A, C, G and T, have byte values of 65, 67, 71 and 84, respectively, for a text being a genomic sequence we take their second and third lowest bits, which are respectively 00, 01, 11 and 10. The actual criterion is such that minimizes $\sum_{h=1}^{4} (n/4 - |\{T[j] : j \in \{1, \ldots, n\} \wedge hash(T[j]) = h\}|)$.

In the search phase, if the minimizer for $P[1 \ldots q]$ is at some position $1 \leq \ell \leq q - p + 1$, we compare the sketches for $P[\max(\ell - u, 1) \ldots \ell - 1]$ and $T[SA'[i] - \min(u, \ell - 1) \ldots SA'[i] - 1]$, and only if they are equal we refer to the "full" string comparison reading symbols from T. Thanks to storing the sketches together with the sampled offsets, in many (negative) cases we avoid extra cache misses, which are almost certain to happen when the verification needs to read symbols from the text. This is especially important for short patterns, when the expected range of candidate matches is large. Note that the extra space we need is $2 \cdot 8n'$ bits, i.e., $2n'$ bytes. Of course, if $q < 8$ we can decrease u e.g. to 4.

2.5 SamSAMi-hash

In [12] we showed how to augment the standard suffix array with a hash table (HT), to start the binary search from a much more narrow interval. The start and end position in the suffix array for each range of suffixes having a common prefix of length k was inserted into the HT, with the hash function calculated for the prefix string. The same function was applied to the pattern's prefix and after a HT lookup the binary search was continued with reduced number of steps. The mechanism requires $m \geq k$. To estimate the space needed by the extra table, the reader may to look at Table 2 in [12] presenting the number of distinct q-grams in five 200 MB datasets from the Pizza & Chili corpus. For example for english the number of distinct 8-grams is 20.8M, which is about 10% of the text length. This needs to be multiplied by 6.67 in our implementation (open addressing with linear probing and 90% load, and 6 bytes per entry, in a dense variant which uses 4 bytes for the left interval boundary and 2 bytes for the corresponding right boundary [13]), which results in less than $0.67n$ bytes overhead.

We can adapt this idea to SamSAMi. Again, the hashed keys will be k-long prefixes, yet now each of the sampled suffixes starts with some minimizer (or its prefix). We can thus expect a smaller overhead. Its exact value for a particular dataset depends on three parameters, k, q and p. Note however that now the pattern length m must be at least $\max(q - p + k, q)$.

3   Sparse Suffix Sorting in $O(n)$ Time and $O(n')$ Space

In this section we solve the well-known problem of sorting n' suffixes arbitrarily chosen from a text T of length n in $O(n)$ time using $O(n')$ extra space. Our algorithm is deterministic and works for an integer alphabet of size $\sigma = n^{O(1)}$. The best known solutions for this problem are two randomized algorithms given by I et al. [15]: a Monte Carlo algorithm running in $O(n)$ time and using $O(n' \log n')$ space and a Las Vegas one with $O(n \log n')$ time and $O(n')$ space.

We make use of the Ferragina and Fischer [7] technique, devised for sorting word-based suffixes, and a lemma from a recent work of Fischer et al. [9].

Lemma 1 ([9]). *One can lexicographically sort strings $P_1, \ldots, P_{n'}$ of total length n in $O(n + \sigma^\varepsilon)$ time using $O(n')$ space, for any $\varepsilon > 0$.*

We present our procedure in the five following steps.

1. The input text T is (conceptually) partitioned into n' consecutive strings P_i, $1 \leq i \leq n'$, where each string is accompanied with its start position in T, stored in the array I as $I[i]$. The strings P_i, together with their i indices, are lexicographically sorted using Lemma 1. Note that the $O(n + \sigma^\varepsilon)$ time remains $O(n)$ for $\sigma = n^{O(1)}$, with a properly chosen constant $\varepsilon > 0$.
2. After the sort, we scan over the strings and assign to them bucket labels. To this end, the first (lexicographically smallest) string gets label 1, and if the jth sorted string, $j > 1$, is equal to the $(j - 1)$th string, it belongs to the

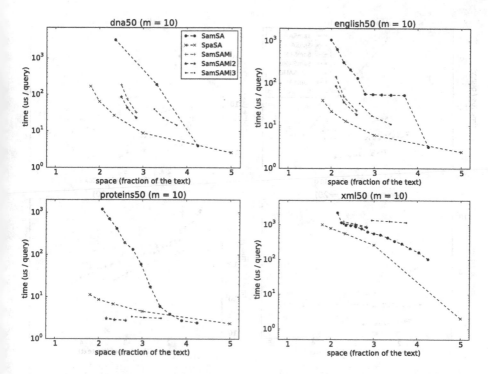

Fig. 1. Pattern search time (count query). All times are averages over 500K random patterns of length 10. The patterns were extracted from the texts. Times are given in microseconds. The index space is a multiple of the text size, including the text.

bucket of its predecessor, otherwise the bucket label is incremented by one and the jth string is assigned to a new bucket. Additionally, we create an array $B[1 \dots n']$, setting $B[i] = k$, where k is the bucket number to which the current string is sent and i is the index assigned to the string in step 1. These operations thake $O(n)$ time.

3. In $O(n')$ time we construct a new text T' of length n', using the formula $T'[i] = B[i], 1 \leq i \leq n'$.
4. We sort the suffixes of T' in $O(n')$ time and using $O(n')$ space, using any linear-time suffix sorting algorithm; as a result the suffix array SA for T' is obtained.
5. Finally, in $O(n')$ time we create the word suffix array A from SA, scanning it from left to right and writing the corresponding suffix to A: $A[i] = I[SA[i]]$.

The last three steps are analogous to the final steps in the Ferragina and Fischer [7] algorithm. We point out, however, that the first step of their algorithm, sorting the sampled suffixes according to their first *word* only, is performed with the MSD radix sort, which splits buckets of suffixes recursively. As the number of resulting buckets is n' (being the number of space-separated words in the text in their scenario), collecting them from a bucket array of size σ may be a bottleneck if $\sigma = \omega(1)$. Using Lemma 1 allows to overcome this limitation.

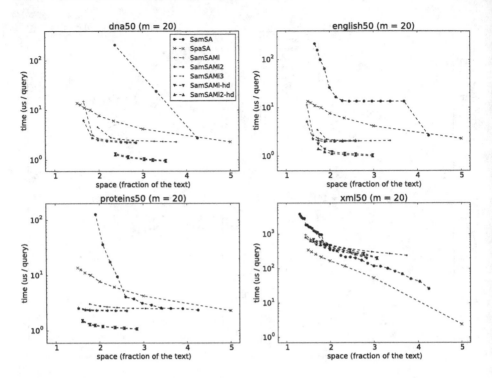

Fig. 2. Pattern search time (count query). All times are averages over 500K random patterns of length 20. The patterns were extracted from the texts. Times are given in microseconds. The index space is a multiple of the text size, including the text.

As Lemma 1 does not require string terminators, the described procedure also works for an arbitrarily sampled set of n' suffixes. We thus obtain the following theorem.

Theorem 2. *Any set of n' suffixes of a string of length n over an alphabet of size $\sigma = n^{O(1)}$ can be sorted in $O(n)$ time using $O(n')$ words of space.*

4   Experimental Results

We have implemented three variants of the SamSAMi index: the basic one (denoted as SamSAMi on the plots), the one with reduced verifications due to storing previous minimizer positions (SamSAMi2) and the one combining SamSAMi2 with the sketches from Sect. 2.4, denoted as SamSAMi3. These algorithms are additionally tested in a variant augmented with a hash table, in its dense version (with the -hd suffix in the name); only the variant SamSAMi3-hd is not shown on the charts as it was not competitive. We compared them against the sparse suffix array (SpaSA), in our implementation, and the sampled suffix array (SamSA) [3], using the code provided by its authors.

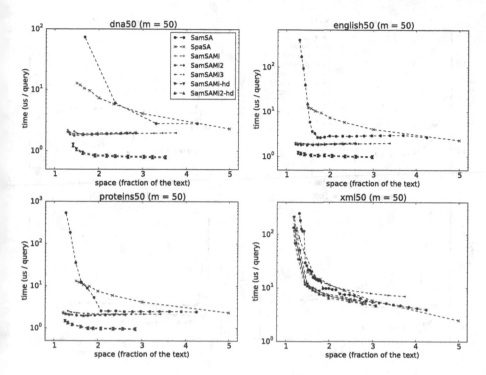

Fig. 3. Pattern search time (count query). All times are averages over 500K random patterns of length 50. The patterns were extracted from the texts. Times are given in microseconds. The index space is a multiple of the text size, including the text.

All experiments were run on a computer with an Intel i7-4930K 3.4 GHz CPU, equipped with 64 GB of DDR3 RAM and running Ubuntu 14.04 LTS 64-bit. All codes were written in C++ and compiled with g++ 4.8.2 with -O3 option.

Pattern searches were run for $m \in \{10, 20, 50, 100\}$, and for each dataset and pattern length 500,000 randomly extracted patterns from the text were used. Figs 1–4 present average search times with respect to varying parameters. For SpaSA we changed its parameter k from 1 (which corresponds to the plain suffix array) to 8. For SamSAMi we varied q from $\{4, 5, 6, 8, 10, 12, 16, 24, 32, 40, 64, 80\}$ setting the most appropriate p (up to 3 or 4) to obtain the smallest index, according to the statistics from Table 1. Obviously, q was limited for $m < 100$; up to 6 for $m = 10$, up to 16 for $m = 20$, and up to 40 for $m = 50$.

We note that SamSAMi is rather competitive against the sparse suffix array, with two exceptions: short patterns ($m = 10$) and the XML dataset (for $m = 10$ and $m = 20$). In most cases, SamSAMi is also competitive against the sampled suffix array, especially when aggressive suffix sampling is applied. (For a honest comparison one should also notice that our implementation uses 32-bit suffix indexes while the Claude et al. scheme was tested with $\lceil \log_2 n \rceil$ bits per index, which is 26 bits for the used datasets.)

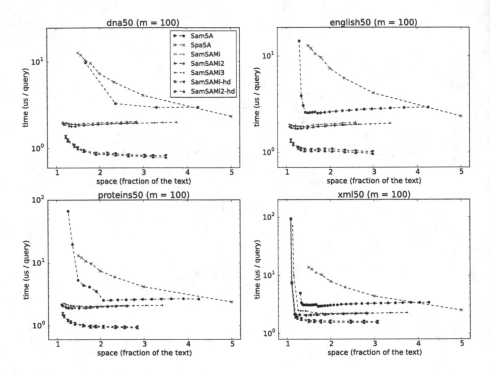

Fig. 4. Pattern search time (count query). All times are averages over 500K random patterns of length 100. The patterns were extracted from the texts. Times are given in microseconds. The index space is a multiple of the text size, including the text.

Unfortunately, the variant with reduced verifications (`SamSAMi2`) is not significantly faster than the original one, only in rare cases, with a large value of the used q, the search time can be approximately halved. Also `SamSAMi3` is competitive with other SamSAMi variants only on the English and DNA datasets for $m = 10$. `SamSAMi-hd`, on the other hand, can be an attractive alternative, similarly as SA-hash used as a replacement of the plain SA [12].

5   Conclusions and Future Work

We presented a simple suffix sampling scheme making it possible to search for patterns effectively. The resulting data structure, called a sampled suffix array with minimizers (SamSAMi), achieves interesting time-space tradeoffs; for example, on English50 dataset the search for patterns of length 50 is still over 10% faster than with a plain suffix array when only 5.3% of the suffixes are retained.

Additionally, based on previous works we show that the well-known problem of sorting n' arbitrarily chosen suffixes for a text of length n can be solved in $O(n)$ time and $O(n')$ words of space. This poses an immediate question if building the SamSAMi structure can also be done in $O(n)$ time and $O(n')$ space, which requires finding the minimizers efficiently, in constant amortized time over a window of size q sliding over the text.

In practice we can assume that the minimizer length, p, is a (small) constant and the solution is trivial. It is however more interesting to consider the case of an arbitrary p. Note there exists a related problem of finding the minimizer in a static sequence (i.e., a pattern prefix), as it is required at the beginning of the pattern search.

For the latter problem variant, we may use a rather theoretical option involving linear-time suffix sorting. To this end, we first remap the pattern prefix alphabet to (at most) $\{0, 1, \ldots, q-1\}$, in $O(q \log q)$ time, using a balanced BST. Next we build a suffix array over the remapped string, using a linear-time suffix sorting algorithm, and finally scan over the sorted list of suffixes until the first suffix of length at least p is found. The total time is $O(q \log q)$. The well-known trick of dividing the window into subwindows of size $2p - 1$ with $(p-1)$-long overlaps improves the time to $O(q \log p)$. The remapping may be done even faster, using the fastest deterministic integer sort by Han [14], to obtain $O(q \log \log p)$ overall time. Still, it is possible to get the optimal $O(n)$ and $O(q)$ times, respectively, for these two variants, as was pointed out to us by Paweł Gawrychowski. The idea will be presented in the full version of the paper.

Acknowledgments. We thank Paweł Gawrychowski for sharing the idea(s) on fast minimizer finding, Travis Gagie for interesting discussion on related issues, Kimmo Fredriksson for helpful comments over the whole manuscript and Francisco Claude for sharing with us the sampled suffix array sources. The work was supported by the Polish National Science Centre under the project DEC-2013/09/B/ST6/03117 (both authors).

References

1. Alstrup, S., Brodal, G.S., Rauhe, T.: Pattern matching in dynamic texts. In: SODA, pp. 819–828. Society for Industrial and Applied Mathematics (2000)
2. Chikhi, R., Limasset, A., Jackman, S., Simpson, J.T., Medvedev, P.: On the representation of de Bruijn graphs. Journal of Computational Biology **22**(5), 336–352 (2015)
3. Claude, F., Navarro, G., Peltola, H., Salmela, L., Tarhio, J.: String matching with alphabet sampling. Journal of Discrete Algorithms **11**, 37–50 (2012)
4. Crescenzi, P., Del Lungo, A., Grossi, R., Lodi, E., Pagli, L., Rossi, G.: Text sparsification via local maxima. In: Kapoor, S., Prasad, S. (eds.) FST TCS 2000. LNCS, vol. 1974, pp. 290–301. Springer, Heidelberg (2000)
5. Crescenzi, P., Lungo, A.D., Grossi, R., Lodi, E., Pagli, L., Rossi, G.: Text sparsification via local maxima. Theoretical Computer Science **1–3**(304), 341–364 (2003)
6. Deorowicz, S., Kokot, M., Grabowski, S., Debudaj-Grabysz, A.: KMC 2: fast and resource-frugal k-mer counting. Bioinformatics **31**(10), 1569–1576 (2015)
7. Ferragina, P., Fischer, J.: Suffix arrays on words. In: Ma, B., Zhang, K. (eds.) CPM 2007. LNCS, vol. 4580, pp. 328–339. Springer, Heidelberg (2007)
8. Ferragina, P., González, R., Navarro, G., Venturini, R.: Compressed text indexes: From theory to practice. ACM Journal of Experimental Algorithmics **13**, article 12, 30 (2009)

9. Fischer, J., Gagie, T., Gawrychowski, P., Kociumaka, T.: Approximating LZ77 via small-space multiple-pattern matching. CoRR, abs/1504.06647 (2015)

10. Gog, S., Petri, M.: Optimized succinct data structures for massive data. Software-Practice and Experience 44(11), 1287–1314 (2014)

11. Grabowski, S., Deorowicz, S., Roguski, Ł.: Disk-based compression of data from genome sequencing. Bioinformatics 31(9), 1389–1395 (2015)

12. Grabowski, S., Raniszewski, M.: Two simple full-text indexes based on the suffix array. In: Holub, J., Zdárek, J. (eds.) PSC, pp. 179–191. Faculty of Information Technology, Czech Technical University in Prague, Department of Theoretical Computer Science (2014)

13. Grabowski, S., Raniszewski, M.: Two simple full-text indexes based on the suffix array (2015). Submitted to a journal

14. Han,Y.: Deterministic sorting in $O(n \log \log n)$ time and linear space. In: STOC, pp. 602–608. ACM (2002)

15. Tomohiro, I., Kärkkäinen, J., Kempa, D.: Faster sparse suffix sorting. In: Mayr, E.W., Portier, N. (eds.) STACS. LIPIcs, vol. 25, pp. 386–396. Schloss Dagstuhl - Leibniz-Zentrum fuer Informatik (2014)

16. Kärkkäinen, J., Ukkonen, E.: Sparse suffix trees. In: Cai, J.-Y., Wong, C.K. (eds.) COCOON 1996. LNCS, vol. 1090, pp. 219–230. Springer, Heidelberg (1996)

17. Li, Y., Kamousi, P., Han, F., Yang, S., Yan, X., Suri, S.: Memory efficient minimum substring partitioning. In: VLDB, pp. 169–180. VLDB Endowment (2013)

18. Mehlhorn, K., Sundar, R., Uhrig, C.: Maintaining dynamic sequences under equality tests in polylogarithmic time. Algorithmica 17(2), 183–198 (1997)

19. Navarro, G., Mäkinen, V.: Compressed full-text indexes. ACM Computing Surveys 39(1), article 2 (2007)

20. Puglisi, S.J., Smyth, W.F., Turpin, A.: Inverted files versus suffix arrays for locating patterns in primary memory. In: Crestani, F., Ferragina, P., Sanderson, M. (eds.) SPIRE 2006. LNCS, vol. 4209, pp. 122–133. Springer, Heidelberg (2006)

21. Roberts, M., Hayes, W., Hunt, B.R., Mount, S.M., Yorke, J.A.: Reducing storage requirements for biological sequence comparison. Bioinformatics 20(18), 3363–3369 (2004)

22. Sahinalp, S.C., Vishkin, U.: Symmetry breaking for suffix tree construction. In: STOC, pp. 300–309. ACM (1994)

23. Wood, D.E., Salzberg, S.L.: Kraken: ultrafast metagenomic sequence classification using exact alignments. Genome Biology 15(3), R46 (2014)

Longest Common Prefix with Mismatches

Giovanni Manzini[✉]

Computer Science Institute, University of Eastern Piedmont, Vercelli, Italy
giovanni.manzini@unipmn.it

Abstract. The Longest Common Prefix (LCP) array is a data structure commonly used in combination with the Suffix Array. However, in some settings we are interested in the LCP values *per se* since they provide useful information on the repetitiveness of the underlying sequence.

Since sequences can contain alterations, which can be either malicious (plagiarism attempts) or pseudo-random (as in sequencing experiments), when using LCP values to measure repetitiveness it makes sense to allow for a small number of errors. In this paper we formalize this notion by considering the longest common prefix in the presence of mismatches. In particular, we propose an algorithm that computes, for each text suffix, the length of its longest prefix that occurs elsewhere in the text with at most one mismatch. For a sequence of length n our algorithm uses $\Theta(n \log n)$ bits and runs in $\mathcal{O}(nL_{ave} \log n / \log \log n)$ time where L_{ave} is the average LCP of the input sequence. Although L_{ave} is $\Theta(n)$ in the worst case, recent analyses of real world data show that it usually grows logarithmically with the input size. We then describe and analyse a second algorithm that uses a greedy strategy to reduce the amount of computation and that can be turned into an even faster algorithm if allow an additive one-sided error.

Finally, we consider the related problem of computing the 1-mappability of a sequence. In this problem we are asked to compute, for each length-m substring of the input sequence, the number of other substrings which are at Hamming distance one. For this problem we propose an algorithm that takes $\mathcal{O}(mn \log n / \log \log n)$ time using $\Theta(n \log n)$ bits of space.

1 Introduction

The longest-common-prefix (LCP) array [20] stores the lengths of the longest-common-prefix between two lexicographically adjacent suffixes of a given sequence. The LCP array has been mostly used to speedup algorithms based on the Suffix Array: it is well known that combining a (compressed) Suffix Array with a (compressed) LCP array we get a data structure as powerful as the Suffix Tree [13] using considerably less space. This application motivates the many recent results on the construction and storage of the LCP array [3,5,9,12,16].

However, there are problems in which we are interested in the LCP values *per se*, that is, without reference to the Suffix Array. Indeed, LCP values give us important information on the repetitiveness of the underlying text and they

© Springer International Publishing Switzerland 2015
C. Iliopoulos et al. (Eds.): SPIRE 2015, LNCS 9309, pp. 299–310, 2015.
DOI: 10.1007/978-3-319-23826-5_29

are useful for the analysis of both natural language texts (e.g. plagiarism detection [17]) and biosequences (e.g. the mappability problem discussed in Section 6).

It is also well known that sequences can contain alterations which can be malicious (computer viruses or plagiarism attempts), due to natural causes (DNA replication) or instrumental errors (sequencing experiments). It is therefore interesting to analyze the LCP values when errors are allowed in the sequence. In this context it is natural to define the longest common prefix with k mismatches. Informally, given a sequence $t[0..n]$, the LCP with k mismatches of the suffix $t[i..n]$ is the length of the longest prefix of $t[i..n]$ that appears elsewhere in t with up to k mismatches (see formal definition in Section 2).

When we enter in the realm of inexact matches even simple problems are extremely challenging so in this paper we consider the problem of computing the longest common prefix with a single mismatch. In Section 4 we describe a simple algorithm for this problem that uses $\mathcal{O}(n \log n)$ bits of space and runs in $\mathcal{O}(n L_{ave} \log n / \log \log n)$ time, where L_{ave} is the average LCP (without mismatches!) of the input sequence. Although in the worst case $L_{ave} = \Theta(n)$ the extensive experimental analysis in [19] shows that for real world data, including both natural language and biosequences, L_{ave} usually grows as $\Theta(\log n)$ so we can expect our algorithm to be usable also for large collections.

In Section 5 we show how to improve our basic algorithm using a greedy strategy. Although the greedy strategy provably reduces the overall amount of computation, in some pathological cases the running time is still $\mathcal{O}(n L_{ave} \log n / \log \log n)$. An additional feature of the greedy algorithm is that we can reduce its worst case running time by a factor q at the cost of an additive one-sided error of at most q (see Lemma 7).

Finally, in Section 6 we show how our techniques can be applied to the problem of determining the mappability of a genomic region (see [8] and references therein). Given a sequence t and a parameter m, the k-mappability problem consists in determining, for each length-m substring α of t, the number of substrings of t which are at Hamming distance at most k from α. The mappability information plays an important role in NGS data analysis but for $k > 0$ the bioinformatics community still relies on heuristic tools (see discussion in Section 6). We propose an algorithm for the computation of the 1-mappability that runs in $\mathcal{O}(mn \log n / \log \log n)$ times using $\mathcal{O}(n \log n)$ bits of space.

2 Notation and Problem Definition

Throughout we consider a string $t = t[0..n] = t[0]t[1]\ldots t[n]$ of $n + 1$ symbols drawn from a constant ordered alphabet Σ. We implicitly assume that t has an additional "end-of-string" character $t[n + 1] = \$$ which is distinct from and lexicographically smaller than all the other characters in Σ.

For $i = 0, \ldots, n$ we write $t[i..n]$ to denote the **suffix** of t of length $n - i + 1$, that is $t[i..n] = t[i]t[i + 1] \cdots t[n]$. For convenience we will frequently refer to suffix $t[i..n]$ simply as "suffix i". Similarly, we write $t[0..i]$ to denote the **prefix** of t of length $i + 1$. We write $t[i..j]$ to represent the **substring** $t[i]t[i + 1] \cdots t[j]$ of t that starts at position i and ends at position j.

The **suffix array** of t is an array $SA[0..n]$ which contains a permutation of the integers $0..n$ such that $t[SA[0]..n] < t[SA[1]..n] < \cdots < t[SA[n]..n]$. In other words, $SA[j] = i$ iff $t[i..n]$ is the jth suffix of t in ascending lexicographical order. Similarly, we define the **prefix array** of t as the array $PA[0..n]$ which contains a permutation of the integers $0..n$ such that $PA[j] = i$ iff $t[0..i]$ is the jth prefix of t in ascending lexicographical (right-to-left) order.

The **lcp array** $LCP = LCP[0..n]$ is an array defined by t and SA. Let $lcp(y, z)$ denote the length of the longest common prefix of strings y and z. For every $j \in 1..n$,

$$LCP[j] = lcp(t[SA[j]..n], t[SA[j-1]..n]),$$

that is, $LCP[j]$ is the length of the longest common prefix of suffixes $SA[j]$ and $SA[j-1]$. $LCP[0]$ is undefined. The **permuted lcp array** [15,22,25], denoted by $PLCP[0..n-1]$, has the same contents as LCP but in different order. Let i^- (resp. i^+) denote the starting position of the lexicographic predecessor (resp. successor) of $t[i..n]$. For $i = 0, \ldots, n$ we define

$$PLCP[i] = LCP[SA^{-1}[i]] = lcp(t[i..n], t[i^-..n]),$$

that is, $PLCP[i]$ is the length of the longest common prefix between $t[i..n]$ and its lexicographic predecessor. Since in this paper we are interested in LCP values *per se*, without reference to the suffix array, we will mostly consider LCP values in this alternative ordering.

For any $k \geq 0$ we define $lcp_k(y, z)$ as the largest $\ell \geq 0$ such that $y[0..\ell - 1]$ and $z[0..\ell - 1]$ exist and are at Hamming distance $\leq k$, that is, differ by at most k mismatches. We define the **permuted lcp array with k mismatches** —$PLCP_k[0..n]$— as follows. For $i = 0, \ldots, n$

$$PLCP_k[i] = \max_{j=0..n-1,\ j\neq i} lcp_k(t[i..n], t[j..n]). \tag{1}$$

Note that, although $lcp_0(y, z) = lcp(y, z)$, it is $PLCP_0[i] \geq PLCP[i]$ since for $PLCP_0$ we consider both the lexicographic predecessor and successor, that is, $PLCP_0[i] = \max(lcp(t[i..n], t[i^-..n]), lcp(t[i..n], t[i^+..n]))$.

Let $E[i] = i + PLCP_0[i] - 1$. By definition of $PLCP_0$ the substring $t[i..E[i]]$ occurs at least twice in t, while $t[i..E[i] + 1]$ occurs only once. Hence, the prefix of $t[i..n]$ defining $PLCP_0[i]$ ends in position $E[i]$. With a little abuse of notation we say that $E[i]$ is the **ending point** of $PLCP_0[i]$.

In this paper we consider the problem of computing the array $PLCP_1$, that is, for $i = 0..n$ we want to find the longest prefix of $t[i..n]$ that appears elsewhere in t with at most one mismatch. Clearly it is $PLCP_1[i] \geq PLCP_0[i] + 1$ since we can simply place the mismatch in position $E[i] + 1$, but $PLCP_1[i]$ can be much greater than $PLCP_0[i]$. For example for $t = a^2 b^{n-2}$ it is $PLCP_0[0] = 1$ and $PLCP_1[i] = n - 1$.

3   Auxiliary Data Structures

In this section we describe some auxiliary data structures used in the paper. To keep the description of our algorithms simple, we are very liberal in the use of these data structures. With a careful implementation of our algorithms some of the auxiliary data structures can be discarded or replaced by some alternatives taking less space (see for example the discussion at the end of Section 4). We do not account for the time for the construction of the auxiliary data structures since it is dominated by the time complexity of our algorithms.

One of the basic data structures used in the literature to handle searches in $t[0..n]$ with one mismatch is the two-dimensional grid $[0..n] \times [0..n]$ containing a point in position (i, k) iff there exists a text position $t[j]$ such that $t[0..j - 1]$ is the i-th smallest lexicographic prefix and $t[j + 1..n]$ is the k-th largest lexicographic suffix, i.e., $PA[i] = j - 1$ and $SA[k] = j + 1$. This data structure was introduced more than fifteen years ago [1] but is still widely used [7,14]. Using this two-dimensional grid, problems involving single mismatches (or gaps) can be solved via orthogonal range queries (or related operations, see [4] and references therein). Note there are other, more space consuming, approaches to indexing with mismatches that can be used to solve the problems considered in this paper with different time/space tradeoffs (see [11] and references therein).

For our convenience, we represent the above two-dimensional grid with an array. Formally for $j = 0, \ldots, n$ we define the arrays $PA^*[0..n]$ and $SA^*[0..n]$ as

$$PA^*[PA^{-1}[j - 1]] = SA^{-1}[j + 1], \tag{2}$$
$$SA^*[SA^{-1}[j + 1]] = PA^{-1}[j - 1]. \tag{3}$$

Note that $PA^*[i]$ is the rank in the suffix array of the suffix that starts *two* positions after the end of the prefix that has rank i in the prefix array (that is, $SA[PA^*[i]] - PA[i] = 2$). A symmetric interpretation holds for SA^*.

For the arrays PA^* and SA^* we build the $\mathcal{O}(n \log n)$ bits data structures described in [2] to support the rangeNext and rangePrev queries in $\mathcal{O}(\log n / \log \log n)$ time. Recall that given an array $A[0..n]$ on the domain $[0..n]$, rangeNext(A, i, j, v) (resp. rangePrev(A, i, j, v)) returns the smallest (resp. largest) element in $A[i..j]$ larger (resp. smaller) than v.

Symmetrically to the LCP array we define the **longest common suffix** array LCS such that $LCS[j]$ is the longest common suffix of the prefixes $PA[j]$ and $PA[j - 1]$. We assume we have a $\mathcal{O}(n \log n)$ bits representation of the LCP and LCS arrays supporting the prevLess, nextLess and RMQ queries in $\mathcal{O}(\log n / \log \log n)$ time (see for example [2,10]). Given an array $A[0..n]$ on the domain $[0..n]$, prevLess(A, i, v) (resp. nextLess(A, i, v)) returns the rightmost element in $A[0..i]$ (resp. leftmost element in $A[i..n]$) that is smaller than v, while RMQ(A, i, j) returns the smallest element in $A[i..j]$.

Definition 1. *Given a text $t[0..n]$ and indices $0 \le i \le j \le k \le n$ we write*

$$\langle t[i..j - 1] * t[j + 1..k] \subseteq t \rangle$$

if the substring $t[i..k]$ appears elsewhere in t with possibly a mismatch in position j. Stated more formally, we require that there exists $\delta \ne 0$ with $-i \le \delta \le n - k$ such that $t[i..j - 1] = t[i + \delta..j - 1 + \delta]$ and $t[j + 1..k] = t[j + 1 + \delta..k + \delta]$. □

Lemma 1. *There exists a $\mathcal{O}(n \log n)$ bits data structure such that given a substring $t[i..j-1]$ we can find in $\mathcal{O}(\log n / \log \log n)$ time the largest k such that $\langle t[i..j-1] * t[j+1..k] \subseteq t \rangle$.*

Proof. Given the index $j-1$ we retrive the position $p_j = \mathrm{PA}^{-1}[j-1]$ of the prefix array containing the prefix $t[0..j-1]$. Then, with a prevLess and nextLess query on the LCS array we find the range $[a,b]$ of prefix array entries that have a common prefix of at least $j-i$ with $t[i..j-1]$; these are all the prefixes of t that ends with the substring $t[i..j-1]$.

Let $s_j = \mathrm{SA}^{-1}[j+1]$ denote the position in SA of suffix $t[j+1..n]$, and let

$$s_j^+ = \min(\mathsf{rangeNext}(\mathrm{PA}^*, a, p_j - 1, s_j), \mathsf{rangeNext}(\mathrm{PA}^*, p_j + 1, b, s_j)),$$

$$s_j^- = \max(\mathsf{rangePrev}(\mathrm{PA}^*, a, p_j - 1, s_j), \mathsf{rangePrev}(\mathrm{PA}^*, p_j + 1, b, s_j))$$

By construction s_j^+ is the smaller value in $\mathrm{PA}^*[a..b]$ larger than s_j, and s_j^- is the larger value smaller than s_j (note $\mathrm{PA}^*[p_j] = s_j$). Hence, among all suffixes preceded by the substring $t[i..j-1]*$, with $*$ an arbitrary character, $t[\mathrm{SA}[s_j^+]..n]$ (resp. $t[\mathrm{SA}[s_j^-]..n]$) is the one immediately following (resp. preceding) suffix $t[j+1..n]$ in the lexicographic order. The desired value k is therefore given by

$$k = j + \max(\mathrm{RMQ}(\mathrm{LCP}, s_j^-, s_j), \mathrm{RMQ}(\mathrm{LCP}, s_j + 1, s_j^+)). \qquad (4)$$

since $\mathrm{RMQ}(\mathrm{LCP}, i, j)$ is the length of the longest common prefix between suffixes $t[\mathrm{SA}[i]..n]$ and $t[\mathrm{SA}[j]..n]$. □

Reversing the roles of the suffix array and prefix array and using SA^* instead of PA^* we get

Lemma 2. *There exists a $\mathcal{O}(n \log n)$ bits data structure such that given a substring $t[j+1..k]$ we can find in $\mathcal{O}(\log n / \log \log n)$ time the smallest i such that $\langle t[i..j-1] * t[j+1..k] \subseteq t \rangle$.* □

4 A Simple Algorithm for Computing PLCP_1

In this section we describe a simple algorithm for computing the array PLCP_1. Our starting point is the following characterization of $\mathrm{PLCP}_1[i]$. Recall that $E[i] = i + \mathrm{PLCP}_0[i] - 1$ is the endpoint of $\mathrm{PLCP}_0[i]$ in the sense that the longest prefix of $t[i..n]$ that appears elsewhere in t ends in position $E[i]$. If $E[i] = n-1$ then $\mathrm{PLCP}_1[i] = \mathrm{PLCP}_0[i] + 1$ since the best we can do is to add a mismatch in position $E[i]+1$. In the general case when $E[i] < n-1$, we observe that computing $\mathrm{PLCP}_1[i]$ is equivalent to computing the largest k such that the substring $t[i..k]$ appears with one mismatch somewhere else in t. Notice that the mismatch must appear in a position j with $i \leq j \leq E[i] + 1$, since if it were $j > E[i] + 1$ we would have a second mismatch in $E[i] + 1$. Putting these observations together, to obtain $\mathrm{PLCP}_1[i]$ we need to compute for $j = i, i+1, \ldots, E[i]+1$ the largest k such that $\langle t[i..j-1] * t[j+1..k] \subseteq t \rangle$. This simple strategy is implemented in the loop of Lines 3–5 in the algorithm in Fig. 4.

1: **for** $i = 0$ **to** n **do**
2: $\quad\quad$ $e_i \leftarrow E[i] + 1$
3: $\quad\quad$ **for** $j = i$ **to** $E[i] + 1$ **do**
4: $\quad\quad\quad\quad$ find the largest k such that $\langle t[i..j-1]*t[j+1..k] \subseteq t \rangle$
5: $\quad\quad\quad\quad$ $e_i \leftarrow \max(e_i, k)$
6: $\quad\quad$ $\text{PLCP}_1[i] \leftarrow e_i - i + 1$

Fig. 1. Simple algorithm for computing the PLCP_1 array.

Theorem 3. *Let $\ell_i = \text{PLCP}_0[i]$, and let $L_{ave} = (\sum_{i=0}^{n} \ell_i)/n$ denote the average LCP of the text t. The algorithm in Fig. 4 computes the PLCP_1 array in $\mathcal{O}(nL_{ave} \log n / \log \log n)$ time using $\mathcal{O}(n \log n)$ bits of space.*

Proof. The loop of Lines 3–5 executes $E[i] + 1 - i = \text{PLCP}_0[i]$ iterations in which at Line 4 we compute the value k in $\mathcal{O}(\log n / \log \log n)$ time using Lemma 1. Summing over all positions i we get the desired time bound.

To prove the space bound, in addition to the $\mathcal{O}(n \log n)$ bits data structures of Lemma 1 the algorithm only need the values $E[i]$. Since it is $\text{PLCP}_0[i] \geq \text{PLCP}_0[i-1] - 1$ using the representation of [23] (and constant time binary select) we can store the PLCP_0 array in $2n + o(n)$ bits and have a constant time access to $\text{PLCP}_0[i]$ and hence to $E[i] = i + \text{PLCP}_0[i] - 1$. $\quad\square$

As already discussed in the introduction, in the worst case it is $L_{ave} = \Theta(n)$ but the extensive empirical analysis in [19] suggests that in real world applications L_{ave} grows as $\Theta(\log n)$.

Note that the actual implementation of the above algorithm does not need all the data structures described in Section 3. Using a bidirectional compressed index [18,24] we can maintain in $\mathcal{O}(1)$ amortized time the values $PA^{-1}[j-1]$, $SA^{-1}[j+1]$ and the ranges of PA rows prefixed by $t[i..j-1]$ that are needed in Lemma 1. With additional $\mathcal{O}(n \log \log n)$ bits we can support $\mathcal{O}(\log n / \log \log n)$ time RMQ queries on the LCP values required by (4). The only data structure for which we apparently need to use $\Theta(n \log n)$ bits is the array PA* with the support for the rangeNext and rangePrev operations in $\mathcal{O}(\log n / \log \log n)$ time.

5   A Greedy Algorithm for PLCP_1

The algorithm in the previous section computes each entry of the PLCP_1 array independently. It is natural to try to improve it by considering simultaneously many PLCP_1 entries. To this end, instead of considering for each position i the location of all possible mismatches influencing $\text{PLCP}_1[i]$, we loop only once over all possible mismatch positions j and we update (an estimate of) all $\text{PLCP}_1[i]$ values that can be influenced by a mismatch in position j.

Let $w^{(j)}$ be the smallest index such that $E[w^{(j)}] + 1 \geq j$. As we observed in the previous section a mismatch in position j can influence the values $\text{PLCP}_1[i]$ for $w^{(j)} \leq i \leq j$. Hence, for any given j we should compute for $i = w^{(j)}, \ldots, j$

```
1:    for i = 0 to n do            // Init e_i values
2:        e_i ← E[i] + 1
3:    for j = 0 to n do            // Loop on all possible mismatch positions
4:        g ← j
5:        while g ≤ E[j + 1] do
6:            find the smallest ℓ such that ⟨t[ℓ..j − 1] * t[j + 1..g] ⊆ t⟩
7:            find the largest h such that ⟨t[ℓ..j − 1] * t[j + 1..h] ⊆ t⟩
8:            e_ℓ ← max(e_ℓ, h)
9:            g ← h + 1
10:   for i = 1 to n do            // Compute endpoints
11:       e_i = max(e_i, e_{i−1})
12:   for i = 0 to n do            // Compute PLCP_1 values
13:       PLCP_1[i] = e_i − i + 1
```

Fig. 2. Greedy algorithm for computing the PLCP_1 array.

the largest $k_i^{(j)}$ such that $\langle t[i..j-1] * t[j+1..k_i^{(j)}] \subseteq t\rangle$. From these values the PLCP_1 array can be obtained by the same reasoning as in the previous section.

Lemma 4. *For* $i = 0, \ldots, n$ *it is*

$$\mathrm{PLCP}_1[i] = \left(\max_{i \leq j \leq E[i]+1} k_i^{(j)} \right) - i + 1. \tag{5}$$

Proof. The maximum in (5) coincides with the largest k such that $t[i..k]$ appears elsewhere in t with one mismatch. □

The greedy algorithm in Fig. 5 is based on Lemma 4. It maintains the values $e_0, \ldots, e_{n-1}$ such that e_i is always a lower bound of the maximum in (5), and at the end (Lines 10–12) it outputs the PLCP_1 array. To prove the correctness of the algorithm we need to analyze its crucial part, that is, the *while* loop at Lines 5–9. Consider a fixed mismatch position j, with $0 \leq j \leq n$. As before we define[1] w to be the smallest index such that $E[w] + 1 \geq j$. For $w \leq i \leq j$ let k_i denote the largest value such that $\langle t[i..j-1] * t[j+1..k_i] \subseteq t\rangle$. It is immediate to see that the sequence of values k_i's is non decreasing, that is:

$$k_w \leq k_{w+1} \leq \cdots \leq k_j. \tag{6}$$

In addition, it is $k_j = E[j+1]$, since $t[j..j-1]$ is the empty string hence k_j is the largest index for which $t[j+1..k]$ appears elsewhere in t and this is $E[j+1]$.

To completely determine the sequence (6) it suffices to know the indices v_i and the values k_{v_i} where the sequence is strictly increasing. Formally, we define $v_1, \ldots, v_d$ such that $w = v_1 < v_2 < \cdots < v_d \leq j$ and for $i = 1, \ldots, d-1$

$$k_{v_i} = k_{v_i+1} = \cdots = k_{v_{i+1}-1} < k_{v_{i+1}} \tag{7}$$

and

---

[1] To simplify the notation we drop the superscript j from $w^{(j)}$ and $k_i^{(j)}$.

$$k_{v_d} = k_{v_d+1} = \cdots = k_j = E[j+1].$$

Clearly, knowing the pairs $(v_1, k_{v_1}), \ldots, (v_d, k_{v_d})$, is equivalent to knowing the whole sequence $k_w, k_{w+1}, \ldots, k_j$. The following lemma shows that the *while* loop in the greedy algorithm of Fig. 5 computes precisely these pairs .

Lemma 5. *For all j, the while loop at Lines 5–9 of the greedy algorithm computes the pairs $(v_1, k_{v_1}), \ldots, (v_d, k_{v_d})$ in the sense that the pair of values (ℓ, h) computed at Lines 6–7 at the i-th iteration of the loop coincides with (v_i, k_{v_i}).*

Proof. Let us denote by (ℓ_i, h_i) the pair of values computed at Lines 6 and 7 during the i-th iteration of the *while* loop. To prove the lemma it suffices to show that the values ℓ_i's coincide with the v_i's since for each i both h_i and k_{v_i} are then defined as the largest k such that $\langle t[v_i..j-1] * t[j+1..k] \subseteq t \rangle$.

At the first iteration it is $g = j$ so $t[j+1..g]$ is the empty string and at Line 6 we compute ℓ_1 as the smallest index such that the substring $t[\ell_1..j-1]$ appears somewhere else in t. This clearly coincides with the smallest w such that $E[w] = j-1$ so $\ell_1 = v_1$ as claimed. At iteration $i+1$, the value ℓ_{i+1} is such that $\langle t[\ell_{i+1}..j-1] * t[j+1..h_i+1] \subseteq t \rangle$. Hence at Line 7 we find $h_{i+1} = k_{\ell_{i+1}}$ such that

$$k_{\ell_{i+1}} \geq h_i + 1 > h_i = k_{v_i}.$$

By (7) we know that $\ell_{i+1} \geq v_{i+1}$, but since by construction ℓ_{i+1} is the smallest integer with that property it must be $\ell_{i+1} = v_{i+1}$.

Finally, let $s+1$ denote the iteration at which we exit from the *while* loop since $g+1 = h_s + 1 > E[j+1]$. Since $\langle t[\ell_s..j-1] * t[j+1..h_s] \subseteq t \rangle$ the substring $t[j+1..h_s]$ appear elsewhere in t which implies $h_s \leq E[j+1]$. We conclude that $h_s = E[j+1] = k_{v_d}$ as claimed. $\square$

We call the algorithm in Fig. 5 "greedy" since at each iteration the *while* loop finds the longest substring with the desired property. As a result it only computes the values where the sequence (6) is strictly increasing. The next theorem shows that the algorithm uses these values to correctly compute PLCP_1.

Theorem 6. *The greedy algorithm correctly computes the PLCP_1 array in time $\mathcal{O}(nL_{ave} \log n / \log \log n)$ using $\mathcal{O}(n \log n)$ bits of space.*

Proof. To show the correctness of the algorithm, we prove that after the loop of Lines 10–11 each value e_i contains the maximum in (5). Assume the maximum is achieved for $j = \tilde{j}$. If the value $k_i^{(\tilde{j})}$ is among the one computed in the *while* loop there is nothing to prove. Otherwise we know that at iteration $\tilde{j}$ the algorithm will store the value $k_i^{(\tilde{j})}$ in a variable e_h with $h < i$, so at the end of the *while* loop we will have $e_h \geq k_i^{(\tilde{j})}$. Hence, after the loop of Lines 10–11 we have $e_i \geq k_i^{(\tilde{j})}$.

To see that it is indeed $e_i = k_i^{(\tilde{j})}$, it suffices to notice that at Line 11 we never store in e_i a value larger than $\text{PLCP}_1[i] + i - 1$ since for every j $\langle t[i-1..j-1] * t[j+1..e_{i-1}] \subseteq t \rangle$ implies $\langle t[i..j-1] * t[j+1..e_{i-1}] \subseteq t \rangle$.

We now analyze the running time. The most expensive part of the algorithm is clearly the nested *while* loop. For $j = 0, \ldots, n - 1$, the *while* loop computes the values

$$k_{v_1} < k_{v_2} < \cdots < k_{v_{d-1}} < k_{v_d}. \tag{8}$$

Since $k_{v_1} \geq j + 1$ and $k_{v_d} = E[j + 1]$, the number of iterations is bounded by $E[j + 1] - j = \mathrm{PLCP}_0[j + 1]$. Since each iteration takes $\mathcal{O}(\log n / \log \log n)$ time, summing over j gives the desired time bound.

The space bound follows from the fact that we only use data structures described in Section 3 which overall take $\mathcal{O}(n \log n)$ bits.   □

The greedy algorithm has the same asymptotic worst case running time than the simpler algorithm in Section 4. However, its actual performances depend on the effectiveness of the greedy strategy in computing only a subset of the sequence (6) instead of the whole sequence that has size $\mathrm{PLCP}_0[j + 1]$. From the above proof we see that if the *while* loop never computes more than $d_{\max}$ values, then the overall running would be bounded by $\mathcal{O}(n \, d_{\max} \log n / \log \log n)$.

Unfortunately the following example shows that in the worst case the greedy strategy is not effective and the *while* loop computes $\mathcal{O}(\mathrm{PLCP}_0[j + 1])$ values. Consider the string $t = a^{3m} b^m a^{2m} c a^{2m}$. When the mismatch is in position $j = 6m$ (so $t[j] = c$) at the first iteration of the *while* loop we have $\ell_1 = 4m$, $h_1 = 7m$ since $t[4m..7m] = a^{2m} c a^{m-1}$ appears with one mismatch in $t[0..3m]$. At the second iteration we have $\ell_2 = 4m+1, h_2 = 7m+1$ since $t[4m+1..7m+1] = a^{2m-1} c a^m$ appears with one mismatch in $t[0..3m]$. Similarly, at the i-th iteration it is $\ell_i = 4m + i - 1$, $h_i = 7m + i - 1$, and there are $m = (\mathrm{PLCP}_0[j + 1])/2$ iterations overall.

We conclude this section observing that the greedy algorithm can be transformed into an approximation algorithm with a one-sided additive error.

Lemma 7. *If in the algorithm of Fig. 5 we replace Line 9 with $g \leftarrow h + q$, the resulting algorithm runs in $\mathcal{O}((n L_{ave}/q) \log n / \log \log n)$ time and the resulting array PLCP_1' is such that for all i $\mathrm{PLCP}_1[i] - q < \mathrm{PLCP}_1'[i] \leq \mathrm{PLCP}_1[i]$.*

Proof. The time bound follows observing that the number of iterations of the *while* loop is now bounded by $\mathrm{PLCP}_0[j+1]/q$. To prove the error bound observe that by setting $g \leftarrow h + q$ the *while* loop will only compute a subset of the sequence (8). The crucial point is that if k_{v_r} is computed then the algorithm will not compute the values k_{v_i} such that $k_{v_r} < k_{v_i} < k_{v_r} + q$ but it will certainly compute the smallest k_{v_i} such that $k_i \geq k_{v_r} + q$. In other words, the algorithm will only skip a value k_{v_s} if a previous value k_{v_r} with $k_{v_s} - k_{v_r} < d$ was not skipped. Hence, if a variable e_ℓ is not updated with a value k_{v_s}, because k_{v_s} was not computed, we know that there exists $\ell' < \ell$ such that when the loop of Lines 10–11 begins it is $e_{\ell'} \geq k_{v_r}$. At the end of the loop we will have $e_\ell \geq k_{v_r} > k_{v_s} - d$ and the thesis follows.   □

6   Computation of 1-mappability

Given a text t and a parameter m, the mappability problem consists in determining, for each length-m substring of t, the number of times this substring appears in t. Formally, for $i = 0..n - m$ let $\mathrm{FQ}_0^m[i]$ denote the number of occurrences in t of the substring $t[i..i + m - 1]$. The mappability at position i is defined as the reciprocal of $\mathrm{FQ}_0^m[i]$.

Mappability is an important information in the analysis of Next Generation Sequencing data. In a NGS experiment reads of length m deriving from regions with $\mathrm{FQ}_0^m[i] = 1$ (high mappability) will likely be aligned in a single position of the reference, while reads deriving from regions with low mappability will be more problematic since they will align in multiple locations of the reference. In quantitative studies, for example of binding affinity in ChIP-Seq experiments, the mappability of the reference is a critical normalization factor (see [8] and references therein).

The array FQ_0^m can be easily computed with a single scan of the LCP and suffix arrays. However, because of sequencing errors and individual variations, NGS tools usually allow for a small number of mismatches. It is therefore important to measure the so called k-mappability, that is the number of substrings which are at Hamming distance k from any given length-m substring of t. When working at the scale of mammalian genomes this is a non trivial problem: the best available tool [8] takes hours or days to compute the exact mappability of the Human genome using 8 cores (there is also a faster variant that computes a reasonable approximation of the true mappability). Not surprisingly, the running time of the algorithm in [8] heavily depends on the parameters m and k. Since the algorithm is based on an exhaustive matching tool [21], for a given number of mismatches the running time *increases* when the length of the region m decreases. Indeed, for smaller m the number of substrings at Hamming distance k from a given string increases and so does the overall running time. In this section we show how to compute the 1-mappability in $\mathcal{O}(mn \log n / \log \log n)$ time. The interest of this approach is that it is efficient for small m and therefore could be a starting point for the development of a tool complementary to [8].

For a given substring length m, let $\mathrm{FQ}_1^m[i]$ denote the number of substrings of t which are at Hamming distance at most 1 from $t[i..i + m - 1]$. Formally, the 1-mappability at position i is defined as the reciprocal of $\mathrm{FQ}_1^m[i]$. To compute a single value $\mathrm{FQ}_1^m[i]$ we consider separately the m possible positions for the mismatch. For $j = i, \ldots, i + m - 1$ we compute the value $\mathrm{Occ}(t[i..j - 1] * t[j + 1..i + m - 1])$, that is, the number of occurrences in t of the strings of the form $t[i..j - 1] * t[j + 1..i + m - 1]$ where $*$ stands for an arbitrary character. The resulting algorithm is shown in Figure 3, where we also take care of the fact that among the $\mathrm{Occ}(t[i..j - 1] * t[j + 1..i + m - 1])$ occurrences there are also the $\mathrm{FQ}_0^m[i]$ occurrences of substrings which are equal to $t[i..i + m - 1]$.

The key to the efficient computation of $\mathrm{Occ}(t[i..j - 1] * t[j + 1..i + m - 1])$ is the following result, originally stated in [6, Lemma 6] in terms of orthogonal range search, and here rephrased in terms of simple arrays.

1:    Compute FQ_0^m with a scan of the LCP array
2:    **for** $i = 0$ **to** $n - m$ **do**
3:         $FQ_1^m[i] \leftarrow FQ_0^m[i]$ // *account for substrings at distance 0*
4:         **for** $j = i$ **to** $i + m - 1$ **do**
5:             // *account for substrings with a mismatch in position j*
6:             $FQ_1^m[i] \mathrel{+}= \mathrm{Occ}(t[i..j - 1] * t[j + 1..i + m - 1]) - FQ_0^m[i]$

Fig. 3. Algorithm for computing the FQ_1^m array.

Lemma 8. *We can represent an array $S[0..n]$ of elements in $[0..n]$ in $n \log n + o(n)$ bits so that given a range $[a..b]$ of positions and a range $[c..d]$ of values we can compute in $\mathcal{O}(\log n / \log \log n)$ time the number of entries in $S[a..b]$ whose values are in the range $[c..d]$.*   □

Theorem 9. *The algorithm in Fig. 3 computes FQ_1^m in $\mathcal{O}(mn \log n / \log \log n)$ time using $\mathcal{O}(n \log n)$ bits of space.*

Proof. Line 1 takes $\mathcal{O}(n)$ time. We show that each iteration of the loop at Lines 2–6 takes $\mathcal{O}(m \log n / \log \log n)$ time assuming that we represent the array PA* defined in (2) using Lemma 8.

For any position i, with a bidirectional compressed index [18,24] we compute in overall $\mathcal{O}(m)$ time the ranges $[a_0, b_0], \ldots, [a_{m-1}, b_{m-1}]$ such that $[a_k, b_k]$ is the range of prefix array entries containing the prefixes ending with $t[i..i + k - 1]$. Similarly we compute the ranges $[c_0, d_0], \ldots, [c_{m-1}, d_{m-1}]$ such that $[c_k, d_k]$ is the range of suffix array entries containing the suffixes starting with $t[i + k + 1..i + m - 1]$. To complete the proof it suffices to show that for $j = 0, \ldots, m - 1$ the $\mathrm{Occ}(t[i..j - 1] * t[j + 1..i + m - 1])$ coincides with the number of entries in PA*$[a_{j-i}..b_{j-i}]$ which are in the range $[c_{j-i}..d_{j-i}]$. Indeed, we have PA*$[x] = y$ with $a_{j-i} \le x \le b_{j-i}$ and $c_{j-i} \le y \le dj - i$ iff the prefix $t[0..PA*[x]]$ ends with $t[i..j - 1]$ and the suffix $t[PA*[x] + 2..n]$ starts with $t[j + 1..n]$. Hence there is a bijection between the substrings of the form $t[i..j - 1] * t[j + 1..i + m - 1]$ and the entries in PA*$[a_{j-i}..b_{j-i}]$ which are in the range $[c_{j-i}..d_{j-i}]$; to count the former we simply count the latter in $\mathcal{O}(\log n / \log \log n)$ time using Lemma 8.   □

References

1. Amir, A., Keselman, D., Landau, G., Lewenstein, M., Lewenstein, N., Rodeh, M.: Text indexing and dictionary matching with one error. J. Algorithms **37**, 309–325 (2000)
2. Barbay, J., Claude, F., Navarro, G.: Compact binary relation representations with rich functionality. Information and Computation **232**, 19–37 (2013)
3. Beller, T., Gog, S., Ohlebusch, E., Schnattinger, T.: Computing the longest common prefix array based on the Burrows-Wheeler transform. J. Discrete Algorithms **18**, 22–31 (2013)
4. Bille, P.: Gørtz, I.L.: Substring range reporting. Algorithmica **69**, 384–396 (2014)

5. Bingmann, T., Fischer, J., Osipov, V.: Inducing suffix and LCP arrays in external memory. In: Proc. 15th Meeting on Algorithm Engineering and Experiments (ALENEX 2013), pp. 88–102. SIAM (2013)
6. Bose, P., He, M., Maheshwari, A., Morin, P.: Succinct Orthogonal Range Search Structures on a Grid with Applications to Text Indexing. In: Dehne, F., Gavrilova, M., Sack, J.-R., Tóth, C.D. (eds.) WADS 2009. LNCS, vol. 5664, pp. 98–109. Springer, Heidelberg (2009)
7. Crochemore, M., Langiu, A., Rahman, M.S.: Indexing a sequence for mapping reads with a single mismatch. Phil. Trans. R. Soc. A 372 (2014)
8. Derrien, T., Estell, J., Marco-Sola, S., Knowles, D.G., Raineri, E., Guig, R., Ribeca, P.: Fast computation and applications of genome mappability. PLoS One 7 (2012)
9. Fischer, J.: Inducing the LCP-Array. In: Dehne, F., Iacono, J., Sack, J.-R. (eds.) WADS 2011. LNCS, vol. 6844, pp. 374–385. Springer, Heidelberg (2011)
10. Fischer, J., Heun, V.: Space-efficient preprocessing schemes for range minimum queries on static arrays. SIAM J. Comput. **40**, 465–492 (2011)
11. Gawrychowski, P., Lewenstein, M., Nicholson, P.K.: Weighted Ancestors in Suffix Trees. In: Schulz, A.S., Wagner, D. (eds.) ESA 2014. LNCS, vol. 8737, pp. 455–466. Springer, Heidelberg (2014)
12. Gog, S., Ohlebusch, E.: Compressed suffix trees: Efficient computation and storage of LCP-values. ACM Journal of Experimental Algorithmics 18 (2013)
13. Gusfield, D.: Algorithms on Strings, Trees, and Sequences : Computer Science and Computational Biology. Cambridge University Press, Cambridge (1997)
14. Iliopoulos, C.S., Rahman, M.S.: Indexing factors with gaps. Algorithmica **55**, 60–70 (2009)
15. Kärkkäinen, J., Manzini, G., Puglisi, S.J.: Permuted Longest-Common-Prefix Array. In: Kucherov, G., Ukkonen, E. (eds.) CPM 2009 Lille. LNCS, vol. 5577, pp. 181–192. Springer, Heidelberg (2009)
16. Kärkkäinen, J., Kempa, D.: LCP Array Construction in External Memory. In: Gudmundsson, J., Katajainen, J. (eds.) SEA 2014. LNCS, vol. 8504, pp. 412–423. Springer, Heidelberg (2014)
17. Khmelev, D.V., Teahan, W.J.: A repetition based measure for verification of text collections and for text categorization. In: Proc. 26th Int. Conference on Research and Development in Information Retrieval (SIGIR), pp. 104–110. ACM (2003)
18. Lam, T.W., Li, R., Tam, A., Wong, S., Wu, E., Yiu, S.M.: High throughput short read alignment via bi-directional BWT. In: Proc. IEEE International Conference on Bioinformatics and Biomedicine (BIBM 2009), pp. 31–36 (2009)
19. Léonard, M., Mouchard, L., Salson, M.: On the number of elements to reorder when updating a suffix array. J. Discrete Algorithms **11**, 87–99 (2012)
20. Manber, U., Myers, G.W.: Suffix arrays: a new method for on-line string searches. SIAM Journal on Computing **22**, 935–948 (1993)
21. Marco-Sola, S., Sammeth, M., Guigó, R., Ribeca, P.: The GEM mapper: Fast, accurate and versatile alignment by filtration. Nature Methods **9**, 1185–1188 (2012)
22. Sadakane, K.: Succinct representations of LCP information and improvements in the compressed suffix arrays. In: Proc. 13th Symposium on Discrete Algorithms (SODA 2002), pp. 225–232. ACM/SIAM (2002)
23. Sadakane, K.: Compressed suffix trees with full functionality. Theory Comput. Syst. **41**, 589–607 (2007)
24. Schnattinger, T., Ohlebusch, E., Gog, S.: Bidirectional search in a string with wavelet trees and bidirectional matching statistics. Inf. Comput. **213**, 13–22 (2012)
25. Sirén, J.: Sampled Longest Common Prefix Array. In: Amir, A., Parida, L. (eds.) CPM 2010. LNCS, vol. 6129, pp. 227–237. Springer, Heidelberg (2010)

Evaluating Geographical Knowledge Re-Ranking, Linguistic Processing and Query Expansion Techniques for Geographical Information Retrieval

Daniel Ferrés[✉] and Horacio Rodríguez

TALP Research Center, Universitat Politècnica de Catalunya,
Jordi Girona 1-3, 08034 Barcelona, Spain
{dferres,horacio}@cs.upc.edu

Abstract. This paper describes and evaluates the use of Geographical Knowledge Re-Ranking, Linguistic Processing, and Query Expansion techniques to improve Geographical Information Retrieval effectiveness. Geographical Knowledge Re-Ranking is performed with Geographical Gazetteers and conservative Toponym Disambiguation techniques that boost the ranking of the geographically relevant documents retrieved by standard state-of-the-art Information Retrieval algorithms. Linguistic Processing is performed in two ways: 1) Part-of-Speech tagging and Named Entity Recognition and Classification are applied to analyze the text collections and topics to detect toponyms, 2) Stemming (Porter's algorithm) and Lemmatization are also applied in combination with default stopwords filtering. The Query Expansion methods tested are the Bose-Einstein (Bo1) and Kullback-Leibler term weighting models. The experiments have been performed with the English Monolingual test collections of the GeoCLEF evaluations (from years 2005, 2006, 2007, and 2008) using the TF-IDF, BM25, and InL2 Information Retrieval algorithms over unprocessed texts as baselines. The experiments have been performed with each GeoCLEF test collection (25 topics per evaluation) separately and with the fusion of all these collections (100 topics). The results of evaluating separately Geographical Knowledge Re-Ranking, Linguistic Processing (lemmatization, stemming, and the combination of both), and Query Expansion with the fusion of all the topics show that all these processes improve the Mean Average Precision (MAP) and RPrecision effectiveness measures in all the experiments and show statistical significance over the baselines in most of them. The best results in MAP and RPrecision are obtained with the InL2 algorithm using the following techniques: Geographical Knowledge Re-Ranking, Lemmatization with Stemming, and Kullback-Leibler Query Expansion. Some configurations with Geographical Knowledge Re-Ranking, Linguistic Processing and Query Expansion have improved the MAP of the best official results at GeoCLEF evaluations of 2005, 2006, and 2007.

Keywords: Information retrieval · Geographical gazetteers · Natural language processing · Toponym disambiguation · Query expansion · Efectiveness measures

© Springer International Publishing Switzerland 2015
C. Iliopoulos et al. (Eds.): SPIRE 2015, LNCS 9309, pp. 311–323, 2015.
DOI: 10.1007/978-3-319-23826-5_30

1   Introduction

Geographical Information Retrieval (GIR) is the task of retrieving a set of relevant documents given a user query need with geographical restrictions expressed in natural language (e.g. "Shark attacks in California"). Geographical queries are normally defined by a triplet < theme, spatial relationship, location > [7]. As an example, the previous query will be treated in the following way: 1) a theme ("shark attacks"), 2) a location ("California), 3) a spatial relationship ("in") between the theme and the location. Current state-of-the-art Information Retrieval (IR)) algorithms treat geographical terms from queries as simple textual tokens without having into account its geographical meaning and the possible geographical restrictions that these terms can imply. As an example, the previous example of geographical query could led to find documents that mention "shark attacks" in California by matching only the geographical token "California" with all the indexed documents. In this way the IR system will not return or will return without enough ranking documents that could report "shark attacks" in places of California but not mentioning California (e.g. "Shark attacks in Santa Barbara"). Theoretically, the treatment and automatic understanding of geographical terms appearing in user queries and indexed documents from IR systems (and major search engines) should provide an improvement of the results by retrieving documents that match the geographical restrictions in the query. The system and the experiments presented in this paper are focused to evaluate how to treat effectively these geographical restrictions in the queries using existing Geographical Knowledge Bases in combination with some conservative Toponym Disambiguation Heuristics. Two kind of of geographical terms are detected and disambiguated in topics and collections: 1) toponyms, 2) geographical feature types. These kind of terms are used in a Geographical Knowledge Re-Reranking process that boosts the ranking of the geographically relevant documents. In addition Linguistic Processing and Query Expansion are also investigated for GIR. This system was initially designed for the GeoCLEF 2007 evaluation in which achieved the best MAP using the TF-IDF algorithm [4].

The GeoCLEF test collections [9] have been used to evaluate the topics. The GeoCLEF GIR evaluation forum took place during 4 years (1 as a pilot task) between 2005 and 2008 in the framework of the CLEF conferences[1]. The test collections are composed of 100 topics (25 topics per year). The GeoCLEF English document collection consists of 169,477 documents composed by stories from the British newspaper *The Glasgow Herald* (1995) and the American newspaper *Los Angeles Times* (1994). In [10] the different kind of geographical topics at GeoCLEF GIR evaluations are reported:

- Feature types with non-geographic restrictions (e.g. rivers with vineyards).
- Feature type with geographical place restriction (e.g. cities in Germany).
- Thematic subject associated to a toponym (e.g. independence of Quebec).
- Topics with a non-geographic subject that is a complex function of place (e.g., European football cup matches).

---

[1] http://www.clef-initiative.eu

```
<title>Whisky making in the Scottlsh Islands</title>
<desc> To be relevant, a document must describe a whisky made, or
a whisky distillery located, on a Scottish island.</desc>
<narr> Relevant islands are Islay, Skye, Orkney, Arran, Jura, Mull.;
Relevant whiskys are Arran Single Malt; Highland Park Single Malt; Scapa;
Isle of Jura; Talisker; Tobermory;  Ledaig; Ardbeg; Bowmore; Bruichladdich;
 Bunnahabhain; Caol Ila; Kilchoman;Lagavulin; Laphroaig  </narr>
```

Fig. 1. Example of a topic of the GeoCLEF 2007 edition.

- Vague topics (e.g., Sub-Saharan Africa).
- Geographical relations among toponyms (e.g., Oil and gas extraction found between the UK and the Continent)
- Geographical relations among events (e.g., F1 circuits where Ayrton Senna competed in 1994).
- Relations between events in specific toponyms (e.g., Casualties in fights in Nagorno- Karabakh).

2   Related Work

GIR systems have very specific issues due to its restricted domain (geography) specificity. Some of these issues have been detailed in the GIR literature [6]:

- geographical names detection (e.g. detecting "Washington" as a possible place name and disambiguate it as a location instead a geo-political entity or a person.)
- spatial natural language qualifiers detection (e.g. north, south of, near, close by,...)
- toponyms disambiguation (e.g. Paris, Texas (USA) vs Paris (France))
- vague place names detection and interpretation. (e.g. Scottish Trossachs, Midlands,...)
- thematic and geospatial indexing and retrieval.

Approaches at GIR used different strategies to perform: 1) stand-alone probabilistic models [8], 2) combination of textual and geographical search [11], 3) filtering or reranking the documents with geographical knowledge [13], 4) geographical query expansion [3] [16], and 5) machine learning for re-ranking [10]. Berkeley 2 group participating at GeoCLEF 2005 used a logistic regression algorithm with the following features: stopwords filtering, Stemming (English Muscat stemmer) and blind feedback with the 30 top-ranked terms from the top 20 ranked documents [8]. Their system achieved the highest result with a MAP of 0.3936 in a run that used the spatial tags included in the topics. Martins et al. [11] presented a GIR sytem at GeoCLEF 2006 that used a geographical ontology of about 12,654 concepts, that include place names, feature types, relationships among places, demographics data, ocurrence statistics of toponyms in corpora, spatial coordinates and bounding boxes. They used this ontology combined with a graph-ranking approach to detect scope of documents and topics and a relevance ranking that combined BM25 and a geographical similarity function for scopes. Their approach did not outperform the baseline with BM25 and manual expansion (that achieved the best MAP at GeoCLEF2006 with 0.3034). Wang and Neumann [16] applied and approach that, besides including geographical knowledge, also included knowledge of natural and human events mined from Wikipedia. They use Query Expansion with ontologies both for events and geographic terms. Their system achieved the best MAP at Geo-CLEF2008 with a 0.3037 with a run with manual work and a MAP of 0.2924 in an automatic run. Buscaldi and Rosso [3] applied the GeoCLEF (2005-2008) topics to test diversity in GIR. They reformulated queries using the meronyms of the places contained in the original queries (using only the title field), with the help of a geographical ontology. They reported that a theoretical improvement is possible. Perea-Ortega et al. [13] using the GeoCLEF data showed that in each evaluation a re-ranking based on the combination of geographical similarity and textual similarity outperforms the baseline (textual based IR). They used POS tagging (TreeTager), stopwords filtering and Snowball Stemmer to process a thematic index of the collection. A geographical index was built with Geo-NER to recognize geographical entities. The textual index uses the stemmed and stopwords filtered text and the geographical entities in its original word form. They applied Lemur[2], Terrier[3] and Lucene[4] for the IR process. Lemur was applied with BM25 with Pseudo Relevance Feedback, Lucene with BM25 and Query Expansion, and Terrier with InL2 and Bo1 (Query Expansion). Their best results were obtained with Terrier InL2 and QE (Bo1) combined with geographical re-ranking with competive (above the average of participants and close to the top ranked) MAP values of 0.3874 (GeoCLEF 2005). 0.2733 (GeoCLEF 2006), 0.2600 (GeoCLEF 2007), 0.2973 (GeoCLEF 2008).

3   System Description

The system is composed of two main phases: 1) Textual and Geographical Indexing, 2) Geographical Information Retrieval. The IR software used in both indexing and retrieval phases is Terrier (version 4.0) [12]. We used the TF-IDF. BM25, and InL2 IR algorithms implemented in the Terrier IR engine. Stopwords filtering is applied by our system using the stopwords list provided in the Terrier IR engine. The baseline system uses all the terms from the topics. This means that no separation between thematic and geographical terms and themes or events is performed by the textual search.

3.1   Textual and Geographical Indexing

We pre-processed the English document collections: Glasgow Herald 1995 (GH95) and Los Angeles Times 1994 (LAT94) with linguistic processing tools

---

[2] www.lemurproject.org/
[3] http://www.terrier.org
[4] http://lucene.apache.org

(described in the next subsection) to mark the part-of-speech (POS) tags, lemmas and Named Entities (NE). After this process the collection is analyzed with a Geographical Knowledge Base and conservative Toponym Disambiguation heuristics (both components are described in the next sub-section). This information was used to built two types of indexes:

- Geographical Index. This is a custom-build index that contains the geographical information of the documents. For each toponym in the document (detected with the NE detector) the feature type, GeoKB ontology information and coordinates are stored in the index. Even if the place is ambiguous all the possible geographical referents are indexed.
- Textual Indexes. These are Terrier based indexes that store the original or the linguistically processed information of the document. Note that in all these indexes geographical entities (toponyms) have been indexed without linguistic processing with exception of the stemmed indexes. The following indexes have been created: 1) *original index with word forms*, 2) *lemmatized index*, 3) *stemmed index* (using the Porter Stemmer, and 4) *lemmatized and stemmed index* (the Porter Stemmer applied over the lemmatized content).

3.2   Geographical Information Retrieval

The retrieval system has four phases performed sequentially: 1) a Linguistic and Geographical Processing of the topics, 2) a textual Document Retrieval with Terrier, 3) a Geographical Document Retrieval with Geographical Knowledge Bases (GKBs), and 4) a Geographical Re-Ranking phase.

Linguistic and Geographical Knowledge Processing of the Topics. The goal of this phase is to extract all the relevant keywords (with its analysis) from the topics. These keywords are then used by the Textual and Geographical Document Retrieval phases. The Topic Analysis phase has two main sub-phases: a Linguistic Analysis and a Geographical Analysis. The Linguistic Analysis sub-phase extracts lexico-semantic and syntactic information using the following set of Natural Language Processing (NLP) tools: 1) *TnT* an statistical POS tagger [2], 2) *WordNet lemmatizer* (version 2.0), 3) A Maximum Entropy based Named Entities Recognizer and Classifier (NERC) trained with the CONLL-2003 shared task English data set, 4) a list of demonyms relationships for each country (e.g. Japanese - Japan). The Geographical Analysis is applied to the Named Entities from the Title and Description and Narrative tags of the topics that have been classified as LOCATION or ORGANIZATION by the NERC module. This analysis uses a Geographical Knowledge Base that has two main components: 1) a Geographical Thesaurus, 2) Feature type thesaurus. The Geographical Thesaurus has been built joining four gazetteers that contain entries with places and their geographical class, coordinates, part-of relationships and other information:

1. NGA GEOnet Names Server (GNS)[5]: a gazetteer covering worldwide excluding the United States and Antarctica, with 5.3 million entries.
2. Geographic Names Information System (GNIS)[6], contains 2.0 million entries about geographic features of the United States and its territories. We used a subset of 39,906 entries of the most important geographical names.
3. *GeoWorldMap*[7] *World Gazetteer*: a gazetteer with 40,594 entries of the most important countries, regions, and cities of the world.
4. *World Gazetteer*[8]: a gazetteer with 171,021 entries of towns, administrative divisions and agglomerations with their features and current population. From this gazetteer we added only the 29,924 cities with more than 5,000 unhabitants.

Each one of these gazetteers has a different set of classes that have been mapped to the ADL Feature Type Thesaurus (ADLFTT) with a resulting set of 575 geographical types. The ADL Feature Type Thesaurus is a hierarchical collection of geographical terms used to type named geographic places in English [5]. Our GNIS mapping is similar to the one exposed by Hill [5]. The following Toponym Disambiguation heuristics are applied using the information from the GeoKB:

- *H1. Hierarchical ranked ontology of feature types.* The ranked hierarchy of the feature types ontology is applied when a toponym can refer to several kinds of feature types (e.g. Africa (the continent) vs Africa, Mexico). The following list of ordered priorities for the different feature types is used: 1) continent, 2) subcontinent (e.g. South America), 3) country capital, 4) country, 5) first order administrative divisions (e.g. states), 6) sea, 7) summit, 8) river, 9) county, 10) important city , 11) other place (can include less important cities and other types).
- *H2. Important places are disambiguated excluding other places with the same name.* GeoWorldMap and Word Gazetteer have priority to disambiguate places because contain less but important places compared with GNIS and GNS.
- *H3. Treatment of toponym vs person name type of Geo/Non-Geo ambiguity when the toponym has the lowest priority (11).* A list of common first and last names is used to filter out Named Entities erroneously recognized as toponyms.
- *H4. Small places are not taken into account (only for USA).* Due to the high amount of places in the GNIS gazetteer, only a small part of its data is used (the US concise gazetteer).
- *H5. Lowest priority toponyms are not disambiguated.* Toponyms with the lowest priority in the hierarchy are not disambiguated and all the possible geographical referents are taken into account in the collection processing and indexing, and the topic analisys phases.

---

[5] NGA GNS. http://geonames.nga.mil/gns/html/namefiles.html
[6] GNIS. http://geonames.usgs.gov/domestic/download_data.htm
[7] Geobytes Inc.: Geoworldmap database. http://www.geobytes.com/
[8] World Gazetteer is not available from its original site. But a copy can be found in this link. http://biit.cs.ut.ee/biodc/dataen.zip

These processes are applied to the topics but have been applied also to the entire document collection before indexing. The GeoKB and the Toponym Disambiguation processes take into account the part-of relationships of the toponyms detected and are used in the retrieval and indexing process (e.g. the toponym "United States" is indexed as $America@North\_America@United\_States$). Geographical coordinates (point-based) for each toponym are also included in the index with exception to the continent and subcontinent feature types. The feature types of each toponym disambiguated is also detected and stored (e.g the toponym "United States" will have the following feature type associated $administrative\_areas@political\_areas@countries$).

Textual Document Retrieval. The textual IR phase is performed retrieving the top 10,000 documents related to the topic using the TF-IDF, BM25 or InL2 algorithms. The default stopwords in English of the IR engine Terrier are used. This phase can perform Stemming (Porter's algorithm) and automatic Query Expansion (QE) using two state-of-the art Query Expansion models based on Divergence From Randomness: Bose-Einstein 1 (Bo1) and Kullback-Leibler (KL) [1]. This pseudo-relevance feedback option extracts the 40 most informative terms from the 10 top-returned documents in first-pass retrieval as the expanded query terms.

Geographical Document Retrieval. Our Geographical Knowledge Base is used to retrieve geographically relevant documents using the following types of geographical terms from GIR queries: 1) toponyms (e.g. places names such as "United States"), 2) feature types (e.g. "cities", "countries"). The GeoKB uses a search method over toponyms and feature types that allows to retrieve all the documents that have a token that matches totally or partially the toponyms or the feature types. As an example for the case of toponyms, the keyword $America@Northern\_America@United\_States$ will retrieve U.S. places like Los Angeles, CA, USA and Baltimore, MD, USA (see Table 1). In addition, each geographical feature type in the query can be expanded using a set of feature type synonyms and related words that has been manually extracted from the GNIS feature types.

Geographical Knowledge Re-Ranking. This component re-ranks the documents retrieved by Terrier using the set of geographically relevant documents

Table 1. Example of full and partial disambiguation.

toponym	disambiguation (full or partial)
Los Angeles	$administrative\_areas@populated\_places@cities$
	$America@Northern\_America@United\_States@California@Los\_Angeles$
Baltimore	$administrative\_areas@@populated\_places@cities$
	$America@Northern\_America@Canada@Ontario@Baltimore$
	$America@Northern\_America@United\_States@Maryland@Baltimore$
	$America@Northern\_America@United\_States@Ohio@Baltimore$

detected by the Geographical Document Retrieval module and returns a set of 1,000 documents. First, the top-scored documents retrieved by Terrier that appear in the document set retrieved by the Geographical Document Retrieval module are selected. Then, if the set of selected documents is less than 1,000, the top-scored documents retrieved by Terrier that not appear in the document set of Geographically Relevant documents are used to complete the retrieved set (changing its ranking and score).

4   Experiments

Several experiments with the full collection of GeoCLEF[9] (100 topics) have been designed to evaluate the relative impact of different features (alone and in combination among them) in GIR over some state-of-the-art effectiveness measures. These experiments will be evaluated with the binary relevance assessments collected with pooling during the GeoCLEF forums (see Table 2 for details about the relevance assesments).

Table 2. Relevance assesment information about GeoCLEF evaluations

	2005	2006	2007	2008
#topics	25	25	25	25
#relevant_documents	1,028	378	650	747
#judged_documents	14,546	17,964	15,637	14,528
#considered_documents	18,000	18,000	18,000	18,000

The baselines to compare are the IR algorithms TF-IDF, BM25, and InL2 with word forms in the indexed collection and the set of queries (topics). These experiments have been performed with three possible uses of the topics metadata: a) title (T), b) title and description (TD), c) title, description and narrative (TDN) . Several experiments have been performed with the full GeoCLEF collection (100 topics) to evaluate the following system components alone or in combination: 1) Linguistic Processing features evaluated in isolation or in combination: a) Lemmatization, b) Stemming, c) Lemmatization + stemming, 2) Automatic Query Expansion: the Bose-Einstein (Bo1) and Kullback-Leibler QE term weighting models, 3) GeoKR, 4) Linguistic Processing with GeoKR, 5) Linguistic Processing, QE and GeoKR combined. The effectiveness measures chosen to evaluate the full collection experiments have been the following: Mean Average Precision, R-Precision. MAP computes the arithmetic mean of average precisions of all topics. The average precision of each topic is the mean of precisions computed at the rank position of each relevant document retrieved. R-Precision is a measure that computes the arithmetic mean of precision at R documents for each

---

[9] The GeoCLEF test topics, relevance assesments and the official experiments performet at GeoCLEF from 2005 to 2008 can be downloaded at http://direct.dei.unipd.it/

topic, being R the number of relevant documents for the topic. Moreover, Precision at N(5,10,15,20,30,100,200,500,100) plots have been used to show a more detailed evaluation of the main features in the best system. All these measures have been applied over the 1,000 top-ranked retrieved documents. Significance testing has been performed using the following tests: two-tailed t-test [14], and Fisher's two-sided paired randomization test [15]. Finally, a set of experiments has been done with the individual GeoCLEF collections of years 2005, 2006, 2007 and 2008 to compute the performance in MAP of the best configurations of the full collection. These experiments will be compared with the best run of each GeoCLEF task.

5 Results

The results of the full GeoCLEF collection experiments are shown in Table 3 and Figure 2. The results of evaluating separately Geographical Knowledge Re-Ranking, Linguistic Processing (lemmatization, stemming, and the combination of both), and Query Expansion show that all these processes improve the Mean Average Precision (MAP) and R-Precision in all the experiments and show statistical significance over the baselines in most of them (see Table 3). All the experiments that use only the title (T) field show statistical significance (p-value < 0.01) in MAP and R-Precision. The experiments with title and description (TD) obtained statistical significance (p-value < 0.01) in MAP (including R-Prec statistical significance with the ones that used the TF-IDF). MAP and RPrecision also show statistical significance (p-value < 0.01) in all the experiments that combine Lemmatization with stemming, GeoKB and Query Expansion. The best results in MAP (0.3116) and R-Precision (0.3142) are obtained with the InL2 algorithm with Title and Description, and using the following techniques: GeoKR, Lemmatization with Stemming, and Kullback-Leibler Query Expansion. This configuration and each method tested alone with respect to the baseline show improvements in Precision at @(5,10,15,20,30,100,200,500,1000) in the majority of the experiments (see Figure 2). Some configurations with GeoKR, Linguistic Processing and Query Expansion have improved the MAP of the best official results at GeoCLEF evaluations of 2005, 2006, and 2007 (see Table 4). In the evaluation with the GeoCLEF 2008 topics a huge drop in MAP (with respect to the use of only TD) is found when using the TDN tags. The textual baseline GeoCLEF 2008 results show a MAP with TDN of 0.1978 which is significantly lower than with T (0.2517) or TD (0.2448).

The narrative terms of the GeoCLEF 2008 topics do not help to improve the MAP with respect to the T and TD experiments while the use of TD and T is not affected. This fact lead us to try experiments using TD for textual retrieval and TDN for GeoKR. This new configuration with improved the MAP and R-Precision of the best MAP experiment in Table 3 from 0.3116 to 0.3198 (MAP) and from 0.3095 to 0.3236 (R-Precision).

Table 3. Results in MAP and R-Precision with the 100 topics of all GeoCLEF collections using the Title (T), the Title and Description (TD), and the Title, Description, and Narrative (TDN) fields of the topics. Results in bold font mark the best results by field tag for each IR algorithm. Underlined results mark the best ones of each kind of field tag. Results in dark grey mark the best effectiveness measure among all field types and IR algorithms. The results marked with * and ** have statistical significance for t-test and randomization tests with p-values < 0.05 and p-values <0.01 respectively.

Configuration	MAP			RPrec		
	T	TD	TDN	T	TD	TDN
TF-IDF (baseline)	0.1938	0.2238	0.2386	0.2040	0.2335	0.2444
+Stemming (S)	0.2642**	0.2740**	0.2742**	0.2678**	0.2811**	0.2707*
+Lemmatization (L)	0.2333**	0.2573**	0.2619*	0.2379**	0.2621**	0.2630
+L+S	0.2631**	0.2726**	0.2728**	0.2680**	0.2792**	0.2712 *
+Bo1	0.2372**	0.2541**	0.2692*	0.2462**	0.2647**	0.2644
+KL	0.2339**	0.2531**	0.2723**	0.2430**	0.2620**	0.2638
+GeoKB	0.2088**	0.2307**	0.2485**	0.2313*	0.2520*	0.2553**
+S+Bo1	0.2926**	**0.3007****	0.2908**	0.2942**	0.3030**	0.2779*
+L+S+Bo1	0.2869**	0.2977**	0.2959**	0.2865**	0.2997**	0.2845**
+L+S+Bo1+GeoKB	0.2899**	0.2988**	**0.3082****	**0.2957****	**0.3066****	0.3050**
+L+S+GeoKB	0.2647**	0.2735**	0.2833**	0.2700**	0.2881**	0.2877*
+S+KL	**0.2954****	0.3001**	0.2906**	0.2900**	0.3018**	0.2780*
+L+S+KL	0.2893**	0.2987**	0.2936**	0.2836**	0.2967**	0.2902**
+L+S+KL+GeoKB	0.2898**	0.2978**	0.3066**	0.2922**	0.3055**	**0.3092****
BM25 (baseline)	0.1935	0.2237	0.2390	0.2030	0.2360	0.24632
+Stemming (S)	0.2653**	0.2756**	0.2748**	0.2678**	0.2835**	0.2767**
+Lemmatization (L)	0.2353**	0.2589**	0.2624*	0.2383**	0.2626*	0.2655
+L+S	0.2643**	0.2752**	0.2744**	0.2702**	0.2800**	0.2755**
+Bo1	0.2384**	0.2635**	0.2718**	0.2405**	0.2640*	0.2650*
+KL	0.2399**	0.2676**	0.2743**	0.2403**	0.2709**	0.2630*
+GeoKB	0.2086**	0.2312**	0.2481**	0.2320	0.2534**	0.2571**
+S+Bo1	0.2898**	0.2997**	0.2908**	0.2933**	0.2962**	0.2836*
+L+S+Bo1	0.2854**	0.2951**	0.2943**	0.2850**	0.2908**	0.2880**
+L+S+Bo1+GeoKB	0.2906**	0.2983**	**0.3062****	**0.2995****	0.3037**	0.3084**
+L+S+GeoKB	0.2661**	0.2755**	0.2826**	0.2715**	0.2875**	0.2943
+S+KL	**0.2940****	0.2991**	0.2907**	0.2949**	0.2986**	0.2853
+L+S+KL	0.2899**	0.2962**	0.2916**	0.2861**	0.2930**	0.2910**
+L+S+KL+GeoKB	0.2939**	**0.3002****	0.3044**	0.2993**	**0.3084****	**0.3115** **
InL2 (baseline)	0.1939	0.2240	0.2387	0.2002	0.2348	0.2466
+Stemming (S)	0.2649**	0.2745**	0.2753**	0.2698**	0.2829**	0.2739**
+Lemmatization (L)	0.2370**	0.2612**	0.2613 *	0.2406**	0.2741**	0.2607
+L+S	0.2646**	0.2749**	0.2750**	0.2705**	0.2789**	0.2724*
+Bo1	0.2388**	0.2595**	0.2732**	0.2469**	0.2612*	0.2682
+KL	0.2384**	0.2592**	0.2764**	0.2454**	0.2658**	0.2698*
+GeoKB	0.2078**	0.2307**	0.2478**	0.2310**	0.2538*	0.2536**
+S+Bo1	0.2969**	0.3052**	0.2947**	0.2948**	0.2995**	0.2835*
+L+S+Bo1	0.2949**	**0.3067****	0.2967**	0.2933**	0.3010**	0.2884**
+L+S+Bo1+GeoKB	0.2974**	0.3052**	0.3092**	0.3029**	0.3106**	0.3060**
+L+S+GeoKB	0.2663**	0.2745**	0.2830**	0.2701**	0.2875**	0.2893
+S+KL	**0.3001****	0.3041**	0.2973**	0.2948**	0.3029**	0.2882**
+L+S+KL	0.2978**	0.3061**	0.2987**	0.2988**	0.3109**	0.2904**
+L+S+KL+GeoKB	0.2976**	0.3047**	**0.3116****	**0.3037****	**0.3142****	**0.3085****

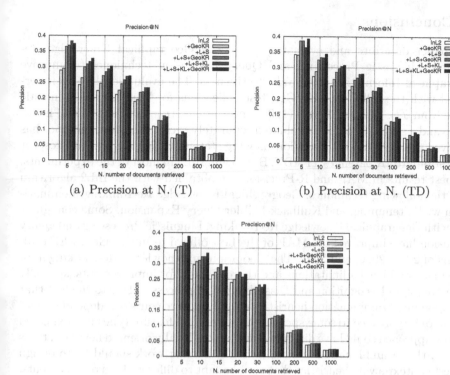

(a) Precision at N. (T)   (b) Precision at N. (TD)

(c) Precision at N (TDN).

Fig. 2. Precision at N plots of the InL2 IR algorithm with different sets of features and the GeoCLEF collection (100 topics) using the Title (T), the Title and Description (TD), and the Title, Description, and Narrative (TDN) field tags of the topics.

Table 4. MAP at 1,000 documents with the best configurations for the full collection applied to each GeoCLEF Monolingual English task. Includes the best official results (in MAP) at GeoCLEF evaluations.

Base Configuration InL2+S+L+GeoKR	MAP			
	GeoCLEF 2005	GeoCLEF 2006	GeoCLEF2007	GeoCLEF2008)
best official results	0.3936 [8]	0.3034 [11]	0.2850 [4]	**0.3037** [17]
+Bo1(T)	0.3823	0.2573	<u>0.2875</u>	0.2624
+KL(T)	0.3881	0.2555	<u>0.2853</u>	0.2616
+Bo1(TD)	0.3863	0.2797	0.2843	0.2710
+KL(TD)	0.3898	0.2781	0.2809	0.2697
+Bo1(TDN)	0.3921	<u>0.3303</u>	**0.2937**	0.2208
+KL(TDN)	**<u>0.3974</u>**	**<u>0.3390</u>**	**0.2924**	0.2178

6 Conclusions

This paper describes and evaluates the use of Geographical Knowledge Re-Ranking, Linguistic Processing, and Query Expansion techniques to improve Geographical Information Retrieval effectiveness. The evaluation has been performed with the full GeoCLEF GIR test collections for English, which include stories from *The Glasgow Herald* (1995) and the *Los Angeles Times* (1994) newspapers and a set of 100 topics. Evaluated separately each one of these methods has improved the MAP and R-Precision showing statistical significance with respect to the standard IR baselines TF-IDF, BM25 and InL2 in most of the experiments. The best results in MAP and R-Precision are obtained with the InL2 algorithm using the following techniques: Geographical Knowledge Re-Ranking, Lemmatization with Stemming, and Kullback-Leibler Query Expansion. Some configurations with Geographical Knowledge Re-Ranking, Linguistic Processing and Query Expansion have improved the MAP of the best official results at GeoCLEF evaluations of 2005, 2006, and 2007. The Geographical Knowledge Re-Ranking approach presented has its limitations and there is room for improvements, specially in the Toponym Recognition and Disambiguation processes. Due to the fact that the Toponym Disambiguation heuristics employed were context independent and gave more importance to some specific toponyms and feature types it is expected that the approach could have difficulties with more locally oriented news and texts in which the disambiguation is more difficult. Further work should be to design and test context aware heuristics that could adapt to different kind of documents. Regarding the adaptability of the techniques to other languages, it is expected that these techniques will work with the same kind of texts but the coverage of the gazetteer for the new language should be checked. Further work also includes the change of the NLP and NERC phases for a Geonames Gazetteer lookup of tokens and evaluate the performance of both methods.

Acknowledgments. Work supported by the Spanish research project SKATER (TIN2012-38584-C06-01).

References

1. Amati, G.: Probability Models for Information Retrieval Based on Divergence From Randomness. Ph.D. thesis, University of Glasgow (2003)
2. Brants, T.: TnT: A Statistical Part-of-speech Tagger. In: Proceedings of the Sixth Conference on Applied Natural Language Processing, ANLC2000, pp. 224–231. Association for Computational Linguistics, Stroudsburg (2000). http://dx.doi.org/10.3115/974147.974178
3. Buscaldi, D., Rosso, P.: Explicit Query Diversification for Geographical Information Retrieval. In: The 33rd European Conference on Information Retrieval, ECIR 2011, Ireland, pp. 73–80. (April 2011). https://hal.archives-ouvertes.fr/hal-00596899
4. Ferrés, D., Rodríguez, H.: TALP at GeoCLEF 2007: Results of a Geographical Knowledge Filtering Approach with Terrier. In: Peters, C., Jijkoun, V., Mandl, T., Müller, H., Oard, D.W., Peñas, A., Petras, V., Santos, D. (eds.) CLEF 2007. LNCS, vol. 5152, pp. 830–833. Springer, Heidelberg (2008)

5. Hill, L.L.: Core Elements of Digital Gazetteers: Placenames, Categories, and Footprints. In: Borbinha, J.L., Baker, T. (eds.) ECDL 2000. LNCS, vol. 1923, pp. 280–290. Springer, Heidelberg (2000)

6. Jones, C.B., Purves, R.S.: Geographical Information Retrieval. International Journal of Geographical Information Science **22**(3), 219–228 (2008). http://dx.doi.org/10.1080/13658810701626343

7. Jones, R., Zhang, W.V., Rey, B., Jhala, P., Stipp, E.: Geographic Intention and Modification in Web Search. Int. J. Geogr. Inf. Sci. **22**(3), 229–246 (2008). http://dx.doi.org/10.1080/13658810701626186

8. Larson, R.R., Gey, F.C., Petras, V.: Berkeley at GeoCLEF: Logistic Regression and Fusion for Geographic Information Retrieval. In: Peters, C., et al. (eds.) CLEF 2005. LNCS, vol. 4022, pp. 963–976. Springer, Heidelberg (2006)

9. Mandl, T., Gey, F.C., Nunzio, G.M.D., Ferro, N., Sanderson, M., Santos, D., Womser-Hacker, C.: An Evaluation Resource for Geographic Information Retrieval. In: Proceedings of the International Conference on Language Resources and Evaluation, LREC 2008, May 26-June 1, Marrakech, Morocco. European Language Resources Association (2008). http://www.lrec-conf.org/proceedings/lrec2008/summaries/8.html

10. Martins, B., Calado, P.: Learning to Rank for Geographic Information Retrieval. In: Purves, R., Clough, P.D., Jones, C.B. (eds.) Proceedings of the 6th Workshop on Geographic Information Retrieval, GIR 2010, Zurich, Switzerland, February 18–19. ACM (2010). http://doi.acm.org/10.1145/1722080.1722107

11. Martins, B., Cardoso, N., Chaves, M.S., Andrade, L., Silva, M.J.: The University of Lisbon at GeoCLEF 2006. In: Peters, C., Clough, P., Gey, F.C., Karlgren, J., Magnini, B., Oard, D.W., de Rijke, M., Stempfhuber, M. (eds.) CLEF 2006. LNCS, vol. 4730, pp. 986–994. Springer, Heidelberg (2007)

12. Ounis, I., Amati, G., Plachouras, V., He, B., Macdonald, C., Lioma, C.: Terrier: A High Performance and Scalable Information Retrieval Platform. In: Proceedings of ACM SIGIR 2006 Workshop on Open Source Information Retrieval (OSIR 2006) (2006)

13. Perea-Ortega, J.M., García-Cumbreras, M.A., Ureña-López, L.A., García-Vega, M.: Geo-Textual Relevance Ranking to Improve a Text-Based Retrieval for Geographic Queries. In: Muñoz, R., Montoyo, A., Métais, E. (eds.) NLDB 2011. LNCS, vol. 6716, pp. 278–281. Springer, Heidelberg (2011)

14. Sakai, T.: Statistical Reform in Information Retrieval? SIGIR Forum **48**(1), 3–12 (2014). http://doi.acm.org/10.1145/2641383.2641385

15. Smucker, M.D., Allan, J., Carterette, B.: A Comparison of Statistical Significance Tests for Information Retrieval Evaluation. In: Proceedings of the Sixteenth ACM Conference on Conference on Information and Knowledge Management, CIKM 2007, pp. 623–632. ACM, New York (2007). http://doi.acm.org/10.1145/1321440.1321528

16. Wang, R., Neumann, G.: Ontology-Based Query Construction for GeoCLEF. In: Peters, C., Deselaers, T., Ferro, N., Gonzalo, J., Jones, G.J.F., Kurimo, M., Mandl, T., Peñas, A., Petras, V. (eds.) CLEF 2008. LNCS, vol. 5706, pp. 880–884. Springer, Heidelberg (2009)

17. Wang, R., Neumann, G.: Ontology-Based Query Construction for GeoCLEF. In: Peters, C., et al. (eds.) CLEF 2008. LNCS, vol. 5706, pp. 880–884. Springer, Heidelberg (2009)

Improved Practical Compact Dynamic Tries

Andreas Poyias$^{(\boxtimes)}$ and Rajeev Raman

University of Leicester, Leicester, UK
{ap480,r.raman}@le.ac.uk

Abstract. We consider the problem of implementing a *dynamic trie* with an emphasis on good practical performance. For a trie with n nodes with an alphabet of size σ, the information-theoretic lower bound is $n \log \sigma + O(n)$ bits. The Bonsai data structure [1] supports trie operations in $O(1)$ expected time (based on assumptions about the behaviour of hash functions). While its practical speed performance is excellent, its space usage of $(1 + \epsilon)n(\log \sigma + O(\log \log n))$ bits, where ϵ is any constant > 0, is not asymptotically optimal. We propose an alternative, *m-Bonsai*, that uses $(1 + \epsilon)n(\log \sigma + O(1))$ bits in expectation, and supports operations in $O(1)$ expected time (again based on assumptions about the behaviour of hash functions). We give a heuristic implementation of m-Bonsai which uses considerably less memory and is slightly faster than the original Bonsai.

1 Introduction

In this paper, we consider *practical* approaches to the problem of implementing a *dynamic trie* in a highly space-efficient manner. A dynamic trie (also known as a dynamic *cardinal* tree [2]) is a rooted tree, where each child of a node is labelled with a distinct symbol from an alphabet $\Sigma = \{0, \dots, \sigma - 1\}$. We consider dynamic tries that support the following operations:

create(): Create a new empty tree.
getRoot(): return the root of the current tree.
getChild(v, i): return child node of node v with symbol i, if any (and return -1 if no such child exists).
addChild(v, i): add new child with symbol i and return the newly created node.
getParent(v): return the parent of node v.

We do not discuss deletions explicitly, but do indicate what is possible with regards to deletions. A trie is a classic data structure (the name dates back to 1959) and has numerous applications in string processing. A naive implementation of tries uses pointers. Using this approach, each node in an n-node binary trie uses 3 pointers for the navigational operations. A popular alternative for larger alphabets is the *ternary search tree (TST)* [3], which uses 4 pointers (3 plus a parent pointer), in addition to the space for a symbol. Other approaches include the *double-array trie (DAT)*, which uses a minimum of two integers per node, each of magnitude $O(n)$. Since a pointer must asymptotically use $\Omega(\log n)$ bits

© Springer International Publishing Switzerland 2015
C. Iliopoulos et al. (Eds.): SPIRE 2015, LNCS 9309, pp. 324–336, 2015.
DOI: 10.1007/978-3-319-23826-5_31

of memory, the asymptotic space bound of TST (or DAT) is $O(n(\log n + \log \sigma))$ bits. However, the information-theoretic space lower bound of $n \log \sigma + O(n)$ bits (see e.g. [2]) corresponds to one symbol and $O(1)$ *bits* per node. Clearly, if σ is small, both TST and DAT are asymptotically non-optimal. In practice, $\log \sigma$ is a few bits, or one or two bytes at most. An overhead of 4 pointers, or $32n$ bytes on today's machines, makes it impossible to hold tries with even moderately many nodes in main memory. Although tries can be *path-compressed* by deleting nodes with just one child and storing paths explicitly, this approach (or even more elaborate ones like [4]) cannot guarantee a small space bound.

Motivated by this, a number of space-efficient solutions were proposed [2,5–8], which represent *static* tries in information-theoretically optimal space, and support a wide range of operations. A number of asymptotic worst-case results were given in [9–12]. As our focus is on practical performance, we do not discuss all previous results in detail and refer the reader to e.g. [11] for a comparison. For completeness, we give a summary of some the results of [11,12]. The first uses almost optimal $2n + n \log \sigma + o(n \log \sigma)$ bits, and supports trie operations in $O(1)$ time if $\sigma = \text{polylog}(n)$ and in $O(\log \sigma / \log \log \sigma)$ time otherwise. The second [12, Theorem2] uses $O(n \log \sigma)$ bits and supports individual dynamic trie operations in $O(\log \log n)$ amortized expected time, although finding the longest prefix of a string in the trie can be done in $O(\log k / \log_\sigma n + \log \log n)$ expected time. Neither of these has been fully implemented, although a preliminary attempt (without memory usage measurements) was presented in [13]. Finally, we mention the *wavelet trie* [14] which is a data structure for a sequence of strings, and in principle can replace tries in many applications. Although in theory it is dynamic, we are not aware of any implementation of a dynamic wavelet trie.

Predating most of this work, Darragh et al. [1] proposed the *Bonsai* data structure, which uses a different approach to support the above dynamic trie operations in $O(1)$ expected time (based on assumptions about the behaviour of hash functions). While its practical speed performance is excellent, we note here that the asymptotic space usage of the Bonsai data structure is $(1 + \epsilon)n(\log \sigma + O(\log \log n))$ bits, where ϵ is any constant > 0, which is not asymptotically optimal due to the addition of $O(\log \log n)$ term. The additive $O(n \log \log n)$ bits term can be significant in many practical applications where the alphabet size is relatively small, including one involving mining frequent patterns that we are considering. The Bonsai data structure also has a certain chance of failure: if it fails then the data structure may need to be rebuilt, and its not clear how to do this without affecting the space and time complexities.

In this paper, we introduce m-Bonsai[1], a variant of Bonsai. Again, based upon the same assumptions about the behaviour of [1], our variant uses $(1+\epsilon)n(\log \sigma + O(1))$ bits of memory in expectation, where ϵ is any constant > 0, which is asymptotically optimal, and operations take $O(1)$ expected time. We give two practical variants of m-Bonsai: m-Bonsai (γ) and m-Bonsai (recursive). Our implementations and experimental evaluations show that m-Bonsai (recursive) is consistently a bit faster than the original Bonsai and significantly more

---

[1] This could be read as mame-bonsai, a kind of small bonsai plant, or mini-bonsai.

space-efficient than the original, while m-Bonsai (γ) is even more space efficient but rather slower. Of course, all Bonsai variants use at least 20 times less space than TSTs for small alphabets and compare well in terms of speed with TSTs. We also note that our experiments show that the hash functions used in Bonsai appear to behave in line with the assumptions about their behaviour. Finally, for both Bonsai and m-Bonsai, we believe it is relatively easy to remove the $(1 + \epsilon)$ multiplicative factor from the $n \log \sigma$ term, but since this is not our primary interest is robust practical performance, we have not pursued this avenue.

The rest of this paper is organized as follows. In Section 2, we talk about the asymptotics of Bonsai [1] and give a practical analysis. Section 3 summarizes m-Bonsai approach which is followed by Section 4 the experimental evaluation.

2 Preliminaries

Bit-vectors. Given a bit string $x_1, \ldots, x_n$, we define the following operations:

$select_1(x, i)$: Given an index i, return the location of i_{th} 1 in x.
$rank_1(x, i)$: Return the number of 1s upto and including location i in x.

Lemma 1 (Pătraşcu [15]). *A bit string can be represented in $n + O(n/(\log n)^2)$ bits such that $select_1$ and $rank_1$ can be supported in $O(1)$ time.*

Asymptotics of Bonsai. We now sketch the Bonsai data structure, focussing on asymptotics. It uses an array Q of size M to store a tree with $n = \lfloor \alpha M \rfloor$ nodes for some $0 < \alpha < 1$ (we assume that n and M are known at the start of the algorithm). We refer to α as the *load factor*. The Bonsai data structure refers to nodes via a unique *node ID*, which is a pair $\langle i, j \rangle$ where $0 \leq i < M$ and $0 \leq j < \lambda$, where λ is an integer parameter that we discuss in greater detail below. If we wish to add a child w with symbol $c \in \Sigma$ to a node v with node ID $\langle i, j \rangle$, then w's node ID is obtained as follows: We create the *key* of w using the node ID of v, which is a triple $\langle i, j, c \rangle$. We evaluate a hash function $h : \{0, \ldots, M \cdot \lambda \cdot \sigma - 1\} \mapsto \{0, \ldots, M - 1\}$ on the key of w. If $i' = h(\langle i, j, c \rangle)$, the node ID of w is $\langle i', j' \rangle$ where $j' \geq 0$ is the lowest integer such that there is no existing node with a node ID $\langle i', j' \rangle$; i' is called the *initial address* of w.

In order to check if a node has a child with symbol c, keys are stored in Q using open addressing and linear probing[2]. The space usage of Q is kept low by the use of *quotienting* [16]. The hash function has the form $h(x) = (ax \bmod p) \bmod M$ for some prime $p > M \cdot \lambda \cdot \sigma$ and multiplier a, $1 \leq a \leq p - 1$. Q only contains the *quotient* value $q(x) = \lfloor (ax \bmod p)/M \rfloor$ corresponding to x. Given $h(x)$ and $q(x)$, it is possible to reconstruct x to check for membership. While checking for membership for x, one needs to know $h(y)$ for all keys y encountered during the search, which is not obvious since keys may not be stored at their initial address due to collisions. The Bonsai approach is to keep all keys with the same initial address in consecutive locations in Q (this means that keys may be moved

---

[2] A variant, *bidirectional* probing, is used in [1], but we simplify this to linear probing

after they have been inserted) and to use two bit-vectors of size M bits to effect the mapping from a node's initial address to the position in Q containing its quotient, for details see [1]. Clearly, being able to search for, and insert keys allows us to support *getChild* and *addChild*; for *getParent(v)* note that the key of v encodes the node ID of its parent.

Asymptotic space usage. In addition to the two bit-vectors of M bits each, the main space usage of the Bonsai structure is Q. Since a prime p can be found that is $< 2 \cdot M \cdot \lambda \cdot \sigma$, it follows that the values in Q are at most $\lceil \log_2(2\sigma\lambda + 1) \rceil$ bits. The space usage of Bonsai is therefore $M(\log \sigma + \log \lambda + O(1))$ bits.

Since the choice of the prime p depends on λ, λ must be fixed in advance. However, if more than λ keys are hashed to any value in $\{0, \ldots, M-1\}$, the algorithm is unable to continue[3]. Thus, λ should be chosen large enough to reduce the probability of more than λ keys hashing to the same initial address to acceptable levels. In [1] the authors, assuming the hash function has full independence and is uniformly random, argue that choosing $\lambda = O(\log M / \log \log M)$ reduces the probability of error to at most M^{-c} for any constant c (choosing asymptotically smaller λ causes the algorithm almost certainly to fail). As the optimal space usage for an n-node trie on an alphabet of size σ is $O(n \log \sigma)$ bits, the additive term of $O(M \log \lambda) = O(n \log \log n)$ makes the space usage of Bonsai non-optimal for small alphabets.

However, even this choice of λ is not well-justified from a formal perspective, since the hash function used is quite weak—it is only 2-universal [17]. For 2-universal hash functions, the maximum number of collisions can only be bounded to $O(\sqrt{n})$ [18] (note that it is not obvious how to use more robust hash functions, since quotienting may not be possible). Choosing λ to be this large would make the space usage of the Bonsai structure asymptotically uninteresting.

Practical Analysis. In practice, we note that choosing $\lambda = 32$, and assuming complete independence in the hash function, the error probability for M up to 2^{64} is about 10^{-19} for $\alpha = 0.8$, using the formula in [1]. Choosing $\lambda = 16$ as suggested in [1] suggests a high failure probability for $M = 2^{56}$ and $\alpha = 0.8$. Also, in practice, the prime p is not significantly larger than $M\lambda\sigma$ [19, Lemma5.1]. The space usage of the Bonsai structure therefore is taken to be $(\lceil \log \sigma \rceil + 7)M$ bits for the tree sizes under consideration in this paper.

3   m-Bonsai

3.1   Overview

In our approach, each node again has an associated key that needs to be searched for in a hash table, again implemented using open addressing with linear probing and quotienting. However, the ID of a node x in our case is a

---

[3] Particularly for non-constant alphabets, it is not clear how to rebuild the data structure without an asymptotic penalty.

number from $\{0, \ldots, M - 1\}$ that refers to the index in Q that contains the quotient corresponding to x. If a node with ID i has a child with symbol $c \in \Sigma$, the child's key, which is $\langle i, c \rangle$, is hashed using a multiplicative hash function $h : \{0, \ldots, M \cdot \sigma - 1\} \mapsto \{0, \ldots, M - 1\}$, and an initial address i' is computed. If i'' is the smallest index $\geq i'$ such that $Q[i'']$ is vacant, then we store $q(x)$ in $Q[i'']$. Observe that $q(x) \leq \lceil 2\sigma \rceil$, so Q takes $M \log \sigma + O(M)$ bits. In addition, we have a *displacement* array D, and set $D[i''] = i'' - i'$. From the pair $Q[l]$ and $D[l]$, we can obtain both the initial hash address of the key stored there and its quotient, and thus reconstruct the key. The key idea is that in expectation, the average value in D is small:

Proposition 1. *Assuming h is fully independent and uniformly random, the expected value of $\sum_{i=0}^{M-1} D[i]$ after all $n = \alpha M$ nodes have been inserted is $\approx M \cdot \frac{\alpha^2}{2(1-\alpha)}$.*

Proof. The average number of probes, over all keys in the table, made in a successful search is $\approx \frac{1}{2}(1 + \frac{1}{1-\alpha})$ [16]. Multiplying this by $n = \alpha M$ gives the total average number of probes. However, the number of probes for a key is one more than its displacement value. Subtracting αM from the above and simplifying gives the result.

Thus, encoding D using variable-length encoding could be very beneficial. For example, coding D in unary would take $M + \sum_{i=1}^{M} D[i]$ bits; by Proposition 1, and plugging in $\alpha = 0.8$, the expected space usage of D, encoded in unary, should be about $2.6M$ bits, which is smaller than the overhead of $7M$ bits of the original Bonsai. As shown in Table 1, predictions made using Proposition 1 are generally quite accurate. Table 1 also suggests that encoding each $D[i]$ using the γ-code, we would come down to about $2.1M$ bits for the D, for $\alpha = 0.8$.

Table 1. Average number of bits per entry needed to encode the displacement array using the unary, γ and Golomb encodings. For the unary encoding, Proposition 1 predicts 1.816, 2.6 and 5.05 bits per value. For file details see Table 2.

	unary			γ			Golomb		
Load Factor	0.7	0.8	0.9	0.7	0.8	0.9	0.7	0.8	0.9
Pumsb	1.81	2.58	5.05	1.74	2.11	2.65	2.32	2.69	3.64
Accidents	1.81	2.58	5.06	1.74	2.11	2.69	2.33	2.69	3.91
Webdocs	1.82	2.61	5.05	1.75	2.11	2.70	2.33	2.70	3.92

3.2 Representing the Displacement Array

We now describe how to represent the displacement array. A *write-once dynamic array* is a data structure for a sequence of supports the following operations:

create(n): Create an array A of size n with all entries initialized to zero.
set(A, i, v): If $A[i] = 0$, set $A[i]$ to v (assume $0 < v \leq n$). If $A[i] \neq 0$ then $A[i]$ is unchanged.
get(A, i): Return $A[i]$.

The following lemma shows how to implement such a data structure. Note that the apparently slow running time of *set* is enough to represent the displacement array without asymptotic slowdown: setting $D[i] = v$ means that $O(v)$ time has already been spent in the hash table finding an empty slot for the key.

Lemma 2. *A write-once dynamic array A of size n containing non-negative integers can be represented in space $\sum_{i=1}^{n} |\gamma(A[i]+1)| + o(n)$ bits, supporting get in $O(1)$ time and set(A, i, v) in $O(v)$ amortized time.*

Proof. We divide A into contiguous blocks of size $b = (\log n)^{3/2}$. The i-th block $B_i = A[bi..bi+b-1]$ will be stored in a contiguous sequence of memory locations. There will be a pointer pointing to the start of B_i. Let $G_i = \sum_{j=bi}^{bi+b-1} |\gamma(A[j]+1)|$.

We first give a naive representation of a block. All values in a block are encoded using γ-codes and concatenated into a single bit-string (at least in essence, see discussion of the *get* operation below). A *set* operation is performed by decoding all the γ-codes in the block, and re-encoding the new sequence of γ-codes. Since each γ-code is $O(\log n)$ bits, or $O(1)$ words, long, it can be decoded in $O(1)$ time. Decoding and re-encoding an entire block therefore takes $O(b)$ time, which is also the time for the *set* operation. A *get* operation can be realized in $O(1)$ time using the standard idea of to concatenating the unary and binary portions of the γ-codes separately into two bit-strings, and to use *select*$_1$ operations on the unary bit-string to obtain, in $O(1)$ time, the binary portion of the i-th γ-code. The space usage of the naive representation is $\sum_i G_i + O((\sum_i G_i)/(\log n)^2) + (n \log n)/b)$ bits: the second term comes from Lemma 1 and the third accounts for the pointers and any unused space in the "last" word of a block representation. This adds up to $\sum_i G_i + o(n)$ bits, as required.

Since at most b *set* operations can be performed on a block, if any value in a block is set to a value $\geq b^2$, we can use the $\Omega(b^2)$ time allowed for this operation to re-create the block in the naive representation, and also to amortize the costs of all subsequent *set* operations on this block. Thus, we assume wlog that all values in a block are $< b^2$, and hence, that γ-codes in a block are $O(\log b) = O(\log \log n)$ bits long. We now explain how to deal with this case. We divide each block into *segments* of $\ell = \lceil c \log n / \log \log n \rceil$ values for some sufficiently small constant $c > 0$, which are followed by an *overflow* zone of at most $o = \lceil \sqrt{\log n} \log \log n \rceil$ bits. Each segment is represented as a bit-string of concatenated γ-codes. All segments, and their overflow zones, are concatenated into a single bit-string. The bit-string of the i-th block, also denoted B_i, has length at most $G_i + (b/\ell) \cdot o = G_i + O(\log n (\log \log n)^2)$. As we can ensure that a segment is of size at most $(\log n)/2$ by choosing c small enough, we can decode an individual γ-code in any segment in $O(1)$ time using table lookup. We can also support a *set* operation on a segment in $O(1)$ time, by overwriting the sub-string of B_i that represents this segment, provided the overflow zone is large enough to accommodate the new segment.

If the overflow zone is exhausted, the time taken by the *set* operations that have taken place in this segment alone is $\Omega(\sqrt{\log n} \log \log n)$. Since the length of B_i is at most $O(\sqrt{\log n} \log \log n)$ words, when any segment overflows, we can simply copy B_i to a new sequence of memory words, and while copying, use table lookup again to rewrite B_i, ensuring that each segment has an overflow zone of exactly o bits following it (note that as each segment is of length $\Omega(\log n/\log\log n)$ bits and the overflow zones are much smaller, rewriting a collection of segments that fit into $O(\log n)$ bits gives a new bit-string which is also $O(\log n)$ bits).

One final component is that for each block, we need to be able to find the start of individual segments. As the size of a segment and its overflow zone is an integer of at most $O(\log\log n)$ bits, and there are only $O(\sqrt{\log n}\log\log n)$ segments in a block, we can store the sizes of the segments in a block in a single word and perform the appropriate prefix sum operations in $O(1)$ time using table lookup, thereby also supporting get in $O(1)$ time. This proves Lemma 2.

Theorem 1. *For any given integers M and σ and constant $0 < \alpha < 1$, there is a data structure that represents a trie on an alphabet of size σ with n nodes, where $n \leq \alpha M$, using $M\log\sigma + O(M)$ bits of memory in expectation, and supporting create() in $O(M)$ time, getRoot and getParent in $O(1)$ time, and addChild and getChild in $O(1)$ expected time. The expected time bounds are based upon the assumption that the hash function has full randomness and independence.*

Proof. Follows directly from Proposition 1 and Lemma 2, and from the observation that $|\gamma(x+1)| \leq x + 2$ for all $x \geq 0$.

Remark 1. In the Bonsai (and m-Bonsai) approaches, deletion of an internal node is in general not $O(1)$ time, since the node IDs of all descendants of a node are dependent on its own node ID. It is possible in m-Bonsai to delete a leaf, taking care (as in standard linear probing) to indicate that a deleted location in Q previously contained a value, and extending Lemma 2 to allow a *reset(i)* operation, which changes $A[i]$ from its previous value v to 0 in $O(v)$ time.

3.3 Alternate Representation of the Displacement Array

The data structure of Lemma 2 appears to be too complex for implementation, and a naive approach to representing the displacement array (as in Lemma 2) may be slow. We therefore propose a practical alternative, which avoids any explicit use of variable-length coding.

The displacement array is stored as an array D_0 of *fixed-length* entries, with each entry being Δ_0 bits, for some integer parameter $\Delta_0 \geq 1$. All displacement values $\leq 2^{\Delta_0} - 2$ are stored as is in D_0. If $D[i] > 2^{\Delta_0} - 2$, then we set $D_0[i] = 2^{\Delta_0} - 1$, and store the value $D'[i] = D[i] - 2^{\Delta_0} + 1$ as satellite data associated with the key i in a second hash table.

This second hash table is represented using the original Bonsai representation, using a value $M' \sim \alpha' n'$, where n' is the number of keys stored in the second hash table, and α' is the load factor of this secondary hash table. The satellite data for this second hash table are also stored in an array of size M' with fixed-length entries of size Δ_1, where Δ_1 is again an integer parameter.

If $D'[i] \leq 2^{\Delta_1} - 2$, it is stored explicitly in the second-level hash table. Yet larger values of D are stored in a standard hash table. The values of α', Δ_0 and Δ_1 are currently chosen experimentally, as described in the next section.

In what follows, we refer to m-Bonsai with the displacement array represented as γ-codes as m-Bonsai (γ) and the representation discussed here as m-Bonsai (recursive), respectively.

4   Experimental Evaluation

4.1   Implementation

We implemented m-Bonsai (recursive), m-Bonsai (γ) and Bonsai in C++, and compared these with Bentley's C++ TST implementation [3]. The DAT implementation of [20] was not tested since it apparently uses 32-bit integers, limiting the maximum trie size to 2^{32} nodes, which is not a limitation for the Bonsai or TST approaches. The tests of [20] suggest that even with this "shortcut", the space usage is only a factor of 3 smaller than TST (albeit it is ~ 2 times faster).

Both Bonsai implementations used the `sdsl-lite` library [21]. The original Bonsai data structure mainly comprises three `sdsl` containers: firstly, the `int_vector<>`, which uses a fixed number of bits for each entry, is used for the Q array (also in m-Bonsai). In addition, we use two `bit_vectors` that to distinguish nodes in collision groups as in [1]. In m-Bonsai (γ), D is split into consecutive blocks of 256 values (initially all zero) each, which are stored as a concatenation of their γ-codes. We used `sdsl`'s `encode` and `decode` functions to encode and decode each block for the *set* and *get* operations.

The m-Bonsai (recursive) uses an alternative approach for the displacement array. D_0 has fixed length entries of Δ_0-bits, thus `int_vector<>` is the ideal container. If a displacement value is larger than Δ_0, we store it as a satellite data in a Bonsai data structure. The satellite data is stored again in an `int_vector<>` of Δ_1-bit entries. Finally, if the displacement value is even larger, then we use the standard C++ `std::map`. In Figure 1, we show how we chose the parameters for this approach. The three parameters α', Δ_0 and Δ_1 are selected given the trade-off of runtime speed and memory usage. For this example we have $\alpha' = 0.8$. Each line represents a different Δ_0 value in bits. The y-axis shows the total bits required per displacement value and the x-axis shows the choice of Δ_1 sizes in bits. As shown, there is a curve formed where its minimum point is when $\Delta_1 = 7$ for any Δ_0 values. $\Delta_0 = 3$ is the parameter with the lower memory usage. $\Delta_0 = 4$ uses relatively more memory and even though $\Delta_0 = 2$ is closer to $\Delta_0 = 3$ in terms of memory, it is slower in terms of runtime speed. This happens since less values are accessed directly from D_0 when $\Delta_0 = 2$, therefore we chose $\Delta_0 = 3$ and $\Delta_1 = 7$. Finally, we consider $\alpha' = 0.8$ as a good choice to have competitive runtime speed and at the same time good memory usage.

4.2   Experimental Analysis

The machine used for the experimental analysis is an Intel Pentium 64-bit machine with 8GB of main memory and a G6950 CPU clocked at 2.80GHz

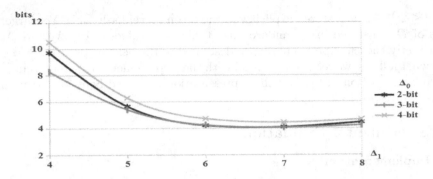

Fig. 1. This graph is an example based on Webdocs8 used in m-Bonsai (recursive) data structure with $\alpha = 0.8$. The y-axis shows the bits per M in the displacement array. The x-axis shows parameter Δ_1 and each line is based on parameter Δ_0.

Table 2. The average bits per node for datasets used for different purposes. In some cases the TST processes were unable to finish execution due to large memory required.

Datasets	Node Number	σ	m-Bonsai (r)	m-Bonsai(γ)	Bonsai	TST
Pumsb	1125375	7117	20.45	18.91	24	390.87
Accidents	4242318	442	15.65	14.12	19.2	388.26
Webdocs8	63985704	5577	20.45	18.91	24	386.79
Webdocs	231232676	5267657	27.04	30.91	36	385.1
splitPumsb	6702990	5	8.45	6.75	12	383.92
splitAccidents	17183926	5	8.45	6.75	12	387.07
splitWebdocs8	333444484	5	8.45	6.76	12	
splitWebdocs	1448707216	5	8.45	6.78	12	
SRR034939.fastq	3095560	5	8.45	6.73	12	385.88
SRR034944.fastq	21005059	5	8.45	6.76	12	385.76
SRR034940-1.fastq	1556235309	5	8.45	6.68	12	
SRR034945-1.fastq	1728553810	5	8.45	6.68	12	

with 3MB L2 cache, running Ubuntu 12.04.5 LTS Linux. All the code was compiled using g++ 4.7.3 with optimization level 6. To measure the resident memory (RES), /proc/self/stat was used. For the speed tests we measured wall clock time using std::chrono::duration_cast.

We use benchmark datasets arising arising in frequent pattern mining [22], where each "string" is a subset of a large alphabet (up to tens of thousands). In some frequent pattern mining algorithms such as [23], these strings need to be traversed in sorted order, which takes a slow $O(n\sigma)$ time in all Bonsai variants because they do not support the next-sibling operation. To get around this, we divide each symbol into 3-bit values, which optimizes the trade-off in runtime

Table 3. The wall clock time in seconds for the construction of the Trie. Note that the TST was affected by thrashing in Webdocs and splitWebdocs8.

Datasets	m-Bonsai (r)	m-Bonsai(γ)	Bonsai	TST
Pumsb	0.55	5.97	0.86	0.64
Accidents	2.06	21.70	3.12	2.33
Webdocs8	27.03	255.25	35.13	19.38
Webdocs	110.35	886.17	125.06	608.91
splitPumsb	3.30	37.03	5.21	2.29
splitAccidents	7.72	82.95	10.92	5.69
splitWebdocs8	128.88	1287.49	173.25	1862.49
splitWebdocs	626.20	5439.71	832.8	
SRR034939.fastq	0.561	9.82	0.74	0.61
SRR034944.fastq	6.041	72.38	6.84	4.39
SRR034940-1.fastq	746.005	5801.6	936.88	
SRR034945-1.fastq	851.164	6456.18	1054.43	

speed and memory usage. Finally, we used sets of short read genome strings given in the standard FASTQ format.

Memory Usage: For the memory usage experiments we set $\alpha = 0.8$ for all Bonsai data structures. Then, we insert all the strings of each dataset in the trees and we measure resident memory. Table 2 is showing the average bits per n. It is obvious that the Bonsai data structure is quite badly affected on datasets with low σ. By converting the values of Table 2 in scale of bits per M (as explained in Section 3.1 $n = \alpha M$), we prove the practical analysis of Section 2, showing that Bonsai requires $10M$-bits for the FASTQ sequences out of which $7M$-bits are used only to map the nodes in Q array. The m-Bonsai (γ) performance is very good which needs more than 40% less memory than Bonsai on lower σ datasets. The m-Bonsai (recursive) is also performing better than Bonsai and it is obvious that as σ gets lower the recursive approach becomes more efficient by avoiding the relatively big overhead of Bonsai.

Tree construction (Runtime speed): In Table 3 we show the wall clock time in seconds for the construction of the Tree. The m-Bonsai (recursive) is proved to be competitively fast and even faster than TST for some cases like Pumsb and Accidents. This happens since m-Bonsai (recursive) is able to fit a big part of data structure in cache memory. However, when both data structures use more heavily the main memory (Webdocs8), the pointer-based TST is 1.4 times faster. The Bonsai implementation is consistently slower than TST and m-Bonsai (recursive). Since the m-Bonsai (recursive) has a write once linear probing approach, when inserting a node in empty location $Q[i]$, we know that $D[i]$ is free for insertions. Now, if $D[i]$ is supposed to be zero then we don't even need to access D as it is already initialised to zeros[4]. However, Bonsai always needs to

---

[4] Approximately 48% of the nodes have 0 displacement value at $\alpha = 0.8$.

Table 4. The wall clock time in nanoseconds per successful search operations.

Datasets	m-Bonsai (r)	m-Bonsai(γ)	Bonsai	TST
Pumsb.search	237	1345	358	105
Webdocs8.search	332	1672	608	117
splitWebdocs.search	416	2037	657	
SRR034940-1.search	403	1932	658	

access at least one more bit-vector to reassure and mark the empty location. Additionally, in case of collision Bonsai requires to swap elements in Q and one of the bit-vectors, to make space for the new node at a matching location. Also, if any satellite data(not included in this experiment) is required, it has to move to match location as well thus potentially impacting the runtime performance. Finally, the compact m-Bonsai (γ) is about ten times slower. This is due to the O(b) time required to access each value as explained in Section 3.1 $n = \alpha M$.

Successful search runtime speed: For this experiment we designed our own .*search* datasets, where we randomly picked 10% of the strings from each dataset. As shown in Table 4 we selected some datasets from our repository mainly due to space limit. After the tree construction, we measured the time needed in nanoseconds per successful search operation. It is obvious that TST is the fastest approach. However, m-Bonsai (recursive) remains competitive with TST and consistently faster than Bonsai by at least 1.5 times, whereas m-Bonsai (γ) in the slowest. Note that there is an increase in runtime speed per search operation for all Bonsai data structures as the datasets get bigger. However, we can't prove this for TST, since it is not able to process the larger datasets.

5   Conclusion

We have demonstrated a new variant of the Bonsai approach to store large tries in a very space-efficient manner. Not only have we (re)-confirmed that the original Bonsai approach is very fast and space-efficient on modern architectures, both m-Bonsai variants we propose are significantly smaller (both asymptotically and in practice) and and one of them is a bit faster than the original Bonsai. In the near future we intend to investigate other variants, to give a less ad-hoc approach to m-Bonsai (recursive), and to compare with other trie implementations.

Neither of our approaches is very close to the information-theoretic lower bound of $(\sigma \log \sigma - (\sigma - 1) \log(\sigma - 1))n - O(\log(kn))$ bits [2]. For example, for $\sigma = 5$, the lower bound is $3.61n$ bits, while m-Bonsai (γ) takes $\sim 5.6M \sim 7n$ bits. Closing this gap would be an interesting future direction. Another interesting open question is to obtain a practical compact dynamic trie that has a wider range of operations, e.g. being able to navigate directly to the sibling of a node.

References

1. Darragh, J.J., Cleary, J.G., Witten, I.H.: Bonsai: a compact representation of trees. Softw., Pract. Exper. **23**(3), 277–291 (1993)
2. Benoit, D., Demaine, E.D., Munro, J.I., Raman, R., Raman, V., Rao, S.S.: Representing trees of higher degree. Algorithmica **43**(4), 275–292 (2005)
3. Bentley, J., Sedgewick, B.: Ternary search trees (1998). http://www.drdobbs.com/database/ternary-search-trees/184410528
4. Nilsson, S., Tikkanen, M.: An experimental study of compression methods for dynamic tries. Algorithmica **33**(1), 19–33 (2002)
5. Jacobson, G.: Space-efficient static trees and graphs. In: Proc. 30th Annual Symposium on Foundations of Computer Science, pp. 549–554. IEEE Computer Society (1989)
6. Raman, R., Raman, V., Satti, S.R.: Succinct indexable dictionaries with applications to encoding k-ary trees, prefix sums and multisets. ACM Transactions on Algorithms **3**(4) (2007)
7. Farzan, A., Munro, J.I.: A uniform paradigm to succinctly encode various families of trees. Algorithmica **68**(1), 16–40 (2014)
8. Farzan, A., Raman, R., Rao, S.S.: Universal succinct representations of trees? In: Albers, S., Marchetti-Spaccamela, A., Matias, Y., Nikoletseas, S., Thomas, W. (eds.) ICALP 2009, Part I. LNCS, vol. 5555, pp. 451–462. Springer, Heidelberg (2009)
9. Munro, J.I., Raman, V., Storm, A.J.: Representing dynamic binary trees succinctly. In: Kosaraju, S.R. (ed.) Proc. 12th Annual Symposium on Discrete Algorithms, pp. 529–536. ACM/SIAM (2001)
10. Raman, R., Rao, S.S.: Succinct dynamic dictionaries and trees. In: Baeten, J.C.M., Lenstra, J.K., Parrow, J., Woeginger, G.J. (eds.) ICALP 2003. LNCS, vol. 2719, pp. 357–368. Springer, Heidelberg (2003)
11. Arroyuelo, D., Davoodi, P., Satti, S.: Succinct dynamic cardinal trees. Algorithmica, 1–36 (2015) (online first)
12. Jansson, J., Sadakane, K., Sung, W.: Linked dynamic tries with applications to lz-compression in sublinear time and space. Algorithmica **71**(4), 969–988 (2015)
13. Takagi, T., Uemura, T., Inenaga, S., Sadakane, K., Arimura, H.:Applications of succinct dynamic compact tries to some stringproblems (presented at WAAC 2013). http://www-ikn.ist.hokudai.ac.jp/~arim/papers/waac13takagi.pdf
14. Grossi, R., Ottaviano, G.: The wavelet trie: maintaining an indexed sequence of strings in compressed space. In: PODS, pp. 203–214 (2012)
15. Patrascu, M.: Succincter. In: 49th Annual IEEE Symp. Foundations of Computer Science, pp. 305–313. IEEE Computer Society (2008)
16. Knuth, D.E.: The Art of Computer Programming. Sorting and Searching, vol. 3, 2nd edn. Addison Wesley Longman (1998)
17. Carter, L., Wegman, M.N.: Universal classes of hash functions. J. Comput. Syst. Sci. **18**(2), 143–154 (1979)
18. Fredman, M.L., Komlós, J., Szemerédi, E.: Storing a sparse table with 0(1) worst case access time. J. ACM **31**(3), 538–544 (1984)
19. Pagh, R.: Low redundancy in static dictionaries with constant query time. SIAM J. Comput. **31**(2), 353–363 (2001)
20. Yoshinaga, N., Kitsuregawa, M.: A self-adaptive classifier for efficient text-stream processing. In: COLING 2014, 25th International Conference on Computational Linguistics, Proceedings of the Conference: Technical Papers, August 23–29, 2014, Dublin, Ireland, pp. 1091–1102 (2014)

21. Gog, S., Beller, T., Moffat, A., Petri, M.: From theory to practice: plug and play with succinct data structures. In: Gudmundsson, J., Katajainen, J. (eds.) SEA 2014. LNCS, vol. 8504, pp. 326–337. Springer, Heidelberg (2014)
22. Goethals, B.: Frequent itemset mining implementations repository. http://fimi.ua. ac.be/
23. Schlegel, B., Gemulla, R., Lehner, W.: Memory-efficient frequent-itemset mining. In: Proceedings of the 14th International Conference on Extending Database Technology, EDBT 2011, Uppsala, Sweden, March 21–24, 2011, pp. 461–472 (2011)

ShRkC: Shard Rank Cutoff Prediction for Selective Search

Anagha Kulkarni[✉]

San Francisco State University, 1600 Holloway Ave, San Francisco, CA 94132, USA
ak@sfsu.edu

Abstract. In search environments where large document collections are partitioned into smaller subsets (*shards*), processing the query against only the relevant shards improves search efficiency. The problem of ranking the shards based on their estimated relevance to the query has been studied extensively. However, a related important task of identifying *how many* of the top ranked relevant shards should be searched for the query, so as to balance the competing objectives of effectiveness and efficiency, has not received much attention. This task of *shard rank cutoff estimation* is the focus of the presented work. The central premise for the proposed solution is that the number of top shards searched should be dependent on – 1. the query, 2. the given ranking of shards, and 3. on the type of *search need* being served (precision-oriented versus recall-oriented task). An array of features that capture these three factors are defined, and a regression model is induced based on these features to learn a query-specific shard rank cutoff estimator. An empirical evaluation using two large datasets demonstrates that the learned shard rank cutoff estimator provides substantial improvement in search efficiency as compared to strong baselines without degrading search effectiveness.

1 Introduction

To facilitate distributed and parallel query processing, large document collections are often partitioned into smaller subsets, referred to as *shards* [1–5]. Search efficiency can be further improved by processing the query against only a few selected shards that are likely to contain relevant documents [2,4–6]. A rich line of work, mainly originating in federated search, has extensively investigated the problem of ranking shards based on their estimated relevance to the query [7–15][1]. However, a related research problem of determining *how many* of the top ranked shards should be searched for the query has been largely understudied (exceptions are [10,16,17]. We refer to this task as that of estimating the *shard rank cutoff* in a given ranking of shards for the query.

Most prior work a adopts query-agnostic approach to set the shard rank cutoff where a preset number of top ranked shards are searched for every query. However, searching a fixed number of top shards for every query often degrades

---

[1] Prior work in federated search has referred to shards as *resources*, however, for the sake of consistency we use the newer term, shards, throughout this paper.

© Springer International Publishing Switzerland 2015
C. Iliopoulos et al. (Eds.): SPIRE 2015, LNCS 9309, pp. 337–349, 2015.
DOI: 10.1007/978-3-319-23826-5_32

either the search effectiveness or search efficiency. For instance, the fixed shard rank cutoff of 1 would lead to poor search effectiveness for more than one fourth of the queries, and a fixed cutoff of 5 would be excessive for more than half of the queries for one of the datasets used in this work. Since searching fewer shards improves efficiency, the optimal cutoff value for a ranking of shards is the smallest (or earliest) rank which maximizes search effectiveness. This optimal shard rank cutoff may be different for every query because of the inherent differences in queries. For instance, a query with a broader information-need requires larger search budget than one with a focused information-need. The other factor that directly influences the number of top shards that ought to be searched for a query is the type of *search need* – a few top shards might be sufficient for precision-oriented tasks where effectiveness at early ranks is important, but for recall-oriented tasks, a higher shard rank cutoff value might be necessary to optimize effectiveness at deeper ranks.

Based on these observations we hypothesize that a shard rank cutoff estimator that is: (1) optimized for a specific search need, and (2) explicitly models query and dataset-specific properties, is needed to balance the competing objectives of minimizing search cost and maximizing search effectiveness. To test this hypothesis we propose a shard rank cutoff estimator (ShRkC) which learns a regression function based on features that are designed to capture the salient aspects of the query and the dataset. Also, a separate estimator is learned for different search needs so as to model the differences in requirements.

The contributions of this work are: first, we formalize the shard rank cutoff estimation problem, especially as an independent research task from the shard ranking problem. Second, we propose a novel shard rank cutoff estimator that leverages three different sources of information to model the quintessential query and dataset properties. Lastly, the empirical evaluation using some of the largest datasets demonstrates that the proposed estimator provides the most cost-effective search setup.

2   Related Work

Although shard ranking has been studied extensively [17–19], the problem of shard rank cutoff estimation, especially as an independent task, has been investigated only in few studies. One of the early estimators was proposed as part of the SUSHI algorithm which also infers shard ranking. For both tasks, SUSHI uses a data structure called *central sample index* (CSI), which is a compilation of documents sampled from each shard that serve as representatives of the complete shard contents. The query is executed against the CSI and the top 50 retrieved documents are used as follows. For each shard represented in these documents, a curve is fitted using the document scores and rank. Three types of curves, linear, logarithmic, and exponential are attempted, and the one with best fit is used to interpolate scores for the top m ranks for each shard. The interpolated points from all shards are merged into a single ranking, P, and the scores are aggregated based on their shard membership. The resulting shard scores are then used to rank the shards. The number of unique shards present in the top R documents of the consolidated ranking P is predicted as the shard rank cutoff for P@R metric. Notice that SUSHI's prediction for the rank cutoff takes into consideration both, the query and the metric of interest.

The Rank-S algorithm [16] is similar to SUSHI in that it uses CSI as the basis for inferring the shard ranking, and also for estimating the shard rank cutoff value. More specifically, Rank-S algorithm uses the CSI results obtained for the query to compute a vote that each document assign to its parent shard as follows: $vote(d_i) = weight(d_i) \cdot B^{-i}$, where d_i is the document at rank i in the CSI results, $weight(d_i)$ is typically the relevance score assigned to the document d_i by the retrieval algorithm, and B is the base of the exponential decay function used to suppress the votes assigned by documents further down the result list. The votes are aggregated based on the parent shard of the documents and then used to rank the shards. The number of unique shards represented until the rank at which the vote converges to zero is predicted as the shard rank cutoff value. Although, Rank-S makes query-specific rank cutoff predictions, it cannot optimize for the metric of interest.

More recently, Markov and Crestani [17] compared some of the popular shard ranking algorithms, all of which use CSI, and observed that the number of shards searched for a query is determined by *parameter k* which is the number of retrieved CSI documents that are used for ranking the shards. They propose an unsupervised approach, named *adaptive k*, for choosing a query-specific value for k. This approach uses the number of documents retrieved from the CSI that contain all the query terms, as the value for k. For queries which yield fewer than three such documents, the documents containing one less query term are used to set the parameter. Empirical evaluation demonstrates that when applied to a variant of ReDDE, the adaptive k is as effective as the best performing fixed k approach. The adaptive k approach although query-specific, it does not adapt to different search needs (that is, metric of interest). Also, as proposed, the adaptive k estimator can only work in setups where CSI is available.

Next we describe a widely used and adapted shard ranking algorithm, ReDDE, that we use in this work as the source of shard ranking. ReDDE uses the central sample index (CSI) as a proxy for the complete collection (all shards). ReDDE first executes the user query against the CSI, and assumes that the top n retrieved documents are relevant (This is the same as parameter k in Markov and Crestani, 2014). If n_R is the number of documents in n that are mapped to shard R then a score s_R for each R is computed as: $\theta_R = n_R * w_R$, where the shard weight w_R is the ratio of size of the shard ($|R|$) and the size of its sample ($|S_R|$). The shard scores θ_R are then normalized to obtain a valid probability distribution which is used to rank the shards. Although, ReDDE provides effective shard ranking, it does not estimate the number of top shards to search for a query. The typical approach is to use a fixed cutoff value for all the queries.

3   Shard Rank Cutoff Prediction (ShRkC)

Shard ranking algorithms, such as, ReDDE [9], infer an ordering of the shards based on their estimated relevance to the query. Given such a ranking of shards for a query, our goal is to estimate the minimum number of top shards that should be searched to maximize effectiveness. More specifically, for each query q, we predict a metric-dependent shard rank cutoff as follows.

$$T_q^{met} = \text{floor}(f(\vec{X})) \tag{1}$$

where: T_q^{met}: The optimal shard rank cutoff for query q and metric met.

f: The learned regression function. $\vec{X}$: Set of features derived from S_q^n, R_q^n, CSI, and from collection statistics gathered from all shards, where: S_q^n is n ranked shards for query q, R_q^n is the corresponding normalized relevance scores for the shards at ranks 1 to n (Is a valid probability distribution), and CSI is the central sample index used to infer the shard ranking and the relevance scores for every query. The set of features, $\vec{X}$, used by the the regression function f are described in Section 4.

Based on preliminary experiments that compared three regression techniques: linear regression, SVM-regression, and random forest (RF), we choose the RF algorithm (RF) [20] to learn the regression function f due to its higher performance. RF is a type of an ensemble learning approach that combines multiple decision trees to improve the overall prediction performance. An open-source R package[21] was employed for the actual implementation of Random Forest. A separate regression model is learned for every metric of interest. In order to obtain reliable evaluation of the learned model, 10-fold cross-validation setup was implemented. Also, in order to account for the variability introduced by the random factors of the RF algorithm, 10 runs of each experiment were conducted and the average values are reported in the result tables. The RF algorithm contains two parameters that can be tuned, $mtry$ (the number of features to sample at each split in the learning process), and $ntree$ (number of trees to fit). For all the experiments in this work we used the default settings for both the parameters ($mtry$: $p/3$ where p is the total number of features, and $ntree$: 500). Section 6 explores the effects of parameter tuning on search performance.

Ground Truth: The definition of the ground truth for the shard rank cutoff prediction task reflects the two-part goal of minimizing the number of shards searched per query, and maximizing the search effectiveness for each query. Given a ranking of shards for a query, the optimal cutoff is the smallest rank T at which the search effectiveness is the maximum it can be for the given shard ranking. The search effectiveness is quantified by one of the standard IR evaluation metrics. Thus the ground truth T is defined for each <query, metric> pair.

4   ShRkC Estimator Features

We use three sources of information to define the features used by the learner: the ranking of shards for the query, the central sample index (CSI), and the collection statistics gathered from all the shards. The features were designed with the goal to model various signals, such as, *query difficulty*, where the premise is that *harder* queries need to search more shards while *easy* queries don't.

4.1   Shard Ranking Based Features

The features in this category make use of only two types of inputs: the ranking of shards produced by an algorithm, such as, ReDDE, for the query (S_q^n), and the corresponding shard relevance scores (R_q^n).

Number of Shards Ranked: This feature simply captures the number of shards that the shard ranking algorithm was able to order for the query (n). Queries with fewer ranked shards would indicate less diffusion of relevant documents across shards, and thus suggest smaller cutoff, T value.

Geometric Distribution Parameter: Queries for which the relevance scores decay rapidly down the rank suggest smaller cutoff values. Whereas, a flat distribution of relevance scores indicates larger cutoff value. To capture this trend we fit a geometric distribution to the ordered set of relevance scores for each query. The distribution parameter, p, is used as a feature in the regression function. For parameter estimation the method of moments is used as follows: $p = 1/E(x) = 1/\sum_{i=1}^{n} x_i \cdot P(x_i) = 1/\sum_{i=1}^{n} x_i \cdot R_q^i$, where x_i is simply the rank i, and R_q^i is the normalized relevance score of the shard at rank i.

Entropy: Shannon's entropy is often used to quantify the information content of a data source. The entropy of a random variable with uniform distribution is high, and that with a skewed distribution is low. As such, high entropy can characterize large T value, and vice-versa. We compute the entropy of the variable x with the probability distribution R_q^n as follows: $H(X) = -\sum_{i=1}^{n} R_q^i \cdot log_2(R_q^i)$. The entropy value, H(X), is used as a feature for the regression model.

Relative Entropy: For this feature we compare the shard relevance score distribution with a reference distribution by computing their relative entropy. The reference distribution is defined as: $Y \sim Uniform(1, n), \therefore p(y) = 1/n$, where n is the number of shards ranked for the query. The relative entropy is calculated as follows, and the resulting value is used as a feature for the regression function:

$$H(X|Y) = \sum_{i=1}^{n} R_q^i \cdot ln\left(\frac{R_q^i}{p(y)}\right)$$

Compared to Reference Distribution: This feature contrasts the shard relevance score distribution with a reference distribution to identify a crossover point. The reference distribution is same as before. The rank i at which $R_q^i <= p(y)$ is used as a feature for the regression function. We expect the ideal shard rank cutoff, T value, to be in the neighborhood of the cross-over point.

Cumulative Moving Average Compared to Reference Distribution: This feature is an extension of the previous feature where the cumulative moving average of the shard relevance scores is compared with the reference distribution to locate a cross-over point. $f = k$ if $\left(\frac{\sum_{i=1}^{k} R_q^i}{k} <= 2 \cdot p(y) \right)$, where $k \in \{1, 2, ..., n\}$. The cross-over point, f, is used as a feature for the regression function.

4.2   CSI Based Features

The features in this category use two sources of information: the shard ranking and relevance scores, and the central sample index (CSI).

Intersection: The number of CSI documents containing all query terms is used as a feature. This and the next two features model the generality (or the specificity) of the query terms, and the coverage of the query topic in the CSI.

Union: The number of CSI documents that contain at least one of the query terms is used as a feature.

Ratio of Intersection and Union: The ratio of the previous two features. If the ratio is close to 1 then the query terms are strongly related to a particular topic, indicating a focused information-need.

Average Inverse Document Frequency for CSI: This and the next two features are inspired by the query performance predictors proposed by He and Ounis [22]. The average inverse document frequency for CSI is computed as: $\text{AvgIDF}_{\text{CSI}}(q) = log_2 \prod_{j=1}^{m} \frac{(|CSI|/df_{CSI}^{q_j})}{m}$, where m is query length, $|CSI|$ is CSI size in documents, and $df_{CSI}^{q_j}$ is document frequency in CSI for query term q_j.

Standard deviation of CSI-IDFs: The inverse document frequency of a term is a measure of its discriminatory power. A query consisting of high IDFs terms needs a different search budget than one where IDFs have high variance, indicating that some are general term while others are on-topic. We compute the standard deviation of query term IDFs, specific to the CSI, to use as a feature.

Simplified Query Clarity: Query clarity was introduced by Cronen-Townsend et al. [23] as a measure of query ambiguity. We expect that larger search budget (larger T) is needed for ambiguous queries because shard ranking is often more erroneous. The following definition of query clarity is used as a feature: $\text{clarity}_{\text{CSI}}(q) = \sum_{j=1}^{m} \frac{qtf_j}{m} \cdot log_2 \frac{(qtf_j/m)}{df_{\text{CSI}}^{q_j}/|\text{CSI}|}$, where qtf_j is the frequency of term q_j in the query, $df_{CSI}^{q_j}$ is the document frequency for q_j in CSI.

AvgIDFs Compared to Reference Distribution: The subset of CSI documents that were sampled from the shard at rank i are identified. The IDF for each query term, specific to this subset, is computed. The average of these IDF values ($avgIDF$) is computed, and this process is repeated for every rank. The sum of $avgIDF$ across all the ranks is used to define a uniform distribution with parameter $p = \frac{\sum_i avgIDF_i}{n}$, as the reference distribution. The rank at which the corresponding $avgI\hat{D}F$ value becomes less than or equal to p is used as a feature.

4.3   Collection Statistics Based Features

This category of features use all three sources of information: shard ranking, CSI, and statistics from all shards. For brevity we refer to *all shards* as *collection.*

Average Inverse Document Frequency for Collection: This feature is an extension of the $AvgIDF_{CSI}$ where the IDF is computed over all the shards.

Standard Deviation of IDFs: This feature is also an adaptation of the earlier feature that incorporates collection level statistics.

Simplified Query Clarity: This feature is also similar to the earlier feature but uses collection level statistics.

Query scope: The concept of query scope models the specificity or the generality of the query, and is used as a feature. $scope(q) = -\log(\frac{|u_q|}{N})$, where $|u_q|$ is the set size of the documents that contain at least one of the query terms, and N collection size in documents.

Shard Prevalence w.r.t. Collection Prevalence: This feature contrasts the *prevalence* of query topic in a shard, with its prevalence in the collection. For every shard in the given ranking a prevalence score is computed as follows: $prevalence_{Coll}(S[i]) = \sum_{j=1}^{m} \frac{df_{S[i]}^{q_j}}{df_C^{q_j}}$, where $S[i]$ is the shard at rank $i \in \{1,..,n\}$, $df_{S[i]}^{q_j}$ is the document frequency of the query term, q_j, in shard $S[i]$, and $df_C^{q_j}$ is the document frequency of q_j in the complete collection. The computed scores are normalized to obtain a valid probability distribution. A reference distribution is defined by uniformly distributing the normalized prevalence scores, and the rank at which $prevalence(S[i])$ becomes less than or equal to the reference distribution probability is used as a feature.

Spread of Query Topic in the Collection: This feature is similar to the previous feature with the exception of the denominator in the following formulation: $spread(S[i]) = \sum_{j=1}^{m} \frac{df_{S[i]}^{q_j}}{|C|}$, where $|C|$ is the size of the collection in terms of documents. The feature value is the rank at which the normalized spread score is smaller or equal to the reference distribution's probability.

Shard Purity: This feature models the *purity* of a shard, in terms of the query topic. $purity(S[i]) = \sum_{j=1}^{m} \frac{df_{S[i]}^{q_j}}{|S[i]|}$, where $|S[i]|$ is the size of the shard at rank i in terms of documents. A reference distribution is defined by uniformly dividing the normalized purity scores. The feature is the rank at which the normalized purity scores becomes less than or equal to the reference distribution probability.

Table 1. Results for ShRkC and three baseline approaches. Dataset: GOV2. Percentage difference w.r.t. ShRkC are in round brackets, and the statistically significant differences (T-test with $p < 0.05$) are underlined. Search cost values are in million documents. Exhaustive search: P@30=0.483 (+5), C_{RES}=3.41 (+543), C_{LAT}=0.31 (+55), MAP=0.29 (+6), C_{RES}=3.41 (+252), C_{LAT}=0.31 (+48)

| | | Search Cost | | | Search Cost | |
	P@30	C_{RES}	C_{LAT}	**MAP**	C_{RES}	C_{LAT}
ShRkC	0.459	0.53	0.20	0.268	0.97	0.21
Fxd T=1	0.313 (-32)	0.15 (-72)	0.15 (-25)	0.130 (-52)	0.15 (-85)	0.15 (-29)
Fxd T=3	0.440 (-4)	0.42 (-21)	0.19 (-5)	0.220 (-18)	0.42 (-57)	0.19 (-10)
Fxd T=5	0.459 (0)	0.65 (+23)	0.21 (+5)	0.241 (-10)	0.65 (-33)	0.21 (+0)
Fxd T=10	0.473 (+3)	1.15 (+117)	0.22 (+10)	0.269 (+0)	1.15 (+19)	0.22 (+5)
Rank-S	0.467 (+2)	0.74 (+40)	0.21 (+5)	0.256 (+4)	0.74 (-24)	0.21 (+0)
SUSHI	0.411 (-11)	0.66 (+25)	0.19 (-5)	0.222 (-17)	0.81 (-16)	0.19 (-10)

5   Experimental Methodology

Data: Two widely used large document collections, GOV2 (Size: 400GB, No. of documents: 25 million) and CW09-Eng (Size: 15TB, No. of documents: 503 million), were used for the empirical evaluation of the proposed estimator. Each dataset was partitioned into topic-based shards using the approach proposed in Kulkarni and Callan [6]. GOV2 was clustered into 50 shards, and CW09-Eng into 1000. Thus, on an average each shard contains about half a million documents, for both the datasets. The evaluation query sets consisting of 150 topics (Avg no. of relevant documents per query: 182 ($\pm$ 149)) from 2004 and 2005 TREC Terabyte track were used with GOV2, and 200 queries (Avg no. of relevant documents per query: 107 ($\pm$ 69)) from the TREC Web tracks 2009 through 2012 were used with CW09-Eng.

Setup: For each dataset, a central sample index (CSI) is created by sampling 0.5% of documents from each shard using simple random sampling. For every query, the ReDDE algorithm (Section 2), is used to rank the shards based on their estimated relevance to the query[2].

Given this ranking of shards, the ShRkC estimator is used to predict the shard rank cutoff (T), the query is run against the top T shards using Indri, an inference network and language modeling based document retrieval algorithm, and the T search results are merged. We assume that the document scores assigned at the T shards are comparable, and thus the T result lists can be merged as is.

Evaluation Metrics: The search effectiveness is quantified using three standard IR evaluation metrics: P@30, P@100, and MAP. We model the corresponding

---

[2] Any other sample-based shard ranking algorithm can be used instead of ReDDE.

search costs using two metrics proposed by Aly et al. [15]: $C_{RES}(q) = \sum_{i=1}^{T} |D_{S[i]}^q| + |D_{CSI}^q|$ provides an upper bound on the number of documents evaluated for the query q, and the second metric $C_{LAT}(q) = \max_{1 \le i \le T} |D_{S[i]}^q| + |D_{CSI}^q|$ quantifies the longest execution path for the query, assuming a distributed query processing framework. Here $|D_{S[i]}^q|$ is the number of documents in the shard at rank i that contain at least one of the query terms, and $|D_{CSI}^q|$ is the number of documents in CSI that contain at least one of the query terms. Both these metrics use the number of documents evaluated as a proxy for query response time, which is supported by results demonstrated by MacDonald et el. [24] where the number of documents scored for a query is strongly correlated with the query response time. The result tables report the average search costs across all the evaluation queries.

Table 2. Results for ShRkC and three baseline approaches. Dataset: CW09-Eng. Percentage difference w.r.t. ShRkC are in round brackets, and the statistically significant differences (T-test with $p < 0.05$) are underlined. Search cost values are in million documents. Exhaustive search: P@10=0.106 (-12), C_{RES}=57.61 (+4105), C_{LAT}= 1.00 (+59), MAP=0.062 (+1), C_{RES}=57.61 (+1461), C_{LAT}=1.00 (+32)

	P@30	**Search Cost** C_{RES}	C_{LAT}	**MAP**	**Search Cost** C_{RES}	C_{LAT}
ShRkC	0.121	1.37	0.63	0.062	3.69	0.76
Fxd T=1	0.108 (-11)	0.58 (-58)	0.58 (-8)	0.041 (-33)	0.58 (-84)	0.58 (-24)
Fxd T=3	0.115 (-5)	1.55 (+13)	0.66 (+5)	0.054 (-11)	1.55 (-58)	0.66 (-13)
Fxd T=5	0.116 (-4)	2.50 (+82)	0.73 (+16)	0.056 (-8)	2.50 (-32)	0.73 (-4)
Fxd T=10	0.115 (-5)	4.72 (+245)	0.80 (+27)	0.058 (-5)	4.72 (+28)	0.80 (+5)
Rank-S	0.116 (-4)	1.46 (+7)	0.66 (+5)	0.052 (-15)	1.46 (-60)	0.66 (-13)
SUSHI	0.113 (-7)	4.63 (+238)	0.77 (+22)	0.058 (-5)	6.19 (+68)	0.81 (+7)

Baselines: The proposed estimator is compared with three baseline approaches: fixed rank cutoffs at $\{1, 3, 5, 10\}$, Rank-S, and SUSHI. The fixed rank cutoff approach is both, query- and metric-agnostic. The choice of the cutoff is often driven by external factors such as the availability of computing resources, or the system administrator's knowledge about the dataset. The Rank-S algorithm provides query-specific estimation of the rank cutoff, but does not optimize for a particular type of search need (metric). The last baseline, SUSHI, is most similar to the proposed approach, in that, its estimates are both, query and metric-specific. Two-fold cross-validation was employed to set the parameters of the baseline algorithms.

6   Results and Discussion

The results for the two datasets, GOV2 and CW09-Eng, are provided in Tables 1 and 2, respectively. One consistent trend is that the fixed cutoff with small T

values provide higher search efficiency (lower search costs) than ShRkC, however, the corresponding search effectiveness is substantially lower than that with ShRkC. Certain fixed cutoff values provide comparable search effectiveness to that with ShRkC, however, the corresponding search costs are significantly higher than those with ShRkC. For example, the Fxd $T=5$ and ShRkC support comparable P@30 values for the GOV2 dataset, but the search costs with the Fxd $T=5$ are higher (23% and 5%) than with ShRkC. This trend is even more pronounced for the larger dataset (CW09-Eng). These results support the hypothesis that the query- and metric-agnostic fixed shard rank cutoff approach cannot provide the best tradeoff between search effectiveness and efficiency. With small fixed T values search effectiveness is compromised, and the large fixed T values waste search effort.

When comparing ShRkC with the stronger baselines, Rank-S and SUSHI, several trends emerge. For the smaller dataset, GOV2, the Rank-S approach is able to balance search efficiency and effectiveness for the MAP metric, however, not for the other two metric (P@30 and P@100). This demonstrates the limitation of this metric-agnostic approach. SUSHI, the approach that is most similar to ShRkC in its goal of optimizing for both, query and metric, unfortunately performs poorly across the board, and illustrates the difficulty of balancing the competing objectives of search effectiveness and efficiency.

The exhaustive search approach, which processes the query against all shards, is expected to provide the upper bound on search effectiveness. In most cases, the search effectiveness with exhaustive search is comparable or better than with ShRkC. The corresponding search costs are naturally higher with exhaustive search than with any of the other approaches. For the larger dataset, CW09-Eng, ShRkC even supports higher P@30 value than exhaustive search. This suggests that searching a subset of the shards reduces the non-relevant documents (false-positives) in the retrieved results. Overall, these results demonstrate the unique ability of ShRkC to adapt to different definitions of effectiveness as well as different query requirements, thus providing the most cost-effective search setup.

Parameter Space of ShRkC: The ShRkC algorithm employs random forests (RF) algorithm to learn a regression function based on the supplied features. There are two parameters that need to be set for the RF algorithm. First, *mtry*: number of features to sample at each split in the learning process. Second, *ntree*: number of trees to fit. For the earlier experiments we used the default settings for both the parameters (mtry:$p/3$ where p is the total number of features, ntree:500). In this section we explore a larger parameter space. We also do the same for the baselines: fixed T, and Rank-S. In the interest of space we restrict this investigation to the metric MAP. Figure 1 plots the search effectiveness versus cost trends with different parameter settings. The corresponding plots for the GOV2 dataset, which exhibit similar trends, are omitted in the interest of space. One of the prominent trends is that ShRkC algorithm have much less variance than the other approaches. This suggests stability and less sensitivity to the parameter settings in case of ShRkC algorithm. The top left corner of

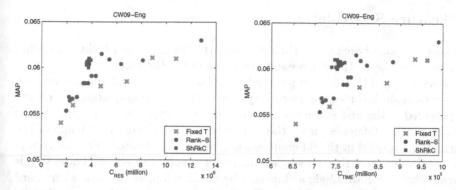

Fig. 1. Exploring parameter space of Fixed T, Rank-S, and ShRkC. Fixed T $\in$ {1,3,5,10,15,20,25}, Rank-S B $\in$ {1.1,..2.0,2.5,3.0,3.5,4.0,5.0,10.0}, and ShRkC mtry $\in$ {3,6,9,12,15,18} and ntree $\in$ {100, 500, 1000}. All the combinations of the mtry and ntree values were investigated.

Table 3. Search effectiveness with ShRkC and different feature sets. r: shard ranking features, csi: CSI features, and coll: collection features. Feature type inclusion is 1, and exclusion is 0. ∇ indicates statistically significant degradation as compared with ShRkC ($p < 0.05$).

ShRkC	GOV2			CW09-Eng		
	P@30	P@100	MAP	P@30	P@100	MAP
r=1 csi=1 coll=1	0.453	0.358	0.268	0.121	0.100	0.061
Leave-one-out						
r=1 csi=1 coll=0	0.459	0.354	0.269	0.122	0.100	0.061
r=1 csi=0 coll=1	0.451	0.356	0.268	0.115	0.100	0.060
r=0 csi=1 coll=1	0.445	0.359	0.268	0.119	0.100	0.059
Leave-one-in						
r=1 csi=0 coll=0	0.464	0.351	∇0.261	∇0.112	0.100	∇0.057
r=0 csi=1 coll=0	∇0.435	0.350	0.265	0.121	0.098	∇0.057
r=0 csi=0 coll=1	∇0.436	0.353	0.264	0.119	0.098	∇0.058

these plots is the desired region where effectiveness is maximized and the cost minimized. The ShRkC algorithm's performance is often the closest to that region. If the search system's need is to balance effectiveness and cost then often ShRkC is the best choice. Better search effectiveness can be obtained with Rank-S, although at a higher search cost.

7   Feature Set Ablation

The contribution of each of the three feature types used by the ShRkC algorithm is investigated here. Table 3 presents the results for the *leave-one-out* (1Out), and *leave-one-in* (1In) feature set analyses. The leave-one-out mode excludes, and leave-one-in includes one feature type at a time. The *all-features-in* results are provided in the first row for reference. The 1Out experiments do not exhibit any substantial differences from the reference run, indicating that some of the information encoded in the different feature types is duplicated. The 1In results however demonstrate that no single feature set is sufficient either. These trends indicate that the ShRkC algorithm can be easily adapted to work with shard ranking algorithms such as CORI, and Taily, that do not use CSI. Since collection based features are correlated with the CSI features, the exclusion of CSI features should not substantially degrade the search effectiveness.

8   Conclusions

This work studied the problem of shard rank cutoff estimation where the goal is to identify the minimum number of top ranked shards that should be searched for the query to maximize search effectiveness. For this task we develop a novel estimator, named ShRkC, that learns a regression model using a set of features that we define over three sources of information: the shard ranking, CSI, and the collection statistics. The results from an empirical evaluation using two large datasets demonstrate that ShRkC provides competitive search effectiveness while substantially lowering the search costs, as compared to three strong baselines. Its ability to adapt to different search needs, and to individual query requirements makes ShRkC-based search highly cost-effective. Balancing the computational load on the shard-servers can be a challenge in search environments where popularity of query topics go through boom-and-burst cycles. Thus a worth-while future direction would be to develop a shard rank cutoff estimator that incorporates the current computational load of the top shards in order to mitigate the load imbalance problem.

References

1. Barroso, L.A., Dean, J., Hölzle, U.: Web Search for a Planet: The Google cluster architecture. IEEE Micro **23**, 22–28 (2003)
2. Baeza-Yates, R., Murdock, V., Hauff, C.: Efficiency trade-offs in two-tier Web search systems. In: Proceedings of the ACM SIGIR Conference, pp. 163–170 (2009)
3. Chowdhury, A., Pass, G.: Operational requirements for scalable search systems. In: Proceedings of the CIKM, pp. 435–442 (2003)
4. Risvik, K.M., Aasheim, Y., Lidal, M.: Multi-tier architecture for Web search engines. In: Proceedings of the First Latin American Web Congress, pp. 132–143 (2003)
5. Baeza-Yates, R., Castillo, C., Junqueira, F., Plachouras, V., Silvestri, F.: Challenges on distributed Web retrieval. In: Proceedings of the ICDE, pp. 6–20 (2007)

6. Kulkarni, A., Callan, J.: Document allocation policies for selective searching of distributed indexes. In: Proceedings of the ACM CIKM, pp. 449–458 (2010)
7. Gravano, L., García-Molina, H., Tomasic, A.: GlOSS: text-source discovery over the internet. ACM Transactions on Database Systems **24**, 229–264 (1999)
8. Callan, J.P., Lu, Z., Croft, W.B.: Searching distributed collections with inference networks. In: Proceedings of the ACM SIGIR Conference, pp. 21–28 (1995)
9. Si, L., Callan, J.: Relevant document distribution estimation method for resource selection. In: Proceedings of the ACM SIGIR Conference, pp. 298–305 (2003)
10. Thomas, P., Shokouhi, M.: Sushi: Scoring scaled samples for server selection. In: Proceedings of the ACM SIGIR Conference, pp. 419–426 (2009)
11. Shokouhi, M.: Central-Rank-Based Collection Selection in Uncooperative Distributed Information Retrieval. In: Amati, G., Carpineto, C., Romano, G. (eds.) ECiR 2007. LNCS, vol. 4425, pp. 160–172. Springer, Heidelberg (2007)
12. Ipeirotis, P.G., Gravano, L.: Distributed search over the hidden Web: Hierarchical database sampling and selection. In: Proceedings of the VLDB Conference, pp. 394–405 (2002)
13. Arguello, J., Callan, J., Diaz, F.: Classification-based resource selection. In: Proceedings of the ACM CIKM, pp. 1277–1286 (2009)
14. Puppin, D., Silvestri, F., Perego, R., Baeza-Yates, R.: Tuning the capacity of search engines: Load-driven routing and incremental caching to reduce and balance the load. ACM Transactions on Information Systems **28**, 5:1–5:36 (2010)
15. Aly, R., Hiemstra, D., Demeester, T.: Taily: Shard selection using the tail of score distributions. In: Proceedings of the ACM CIKM, pp. 673–682 (2013)
16. Kulkarni, A., Tigelaar, A., Hiemstra, D., Callan, J.: Shard ranking and cutoff estimation for topically partitioned collections. In: Proceedings of the ACM CIKM, pp. 555–564 (2012)
17. Markov, I., Crestani, F.: Theoretical, qualitative, and quantitative analyses of small-document approaches to resource selection. ACM Transactions on Information Systems **32**, 9:1–9:37 (2014)
18. Callan, J.: Distributed information retrieval, 127–150 (2000)
19. Shokouhi, M., Si, L.: Federated search. Foundations and Trends in Information Retrieval **5**, 1–102 (2011)
20. Breiman, L.: Random forests. Machine Learning **45**, 5–32 (2001)
21. Liaw, A., Wiener, M.: Classification and regression by randomforest. R News **2**, 18–22 (2002)
22. He, B., Ounis, I.: Query performance prediction. Information Systems **31**, 585–594 (2006)
23. Cronen-Townsend, S., Zhou, Y., Croft, W.B.: Predicting query performance. In: Proceedings of the ACM SIGIR Conference, pp. 299–306. ACM (2002)
24. Macdonald, C., Tonellotto, N., Ounis, I.: Learning to predict response times for online scheduling. In: Proceedings of the ACM SIGIR Conference, pp. 621–630 (2012)

Range LCP Queries Revisited

Amihood Amir[1,2], Moshe Lewenstein[1]([✉]), and Sharma V. Thankachan[3]

[1] Department of Computer Science, Bar-Ilan University, 52900 Ramat-Gan, Israel
{amir,moshe}@cs.biu.ac.il
[2] Department of Computer Science, Johns Hopkins University,
Baltimore, MD 21218, USA
[3] College of Computing, Georgia Institute of Technology,
801 Atlantic Drive, Atlanta, GA 30318, USA
sharma.thankachan@gatech.edu

Abstract. The Range LCP problem is to preprocess a string $S[1 \ldots n]$, to enable efficient solutions of the following query: given a range $[l, r]$ as the input, report $\max_{i,j \in \{l,\ldots,r\}} |\mathsf{LCP}(S_i, S_j)|$. Here $\mathsf{LCP}(S_i, S_j)$ is the longest common prefix of the suffixes of S starting at locations i and j and $|\mathsf{LCP}(S_i, S_j)|$ is its length. We study a natural extension of this problem, where the query consists of two ranges. Additionally, we allow a bounded number (say $k \geq 0$) of mismatches in the LCP computation. Specifically, our task is to report the following when two ranges $[\ell_1, r_1]$ and $[\ell_2, r_2]$ comes as input:

$$\max_{\{\ell_1 \leq i \leq r_1, \ell_2 \leq j \leq r_2\}} |\mathsf{LCP}_k(S_i, S_j)|$$

Here $\mathsf{LCP}_k(S_i, S_j)$ is the longest prefix of S_i and S_j with at most k mismatches allowed. We show that the queries can be answered in $O(k)$ time using an $O(n^2/w)$ space data structure, where w is the word size. We also present space efficient data structures for $k = 0$ and $k = 1$. For $k = 0$, we obtain a linear space data structure with query time $O(\sqrt{n/w} \log^\epsilon n)$, where w is the word size and $\epsilon > 0$ is an arbitrarily small constant.

For the case $k = 1$ we obtain an $O(n \log n)$ space data structure with query time $O(\sqrt{n} \log n)$.

Finally, we give a reduction from Set Intersection to Range LCP queries, suggesting that it will be very difficult to improve our upper bound by more than a factor of $O(\log^\epsilon n)$.

1 Introduction

The Longest Common Prefix (LCP) has been historically an important tool in Combinatorial Pattern Matching:

A. Amir—Partly supported by ISF grant 571/14.
M. Lewenstein—Partly supported by GIF 1147/2011 and BSF 2010437.

© Springer International Publishing Switzerland 2015
C. Iliopoulos et al. (Eds.): SPIRE 2015, LNCS 9309, pp. 350–361, 2015.
DOI: 10.1007/978-3-319-23826-5_33

1. The connection between Edit Distance and Longest Common Prefix (LCP) calculation has been shown and exploited in the classic Landau-Vishkin paper in 1989 [13]. It was shown in that paper that computing *mismatches* and LCPs is sufficient for computing the Edit Distance.
2. The LCP is the main tool in various bioinformatics algorithms for finding maximal repeats in a genomic sequence.
3. The LCP plays an important role in compression. Its computation is required in order to compute the Ziv-Lempel compression, for example [15].

Therefore, the LCP has been amply studied and generalized versions of the problem are of interest. A first natural generalization of the LCP problem is a "range" version. Such a problem was considered by Cormode and Muthukrishnan [6] in the context of data compression. They called it the *Interval Longest Common Prefix* (ILCP) Problem. In that version, the maximum LCP between a given suffix $S[p \dots n]$ and all suffixes in a given interval $[l, r]$, is sought. Cormode and Muthukrishnan provide an algorithm whose preprocessing time is $O(n \log^2 n \log \log n)$, and whose query time is $O(\log n \log \log n)$. This result was then improved by Keller et al. [12] to $O(n \log n)$ preprocessing time and $O(\log n)$ query time. Also see [19] for a recent linear space and $O(\log^\epsilon n)$ query time result.

Another generalization, called Range-LCP (RLCP), was considered by Amir et al. [1]. They wanted to solve the following problem. Preprocess a string S, of length n, to enable efficient solutions of the following query:

Given $[l, r]$, $0 < l \le r \le n$, compute $\max_{i,j \in \{l,\dots,r\}} |\mathsf{LCP}(S_i, S_j)|$, where $\mathsf{LCP}(S_i, S_j)$ is the length of the longest common prefix of the suffixes of S starting at locations i and j. Here $S_i = S[i \dots n]$, the suffix starting at location i. They provided a number of algorithms for the problem:

1. Preprocessing Time: $O(n)$, Space: $O(n)$, Query Time: $O(|r - l| \log \log n)$.
2. Preprocessing Time: no preprocessing, Space: $O(|r-l| \log |r-l|)$, Query Time: $O(|r - l| \log |r - l|)$. However, the query just gives the pairs with the longest LCP, *not* the LCP itself.
3. Preprocessing Time: $O(n \log^2 n)$, Space: $O(n \log^{1+\epsilon} n)$ for arbitrary small constant ϵ, Query Time: $O(\log \log n)$

Patil et al. [19] improve the above linear space solution by describing a linear space data structure that enables an $O(\sqrt{|r - l|} \log^\epsilon(|r - l|))$ query time, for any constant ϵ.

In this paper we study a natural extension of the range LCP problem, where the query input consists of two ranges and we allow a bounded number of mismatches in the LCP computations. The formal definition of the problem is given below.

Problem 1 (Two-RLCP with k Mismatches (2-RLCP$_k$)) *Let* $\mathsf{LCP}_k(S_i, S_j)$ *be the longest common prefix of the suffixes of a string S, starting at locations i and j, with at most k mismatches allowed. The 2-RLCP$_k$ problem is to index S for a fixed k, so as to answer the following query: given two ranges $[\ell_1, r_1]$ and*

$[\ell_2, r_2]$ as the input, report

$$\max_{\{\ell_1 \leq i \leq r_1, \ell_2 \leq j \leq r_2\}} |\mathsf{LCP}_k(S_i, S_j)|$$

Understanding this generalization enhances our understanding of this important tool. In addition, this version is motivated by a natural application. McClintock earned the Nobel prize in 1983 for her discovery of DNA sequences that change or replicate within the genome [18]. These *transposons* are important for many reasons, among them the study of diseases and evolution. It is necessary to be able to find them efficiently [17]. The problem is mitigated by the fact that there are areas in the genome "suspected" of having such transposons. When we have two such areas we seek the largest common DNA sequence between these two areas. This is exactly answered by a case *Two Range LCP*.

The paper is structured as follows. Section 3 shows an $O(k)$ query time algorithm. Section 4 presents an almost optimal linear space data structure algorithm. Section 5 presents an algorithm for interval longest common prefix with one mismatch. Section 6 proves the hardness result, suggesting that our result of Section 4 is within a logarithmic factor of optimal. We conclude with some open problems.

2   Preliminaries: Suffix Arrays, Trees, LCA, and LCP

In this section we present the background needed to understand the solutions presented in this paper. The reader familiar with them can skip this subsection.

Definition 1. *Let $S = S[1 \ldots n]$ be a string over alphabet Σ. Let $\{S_1, \ldots, S_n\}$ be the set of suffixes of S, where $S_j = S[j \ldots n]$, $j = 1, \ldots, n$. Let $S_{i_1}, \ldots, S_{i_n}$ be the suffixes S_j, $j = 1, \ldots, n$, ordered lexicographically in increasing order. The suffix array of S is array SA of length n where $SA[1] = i_1, SA[2] = i_2, \ldots, SA[\ell] = i_\ell, \ldots, SA[n] = i_n$.*

Theorem 1. [Kärkkäinen and Sanders [10]] *For alphabet $\Sigma = \{1, \ldots, n^{O(1)}\}$ the suffix array can be constructed in time $O(n)$. For general alphabets it can be constructed in time $O(n \log \sigma)$, where $\sigma = \min(|\Sigma|, n)$.*

Theorem 2. [Kasai et al. [11]] *The suffix array can be preprocessed in time $O(n)$ to enable constant time Longest Common Prefix (LCP) query computations on the suffixes.*

Definition 2. *Let $S_1, \ldots, S_k$ be strings over alphabet Σ and let $\$ \notin \Sigma$. We assume that every string S_i, $i = 1, \ldots, k$, ends with a $\$$ symbol.*

An uncompacted trie of strings $S_1, \ldots, S_k$ is an edge-labeled tree with k leaves. Every path from the root to a leaf corresponds to a string S_i, with the edges labeled by the symbols of S_i. Strings with a common prefix start at the root and follow the same path of the prefix, and the paths split where the strings differ.

A compacted trie *is a trie where every chain of edges connected by degree-2 nodes is contracted to a single edge whose label is the concatenation of the symbols on the edges of the chain.*

Let $S = S[1 \ldots n]$ *be a string over alphabet* Σ. *Let* $\{S_1, \ldots, S_n\}$ *be the set of suffixes of* S, *where* $S_i = S[i \ldots n]$, $i = 1, \ldots, n$. *A* suffix tree *of* S *is the compacted trie of the suffixes* $S_1, \ldots, S_n$.

Theorem 3. [Weiner [20]] *For constant size alphabet* Σ, *the suffix tree of a length-n string can be constructed in time* $O(n)$. *For general alphabets it can be constructed in time* $O(n \log \sigma)$, *where* $\sigma = \min(|\Sigma|, n)$.

Since we now have our string in a tree data structure, one can exploit tree properties as they translate to strings. A particularly important property is the *Lowest Common Ancestor* (LCA).

Definition 3. *Let* $\mathcal{T}$ *be a tree*, u, v *nodes in* $\mathcal{T}$. *Node* w *is the* Lowest Common Ancestor (LCA) *of* u *and* v, *if* w *is an ancestor of both* u *and* v, *and every other common ancestor of* u *and* v *is also an ancestor of* w.

Landau and Vishkin [14] made the crucial observation that the substring from the root of a suffix tree to the LCA of nodes s and t is the LCP of the substrings on the paths from the root to nodes s and t. The following theorem allows us to compute the LCP efficiently using suffix trees.

Theorem 4. [Harel and Tarjan [9]] *Given an* n *node tree, it can be preprocessed in time* $O(n)$ *allowing subsequent* LCA *queries in constant time.*

3 Data Structure for General k

In this section we present a solution for the 2-RLCP$_k$ problem. We start with a well-known lemma solved by the kangaroo method, see [8,14].

Lemma 1. *Using a suffix tree* $\mathcal{T}$ *of* S, *we can compute* $|\mathsf{LCP}_k(S_i, S_j)|$ *for any* i, j *and* k *in* $O(k)$ *time.*

Next we define a useful problem.

Definition 4. *The* Two-dimensional Range Maximum *(2D-RMQ) problem is that of preprocessing a two-dimensional* $n \times n$ *array* A *of numbers to enable the following queries.*

Query: *given two ranges* $[\ell_1, r_1]$ *and* $[\ell_2, r_2]$, *where* $1 \leq \ell_1 \leq r_1 \leq n$ *and* $1 \leq \ell_2 \leq r_2 \leq n$, *output* $\max\{A[i, j] \mid \ell_1 \leq i \leq r_1, \ \ell_2 \leq j \leq r_2\}$.

Yuan and Atallah [22] showed an algorithm whereby a linear preprocessing allows for constant-time queries. Consequently, Algorithm Constant-Time Query (see Fig. 1) solves the 2-RLCP$_k$ problem with $O(k)$ time queries.

Algorithm Constant-Time Query

Preprocessing:
Preprocess string S for LCP queries.
Construct two-dimensional array A where

$$A[i,j] = \begin{cases} |\mathsf{LCP}_k(S_i, S_j)| & \text{if } i \neq j \\ 0 & \text{if } i = j \end{cases}$$

Preprocess array A for range maximum queries.

Query: Given ranges $[\ell_1, r_1]$ and $[\ell_2, r_2]$.
do a 2D-RMQ of $[\ell_1, r_1]$ and $[\ell_2, r_2]$ on A and obtain the maximum.

Fig. 1. The constant-time query algorithm

Space: Dominated by $O(n^2)$ for array A.
Time: $O(n)$ for the LCP queries preprocessing. Each entry in array A is an LCP_k query, which can be answered in $O(k)$ time, thus $O(n^2 k)$ time for constructing array A, and $O(n^2)$ time for preprocessing array A for range-maximum queries. Thus the total preprocessing time is $O(n^2 k)$. The query time is a range-maximum query, which is done in constant time.

We will now reduce the space by a factor of w, where w is the length of a word. Much effort has been made recently in the context of reducing additional space for various problems. In particular, Brodal et al. [3] showed that the 2D-RMQ problem can be solved using $O(n^2)$ additional bits to the initial input array. In a later paper, Brodal et al. [2] showed a tradeoff for the additional bits needed vs. the query time.

In our 2-RLCP$_k$ problem, the additional space necessary for the RMQ mechanism is t hus $O(n^2/w)$. However, the initial array data can be encoded by $O(n)$ space, since the data of array A is really LCP's of suffixes of S. As mentioned in Lemma 1, these LCPs can be computed in $O(k)$ time.

We conclude:

Theorem 5. *There exists an $O(n^2/w)$ space and $O(k)$ query time data structure for 2-RLCP$_k$ problem.*

4   A Space Efficient Framework

In this section we present a framework by which the 2-RLCP$_k$ problem can be reduced to the ILCP$_k$ problem, where ILCP$_k$ is a generalized version of Interval-LCP problem.

Problem 2 (Interval-LCP with k Mismatches (ILCP$_k$)) *Index a string $S[1 \ldots n]$, so as to answer the following query: given an integer $1 \leq p \leq n$ and a range $[l, r]$, where $1 \leq l \leq r \leq n$, report $\max_{\{l \leq i \leq r\}} |\mathsf{LCP}_k(S_i, S_p)|$.*

In the 2D-orthogonal range successor query data structure problem, one is given a set of n points on an $[n]^2$ grid which need to be preprocessed, to answer queries that are to return the leftmost point in a rectangle. The $ILCP_0$ problem can be reduced to a 2D-orthogonal range successor query data structure, see [16]. Therefore, $ILCP_0$ queries can be answered in $O(\log^\epsilon n)$ time using known $O(n)$ space data structures for the range successor problem. However, we are not aware of any prior result on $ILCP_k$ with $k \geq 1$.

4.1 The Framework

We prove the following result in this section.

Lemma 2. *If there exists an s_k space and t_k query time data structure for $ILCP_k$ queries on string $S[1 \ldots n]$, then there exists an $s_k + O(\frac{n^2}{w\Delta^2})$ space and $O(t_k \cdot \Delta)$ query time data structure for $2\text{-}RLCP_k$ problem, where Δ is a parameter.*

Proof. Recall that A is an $n \times n$ matrix, where $A[i,j] = 0$ if $i = j$ and is $|LCP_k(S_i, S_j)|$ otherwise. We create a new matrix A' of size $n/\Delta \times n/\Delta$ as follows: partition A into horizontal and vertical slabs of size Δ and obtain $(n/\Delta)^2$ $\Delta \times \Delta$ sub-matrixes of A. Then, set $A'[i,j]$ to the maximum element in the sub-matrix in the i-th horizontal and j-th vertical slab. Specifically,

$$A'[i,j] = \max_{1+(i-1)\Delta \leq a \leq i\Delta, 1+(h-1)\Delta \leq b \leq j\Delta} A[a,b]$$

Our data structure only consists of a 2D-RMQ structure over A' and the structure for $ILCP_k$ queries. Notice that the value of any entry in A' can be computed using Δ number of $ILCP_k$ queries. Therefore the total space is $s_k + O(\frac{n^2}{w\Delta^2})$ words.

Query Algorithm: Let $[l_1, r_1]$ and $[l_2, r_2]$ be the input ranges. Clearly, the query can be answered using $\min\{r_1 - l_1 + 1, r_2 - l_2 + 1\}$ $ILCP_k$ queries. Therefore the result follows if either $r_1 - l_1 + 1 \leq \Delta$ or $r_2 - l_2 + 1 \leq \Delta$. If this is not the case, do the following: partition the range $[l_1, r_1]$ into $[l_1, l_1'], [l_1' + 1, r_1'][r_1' + 1, r_1]$ and the range $[l_2, r_2]$ into $[l_2, l_2'], [l_2' + 1, r_2'][r_2' + 1, r_2]$, where $l_1' = \lceil l_1/\Delta \rceil \Delta, l_2' = \lceil l_2/\Delta \rceil \Delta$ and $r_1' = \lfloor r_1/\Delta \rfloor \Delta, r_2' = \lfloor r_2/\Delta \rfloor \Delta$. Now the answer to our query is the maximum among the outputs of the following queries:

a. $2\text{-}RLCP_k$ with input ranges $[l_1, l_1']$ and $[l_2, r_2]$.
b. $2\text{-}RLCP_k$ with input ranges $[r_1' + 1, r_1]$ and $[l_2, r_2]$.
c. $2\text{-}RLCP_k$ with input ranges $[l_2, l_2']$ and $[l_1, r_1]$.
d. $2\text{-}RLCP_k$ with input ranges $[r_2' + 1, r_2]$ and $[l_1, r_1]$.
e. $2\text{-}RLCP_k$ with input ranges $[l_1' + 1, r_1']$ and $[l_2' + 1, r_2']$.

Here (a),(b),(c) and (d) can be reduced to $O(\Delta)$ number of $ILCP_k$ queries as one of the input ranges is guaranteed to be smaller than Δ. The ranges in (e) can be arbitrarily large. However this can also be converted to $O(\Delta)$ number of $ILCP_k$ queries as follows: the region in A corresponding to (e), which is an $(r_1' - l_1') \times (r_2' - l_2')$ sub-matrix, is equivalent to an $(r_1' - l_1')/\Delta \times (r_2' - l_2')/\Delta$ sub-matrix of A'. The cell containing the maximum element in this sub-matrix of A' can be computed in $O(1)$ time using the 2D-RMQ structure. Once that particular cell of A' is identified, the corresponding $\Delta \times \Delta$ sub-matrix of A can also be identified. Notice that the maximum element within this $\Delta \times \Delta$ sub-matrix of A is our answer and it can be computed using Δ $ILCP_k$ queries. In summary, we can answer an 2-$RLCP_k$ query in $O(\Delta)$ number of $ILCP_k$ queries. Therefore, query time is $O(t_k \cdot \Delta)$. ∎

We are now ready to present our results on 2-$RLCP_k$ for $k \geq 1$.

Theorem 6. *Using an $O(n)$ space data structure, 2-$RLCP_0$ queries can be answered in time $O(\sqrt{n/w} \log^\epsilon n)$, where $\epsilon > 0$ is an arbitrarily small constant.*

Proof. The result follows from Lemma 2 with $\Delta = \sqrt{n/w}$ and by choosing linear space and $O(\log^\epsilon n)$ query time structure for $ILCP_0$ queries. ∎

In Section 6 we show that if one insists on a linear space data structure, the query result obtained here is likely to be optimal within an $O(\log^\epsilon n)$ factor.

5   Data Structure for $ILCP_1$ Queries

This section is dedicated to proving the following result.

Theorem 7. *Using an $O(n \log n)$ space structure, we can answer $ILCP_1$ queries on $S[1 \ldots n]$ in $O(\log^2 n)$ time.*

The result is based on the seminal paper by Cole et al. [5] for text indexing with errors. The central structure is a suffix tree $\mathcal{T}$ of S. We use ℓ_i to denote the i-th leftmost leaf of $\mathcal{T}$, $path(u)$ to denote the concatenation of edge labels on that path from root to node u and $size(u)$ to denote the number of leaves in the subtree of u. We now briefly describe the heavy path decomposition of the suffix tree.

Heavy Path and Heavy Path Decomposition: The heavy path of $\mathcal{T}$ is the path starting from the root, where each node v on the path is the child with the largest subtree (ties broken arbitrary) of its parent. The *heavy path decomposition* is the operation where we decompose each off-path subtree of the heavy path recursively. Therefore, the edges in $\mathcal{T}$ will be partitioned into disjoint heavy paths. Each heavy path leads to a unique leaf node (therefore we have n heavy paths) and each internal node is on a unique heavy path. We say that two (arbitrary) paths in a tree *intersect* if they have a node in common. The following lemma is well-known and can easily be proven.

Lemma 3. *The number of heavy paths intersected by any (arbitrary) root-to-leaf path is at most $\log n$.*

We shall call a node u *heavy* if both u and its parent are on the same heavy path. Otherwise u is *light*. The root is always light.

Let H_i be the heavy path that leads to leaf ℓ_i and C_i be the set of leaves whose path to root intersect with H_i. We now define a new problem.

Problem 3 *An* HLCP_1 *query* (p, l, r, i) *asks to report* 0 *if* $\ell_{SA^{-1}[p]} \notin C_i$, *and otherwise report* $\max\{|\text{LCP}_1(S_p, \text{path}(\ell_j))| \mid \ell_j \in C_i, SA[j] \in [l, r]\}$.

Lemma 4. *An* ILCP_1 *query* (p, l, r) *can be decomposed into* $O(\log n)$ *number of* HLCP_1 *queries.*

Proof. First, using Lemma 3, find the heavy paths $H_{i_1}, H_{i_2}, \ldots, H_{i_k}$, where $k = O(\log n)$ that intersect with the root-to-leaf path of the suffix $S[p \ldots n]$ in T. Then, the desired answer is given by the maximum among the answers reported by HLCP_1 queries (p, l, r, i_x) for $x = 1, 2, \ldots, k$. The correctness can be argued as follows: let $\text{path}(\ell_g)$ be the suffix that corresponds to our desired answer. Then, HLCP_1 query on the heavy path that intersects with the node $\text{LCA}(\ell_g, \ell_{SA^{-1}[p]})$ returns our answer and the answers returned by other HLCP_1 queries cannot be more than this. ∎

We now proceed to describe an $O(n \log n)$ space structure for answering HLCP_1 queries in $O(\log^\epsilon n)$ time. Combining this with Lemma 4 gives Theorem 7.

5.1   Structures for HLCP_1 Queries

The structure for answering HLCP_1 queries (p, l, r, i) for a fixed i can be constructed as follows:

1. From the set C_i of leaves intersecting with H_i, create a set Z_i of $2|C_i| - 1$ strings as follows:

$$Z_i = \{\text{path}(\ell_j) \mid j \in C_i\} \cup \{\text{path}'(\ell_j) \mid j \in C_i \text{ and } j \neq i\}$$

   Here $\text{path}'(\ell_j)$ is the string obtained from $\text{path}(\ell_j)$ by replacing its character at position $f = 1 + |\text{LCP}(\ell_i, \ell_j)|$ by the f-th character of $\text{path}(\ell_i)$. Observe that $|\text{LCP}(\text{path}(\ell_i), \text{path}'(\ell_j))| = |\text{LCP}_1(\text{path}(\ell_i), \text{path}(\ell_j))|$.
2. We now map each string (i.e., $\text{path}(\ell_j)$ or $\text{path}'(\ell_j)$) into a 2D point whose x-coordinate is the lexicographic rank of the string among all strings in Z_i and y-coordinate is $SA[j]$. These points are then preprocessed into linear space data structure for answering orthogonal range successor/predecessor search queries. We call the structure D_i.
3. Whenever p and a range $[l, r]$ are given as input, the structure D_i can return the maximum of the length of the longest common prefix of $S[p \ldots n]$ and any string in Z_i using an orthogonal range successor and an predecessor search queries
4. The size of D_i is $O(|C_i|)$ words, therefore the overall space for all D_i's is $O(n \log n)$ words.

5.2   Query Time

In this section, we show that $HLCP_1$ queries can be answered in $O(\log^\epsilon n)$ time using an $O(n \log n)$ space structure.

Let p and $[l, r]$ be the input parameters to the $ILCP_1$ query, then our task is the following: among all leaves ℓ_j, where $SA[j] \in [l, r]$, report the maximum of $|LCP_1(\text{path}(\ell_j), S_p)|$, i.e., report the maximum of $|LCP(\text{path}(\ell_j), S_p))|$ after ignoring the first mismatch, which is at the position $1 + |LCP(\text{path}(\ell_j), S_p)|$. To execute this task efficiently, we maintain several data structures, called D_i's and $E(u)$'s.

5.3   Structure D_i

Let H_i be the heavy path that leads to leaf ℓ_i and C_i be the set of leaves, whose path to root intersects with H_i. Notice that $\sum_i |C_i| \le 2 \log n$, as each leaf node can be a part of at most $\log n$ number of C_i's. For each $i \in [1, n]$, we construct a data structure D_i by preprocessing C_i as follows:

1. We preprocess Z_i into a data structure, that given p and $[l, r]$ as inputs, if $S_p = \text{path}(\ell_{SA^{-1}[p]}) \in Z_i$, then report the length of the longest common prefix of S_p and all strings $\text{path}(\ell_j)$ or $\text{path}'(\ell_j)$ in Z_i with $SA[j] \in [l, r]$. The problem can be reduced to orthogonal range successor query over a set of $|Z_i|$ 2D points. Therefore, we maintain a linear space (i.e., $O(|Z_i|)$ words) range successor query data structure with query time $O(\log^\epsilon n)$ and call it D_i.
2. The overall space of all D_i is $O(n \log n)$ words.

5.4   Structure $E(u)$

With every internal node u, we associate a data structure $E(u)$ as follows: let L_u be the set of leaves in the subtree of u, but not in the subtree of the heavy child v of u. Then, create a set of strings $Z(u) = \{\text{path}'(\ell_j) \mid \ell_j \in L_u\}$. Here $\text{path}'(\ell_j)$ is same as $\text{path}(\ell_j)$ with its character at position $f = 1 + |LCP(\text{path}(\ell_j), \text{path}(v))|$ is changed to $\text{path}(v)[f]$. We preprocess $Z(u)$ into a data structure such that, given p and $[i, j]$ as queries, if $\text{path}'(\ell_{SA^{-1}[p]}) \in Z(u)$, then report the length of the longest common prefix of $\text{path}'(\ell_{SA^{-1}[p]})$ and all other strings $\text{path}'(\ell_j)$ in $Z(u)$ with $SA[j] \in [l, r]$. Again, the problem can be educed to orthogonal range successor query. Hence by maintaining an $O(|Z(u)|)$ space structure (called $E(u)$), the query can be answered in $O(\log^\epsilon n)$ time. Notice that the overall size of all $E(u)$'s is $O(n \log n)$.

5.5   Query Answering

We need to do $\log n$ range successor queries, each will require a binary search, hence the overall query time is $O(\log^2 n)$.

This concludes the proof of Theorem 7.

Corollary 1. *Using an $O(n \log n)$ space data structure, 2-RLCP$_1$ queries can be answered in time $O(\sqrt{n} \log n)$.*

Proof. By Theorem 7, ILCP$_1$ queries can be answered in $O(\log^2 n)$ time using an $O(n \log n)$ space structure. By combining this result with Lemma 2 with $\Delta = \sqrt{n}/\log n$, the result can be obtained. ∎

6   The Hardness Results

Set intersection is believed to be a hard problem. While there are some known lower bounds on the problem of maintaining a collection of sets under insertion and deletion of elements into individual sets (initialized empty) and set-intersection queries [7,21], the following problem has eluded a lower bound.

Definition 5. *Let $S_1, S_2, \ldots$ be a collection of sets of total cardinality n, The Set Emptiness Problem (SEP) is defined as follows.*
Preprocess the collection of sets such that the following queries can be answered efficiently.
Query: *A set emptiness query (i, j) asks "Is $S_i \cap S_j$ empty ?".*

The best known data structure for answering such queries takes $O(n)$ space and $O(\sqrt{n/w})$ query time, where w is the word size [4][1]. We show that the 2-RLCP$_k$ problem is at least as hard as the Set Emptiness problem, even for $k = 0$.

Theorem 8. *The Set Emptiness problem can be solved in the same space and query time complexity as that of the 2-RCLP$_0$ query problem, if the total cardinality of all sets in the first problem is asymptotically equal to the length of the string S in the second problem.*

Proof. Reduce SEP to the 2-RLCP$_0$ problem as follows. Each set S_i is written as a string of all its elements. By concatenating the strings corresponding to all S_i's, we construct a string S of length n. The substring corresponding to any S_i can be represented as a range $[\ell_i, r_i]$. A data structure over S for 2-RLCP$_0$ query can answer a set emptiness query (i, j) as follows: return yes, iff 2-RLCP$_0$ on S with input ranges $[\ell_i, r_i]$ and $[\ell_j, r_j]$ returns 0. ∎

As we have seen, the best known linear space result of set emptiness query requires $O(\sqrt{n/w})$ time, therefore it is unlikely that there exists an $O(n)$-space data structure for 2-RLCP$_0$ queries that yields a $o(\sqrt{m/w})$ query time. Consequently, we conclude that the query time in the result in Theorem 6 is likely to be optimal within an $O(\log^\epsilon n)$ factor.

---

[1] Cohen and Porat achieve space $O(n)$ words and time $O(\sqrt{n})$, but one can easily reduce the time to $O(\sqrt{n/w})$.

References

1. Amir, A., Apostolico, A., Landau, G.M., Levy, A., Lewenstein, M., Porat, E.: Range LCP. J. Comp. and Sys. Sci. **80**(7), 1245–1253 (2014)
2. Brodal, G.S., Davoodi, P., Lewenstein, M., Raman, R., Srinivasa Rao, S.: Two Dimensional Range Minimum Queries and Fibonacci Lattices. In: Epstein, L., Ferragina, P. (eds.) ESA 2012. LNCS, vol. 7501, pp. 217–228. Springer, Heidelberg (2012)
3. Brodal, G.S., Davoodi, P., Rao, S.S.: On Space Efficient Two Dimensional Range Minimum Data Structures. In: de Berg, M., Meyer, U. (eds.) ESA 2010, Part II. LNCS, vol. 6347, pp. 171–182. Springer, Heidelberg (2010)
4. Cohen, H., Porat, E.: Fast set intersection and two-patterns matching. Theor. Comput. Sci. **411**(40–42), 3795–3800 (2010)
5. Cole, R., Gottlieb, L., Lewenstein, M.: Dictionary matching and indexing with errors and don't cares. In: Proc. 36th Annual ACM Symposium on the Theory of Computing (STOC), pp. 91–100. ACM Press (2004)
6. Cormode, G., Muthukrishnan, S.: Substring compression problems. In: Proc. 16th Annual ACM-SIAM Symposium on Discrete Algorithms (SODA), pp. 321–330 (2005)
7. Dietz, P., Mehlhorn, K., Raman, R., Uhrig, C.: Lower bounds for set intersection queries. Algorithmica **14**(2), 154–168 (1995)
8. Galil, Z., Giancarlo, R.: Improved string matching with k mismatches. SIGACT News **17**(4), 52–54 (1986)
9. Harel, D., Tarjan, R.E.: Fast algorithms for finding nearest common ancestor. Computer and System Science **13**, 338–355 (1984)
10. Kärkkäinen, J., Sanders, P.: Simple linear work suffix array construction. In: Baeten, J.C.M. et al., (eds.): ICALP 2003. LNCS, vol. 2719, pp. 943–955. Springer, Heidelberg (2003)
11. Kasai, T., Lee, G.H., Arimura, H., Arikawa, S., Park, K.: Linear-Time Longest-Common-Prefix Computation in Suffix Arrays and Its Applications. In: Amir, A., Landau, G.M. (eds.) CPM 2001. LNCS, vol. 2089, pp. 181–192. Springer, Heidelberg (2001)
12. Keller, O., Kopelowitz, T.: Shir Landau Feibish, and Moshe Lewenstein, Generalized substring compression. Theor. Comput. Sci. **525**, 42–54 (2014)
13. Landau, G.M., Vishkin, U.: Fast parallel and serial approximate string matching. Journal of Algorithms **10**(2), 157–169 (1989)
14. Landau, G.M., Vishkin, U.: Efficient string matching in the presence of errors. In: Proc. 26th IEEE FOCS, pp. 126–126 (1985)
15. Lempel, A., Ziv, J.: On the complexity of finite sequences. IEEE Transactions on Information Theory **22**, 75–81 (1976)
16. Lewenstein, M.: Orthogonal Range Searching for Text Indexing. In: Brodnik, A., López-Ortiz, A., Raman, V., Viola, A. (eds.) Ianfest-66. LNCS, vol. 8066, pp. 267–302. Springer, Heidelberg (2013)
17. Makalowski, W., Pande, A., Gotea, V., Makalowska, I.: Transposable elements and their identification. Methods Mol. Biol. **855**, 337–359 (2012)
18. McClintock, B.: The origin and behavior of multiple loci in maize. Proc. Natl. Acad. Sci. **36**(6), 344–355 (1950)

19. Patil, M., Shah, R., Thankachan, S.V.: Faster Range LCP Queries. In: Kurland, O., Lewenstein, M., Porat, E. (eds.) SPIRE 2013. LNCS, vol. 8214, pp. 263–270. Springer, Heidelberg (2013)
20. Weiner, P.: Linear pattern matching algorithm. In: Proc. 14 IEEE Symposium on Switching and Automata Theory, 1–11 (1973)
21. Yellin, D.: Data structures for set equality-testing. In: Proc. 9th Annual ACM-SIAM Symposium on Discrete Algorithms (SODA), pp. 386–392 (1992)
22. Yuan, H., Atallah, M.J.: Data structures for range minimum queries in multidimensional arrays. In: Proc. 21st ACM-SIAM Symposium on Discrete Algorithms (SODA), pp. 150–160 (2010)

Feasibility of Word Difficulty Prediction

Ricardo Baeza-Yates[1,2]($\boxtimes$), Martí Mayo-Casademont[2], and Luz Rello[3]

[1] Yahoo Labs, New York, USA
rbaeza@acm.org
[2] DTIC, Universitat Pompeu Fabra, Barcelona, Spain
[3] HCI Institute, Carnegie Mellon University, Pittsburgh, USA

Abstract. We present a machine learning algorithm to predict how difficult is a word for a person with dyslexia. To train the algorithm we used a data set of words labeled as easy or difficult. The algorithm predicts correctly slightly above 72% of our instances, showing the feasibility of building such a predictive solution for this problem. The main purpose of our work is to be able to weight words in order to perform lexical simplification in texts read by people with dyslexia. Since the main feature used by the classifier, and the only that is not computed in constant time, is the number of *similar* words in a dictionary, we did a study on the different methods that exist to compute efficiently this feature. This algorithmic comparison is interesting on its own sake and shows that two algorithms can solve the problem in less than a second.

1 Introduction

Text can often be complex and difficult to read, especially for people with cognitive impairments or low literacy skills. The complexity of the word is directly related to its readability and comprehensibility [20]. Readability refers to the ease a text can be read; and comprehensibility refers to the ease a text can be understood. Determining word complexity is crucial for text simplification, that is, a process that reduces the complexity of both wording and structure in a sentence, while retaining its meaning. Hence, determining word complexity automatically can be useful for these target populations to read efficiently, such as on people with autism [9], Down syndrome [23], dyslexia [21], or aphasia [3].

For the reasons above, in this paper we focus on studying the feasibility of predicting through machine learning if a word is *difficult* for a person that has dyslexia. For instance, people with dyslexia read significantly faster when the text contains more frequent or shorter words [13]. As with other learning disabilities, dyslexia is a lifelong challenge that people are born with. This language processing disorder can hinder reading and writing. Dyslexia occurs among people of all economic and ethnic backgrounds and is estimated that 10% of the world population have it.

In this paper we show that we can predict complex words with over 72% accuracy by using machine learning and a small set of features. The main feature is the number of words in a dictionary at distance one. As this was the only

© Springer International Publishing Switzerland 2015
C. Iliopoulos et al. (Eds.): SPIRE 2015, LNCS 9309, pp. 362–373, 2015.
DOI: 10.1007/978-3-319-23826-5_34

feature that took non constant time to compute, we compared four different algorithms, finding out that a particular case of [11] and a variation of [12] were the best choices, although only the former achieves constant time search with respect to the dictionary size.

The rest of this paper is structured as follows. Section 2 gives the background for our problem. In Section 3 we give the machine learning solution and the analysis of it. Then, in Section 4 we present the algorithms that we considered to compute all similar words and the experimental comparison. We end with some conclusions in Section 5.

2   Background

Word complexity depends on various factors such as its length, frequency or and semantic ambiguity [20]. Traditional methods to address text complexity relay in length number of letter such as the Flesch Reading Ease [10] or the Coleman-Liau index [6]. More recently, readability measures have included other type of word complexity features. These can take into consideration lexico-semantic features, such as lexical richness [16]; or psycholinguistics-based lexical features, such as word concreteness [24]. However, these word complexity measures are not enough for people with dyslexia because this target group present difficulties with different kind of words. People with dyslexia have difficulties in recognizing words that are phonetically and orthographically similar [8,25]. Hence, word complexity measures that take into account features such as word similarity (or string similarity), such as Colthearts orthographic neighborhood size metric [7], could be useful to rank word complexity using machine learning.

Regarding finding similar words, to formalize the problem let us consider the following notation:

1. Let Σ be a finite alphabet and Σ^* all the possible words generated by this alphabet. We use $|\Sigma|$ to denote the size of the alphabet.
2. Let $D \subseteq \Sigma^*$ be a set of words that conform the *dictionary*.
3. Let $P \subseteq \Sigma^*$ be a set of words called the *patterns*.
4. Let $k \in \mathbb{N} \cup \{0\}$ be the maximum number of errors (distance) allowed.
5. Let $d : \Sigma^* \times \Sigma^* \to \mathbb{N} \cup \{0\}$ be a distance function.

Now we can define our problem as finding all the words $t \in D$ such that $d(p,t) \leq k$ for $p \in P$. In our case P is the set of words labeled as difficult or easy and D is the dictionary of all possible words. The main approaches to this problem, called *Multiple Pattern Approximate String Matching*, are surveyed in Navarro [17] and Navarro *et al.* [18].

3   A Machine Learning Solution

3.1   Dataset

Our dataset consisted in 995 Spanish words, 666 of them labeled as difficult words and 329 as easy. This dataset is small as it requires to convince several dyslexic volunteers to contribute with texts and then that an expert does a subjective assessment of the difficulty level. Nevertheless, the dataset is large enough for the goal of

Table 1. List of initial features.

Name	Description
d1	How many words there are in a dictionary that are at distance 1 or less of the featured word.
d2	How many words there are in a dictionary that are at distance 2 or less of the featured word.
vowels	How many vowels has the word.
has accent	Binary feature. Does any character of the word have an accent?
[h], [y], [b,v], [b,d], [r,s,l], [m,n,], [s,z,ce,ci,x], [ca,co,cu,q,k]	How many of the characters in brackets does the word have for each class? (eight features)
[gu], [j,ge,gi], [rr,ll], [ctr, str, sfr, scr, spl, xpl, xpr, xtr, nsp]	How many of the strings in brackets does the word have for each class? (four features)
word length	How many characters does the word have.
vowel ratio	Number of vowels divided by word length.
three consonants	Binary feature. Does the word have three consecutive consonants?
two vowels	Does the word have two vowels in a row?
has [x]	Binary feature. Does the word have an 'x'?
alternate	Binary feature. Do the characters in the word alternate between consonants and vowels or not? For example this feature would be true for *vowel* and false for *consonant*.
has [h]	Binary feature. Does the word have an 'h'?
has double letter	Binary feature. Does the word have two consecutive letters equal like for example "ll" or "rr"?
webhits	How many results does Yahoo search return for this word.
has [o] and [u]	Binary feature. Does this word have both and 'o' and a 'u'?
has [ua]	Binary feature. Do the characters 'u' and 'a' appear consecutively in this word?

this study that was to prove that it was possible to predict word difficulty. If feasible, more resources can be invested to label a larger set of words. On the other hand, this is the largest dataset of its kind[1] and has been described and analyzed in [22].

3.2   Feature Selection

The set of initial features are given in Table 1, where the letters in brackets are phonologically and orthographically similar, as they are potentially more challenging for people with dyslexia, plus some characteristics of words. These features are based on our analysis of the dataset [22].

First of all we did some internal testing to discard or tune the features above. A big part of them turned out to be irrelevant to our problem according to our experiments. We consider relevant to stress here that during our internal testing

---

[1] http://grupoweb.upf.es/WRG/resources/DysWebxia/DysList_resource.csv.gz

Table 2. Most correlated features.

Feature pair(s)	Corr.	Feature pair	Corr.
([h], has h)	0.993	([m,n,], length)	0.366
(d1,d2)	0.949	([r,s,l], [rr, ll])	0.360
([rr, ll], has double letter)	0.908	([r,s,l], has double letter)	0.345
(vowels, word length)	0.902	(vowels, [r,s,l])	0.315
([ctr, str, ..], three consonants)	0.716	(vowels, [m,n,])	0.309
(d2, alternate)	0.524	([ca,co,cu,q,k], two vowels)	0.305
(vowel percentage, two vowels)	0.512	(d2, [r,s,l])	-0.302
([r,s,l], length)	0.504	(vowels, alternate)	-0.377
(vowels, two vowels)	0.469	([r,s,l], vowel percentage)	-0.381
(d1,alternate), ([s,z,ce,ci,x], length)	0.466	(d1,vowels)	-0.545
([b,v],[b,d])	0.465	(d2,vowels)	-0.600
([s,z,ce,ci,x], [r,s,l])	0.461	(d1,word length)	-0.602
(vowels, [s,z,ce,ci,x])	0.393	(d2,word length)	-0.661

Table 3. New binary features.

Name	Description
[y] > 0	There is any of the characters in brackets in the word?
[b,v] > 1	There is more than 1 of the characters in brackets in the word?
[ge, gi, j] > 0	There is any of the strings in brackets in the word?
[ca,co,cu,q,k] > 0	There is any of the strings in brackets in the word?
[gu] > 0	There is any of the strings in brackets in the word?

we realized that the correlation between the approximate search attributes (d1, d2) and the easy or difficult labels were working in a different way as we initially expected. We thought that a word might be more difficult if there were a lot of words similar to it and easier otherwise. However it turned out to be the other way around. Our guess here is that difficult words tend to be larger and due to the nature of the language, larger words tend to have less *neighbors* at distance 1 or 2, because we have less long words than short words [14].

Afterwards, in order to shrink this set we computed the correlation matrix of the remaining features to discard those that were not adding information. Table 2 summarizes the most important values of the correlation matrix. We considered that the correlation between two features was significant if was greater than 0.3 or lower than −0.3. According to this we decided to discard the features d2, vowels, alternate, [s,z,ce,ci,x], [r,s,l], [m,n,], two vowels, [b,d], [h], [ctr, str, ..] and [rr,ll]. Based on this analysis, we used the boldfaced features in Table 1 plus the five revised features shown in Table 3.

3.3 Algorithm and Results

Since most machine learning algorithms are sensible to class unbalance, we had to do some sampling on our dataset. That is, we built a dataset made of all the words labeled as easy and as many words labeled as difficult picked at random.

Table 4. Confusion matrix and performance of the ML algorithm.

Prediction / Class	Difficult	Easy	Accuracy
Difficult	214	67	0.723
Easy	115	262	
Class	Recall	Precision	F-Measure
Difficult	0.762	0.650	0.702
Easy	0.695	0.796	0.742

We also used a standard 10-fold cross-validation as testing method. Since our dataset is small we aimed for not very complex algorithms that could make the solution unnecessary sophisticated and thus prone to have overfitting problems.

We tried several approaches and the best result was obtained for Platt's sequential minimal optimization (SMO) algorithm for training a support vector classifier [19]. Indeed, out of 658 instances, 476 were classified correctly giving a 72.34% accuracy. Table 4 gives the complete results and the confusion matrix. SMO not only classifies the words in the two categories but also gives a value between 0 and 1 indicating how likely the word is easy/difficult. This is useful for establishing different degrees of difficulty to perform lexical simplification on those words that show to be harder.

In Table 5 we present the ranking of our features according to the classical information gain measure, that evaluates the quality of an attribute by measuring the information gain with respect to not choosing it. In this table we see that the ranking position obtained by the features in each of the 10 times we ran the algorithm (1 means the most important feature and 10 the least important). Each trial was done in a different test set obtained by sampling our data as explained earlier. In addition, the last column of the table shows the average information gain ratio obtained by the feature in 10 trials.

One can see that the first five attributes are more or less stable in the first 5 positions of the ranking while others, like *webhits* or *vowel percentage*, are most of the time mediocre, but for some particular data sets become more relevant.

Also important is the presence of the *d1* attribute in top of the ranking list in all the trials. This, together with the fact that is the only feature that cannot be obtained in linear time motivated the approximate string matching experimental comparison of the next section.

4   Finding Similar Words

To measure the similarity between words we use the classical edit distance or Levenshtein distance [15], that counts the minimum number of insertions, deletions and substitutions required to make two strings equal. This distance can be easily computed in $O(nm)$ time and space but there are improved algorithms that can have $O(kn/\sqrt{|\Sigma|})$ average time [4].

Table 5. Feature analysis.

Name	1	2	3	4	5	6	7	8	9	10	IG ratio
d1	1	1	1	1	1	1	1	1	1	1	0.144
word length	2	2	2	2	2	2	2	2	2	2	0.100
[ge,gi,j] > 0	3	4	3	3	6	3	4	3	4	4	0.0253
has h	4	3	4	4	4	4	3	5	3	3	0.0248
[y] > 0	5	5	5	5	7	6	5	6	6	5	0.0146
webhits	14	14	14	14	5	5	14	4	5	14	0.00947
[b,v] > 1	6	6	6	9	9	9	12	8	9	7	0.00707
three consonants	8	10	7	6	8	8	6	12	13	8	0.00693
[ca,c0,cu,q,k] > 0	7	11	8	11	11	12	8	7	8	6	0.00628
has ua	10	9	9	10	12	13	7	9	7	9	0.00518
[gu] > 0	12	7	12	7	13	10	9	10	10	11	0.00467
has o and u	9	8	11	8	10	11	10	11	11	13	0.00456
vowel percentage	15	15	15	15	3	15	15	15	15	15	0.00399
has double letter	11	13	10	12	14	7	13	14	14	10	0.00303
has x	13	12	13	13	15	14	11	13	12	12	0.00205

To find similar words we consider the following requirements:

1. **Indexed dictionary**: As opposed to online search, we assume that our dictionary of words can be indexed, as this is more efficient if we need to solve several instances of the problem (large P).
2. **Linearity**: We want our structures to be at most linear, both in size and processing time with respect to the total number of words.
3. **Containing the pattern is not enough**: As opposed to the sibling problem of finding all words in the dictionary that after k operations contain the pattern, we want the words to match exactly the pattern after k operations.
4. **Multiple patterns**: We want to be able to efficiently find the words in the dictionary that are similar to a set of patterns. This differentiates our problem of those that try to find similar words to a single pattern.

4.1 Algorithms

Based on the previous discussion, we consider four different algorithms: a non-indexed brute force approach as a basic baseline, a multi-pattern search algorithm (partial index), a tree-based metric space search algorithm (full index), and a trie-based algorithm (full index). Although theoretically there are solutions of $O(n \log^k n)$ space and $O(m + \log^k n \cdot \log \log n + \#matches)$ query time [5], they are still not practical. For example, this solution will need several GB of space to store just a 10MB word index!

Naive Approach. The naive or brute force approach consists of computing the distance between each pattern and each word of the text, and for those cases where the distance is lower than k, output the text word. The complexity is obviously $O(|P||D|)$ distance computations. For this we use two arrays of words, one containing the patterns and another containing the dictionary.

APATS. Our second technique is the multi pattern approximate string matching algorithm by Fulwider and Mukherjee [12] called APATS. They apply bit-parallelism to multi pattern approximate string search and implement and efficient solution for the case where the pattern is larger than the memory-word of the machine.

However, the problem solved by APATS is a little bit different from ours in two aspects. The first one is that they consider that a word of the dictionary is at distance k from a pattern if with k operations the dictionary word contains the pattern. So for example, using this metric the words *collaboration* and *soup* are at distance 3 because with three operations you can modify *soup* to obtain *coll* which is contained in *collaboration*. The second difference is that they consider the set of patterns as a unity. They output a word from the dictionary as a result if there is at least a pattern that is at distance k or less from the word, although they cannot say which is(are) the pattern(s) responsible of that. Despite these two differences, we consider their solution as an optimistic baseline as can be easily adapted to our problem.

Burkhard-Keller Tree. The Burkhard-Keller (BK) tree [2] is a tree-based data structure designed to quickly find near matches between similar objects in a metric space. In our case the objects are strings and the metric is the Levenshtein distance.

Consider now that we have a string q and a threshold k and we are interested on finding strings whose distance from q is less or equal than k. On top of that suppose we have a test string t at distance d from q. How far from t are the strings we are looking for? The triangle inequality gives us the answer: they are in the range $[d - k, d + k]$ from t.

From here, the construction of the BK-Tree is simple. Every node has an arbitrary number of children, and each edge has a number corresponding to a possible Levenshtein distance. All the nodes in the subtree labeled with the edge x are at distance x from the parent node. To build a tree from a set of words, take any word of the set as the root. Then insert each of the other words one by one. To insert a word compare it with the root node and obtain the distance x between the word and the root. If there is no x edge in the root node, create the subtree having your word as a root. On the other hand, if there is such an edge, compare your word with the root of the subtree corresponding with the x edge and proceed recursively.

To query the tree, compute the distance between your term and the root and recursively query each children numbered between $d - k$ and $d + k$ (inclusive). If the node you are examining is within d of your search term, return it and continue the search.

Although there are other tree-based data structures that are optimized to reduce the number of distance computations, the potential improvement in our case is small as the Levenshtein distance is easy to compute.

Trie based NDFA Algorithm. If we represent our dictionary of words in a trie or digital tree [11], we can find similar words by using a particular case of efficient searching of regular expressions on a trie by Baeza-Yates and Gonnet [1]. That is, we simulate the non deterministic finite automata (NDFA) that finds all possible words at edit distance k to a given query word q over the trie or equivalently, we simulate the standard dynamic programming algorithm to compute the edit distance. The algorithm shown below starts traversing all the children of the root node. If a path exists in the tree where the next letter coincides with the next letter of our query word, we use it. In addition, we take any other path paying a penalty of 1 (we subtract one from k). When k becomes negative in any path, we stop. The algorithm uses the following notation:

1. $T[1..i]$ denotes both the word resulting of following the path from the root to the actual node (without concatenating the substring stored in that node) and the actual node itself, and $T[i + 1..n]$ denotes the substring stored in that node.
2. Similarly, $T[1..i + 1]$ denotes to advance in the tree in one of the children. Among all the possible children, $T[1..i + 1]$ refers to the successor we are dealing with in the loop of line 8 (see Algorithm 1). If the node is a *compressed node, i.e.,* a node with more than one character, we advance in the prefix inside the node one position. This is why we say that a node has a successor if either has children or we are not done with the prefix of the node yet.
3. $T[i]$ denotes the i-th character of the string corresponding to the actual node.

Algorithm 1. Find_within_distance(query[j..n], T[1..i], k)

```
 1: if k < 0 then
 2:     return
 3: end if
 4: if T[1..i] is terminal and T[1..i] not visited and d(query[j..n], T[i+1..n]) ≤ k then
 5:     T[1..i] is visited
 6:     output T[1..i]
 7: end if
 8: for all successors of T[1..i] do
 9:     if query[j] = T[i] then
10:         Find_within_distance(query[j+1..n], T[1..i+1], k)
11:     end if
12:     Find_within_distance(query[j+1..n], T[1..i+1], k-1)  //substitution
13:     Find_within_distance(query[j..n], T[1..i+1], k-1)    //insertion
14:     Find_within_distance(query[j+1..n], T[1..i], k-1)    //deletion
15: end for
```

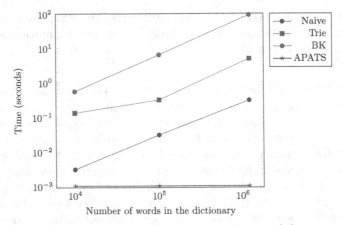

Fig. 1. Building time for a lexicographically sorted dictionary.

By a straightforward but tedious proof by induction it can be shown that the searching cost of this algorithm is in the worst case $O((|w|^{k+1} + k)|\Sigma|^k)$, that is $O((|w||\Sigma|)^k)$.

4.2 Experimental Comparison

Now we compare the performance of the algorithms we considered regarding the time to build the data structure, as well as the time to query the whole set of patterns. All the experiments shown considers the average of five runs. The machine used was a 4-core at 3.30 GHz with 4GB of RAM.

We use our own optimized implementations for all the algorithms but APATS, where we use the code provided by the authors.[2]

Building Time. We consider three different dictionary sizes: 10,000 words, 100,000 words and 1.2 million words. The results can be seen in Figure 1 and do not change if we consider a randomly ordered dictionary, as expected.

We can see that the Naive approach and APATS take almost no time to build their data structures while the Trie requires a few seconds to build the 1.2 million-word index. On the other hand, the BK-Tree is the worst with respect to scalability.

Querying Time. Figure 2 shows the results of querying 1,000 patterns into the largest dictionary. We only run experiments for 0, 1, and 2 errors since we previously found out that the main feature of our machine learning solution was for 1 error. In addition, in practice having 3 or more errors is quite rare and adds noise as the proportion of pairs of correct words that are at distance 3 or more grows fast.

---

[2] The source code can be found in http://www.cs.ucf.edu/~stephen/pats-apats/

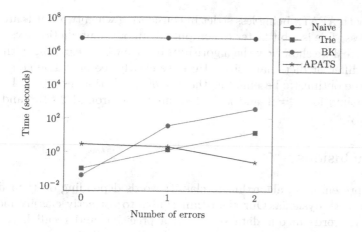

Fig. 2. Querying time for 1,000 words using a dictionary of 1.2 million words.

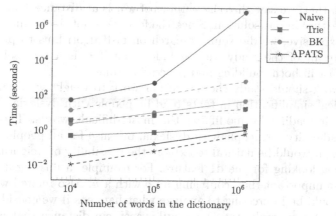

Fig. 3. Querying time for 1,000 words for $k = 1$ and $k = 2$ (dashed) using different dictionary sizes. Naive has the same search time in both cases.

Leaving the naive approach aside, which is clearly unpractical because its processing time is several orders of magnitude higher than the other three contestants, we can see that Trie performs better than BK in all cases except in the exact search case. This is not surprising because BK has a general purpose being able to perform matching and approximate matching in any metric space, not only the one of strings. Indeed, as we already mentioned, BK tries to minimize the number of distance computations performed, which is not that relevant for strings as in this case the distance function is easy to compute (in opposition to other objects such as images). On the other hand, Trie is an *ad hoc* method for the case of strings.

With the comparison against APATS, surprisingly enough, the Trie stands quite well. It is clearly slower when the dictionary is small but as we do experiments with larger dictionaries it improves, specially when looking for words with at most 2 errors. As we have already explained, the output of the two algorithms is different with APATS being less strict in considering when two words are similar. Nevertheless, Trie is a bit faster to compute d1 in our main dictionary.

In Figure 3 we show how the algorithms behave when searching with $k = 1, 2$ errors on different dictionary sizes. Here we clearly see how Trie gets closer to APATS (the optimistic baseline) as the size of the dictionary grows. In fact the crossing points for $k = 1$ and $k = 2$ seems to be around $2 \cdot 10^6$ and $5 \cdot 10^7$, respectively.

5   Conclusions

We have presented an algorithm to classify words depending on their difficulty for people with dyslexia. Our algorithm is able to correctly classify more than 70% of the words in our data sets with a precision and recall between 0.65 and 0.8. Hence, we were successful in proving the feasibility of predicting word difficulty and therefore a larger dataset should be generated.

Added to that, we saw that the number of words at distance 1 was the most important feature in our solution. Since the feature is not linear and computing it is quite expensive, we did some research on fast algorithms to perform this task. As a result of the study we concluded that Trie is a very solid method being very fast in both, building and processing time.

Future work should study the use of our work to weight words in order to perform lexical simplification in texts read by people with dyslexia. This could facilitate their reading by modifying the most difficult words. This not only improves readability for people with dyslexia, but also for all people.

Moreover, it would be natural to explore the possibility of weighting the edit distance when looking for the d1 feature. For example, a missing h in a word might be less important than confusing an o with a u. Also related with the d1 feature, it would be interesting to analyze what happens if we consider another string metric, for example, the DamerauLevenshtein distance that considers a transposition of two adjacent characters as a single error. Obviously, this requires to adapt Trie to this method adding a fourth recursive call in our procedure.

References

1. Baeza-Yates, R., Gonnet, G.H.: Fast text searching for regular expressions or automaton searching on tries. Journal of the ACM **43**(6), 915–936 (1996)
2. Burkhard, W.A., Keller, R.M.: Some approaches to best-match file searching. Communications of the ACM **16**(4), 230–236 (1973)
3. Carroll, J., Minnen, G., Canning, Y., Devlin, S., Tait, J.: Practical Simplification of English Newspaper Text to Assist Aphasic Readers. In: Proc. of AAAI 1998 Workshop on Integrating Artificial Intelligence and Assistive Technology, pp. 7–10 (1998)
4. Chang, W.I., Lampe, J.: Theoretical and empirical comparisons of approximate string matching algorithms. In: Apostolico, A.,Crochemore, M., Galil, Z., Manber, U. (eds.) Combinatorial Pattern Matching. LNCS, vol. 644, pp. 175–184. Springer, Heidelberg (1992)

5. Cole, R., Gottlieb, L.-A., Lewenstein, M.: Dictionary matching and indexing with errors and don't cares. In: Proceedings of the Thirty-Sixth Annual ACM Symposium on Theory of Computing, pp. 91–100. ACM (2004)
6. Coleman, M., Liau, T.L.: A computer readability formula designed for machine scoring. Journal of Applied Psychology 60(2), 283 (1975)
7. Coltheart, M., Davelaar, E., Jonasson, T., Besner, D.: Access to the internal lexicon. Attention and Performance VI, pp. 535–555 (1977)
8. Ellis, A.W.: Reading, writing and dyslexia. Erlbaum, London (1984)
9. Evans, R., Orasan, C., Dornescu, I.: An evaluation of syntactic simplification rules for people with autism. In: Proceedings of the 3rd Workshop on Predicting and Improving Text Readability for Target Reader Populations (PITR) at EACL, pp. 131–140 (2014)
10. Flesch, R.: A new readability yardstick. Journal of Applied Psychology 32(3), 221 (1948)
11. Fredkin, E.: Trie memory. Communications of the ACM 3(9), 490–499 (1960)
12. Fulwider, S., Mukherjee, A.: Multiple pattern matching. In: PATTERNS 2010, The Second International Conferences on Pervasive Patterns and Applications, pp. 78–83 (2010)
13. Hyönä, J., Olson, R.K.: Eye fixation patterns among dyslexic and normal readers: Effects of word length and word frequency. Journal of Experimental Psychology: Learning, Memory, and Cognition 21(6), 1430 (1995)
14. Jurafsky, D., Bell, A., Gregory, M., Raymond, W.D.: Evidence from reduction in lexical production. Frequency and the Emergence of Linguistic Structure 45, 229 (2001)
15. Levenshtein, V.: Binary codes capable of correcting spurious insertions and deletions of ones. Problems of Information Transmission 1(1), 8–17 (1965)
16. Malvern, D., Richards, B.: Measures of lexical richness. The Encyclopedia of Applied Linguistics (2012)
17. Navarro, G.: A guided tour to approximate string matching. ACM Comput. Surv. 33(1), 31–88 (2001)
18. Navarro, G., Baeza-Yates, R., Sutinen, E., Tarhio, J.: Indexing methods for approximate string matching. IEEE Data Eng. Bull. 24(4), 19–27 (2001)
19. Platt, J.: Sequential minimal optimization: A fast algorithm for training support vector machines. Technical Report MSR-TR-98-14, Microsoft Research (1998)
20. Rayner, K., Duffy, S.A.: Lexical complexity and fixation times in reading: Effects of word frequency, verb complexity, and lexical ambiguity. Memory & Cognition 14(3), 191–201 (1986)
21. Rello, L., Baeza-Yates, R., Bott, S., Saggion, H.: Simplify or help? Text simplification strategies for people with dyslexia. In: Proc. W4A 2013, Rio de Janeiro, Brazil (2013)
22. Rello, L., Baeza-Yates, R., Llisterri, J.: A resource of errors written in Spanish by people with dyslexia and its linguistic, phonetic and visual analysis. Language Resources and Evaluation (to appear)
23. Saggion, H., Stajner, S., Bott, S., Mille, S., Rello, L., Drndarevic, B.: Making it Simplext: Implementation and evaluation of a text simplification system for Spanish. ACM Transactions on Accessible Computing (to appear, 2015)
24. Tanaka, S., Jatowt, A., Kato, M.P., Tanaka, K.: Estimating content concreteness for finding comprehensible documents. In: Proceedings of the sixth ACM International Conference on Web Search and Data Mining, pp. 475–484. ACM (2013)
25. Temple, C.M., Marshall, J.C.: A case study of developmental phonological dyslexia. British Journal of Psychology 74(4), 517–533 (1983)

Author Index

Printed in the United States
By Bookmasters